An Introduction to
Conservation
Biology
THIRD EDITION

An Introduction to
Conservation Biology

THIRD EDITION

Anna A. Sher

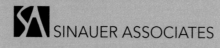

SINAUER ASSOCIATES

NEW YORK   OXFORD
OXFORD UNIVERSITY PRESS

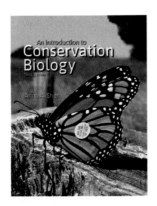

About the Cover

Monarch butterflies (*Danaus plexippus*) travel as much as 6000 km across North America in their annual migration to overwinter in Mexico. Like most species, their primary threat is habitat loss, symbolized here by the cut stump upon which the butterfly rests. The conservation effort of tracking monarchs and monitoring their populations involves gently placing labelled stickers on the wings. The scope of this work has greatly expanded with the help of citizen scientists, who both label butterflies and report recoveries of labels. Monarch Watch, whose label is shown in this picture, distributes more than a quarter million labels to thousands of volunteers each year. Tracking monarchs provides data about their origins, timing and routes of migration, mortality rates, and geographic distribution, which conservation biologists can use to help protect them in the future, in the face of climate change and other threats.

Oxford University Press is a department of the University of Oxford. It furthers the University's objective of excellence in research, scholarship, and education by publishing worldwide. Oxford is a registered trade mark of Oxford University Press in the UK and certain other countries.

Published in the United States of America by Oxford University Press
198 Madison Avenue, New York, NY 10016, United States of America

© 2022, 2020, 2016 Oxford University Press
Sinauer Associates is an imprint of Oxford University Press.

For titles covered by Section 112 of the US Higher Education Opportunity Act, please visit www.oup.com/us/he for the latest information about pricing and alternate formats.

Address editorial correspondence to:
Sinauer Associates
23 Plumtree Road
Sunderland, MA 01375 USA

ACCESSIBLE COLOR CONTENT   Every opportunity has been taken to ensure that the content herein is fully accessible to those who have difficulty perceiving color. Exceptions are cases where the colors provided are expressly required because of the purpose of the illustration.

Library of Congress Cataloging-in-Publication Data

Names: Sher, Anna, author.
Title: An introduction to conservation biology / Anna A. Sher, University of Denver.
Other titles: Conservation biology
Description: Third edition. | New York, NY : Oxford University Press/Sinauer Associates, [2022] | Revised edition of: An introduction to conservation biology / Anna A. Sher (University of Denver), Richard B. Primack (Boston University). Second edition. [2020]. | Includes bibliographical references and index. | Summary: "An Introduction to Conservation Biology is well suited for a wide range of undergraduate courses, as both a primary text for conservation biology courses and a supplement for ecological and environmental science courses. This new edition focuses on engaging students through videos and activities, and includes new pedagogy to scaffold students' learning. Coverage of recent conservation biology events in the news-such as global climate change and sustainable development-keeps the content fresh and current"-- Provided by publisher.
Identifiers: LCCN 2021046178 (print) | LCCN 2021046179 (ebook) | ISBN 9780197564370 (paperback) | ISBN 9780197559079 (ebook)
Subjects: LCSH: Conservation biology. | Biodiversity.
Classification: LCC QH75 .P7523 2022 (print) | LCC QH75 (ebook) | DDC 333.95/16--dc23/eng/20211013
LC record available at https://lccn.loc.gov/2021046178
LC ebook record available at https://lccn.loc.gov/2021046179

9 8 7 6 5 4 3 2 1
Printed in the United States of America

To Dr. Richard B. Primack.
Your past and ongoing work in the discipline inspires and
motivates us to do all we can for the future of biodiversity.

And for Michael Soulé (1936–2020), a father of the field.

Contents

Preface

To say that much has happened since the last edition is, of course, a gross understatement. Climate change has manifested in unprecedented wildfires, floods, and global temperatures. I spent much of the pandemic revising both this text and my other textbook in ecology, which is reflected in several chapters referencing this major historical event. I discuss how COVID-19 opened many people's eyes to the connections between wildlife, conservation, and disease (Chapter 5). A proposed vector of SARS-CoV-2, the virus that causes COVID-19, has been the pangolin, the most trafficked wild animal in the world. There are eight species of pangolin, ranging from vulnerable to critically endangered. When animals such as the pangolin are trapped and sold, the proximity to humans can lead to *zoonosis*, the spread of disease to humans from another species (a new term in this edition). The Chinese government established new protections for pangolins in 2020, and the US government may also increase protections for the pangolin in 2021.

The COVID-19 pandemic facilitated other benefits for conservation. It made people aware of the wildlife that coexists with us, even in cities, but had been rarely seen before we were sequestered in our homes (see Chapter 10 opening photo). Widespread quarantine also provided a decrease, even if short, in industrial emissions that contribute to global warming (Chapter 5). The quarantine and closures due to COVID-19 also provided taxonomists with the opportunity to work on the backlog of field-collected specimens, revealing hundreds of new species (Chapter 2).

Throughout the past two years, the increase in use of social media has helped to spread information and awareness around conservation. This has included illuminating the role African American Civil War soldiers (nicknamed Buffalo Soldiers) played in the protection of Yosemite and Sequoia National Parks under the leadership of Captain Charles Young. Several streamed movies and television series promoted a conservation message, with *David Attenborough: A Life on Our Planet, My Octopus Teacher*, and *Secrets of the Whales* recently among the most popular. Some argue that even the COVID-19 quarantine favorite *Tiger King* helped educate the public about the plight of endangered animals in private zoos. TikTok, Facebook, Instagram, and other social media platforms have become important tools to communicate conservation ideas.

Although the human population is both larger and growing faster than ever before (I had to revise the estimate for 2050 upward by 300 million people), this is the first edition to forecast a future human population decline (Chapter 1). What this may mean for global biodiversity is unclear, but given the association between human population growth and biodiversity loss, I choose to be optimistic. Other positive developments include the progress toward reaching the Aichi Biodiversity Target 11 for protected areas (Chapter 9) and the significant increases in foreign aid to developing countries to protect biodiversity (Chapter 12). We have also seen the power of effective legislation, an example being the dramatic decrease in the trade of wild-caught birds after the European Union banned it (Chapter 5).

Conservation science is necessary to continue progress where possible and to help find new solutions (Chapter 13). We must also capitalize upon the growing public awareness that the fate of biodiversity is connected to our own. It is critical that we not allow the losses we have witnessed overshadow the progress and reasons for hope. Indeed, I believe our very future depends on it.

New to the Third Edition

The following updates maintain *An Introduction to Conservation Biology* as the most current and scientifically accurate textbook of its type. For the third edition, my team's systematic review of every numerical value in the textbook resulted in over 240 changes to text and tables, 275 new citations, and new figures in every chapter. These changes include updating numbers of species, endangerment figures, trade in wildlife, nonprofit and government organization statistics, and many, many other numerical values based on the best information available.

The new citations also reflect my commitment to increase the representation of women and BIPOC in the textbook—an important and ongoing effort to make the text reflect the diversity of scientists in the field. One aspect of this work involved making significant edits and additions regarding the roles and experience of Indigenous People, also known as first people or native people, in the field of conservation biology. These included acknowledging the theft of land to create protected areas (Chapter 9 opening photo), incorporating the importance of traditional ecological knowledge (TEK) (Chapters 10 and 11), expanding coverage of the role of Indigenous People in development and implementation of environmental policy and programs (Chapters 12 and 13), and generally reviewing language used throughout the textbook.

In addition to COVID-19 and other current events, I incorporated several new discoveries and developments from the past two years, including the following:

- The latest understanding of the causes of the Permian extinction
- The UN Decade on Ecosystem Restoration (2021–2030) and its principles to guide restoration
- Connections between human disease and biodiversity conservation

GLOBAL CHANGE
CONNECTION

To highlight content related to the many ways we are changing the Earth, this edition has the added feature of the "Global Change Connection" icon. *Global change* in this context refers to any impact on ecosystems caused by anthropogenic activity that can be considered a global issue, including climate change, invasive species, pollution, and habitat loss.

I have responded to the feedback of expert reviewers, instructors, and student users of the textbook with several edits. Among the dozens of changes are these:

- Additional discussion on the political aspects of climate change and of genetically modified organisms (GMOs)
- An elaboration of the concept of the types of biodiversity, including a refinement of the definition of *species diversity*, with additional examples
- A reorganization of the chapter on restoration ecology, including new graphical representation of the outcomes of restoration and new research on the importance of the human element

In addition to the aforementioned revisions to the text, the new edition includes upgraded digital resources for instructors and students. First, the text is available in a new enhanced e-book format, which includes self-assessment questions following each major section heading. The second digital resource for instructors is a curated list of freely accessible videos from various sources that can be used as teaching aids for key concepts in each chapter.

Acknowledgments

First and foremost, I acknowledge Richard B. Primack, the author of the conservation biology books that were the foundation of this series. I was honored when he asked me to be his successor author for *A Primer of Conservation Biology* and then when it was decided that my integration of that text with Primack's *Essentials of Conservation Biology* warranted a new series, *An Introduction to Conservation Biology*. There are nods to both original textbooks in this edition, the most important being my goal to uphold each book's legacy as a scientifically accurate, relevant, and accessible resource for undergraduate students. Thank you, Richard, for your past mentorship and the privilege of being the sole author of this, the newest edition.

The monumental effort necessary to keep a textbook current is not possible without the support of many people. First, I thank Eduardo González, my academic partner in all things, and Violet Butler, who together did significant research to update this edition, and the many reviewers who helped us make this the best book it can be. I am also grateful to the rest of the Sher Lab and the University of Denver DUEEB group for their general support and feedback on figures.

Thank you to our reviewers of the second edition, including Andrew J. Rassweiler, Janet Steven, Susan Margulis, Elizabeth Freeman, John B. Graham, Pamela A. Morgan, Bibit Traut, Jessica Claxton, Michael Remke, and three anonymous reviewers. Know that some of the more substantial changes among your recommendations will help shape future editions. Thank you to all those who provided data or reviewed sections, including Patrick M. Burchfield.

I am grateful for all the photos that have been donated by colleagues over the years. Wonderful new images for this edition have been provided by Wayne Armstrong, Hector R. Chenge, Scott Dressel-Martin of Denver Botanic Gardens, and Joseph Thomas. As always, a special thanks to Richard Reading for both his excellent photographs and conservation stories.

I cannot write these textbooks without the support of my wife, Fran, and our son, Jeremy. The team at Sinauer Associates and Oxford University Press who produced this book during such a challenging time also deserve special kudos: Senior Acquisitions Editor Jason Noe, Senior Production Editor Alison Hornbeck, Production Manager Joan Gemme, Production Specialist and Book Designer Rick Neilsen, Photo Researcher Mark Siddall, Permissions Supervisor Michele Beckta, Editorial Assistant Sarah D'Arienzo, the excellent copyeditor Lou Doucette, and everyone else who helped make this third edition the best possible book.

Finally, I acknowledge all the scientists and practitioners who do the work of conservation. In particular, I wish to honor the brave conservationists who lost their lives because of their dedication; there were literally dozens in only the past two years. These souls include Homero Gómez González and Raúl Hernández Romero, champions of the monarch butterfly in Mexico; Joannah Stutchbury, a Kenyan known for her work to conserve the Kiambu Forest; Irishman Rory Young, cofounder of the antipoaching organization Chengeta Wildlife; and Kavous Seyed-Emami, an environmentalist and professor studying the rare Asiatic cheetah in Iran. In most if not all cases, the violence was attributed to individuals or groups who felt that conservation work was interfering with their financial interests. May the injustice of their deaths motivate to action those of us with the privilege and capacity to act.

Anna A. Sher
Denver, Colorado
October 21, 2021

About the Author

Anna A. Sher is a Professor of Biology at the University of Denver, where she has taught conservation biology since 2003. She held a joint position as the Director of Research and Conservation at Denver Botanic Gardens from 2003–2010. Dr. Sher has published books and articles for academic, trade, and popular audiences on various topics within conservation biology, including restoration ecology, rare plant conservation, and climate change. She is one of the foremost experts on the ecology of invasive *Tamarix* trees and was the lead editor of the book *Tamarix: A Case Study of Ecological Change in the American West* (Oxford University Press, 2013).

She is also first author of the textbook *Ecology: Concepts and Applications, 9th Ed.* (McGraw-Hill Education). Dr. Sher received her Ph.D. in Biology at the University of New Mexico in 1998 and was a postdoctoral fellow at the University of California, Davis and as a Fulbright Scholar in Israel. Dr. Sher also led scientific study-abroad programs in East Africa, and has been a visiting scholar at the University of Otago in New Zealand. She is an advocate for social justice and is currently leading a campus-wide coalition to support women and increase diversity in STEM academic professions. She and her wife and son live, work, and play in Denver, Colorado.

Media & Supplements to accompany
An Introduction to Conservation Biology Third Edition

oup.com/he/sher3e

For the Instructor

(Instructor resources are available to adopting instructors online. Registration is required. Please contact your Oxford University Press representative to request access.)

The Instructor Resources for *An Introduction to Conservation Biology*, Third Edition offer all the textbook's figures and tables, making it easy for instructors to incorporate visual resources into their lecture presentations and other course materials.

The site includes the following resources:

Figure PowerPoint Slides – All the figures and tables from each chapter, with titles on each slide, and complete captions in the Notes field.

Video Guide – A curated list of freely accessible videos from various sources that can be used as teaching aids for key concepts in each chapter.

Chapter Summaries – Useful aids summarizing each chapter's key concepts.

enhanced e-book (ISBN 978-0-19755-907-9)

An Introduction to Conservation Biology, Third Edition, is available as an e-book via several different e-book providers, including RedShelf and VitalSource. Please visit the Oxford University Press website at **oup.com/us/he** for more information.

An Introduction to
Conservation Biology
THIRD EDITION

Santos Alfonso Bernal Avalos releases Kemp's ridley sea turtles (*Lepidochelys kempii*) off the coast of Tamaulipas, Mexico, in 2020. A binational conservation effort has been successful in rescuing this species from the brink of extinction.

Defining Conservation Biology

Popular interest in protecting the world's **biological diversity**—including its amazing range of species, its complex ecosystems, and the genetic variation within species—has intensified during the last few decades. Studies have found that the number of Google searches on conservation-related topics, including extinction, has dramatically and consistently increased each year (Burivalova et al. 2018; Williams et al. 2020). It has become increasingly evident to both scientists and the general public that we are living in a period of unprecedented losses of **biodiversity**[1] (Pimm et al. 2014). Around the globe, biological ecosystems—the interacting assemblages of living organisms and their environment that took millions of years to develop—are being devastated, including tropical rain forests, coral reefs, temperate old-growth forests, and prairies. Thousands, if not tens of thousands, of species and millions of unique populations are predicted to go extinct in the coming decades (Ceballos et al. 2015). Unlike the mass extinctions in the geologic past, which followed catastrophes such as asteroid collisions with Earth, today's extinctions have a human face.

---

[1] *Biological diversity* is often shortened to *biodiversity* (a term credited to biologist E. O. Wilson in 1992); it includes all species, genetic variation, and biological communities and their ecosystem-level interactions.

During the last 200 years, the human population has exploded. It took more than 160,000 years for the number of *Homo sapiens* to reach 1 billion, an event that occurred sometime around the year 1805. Estimates for 2021 put the number of humans at 7.8 billion, with a projected 9.7 billion by 2050 (https://www.worldometers.info/world-population/; United Nations 2019) (**FIGURE 1.1**). While birth rates have slowed since the 1960s due to women's increasing access to education, birth control, and other opportunities (BBC 2018), at this size, even a modest rate of population increase adds tens of millions of individuals each year. But the global population in 2050 may represent a peak; it is now projected that the world's population may actually decline in the second half of the century (Vollset et al. 2020). However, growth will continue to occur in many areas of tropical Africa, Latin America, and Asia, where the greatest biological diversity is also found.

GLOBAL CHANGE
CONNECTION

The demands of the rapidly increasing human population and its rising material consumption correspond to an acceleration of threats to biodiversity. People deplete natural resources such as firewood, coal, oil, timber, fish, and game, and they convert natural habitats to land dominated by agriculture, cities, housing developments, logging, mining, industrial plants, and other human activities. These changes are not easily reversible, and even aggressive

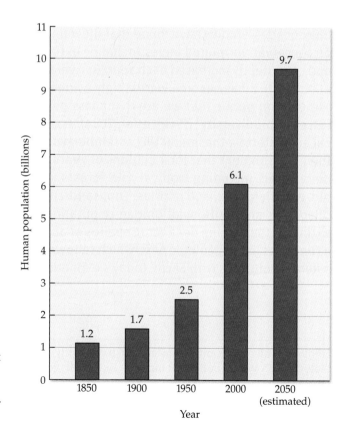

FIGURE 1.1 The human population in 2021 stood at about 7.8 billion. It is notable that the human population is growing even faster than projected; the estimate for 2050 has been revised upward by 300 million people since just 2016. (Data from K. Klein Goldewijk et al. 2016. *Earth Syst Sci Data* 9: 927-953 and World Population Prospects: The 2017 Revision. 2017. United Nations DESA/Population Division.)

programs to slow population growth do not adequately address the environmental problems we have caused (Bradshaw and Brook 2014).

Worsening the situation is the fact that as countries develop and industrialize, the consumption of resources increases. For example, the average citizen of the United States uses more than three times more total energy than the average global citizen (EIA 2018), and more than 10 times the energy of the average Indian (Richie and Roser 2019). In terms of oil, the top 10 countries account for 60% of the world's oil consumption, with the United States accounting for a full third of this (EIA 2020). However, the country with the greatest energy use is now China, reflecting recent development in that country and growing per capita use (**FIGURE 1.2**). The ever-increasing number of human beings and their intensifying use of natural resources have direct and harmful consequences for the diversity of the living world (Tilman et al. 2017).

Threats to biodiversity directly threaten human populations as well, because people depend on the natural environment for raw materials, food, medicines, air, and the water they drink. The poorest people are, of course, the ones who experience the greatest hardship from damaged environments because they have fewer reserves of food and less access to medical supplies, transportation, and construction materials.

GLOBAL CHANGE
CONNECTION

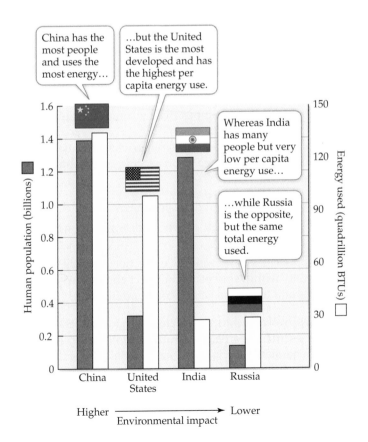

FIGURE 1.2 Population numbers and total energy use by country. Energy use, such as the burning of fossil fuels, is generally associated with negative environmental impacts. Energy use is not simply a function of population size; per capita use (energy/population) has important implications. Even though rapidly developing China's per capita use is lower than that of either the United States or Russia, its large population means that it uses the most energy and thus has the greatest environmental impact. Although India has a much larger population than either the United States or Russia, its energy use is more similar to Russia's because of India's low per capita use. (After US Energy Information Administration. 2018. *International Energy Outlook 2018*, p.2. https://www.eia.gov/outlooks/ieo/india/pdf/india_detailed.pdf.)

1.1 The New Science of Conservation Biology

LEARNING OBJECTIVES

By the end of this section you should be able to:

1.1.1 Trace the history of ideas about the conservation of nature over human history and across cultures.

1.1.2 Contrast the three guiding ethical principles of the field of conservation biology.

1.1.3 Identify the interdisciplinary elements of conservation biology for a case study of species management.

GLOBAL CHANGE
CONNECTION

The avalanche of species extinctions and the wholesale habitat destruction occurring in the world today is devastating, but there is reason for hope. The last several decades have included many success stories, such as that of the American bald eagle (*Haliaeetus leucocephalus*), which was rescued from near extinction due to a combination of scientific inquiry, public awareness, and political intervention. Actions taken—or bypassed—during the next few decades will determine how many of the world's species and natural areas will survive. It is quite likely that people will someday look back on the first half of the twenty-first century as an extraordinarily exciting time, when collaborations of determined people acting locally and internationally saved many species and even entire ecosystems. In a review of nearly 13,000 conservation biology publications, more than half of the research papers measuring conservation outcomes reported positive results (Godet and Devictor 2018). Examples of such conservation efforts and positive outcomes are described throughout this book.

Conservation biology is an integrated, multidisciplinary scientific field that has developed in response to the challenge of preserving species and ecosystems. It has three goals:

1. To document the full range of biological diversity on Earth
2. To investigate human impact on species, genetic variation, and ecosystems
3. To develop practical approaches to prevent the extinction of species, maintain genetic diversity within species, and protect and restore biological communities and their associated ecosystem functions

The first two of these goals involve the dispassionate search for factual knowledge that is typical of scientific research. The third goal, however, defines conservation biology as a **normative discipline**—that is, a field that embraces certain values and attempts to apply scientific methods to achieving those values (Lindenmayer and Hunter 2010). Just as medical science values the preservation of life and health, conservation biology values the preservation of species and ecosystems as an ultimate good, and its practitioners intervene to prevent human-caused losses of biodiversity.

Conservation biology arose in the 1980s, when it became clear that the traditional applied disciplines of resource management alone were not comprehensive enough to address the critical threats to biodiversity. The applied disciplines of agriculture, forestry, wildlife management, and fisheries biology have gradually expanded to include a broader range of species and ecosystem processes. Conservation biology complements those applied disciplines and provides a more general theoretical approach to the protection of biodiversity. It differs from these disciplines in its primary goal of long-term preservation of biodiversity.

Like medicine, which applies knowledge gleaned from physiology, anatomy, biochemistry, and genetics to the goal of achieving human health and eliminating illness, conservation biology draws on other academic disciplines within biology, including taxonomy, evolution, ecology, and genetics. Many conservation biologists have come from these ranks. Others come from backgrounds in the applied disciplines, such as forestry and wildlife management. In addition, many leaders in conservation biology have come from zoos and botanical gardens, bringing with them experience in locating rare and endangered species in the wild and then maintaining and propagating them in captivity.

Conservation biology is also closely associated with, but distinct from, **environmentalism**, a widespread movement characterized by political and educational activism with the goal of protecting the natural environment. Conservation biology is a scientific discipline based on biological research whose findings often contribute to the environmental movement.

> Conservation biology merges applied and theoretical biology by incorporating ideas and expertise from a broad range of scientific fields toward the goal of preserving biodiversity.

Because much of the biodiversity crisis arises from human pressures, conservation biology also incorporates ideas and expertise from a broad range of fields outside of biology (**FIGURE 1.3**). For example, environmental law and policy provide the basis for government protection of rare and endangered species and critical habitats. Environmental ethics provides a rationale for preserving species. Ecological economists provide analyses of the economic value of biological diversity to support arguments for preservation. Climatologists monitor the physical characteristics of the environment and develop models to predict environmental responses to disturbance and climate change. Both physical and cultural geography provide information about the relationships among elements of the environment, helping us understand causes and distributions of biodiversity and how humans interact with it. Social sciences, such as anthropology and sociology, provide methods to involve local people in actions to protect their immediate environment. Conservation education links academic study and fieldwork to solve environmental problems, teaching people about science and helping them realize the value of the natural environment. Because conservation biology draws on the ideas and skills of so many separate fields, it can be considered a truly multidisciplinary discipline.

GLOBAL CHANGE
CONNECTION

FIGURE 1.3 Conservation biology represents a synthesis of many basic sciences (left) that provide principles and new approaches for the applied fields of resource management (right). The experiences gained in the field, in turn, influence the direction of the basic sciences. (After S. A. Temple. 1991. In *Challenges in the Conservation of Biological Resources: A Practitioner's Guide*, D. J. Decker et al. [Eds.], pp. 45–54. Westview Press, Boulder, CO.)

Field experience and research needs

Basic Sciences	Resource Management
Anthropology	Agriculture
Biogeography	Community education
Climatology	and development
Ecology:	Fisheries management
Community ecology	Forestry
Ecosystem ecology	Land-use planning and
Landscape ecology	regulation
Environmental studies:	Management of captive
Ecological economics	populations:
Environmental ethics	Zoos
Environmental law	Aquariums
Ethnobotany	Botanical gardens
Evolutionary biology	Seed banks
Genetics	Management of protected
Population biology	areas
Sociology	Sustainable development
Taxonomy	Wildlife management
Other biological, physical,	Other resource conservation
and social sciences	and management activities

New ideas and approaches

The roots of conservation biology

Religious and philosophical beliefs about the relationship between humans and the natural world are seen by many as the foundation of conservation biology (Singh et al. 2017). Eastern philosophies such as Taoism, Hinduism, and Buddhism revere wilderness for its capacity to provide intense spiritual experiences. These traditions see a direct connection between the natural world and the spiritual world, a connection that breaks down when the natural world is altered or destroyed. Strict adherents to the Jain and Hindu religions in India believe that all killing of animal life is wrong. Islamic, Judaic, and Christian teachings are used by many people to support the idea that people are given the sacred responsibility to be guardians of nature (**FIGURE 1.4**; see Section 3.5 in Chapter 3). Many of the leaders of the early Western environmental movement that helped to establish parks and wilderness areas did so because of strong personal convictions that developed from their Christian religious beliefs. Contemporary religious leaders have pointed out that some of the most profound moments in the Bible occur on mountaintops, in the wilderness, or on the banks of rivers (Korngold 2008). In Native American tribes of the Pacific Northwest, hunters undergo purification rituals in order to be considered worthy, and the Iroquois consider how their actions would affect the lives of their descendants after seven generations. Pawnee Eagle Chief Letakots-Lesa is quoted as saying, "Tirawa, the one above, did not speak directly to man … he showed himself through the beasts, and from them and from the stars, the sun, and the moon should man learn" (from Burlin 1907).

Examples of humans safeguarding nature can be found throughout history and across the globe. For example, an ancient Greek book by Hippocrates, *De aëre, aquis et locis* (*Air, Waters, and Places*), could be considered the earliest surviving European work on the topic of protecting nature. In the early modern era, actions to protect forests include successful advocacy in the 1760s by the Frenchman Pierre Poivre, who had observed the relationship between deforestation and regional climate change (Grove 2002), and the Bishnoi Hindus of Khejarli in India, who gave their lives protecting trees from being felled for the palace of Maharaja of Jodhpur in 1720.

Modern conservation biology in Europe and the United States arose in parallel through individuals and works concerned for landscapes and living organisms, both for their intrinsic value and their utility to humans. In Europe, Romanticism was a movement in the early 1800s that emphasized appreciation of nature, partially in response to the Industrial Revolution. Meanwhile in the United States, nineteenth-century transcendentalist philosophers Ralph Waldo Emerson and Henry David Thoreau wrote about wild nature as an important element in human moral and spiritual development. Emerson (1836) saw nature as a temple in which people could commune with the spiritual world and achieve spiritual enlightenment. Thoreau was both an advocate for nature and an opponent of materialistic society, writing about his ideas and experiences in *Walden*, a book published in 1854 that has influenced many generations of students and environmentalists. Eminent American wilderness advocate John Muir used the transcendental themes of Emerson and Thoreau in his campaigns to preserve natural areas. According to Muir's **preservationist ethic**, natural areas such as forest groves, mountaintops, and waterfalls have spiritual value that is generally superior to the tangible material gain obtained by their exploitation (Muir 1901).

Subsequent leaders paved the way for conservation biology as an applied academic discipline. Gifford Pinchot, the first head of the US Forest Service, developed a view of nature known as the **resource conservation ethic** (Ebbin 2009). He defined natural resources as the commodities and qualities found in nature, including timber, fodder, clean water, wildlife, and even beautiful landscapes (Pinchot 1947). The proper use of natural resources, according to the resource conservation ethic, is whatever will further "the greatest good of the greatest number [of people] for the longest time." From the perspective

Photo by Anna Sher

FIGURE 1.4 Religious convictions combined with a history of traditional relationships with nature motivate the grassroots conservation organization Kakamega Environmental Education Program (KEEP), established in 1998 to protect one of the last remnants of tropical forest left in Kenya, East Africa.

of conservation biology, **sustainable development** is development that best meets present and future human needs while respecting ecosystem function (Mathevet et al. 2018).

A contemporary of Pinchot was biologist Aldo Leopold, who also worked for the U.S. Forest Service and is now considered by many to be a father of both conservation biology and ecology. He published *A Sand County Almanac* in 1949, a highly influential book that illustrated the interrelatedness of living things and their environment, promoting the idea that the most important goal of conservation is to maintain the health of natural ecosystems and ecological processes (Leopold 2004). As a result, Leopold and many others lobbied successfully for certain parts of national forests to be set aside as wilderness areas (Shafer 2001). This idea of considering the ecosystem as a whole, including human populations, is now termed the **land ethic**.

Marine biologist Rachel Carson (**FIGURE 1.5**) is credited with raising public awareness of the complexity of nature with her best-selling books, including *Silent Spring* (1962), which brought attention to the dangers of pesticides and spurred an international environmental movement. A more recent approach known as **ecosystem management**, which combines ideas of Carson, Leopold, and Pinchot, places the highest management priority on cooperation among businesses, conservation organizations, government agencies, private citizens, and other stakeholders to provide for human needs while maintaining the health of wild species and ecosystems.

Depictions of nature in the creative and performing arts have also played an important role in the growing awareness of the value of nature and its preservation in the United States. In the mid-nineteenth century, prolific painters of the Hudson River School were noted for their romantic depictions of "scenes of solitude from which the hand of nature has never been lifted" (Cole 1965). Photographer Ansel Adams (1902–1984) took breathtaking images of wild America, helping to foster public support for its protection (**FIGURE 1.6**). Popular singer-songwriter John Denver (1943–1997) inspired interest in conservation with songs such as his 1975 hit "Rocky Mountain High":

> *Now he walks in quiet solitude the forest and the streams,*
> *seeking grace in every step he takes … His life is filled*
> *with wonder but his heart still knows some fear of a simple*
> *thing he cannot comprehend: why they try to tear the*
> *mountains down … more people, more scars upon the land.*

More recently, wildlife photographer Brian Skerry has captivated audiences with his images and stories about whales; there was even an acclaimed video series made based on his

> Preservation of natural resources, ecosystem management, and sustainable development are major themes in conservation biology.

 GLOBAL CHANGE CONNECTION

FIGURE 1.5   Rachel Carson (1907–1964) was a marine biologist who, through her popular writing, including *Silent Spring* (1962), helped to found both conservation biology and the environmental movement.

FIGURE 1.6   The photographs of Ansel Adams (1902–1984) showed the public the beauty of wild spaces, such as this image of the Snake River and the Tetons in Wyoming, USA.

experiences. Documentaries such as these are as much about beauty as they are about science, and spread messages about species preservation to a very broad audience. Clearly, the arts play a unique role in fostering interest in conservation; it has been argued that, in fact, science and art are inextricable (Bullot et al. 2017).

In Europe, dramatic losses of wildlife caused by the expansion of agriculture and use of firearms stimulated the modern British conservation movement, leading to the founding of the Commons, Open Spaces and Footpaths Preservation Society in 1865, the National Trust for Places of Historic Interest or Natural Beauty in 1895, and the Royal Society for the Protection of Birds in 1899. Altogether, these groups have preserved nearly 1 million hectares of open land (**TABLE 1.1** provides an explanation of the term *hectare* and other measurements). More recently, the formation of the European Union has facilitated conservation in a variety of ways, in part by establishing ambitious objectives for conservation and habitat management. In 2010, it established the EU Biodiversity Strategy to 2020, with a goal of halting species loss. This was supported by the establishment of "Nature Directives" and other legislation to guide resource use and species management. Perhaps the most

TABLE 1.1	Some Useful Units of Measurement
Length	
1 meter (m)	1 m = 39.4 inches = ~3.3 feet
1 kilometer (km)	1 km = 1000 m = 0.62 mile
1 centimeter (cm)	1 cm = 1/100 m = 0.39 inches
1 millimeter (mm)	1 mm = 1/1000 m = 0.039 inches
Area	
1 square meter (m^2)	Area encompassed by a square, each side of which is 1 meter
1 hectare (ha)	1 ha = 10,000 m^2 = 2.47 acres
	100 ha = 1 square kilometer (km^2)
Temperature	
degree Celsius (°C)	°C = 5/9 (°F − 32)
	0°C = 32° Fahrenheit (the freezing point of water)
	100°C = 212° Fahrenheit (the boiling point of water)
	20°C = 68° Fahrenheit ("room temperature")
Energy	
British Thermal Unit (BTU)	The amount of energy it takes to increase the temperature of 1 pound of water 1 degree Fahrenheit

important action of the European Union has been the establishment of a network of protected areas, called the Natura 2000 (see Chapter 9). In contrast to its origins in the United States, biological conservation in Europe has had a more integrated view of human society and ecosystems as a whole, rather than envisioning a dichotomy of man versus nature (Linnell et al. 2015).

Many societies worldwide similarly have strong traditions of nature conservation and land protection. Tropical countries such as Brazil, Costa Rica, and Indonesia have a history of reverence for nature, and their governments have established increasing numbers and areas of national parks. The economic value of these protected areas is constantly increasing because of their importance for tourism and the valuable ecosystem services they provide, such as purifying water and absorbing carbon dioxide (see Chapter 3). Many tropical countries have established agencies to regulate the exploration and use of their biodiversity, and these efforts increasingly involve the Indigenous Peoples who depend on and have unique knowledge of these ecosystems. Hunting and gathering societies, such as the Penan of Borneo, give thousands of names to individual trees, animals, and places in their surroundings to create a cultural landscape that is vital to the well-being of the tribe. Indeed, traditional societies throughout the world have influenced and enriched modern conservation biology.

As demonstrated by the conservation tradition in Europe, habitat degradation and species loss can catalyze long-lasting conservation efforts.

A new science is born

By the early 1970s, scientists throughout the world were aware of an accelerating biodiversity crisis, but there was no central forum or organization to address the issue. Scientist Michael Soulé organized the first International Conference on Conservation Biology in 1978 so that wildlife conservationists, zoo managers, and academics could discuss their common interests. At that meeting, Soulé proposed a new interdisciplinary approach that could help save plants and animals from the threat of human-caused extinctions, which he called *conservation biology* (Soulé 1985). Subsequently, Soulé, along with colleagues including Paul Ehrlich of Stanford University and Jared Diamond of the University of California at Los Angeles, began to develop conservation biology as a discipline that would combine the practical experience of wildlife, forestry, fisheries, and national park management with the theories of population biology and biogeography. In 1985, this core of scientists founded the Society for Conservation Biology. This organization, which continues today with an international membership, sponsors conferences and publishes the scientific journal *Conservation Biology*. That journal is now joined by many peer-reviewed publications that feature research in conservation biology, such as *Biological Conservation* and *Conservation Letters*, expanding our understanding of the importance of biodiversity and how it can be protected.

Public and policymaker awareness of the value of, and threats to, biodiversity greatly increased following the international Earth Summit held in Rio de Janeiro, Brazil, in 1992 (see Chapter 12). At this meeting, representatives of 178 countries formulated and eventually signed the Convention on Biological Diversity (CBD), which obligates countries to protect their biodiversity but also allows them to obtain a share in the profits of new products developed from that diversity. In 2000, the United Nations General Assembly adopted May 22 as International Biodiversity Day to commemorate the conference, and in 2015, the United Nations Climate Change Conference (COP21) was attended by 196 parties and resulted in the Paris Agreement to reduce global greenhouse emissions. Arguably, this increase in understanding and concern would not have been possible without the foundation of a recognized scientific discipline.

> Interdisciplinary approaches, the involvement of local people, and the restoration of important environments and species all attest to progress in the science of conservation biology.

GLOBAL CHANGE
CONNECTION

The interdisciplinary approach: A case study with sea turtles

Throughout the world, scientists are using the approaches of conservation biology to address challenging problems, as illustrated by the efforts to save the Kemp's ridley sea turtle (*Lepidochelys kempii*). The Kemp's ridley is the rarest and smallest of the world's sea turtle species, at 70–100 cm (2–3 feet) long and about 45 kg (100 pounds). It also has the most restricted range, which has contributed to its rarity and risk of extinction (see Chapter 6). However, its numbers have dramatically increased as a result of international conservation efforts and cooperation among scientists, conservation organizations, government officials, and the interested public.

After its discovery as a distinct species in the late 1800s, it took nearly a century for scientists to determine how and where the Kemp's ridley reproduces (Wibbels and Bevan 2015). They discovered that nearly 95% of Kemp's ridley nesting takes place on beaches in the state of Tamaulipas, in the northeastern corner of Mexico, in highly synchronized gatherings of turtles called *arribadas*. The largest arribada ever documented was 40,000 nesting females in 1947 (Wibbels and Bevan 2016). This highly concentrated nesting is unusual among turtles and makes the species particularly vulnerable to intensive harvesting. Over many decades, local people collected an estimated 80% of Kemp's ridley eggs from the nesting beaches for eating, and many eggs were lost to predation by coyotes (*Canis latrans*). Thousands of turtles also drowned in fishing gear, especially when caught accidentally in shrimp nets. By 1985, turtle numbers had declined to a low point of only 702 nests (an estimate of the number of breeding females) worldwide, making the Kemp's ridley the most endangered sea turtle in the world.

Heeding the warning of wildlife biologists that the species was nearing extinction, government officials from Mexico and the United States worked together to help the species recover and establish stable populations. As a first step, nesting beaches were protected as refuges, reserves, and parks (see Chapter 9 for the importance of protected areas). Egg collection was banned. And at sea, shrimp trawlers were required to use turtle excluder devices (TEDs), consisting of a grid of bars with an opening that allows a caught turtle to escape.

In addition to reducing threats, a collaborative group of national and state agencies and conservation organizations in Mexico and the United States has undertaken an ambitious effort to increase nest and hatchling survival and to improve education and appreciation of sea turtle conservation. In the United States, national park authorities began to reestablish a population on Padre Island in Texas, where the species had formerly occurred. From 1978 to 1988, scientists, conservationists, and volunteers collected 22,507 eggs from Mexico, packed them in sand, and transported them to Padre Island National Seashore, which is managed by the US National Park Service (Caillouet et al. 1997). Because most turtles die as hatchlings, the turtles were reared in captivity for 9–11 months to allow them to grow large enough to avoid most predators before being released permanently into the Gulf of Mexico. These types of release programs will be discussed in more detail in Chapter 8.

Now, each year, the staff at Padre Island, many partner organizations, and over a hundred volunteers patrol the beach during the nesting season, searching for Kemp's ridleys and their nests. When they find nests, teams carefully excavate them and bring the eggs to an incubation facility or a large screen enclosure called a corral. When the young hatchlings are released, it is now a public event that doubles as an education tool—the hope is that the people watching each release will become advocates for the turtles' protection. Outside the national seashore, private conservation organizations also help protect the turtles on their feeding grounds (see more on conservation

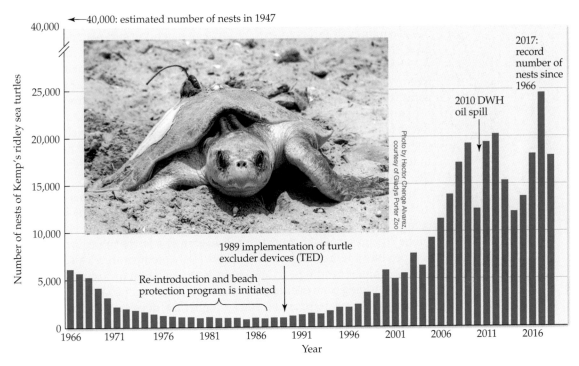

←—40,000: estimated number of nests in 1947

40,000

2017: record number of nests since 1966

2010 DWH oil spill

Number of nests of Kemp's ridley sea turtles

25,000

20,000

15,000

10,000

1989 implementation of turtle excluder devices (TED)

Re-introduction and beach protection program is initiated

5,000

0

1966   1971   1976   1981   1986   1991   1996   2001   2006   2011   2016

Year

Photo by Hector Chenge Alvarez, courtesy of Gladys Porter Zoo

FIGURE 1.7   A Kemp's ridley sea turtle (*Lepidochelys kempii*) with a tracking device mounted on its shell climbs out of the sand. The device transmits signals to a satellite, from which they can be recorded by scientists, allowing the movements of the turtle to be determined over long distances. Populations of these turtles increased after intense conservation efforts on the part of the US and Mexican governments and the implementation of devices that reduced turtle deaths associated with fishing. Despite ongoing egg collection and reintroduction, the population decreased after the *Deepwater Horizon* (DWH) oil spill in 2010. And yet a record number of nests was recorded in 2017. However, this is far from historic numbers, and another population dip in 2018 underlines the necessity of continued research and management. (Data for 1966–2014 from La Comisión Nacional de Áreas Naturales Protegidas, as reported in Gallaway et al. 2016. *Gulf of Mexico Science* 33 [2]. Data for 2015–2021 courtesy of Gladys Porter Zoo, Mexico's Comisión Nacional de Áreas Naturales Protegidas, and the Tamaulipas Commission of Parks and Biodiversity.)

outside of protected areas in Chapter 10). The program has been tremendously successful, and there are now nearly 18,000 nests on the island each year, on average, with over 750,000 hatchlings (**FIGURE 1.7**). Unfortunately, habitat degradation is still a leading threat to this endangered species (discussed in Chapter 4), with climate change not far behind (discussed in Chapter 5). The number of nests and hatchlings dropped by 35% in 2010, the same year as the *Deepwater Horizon* oil spill, which is believed to have killed hundreds of juvenile turtles (Caillouet et al. 2018). Turtles' contact with the oil was confirmed through satellite transmitter tracking (Figure 1.7 inset) as well as chemical analysis of their shells (Reich et al. 2017). Scientists determined that turtles were ingesting oil from the spill by measuring the abundance of certain forms of carbon molecules in their shells, a process called *stable isotope analysis*.

GLOBAL CHANGE
CONNECTION

However, despite the threat from the oil spill, a study of the nests in Mexico found predation rates to be low and hatchling survival high (Bevan et al. 2014), and in 2017 there were 24,586 nests recorded—the highest number in the 40-year history of the Kemp's Ridley Binational Project. Even with such success, periodic population dips since the oil spill raise concerns about food availability and habitat quality, highlighting the need for continued species management and research (Caillouet et al. 2018). Recent research also suggests that Kemp's ridley may be at risk from organic pollutants from pesticides, given the high levels found in the blood of individuals sampled in the Playa Rancho Nuevo Sanctuary in Tamaulipas, Mexico (Montes et al. 2020).

GLOBAL CHANGE CONNECTION

Climate change also poses several risks to this species. Higher than normal temperatures can affect sex ratios of eggs, since sex of the embryo is determined during the second third of incubation and can be changed by subtle changes in nest temperature. The warmer it is, the more likely the egg will produce a female (Bevan et al. 2019). Higher temperatures also threaten survivorship of nestlings; in one study of a population in Texas, nearly 23% of nestlings died from overheating or desiccation (Shaver 2020). Warmer temperatures also mean that sea turtles at higher latitudes aren't signaled to migrate south before the onset of late autumn storms, making them vulnerable to "cold-stunning" that leaves them paralyzed but alive (Griffin et al. 2019).

Even once they have begun to migrate south, they can still be in danger. In February of 2021, the largest cold-stunning event in recorded history in the United States took place: nearly 5000 sea turtles, including green, loggerhead, and Kemp's ridley, washed up on the shore of South Padre Island, Texas, following a rare cold front from the Arctic (Daly National Geographic 2021). Fortunately, residents rallied to help by transporting the cold-stunned turtles to the island's convention center, where the sea turtles could be warmed and revived (**FIGURE 1.8A**).

Sea turtles were expected to benefit from the quarantine associated with COVID-19, since there were fewer people (and their dogs) on beaches to trample nests or disturb hatchlings. However, the decimation of tourism in 2020 also meant decreased revenue that could have funded conservation efforts (CNN 2020). The loss of income and jobs also increased the threat of harvesting of eggs for food or profit. Meanwhile, fewer conservation volunteers in 2020 meant less work done surveying populations and protecting nests. Furthermore, the summer of 2020 was the hottest in recorded history for the Northern Hemisphere (NOAA 2020).

Fortunately, in the case of the Kemp's ridley sea turtle, determined staff followed COVID quarantine guidelines and continued their work. In addition to relocating eggs to nests in fenced enclosures where they could be monitored and protected from physical disturbance, staff also spread shade cloth over the area to protect the nests from overheating (**FIGURE 1.8B**). Despite all of the challenges, 2020 was a banner year for Kemp's ridley sea turtles and a credit to the efforts of many people; over 20,000 nests were recorded, and nearly a million hatchlings were released (see the chapter opening photo) (Burchfield et al. 2020).

(A)

(B)

Courtesy of Hector Raul Chenge Alvarez

Courtesy of Hector Raul Chenge Alvarez

FIGURE 1.8   Sea turtles are at risk from both high and low temperature anomalies.
(A) Here we see volunteers Irving A. Hernandez, David Daniel Barrera, and Robert "DJ"
Lerma rescue a cold-stunned green turtle. It was one of more than 4900 sea turtles
(including Kemp's ridley), a record number, that washed up on the beaches of South
Padre Island, Texas. (B) Technicians cover the area of the nests with black shade cloth
to protect them from record high temperatures that could alter sex ratios of eggs and/or
kill hatchlings. Individual nest *corralitos* are covered with white mosquito net to exclude
parasitic flies.

Scientific scrutiny, international partnerships, and the participation of
volunteers and local communities have brought the Kemp's ridley sea turtle
back from the brink of extinction, and all of those involved will continue
to seek the answers leading to its complete recovery. A variety of scientific
tools and conservation strategies have been useful in the recovery of this and
many other species of sea turtles that have had significant population gains
in recent years (Mazaris et al. 2017).

1.2 The Organizational Values of Conservation Biology

LEARNING OBJECTIVES

By the end of this section you should be able to:

1.2.1   Relate the five organizational values of conservation
biology to its history as a normative discipline.

1.2.2   Identify novel examples of biophilia.

Earlier in the chapter, we mentioned that conservation biology is a normative discipline in which certain values are embraced. The field rests on an

underlying set of principles that is generally agreed on by practitioners of the discipline (Soulé 1985; SCB 2016 [see "Organizational Values"]) and can be summarized as follows:

1. *Biological diversity has intrinsic value.* Species and the biological communities in which they live possess value of their own, regardless of their economic, scientific, or aesthetic value to human society. This value is conferred not just by their evolutionary history and unique ecological role, but also by their very existence. (See Chapter 3 for a more complete discussion of this topic.)

2. *The untimely extinction of populations and species should be prevented.* The ordinary extinction of species and populations as a result of natural processes is an ethically neutral event. In the past, the local loss of a population was usually offset by the establishment of a new population through dispersal. However, as a result of human activity, the loss of populations and the extinction of species have increased more than a hundredfold, with no simultaneous increase in the generation of new populations and species (see Chapter 6).

3. *The diversity of species and the complexity of biological communities should be preserved.* In general, most people agree with this principle simply because they appreciate biodiversity; it has even been suggested that humans may have a genetic predisposition to love biodiversity, called **biophilia** (**FIGURE 1.9**) (Wilson 2017). Many of the most valuable properties of biodiversity are expressed only in natural environments. Although the biodiversity of species may be partially preserved in zoos and botanical gardens, the ecological complexity that exists in natural communities will be lost without the preservation

> There are ethical reasons why people want to conserve biodiversity, such as belief that species have intrinsic value.

FIGURE 1.9 People enjoy seeing the diversity of life, as illustrated by the popularity of planting gardens and of public botanical gardens as tourist destinations. Butchart Gardens in Victoria, British Columbia, Canada, is shown here.

of natural areas (see Chapter 9). Furthermore, biodiversity has been directly linked to ecosystem productivity and stability (Hautier et al. 2015), among other values (see Chapter 3).

4. *Science plays a critical role in our understanding of ecosystems.* It is not enough to simply value diversity and protect natural spaces; objective research is necessary to identify which species and environments are at greatest risk, as well as to understand the nature of these risks and how to mitigate them. Ideally, scientists are involved in all stages of conservation, including implementation of conservation actions and monitoring results.

5. *Collaboration among scientists, managers, policymakers, and the public is important and often necessary.* In order to achieve the goals of reduced extinction rates and preservation of biological communities, high-quality research findings must be shared with those who create the laws and provide the funding for conservation actions, with those who must implement them, and with those who live and work in the areas affected. These actions are most likely to succeed when they are based on scientifically sound information and are supported by local people.

Not every conservation biologist accepts every one of these principles, and there is no hard-and-fast requirement to do so. Individuals or organizations that agree with even two or three of these principles are often willing to support conservation efforts. Current progress in protecting species and ecosystems has been achieved in part through partnerships between traditional conservation organizations, such as The Nature Conservancy, and cattle ranchers, hunting clubs like Ducks Unlimited, and other groups with a vested interest in the health of ecosystems.

1.3 Looking to the Future

The field of conservation biology has set itself some imposing—and absolutely critical—tasks: to describe Earth's biological diversity, to protect what remains, and to restore what is degraded. The field is growing in strength, as indicated by increased governmental participation in conservation activities, increased funding of conservation organizations and projects, and an expanding professional society.

In many ways, conservation biology is a crisis discipline. Decisions about selecting national parks, species management, and other aspects of conservation are made every day under severe time pressure (Martin et al. 2017). As one of the guiding values mentioned above, biologists and scientists in related fields seek to provide the advice that governments, businesses, and the general public need in order to make crucial decisions, but because of time constraints, scientists are often compelled to make recommendations without thorough investigation. Decisions must be made, with or without scientific input, and conservation biologists must be willing to express opinions and take action based on the best available evidence and informed judgment (Garrard et al. 2016). They must also articulate a long-term conservation vision that extends beyond the immediate crisis (Wilhere 2012).

Despite the threats to biodiversity and the limitations of our knowledge, we can detect many positive signs that allow conservation biologists to be cautiously hopeful (Godet and Devictor 2018; Roman et al. 2015). Indeed, optimism may be critical for the field and our ability to motivate positive change (Beever 2020). Per capita energy use in the United States, while still high, has been decreasing; in 2014 it was the lowest it had been since the 1960s (World Bank 2014), and the percent of renewable energy had grown to 11% in 2019 (US Energy Information Administration 2021). As a consequence, the total amount of energy consumed has not grown in the past couple of decades, despite a growing population. The number of protected areas around the globe continues to increase, particularly the number of marine protected areas. The involvement of the public in collecting meaningful data and advocating for conservation continues to rise, facilitated by the ubiquity of smartphones, user-friendly apps, and social media. Technology and science are evolving, providing tools for both the study and preservation of species.

These gains are due in part to action spurred by the public. Increasing numbers of social and religious leaders have rallied for the protection of biodiversity, while others become leaders because of their conservation message. Figures who champion protecting life on Earth have increasingly become more famous. As an example, Greta Thunberg, a teenager from Sweden, spoke out about climate change, became *Time*'s Person of the Year, and made the *Forbes* list of the World's 100 Most Powerful Women in 2019. The optimism and determination of such individuals is powerful. As Ms. Thunberg said to the US Congress:

GLOBAL CHANGE
CONNECTION

> *You must unite behind the science. You must take action. You must do the impossible. Because giving up can never ever be an option.*
>
> – Greta Thunberg, September 2019, Washington, DC

Our ability to protect biodiversity has been strengthened by a wide range of local, national, and international efforts. Many endangered species are now recovering as a result of such conservation measures (IUCN 2019). Effective action has resulted from our continuing expansion of knowledge in conservation science, the developing linkages with rural development and social sciences, and our increased ability to restore degraded environments. All of these advances suggest that progress is being made, despite the enormous tasks still ahead.

Summary

- Human activities are causing the extinction of thousands of species both locally and globally, with threats to species and ecosystems accelerating due to human population growth and the associated demands for resources.
- Conservation biology is a field that combines basic and applied disciplines with three goals: to describe the full range of biodiversity on Earth; to understand human impact on biodiversity; and to develop practical

approaches for preventing species extinctions, maintaining genetic diversity, and protecting and restoring ecosystems.

■ Elements of conservation biology can be found in many cultures, religions, and forms of creative expression, and they began to develop in the United States and Europe in the nineteenth and twentieth centuries. The modern field of conservation biology became a recognized scientific discipline with a professional society and academic journals by the 1980s.

■ Conservation biology rests on a number of underlying assumptions that are accepted by most professionals in the discipline: biodiversity has value in and of itself; extinction from human causes should be prevented; diversity at multiple levels should be preserved; science plays a critical role, and scientists must collaborate with nonscientists to achieve our goals.

■ The conservation of biodiversity has become an international undertaking. Many successful projects, such as the conservation of Kemp's ridley sea turtles, indicate that progress can be made.

For Discussion

1. Explain the connection between human population growth and species loss.

2. How is conservation biology fundamentally different from other branches of biology, such as physiology, genetics, or cell biology? How is it similar to the science of medicine? How is it different from environmentalism?

3. What do you think are the major conservation and environmental problems facing the world today? What are the major problems facing your local community? What ideas for solving these problems can you suggest? (Try answering this question now, and once again when you have completed this book.)

4. Consider the public land management and private conservation organizations with which you are familiar. Do you think their guiding philosophies are closest to the resource conservation ethic, the preservation ethic, or the land ethic? What factors allow them to be successful or limit their effectiveness? Learn more about these organizations through their publications and websites.

5. How would you characterize your own viewpoint about the conservation of biodiversity and the environment? Which of the religious or philosophical viewpoints of conservation biology stated here do you agree or disagree with? How do you, or could you, put your viewpoint into practice?

Suggested Readings

Carson, R. 1962. *Silent Spring*. Houghton Mifflin, Boston. Essays written over a period from 1958 to 1962 on the devastating effects of pesticide use on ecosystems, particularly on birds.

Godet, L., and V. Devictor. 2018. What conservation does. *Trends in Ecology & Evolution* 33: 720–730. A quantitative review of the conservation literature that reveals global patterns and reasons for optimism.

Kloor, K. 2015. The battle for the soul of Conservation Science. *Issues in Science and Technology* 31: 74. The ongoing debate between those who believe that the field should be primarily guided by "nature for nature's sake" and others who believe that human needs should have greater weight.

Leopold, A. 1949. *A Sand County Almanac*. Oxford University Press, New York. Leopold's evocative essays articulate his "land ethic," defining human duty to conserve the land and the living things that thrive upon it.

Mazaris, A. D., et al. 2017. Global sea turtle conservation successes. *Science Advances* 3: e1600730. While many species are doing better, most still require intensive conservation effort to secure their future.

Rutz, C., et al. 2020. COVID-19 lockdown allows researchers to quantify the effects of human activity on wildlife. *Nature Ecology & Evolution* 4: 1156–1159. The pandemic changed our impacts on nature, including both increases and decreases, creating a unique opportunity to study anthropogenic effects.

Soulé, M. E. 1985. What is conservation biology? *BioScience* 35: 727–734. A key early paper defining the field, still relevant today for its emphasis on the intrinsic value of biodiversity.

Tilman, D., et al. 2017. Future threats to biodiversity and pathways to their prevention. *Nature* 546: 73. There are means to address conservation threats. Proactive international efforts could still preserve much of the planet's remaining biodiversity.

WWF (World Wildlife Fund). 2020. *Living Planet Report 2020: Bending the Curve of Biodiversity Loss*. R. E. A. Almond, M. Grooten, and T. Petersen (eds). WWF, Gland, Switzerland. Alarming losses still threaten species and ecosystems in many parts of the world.

KEY JOURNALS IN THE FIELD *Biodiversity and Conservation, Biological Conservation, BioScience, Conservation Biology, Conservation Letters, Ecological Applications, National Geographic, Trends in Ecology and Evolution.*

Biodiversity can be considered at a range of scales, from the genetic diversity among these Indian paintbrush flowers to the diversity of ecosystems. Ecosystems shown here include sub-alpine meadow, forest, and alpine, each with their own unique species assemblages.

What Is Biodiversity?

2

The protection of biological diversity is central to conservation biology. Conservation biologists use the term *biological diversity*, or simply *biodiversity*, to mean the complete range of species and biological communities on Earth, as well as the genetic variation within those species and all ecosystem processes. By this definition, biodiversity must be considered on at least three levels (**FIGURE 2.1**):

1. *Species diversity*   All the species on Earth, including single-celled bacteria and protists as well as the species of the multicellular kingdoms (plants, fungi, and animals) may be included in this level of diversity. As we will see in this chapter, there are several ways that species diversity can be measured for a particular geographic region or ecosystem.

2. *Genetic diversity*   The genetic variation within species, both among geographically separate populations and among individuals within single populations. A **population** is a group of individuals that mate with one another and produce offspring; species may contain one population or many.

3. *Ecosystem diversity*   The different biological communities and their associations with the chemical and physical environment (the **ecosystem**). A **community** is an assemblage of interacting populations of different species living in a particular area.

In ecology, the term *ecosystem* is applied to biological communities at widely different scales, from the bacteria in the gut of an individual mouse to the hundreds of square kilometers of boreal forest across a mountain range. However, ecosystem diversity in a conservation

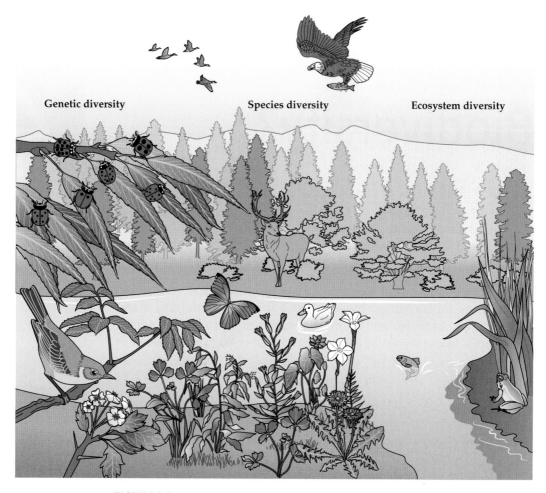

Genetic diversity   Species diversity   Ecosystem diversity

FIGURE 2.1 Biological diversity includes genetic diversity (e.g., the genetic variation found within each species, as shown with the ladybird beetles), species diversity (e.g., the range of species in a given ecosystem), and ecosystem diversity (e.g., the variety of habitat types and ecosystem processes extending over a given region).

context typically refers to the different types of communities and their associated environments within a given geographic area (see chapter opening photo). An example of ecosystem diversity is the wetland, forest, and grassland ecosystems all included within a protected wilderness area (see Chapter 9); diversity at this level would likely be lost if the area were instead converted to farmland.

Note also that this three-tiered list is a simplification that includes only the most often studied and quantified levels; conservation biologists can also be concerned with diversity at other, intermediate scales. For example, genetic diversity can be considered in terms of population diversity, that is, populations with different genetic compositions within a species. Community diversity can refer to unique assemblages of different populations of species, and so on. These different levels of biodiversity are necessary for the continued survival of life as we know it (Methorst et al. 2021; Mori et al. 2017; Xu et al. 2017). All of these levels are also currently facing significant threats, to be discussed in Chapters 4, 5, and 6, although threats to species diversity tend to receive the most attention.

Species diversity reflects the entire range of evolutionary and ecological adaptations of species to particular environments. It provides people with resources and resource alternatives; for example, a tropical rain forest or a temperate swamp with many species produces a wide variety of plant and animal products that can be used as food, shelter, and medicine. **Genetic diversity** is necessary for any species to maintain reproductive vitality, resistance to disease, and the ability to adapt to changing conditions. For example, it was discovered that genetic variability helped rare frog populations survive a deadly fungal disease (Savage et al. 2018). Genetic diversity is also of value in the breeding programs necessary to sustain and improve modern domesticated plants and animals and their disease resistance (Mastretta-Yanes et al. 2018). **Ecosystem diversity** results from the collective response of species to different environmental conditions. Biological communities found in deserts, grasslands, wetlands, and forests support the continuity of proper ecosystem functioning, which provides crucial services to people, such as water for drinking and agriculture, flood control, protection from soil erosion, and filtering of air and water (these different values are discussed further in Chapter 3). We will examine each level of biodiversity in this chapter.

2.1 Species Diversity

LEARNING OBJECTIVES

By the end of this section you should be able to:

2.1.1 Distinguish among different species definitions.

2.1.2 Explain the specific difficulties in identifying what a species is.

2.1.3 Calculate alpha, gamma, and beta species diversity for a sample population/region.

2.1.4 Calculate diversity using the Shannon index for a sample population when relative abundances are provided.

Recognizing and classifying species is one of the major goals of conservation biology. Identifying the process whereby one species evolves into one or more new species is one of the ongoing accomplishments of modern biology. The origin of new species is normally very slow, taking place over hundreds, if not thousands, of generations. The evolution of higher taxa, such as new genera and families, is an even slower process, typically lasting hundreds of thousands or even millions of years. In contrast, human activities are destroying in only a few decades the unique species built up by these slow natural processes.

GLOBAL CHANGE
CONNECTION

Every four years the world's largest conservation event is held: the International Union for Conservation of Nature's World Conservation Congress (for more about the IUCN, see Chapter 7 and Chapter 9). One of the tasks of this international meeting of scientists, policymakers, and others is to come to agreements about the number of species that have currently been **described** (officially identified by science) and, for those species with enough information, how safe or endangered they are. The product of the meeting is a new formal list, representing the result of years of work by thousands of people. One of the reasons it takes so long is because what counts as a unique species is not always clear.

What is a species?

Although seemingly a straightforward concept, how to distinguish a selection of organisms as a species is subject to great scientific discussion, and at least seven different ways of doing this have been proposed (Wiens 2007). Three are most commonly used in conservation biology:

1. *Morphological species*: A group of individuals that appear different from others—that is, that are morphologically distinct. A group that is distinguished exclusively by such visible traits as form or structure may be referred to as a **morphospecies**.

2. *Biological species*: A group of individuals that can potentially breed among themselves in the wild and that do not breed with individuals of other groups.

3. *Evolutionary species*: A group of individuals that share unique similarities in their DNA and hence their evolutionary past.

Because they rely on different methods and assumptions, these three approaches to distinguishing species sometimes do not give the same results. Increasingly, DNA sequences and other molecular markers are being used to identify and distinguish species that look almost identical. For example, a commonly harvested emperor fish in the southwest Indian Ocean that was assumed from morphological characteristics to be a single species was discovered through genetic analysis to actually be two species, with important implications for their management (Healey et al. 2018).

Using morphological and genetic information to identify species is a major activity for taxonomists; accurate identification of a species is a necessary first step in its conservation.

The **morphological definition of species** is the one most commonly used by **taxonomists**, biologists who specialize in the identification of unknown specimens and the classification of species. The **biological definition of species** is widely accepted, but it is problematic for groups of organisms in which different species readily interbreed, or **hybridize**, such as plants. Furthermore, the biological definition of species is difficult to use because it requires a knowledge of which individuals have the potential to breed with one another. Similarly, the **evolutionary definition of species** requires access to expensive laboratory equipment and so cannot be used in the field. As a result, field biologists must rely on observable attributes, and they may name a group of organisms as a morphospecies until taxonomists can investigate them more carefully to determine whether they are a distinct species (e.g., Chan et al. 2015).

Ideally, specimens collected in the field are catalogued and stored in one of the world's >6500 natural history museums (for animals and other organisms) or its >300 major herbaria (for plants and fungi) (**FIGURE 2.2**). Increasingly, these physical records of global biodiversity are being digitized and shared on the World Wide Web for greater access and scientific use. Permanent collections curated by museums and herbaria form the basis of species descriptions and systems of classification. Each species is given a **binomial**—a unique two-part name—such as *Giraffa tippelskirchi* for the Maasai giraffe. The first part of the binomial, *Giraffa*, identifies the genus (the giraffes). The second part, *tippelskirchi*, identifies the smaller group within the genus, the species that is the Maasai giraffe, the iconic animal that lives in East Africa. This naming system both separates the Maasai giraffe from and connects it to similar species—such as *Giraffa camelopardalis*, the northern giraffe, and

(A)

Photograph courtesy of Museum of Comparative Zoology, Harvard University, © President and Fellows of Harvard College

(B)

© Scott Dressel-Martin · Denver Botanical Garden

FIGURE 2.2 (A) An ornithologist at the Museum of Comparative Zoology, Harvard University, classifying collections of orioles: black-cowled orioles (*Icterus prosthemelas*), from Mexico, and Baltimore orioles (*Icterus galbula*), which occur throughout eastern North America. (B) Modern museums, including herbaria, are increasingly using digitization to safeguard data and make it available to a broader global community. At the Kathryn Kalmbach Herbarium of Vascular Plants, high-resolution photos are taken of specimens in a specially designed light box. These images and associated specimen data are put online and shared among numerous digital museum data aggregators, such as the Global Biodiversity Information Facility.

Giraffa reticulata, the reticulated giraffe. In some cases for animals, there may be the same name for both genus and species, such as *Giraffa giraffa*, the southern giraffe. *Bison bison*, the American bison, and *Gorilla gorilla*, the western gorilla, are other examples; such species were likely the first of their genus to be classified.

Problems in distinguishing and identifying species are more common than many people realize. For example, a single species may have several varieties that have observable morphological differences yet are similar enough to be a single biological or evolutionary species. Different varieties of dogs, such as German shepherds, collies, and beagles, all belong to one species; their genetic differences are actually very small, and they readily interbreed. Alternatively, closely related "sibling" species appear very similar in morphology and physiology yet are genetically quite distinct (**FIGURE 2.3**). These are also known as **cryptic species**, which include the emperor fish discussed earlier.

To further complicate matters, individuals of related but distinct species may occasionally mate and produce **hybrids**, intermediate forms that blur the distinction between species. Hybridization is particularly common among plant species in disturbed habitats. Hybridization in both plants and animals frequently occurs when a few individuals of a rare species are surrounded by large numbers of a closely related species. For example, the endangered California tiger salamander (*Ambystoma californiense*) and the introduced

(A)

(B)

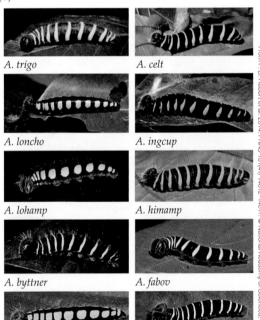

A. trigo   A. celt

A. loncho   A. ingcup

A. lohamp   A. himamp

A. byttner   A. fabov

A. yesenn   A. sennov

FIGURE 2.3 (A) The two-barred flasher butterfly (*Astraptes* spp.) is a cryptic species complex in the skipper butterfly genus. (B) Although the individuals are nearly identical as adults, differences at the larval stage prompted further investigation into differences between populations; a molecular analysis tool called DNA barcoding revealed that they were in fact different species.

Amado Demensa/CC BY-SA 2.0

From P. D. Hebert et al. 2014. *PNAS* 101(41) 14812–14817. © National Academy of Sciences, U.S.A.

From J. R. Johnson et al. 2010. *Funct Ecol* 24: 1073–1080. © 2010 The Authors. Journal compilation © British Ecological Society

FIGURE 2.4 The hybrid tiger salamander (left) is larger than its endangered parent species, the California tiger salamander (right), and is increasing in abundance. Note the much larger head of the hybrid salamander.

barred tiger salamander (*A. mavortium*) are thought to have evolved from a common ancestor 5 million years ago, yet they readily mate in California (**FIGURE 2.4**). The hybrid salamanders have a higher fitness and are better able to tolerate environmental pollution than the native species, *A. californiense*, further complicating the conservation of this endangered species (Ryan et al. 2013).

The inability to clearly distinguish one species from another, whether due to similarities of characteristics or to confusion over the correct scientific name, often slows down efforts at species protection. It is difficult to write precise, effective laws to protect a species if scientists and lawmakers are not certain which individuals belong to which species. At the same time, species are going extinct before they are even described. More than 10,000 new species are being described each year, but even this rate is not fast enough. The key to solving this problem is to train more taxonomists and improve scientific collaboration, especially in the species-rich tropics (Baker et al. 2017).

Those conservation biologists primarily concerned with ecosystem function rather than individual species extinction have argued that a better measure than species diversity is **functional diversity**—that is, the diversity of organisms categorized by their ecological roles or traits rather than their taxonomy (Díaz and Cabido 2001; Gagic et al. 2015). Functional diversity has been found to increase ecosystem resilience against change (de la Riva et al. 2017) and is arguably more important than species diversity for understanding processes on the ecosystem scale (Dawud et al. 2017), such as in response to human interventions (Henry et al. 2021). Because of this, functional diversity is an especially important concept in the context of habitat restoration (see Chapter 11). However, if our goal is to prevent untimely extinctions, we cannot avoid the task of identifying species and measuring species diversity.

GLOBAL CHANGE
CONNECTION

Measuring species diversity

Conservation biologists often want to identify locations of high species diversity. Quantitative definitions of species diversity have been developed by ecologists as a means of comparing the overall diversity of different communities at varying geographic scales (Bhatta et al. 2018).

At its simplest level, species diversity can be defined as the number of species present, called **species richness**. This number can be determined by several methods and at different geographic scales. Three diversity measurements are based on species richness:

- **Alpha diversity** is the number of species found in a given community, such as a lake or a meadow.

- **Gamma diversity** is the number of species at larger geographic scales that include a number of ecosystems, such as a mountain range or a continent.

- **Beta diversity** links alpha and gamma diversity and represents the rate of change of species composition as one moves across a large region. For example, if every lake in a region contained a similar array of fish species, then beta diversity would be low; on the other hand, if the bird species found in one forest were entirely different from the bird species in separate but nearby forests, then beta diversity would be high. There are several ways of calculating beta diversity; a simple measure of beta diversity can be obtained by dividing gamma diversity by alpha diversity.

We can illustrate these three types of diversity with a theoretical example of three mountain ranges (**FIGURE 2.5**). Region 1 has the highest alpha diversity, with more species per mountain on average (6 species) than the other two regions. Region 2 has the highest gamma diversity, with a total of 10 species. Dividing gamma diversity by alpha diversity shows that Region 3 has a higher beta diversity (2.7) than Region 2 (2.5) or Region 1 (1.2) because all but one of its species are found on only one mountain each. In practice, indexes of diversity are often highly correlated. The plant communities of the eastern foothills of the Andes, for instance, show high levels of diversity at alpha, beta, and gamma scales.

Identifying patterns of species diversity helps conservation biologists establish which locations are most in need of protection.

More-complex indexes, such as the **Shannon diversity index** (also called the Shannon-Wiener index), the Simpson index, and the Pielou evenness index, take the relative abundance of different species into account; by these measures, a community dominated by a few species is less diverse than one with a more even distribution of species, even with the same species richness. The Shannon diversity index (H) is calculated as

$$H = -\sum \left[p_i \times \ln(p_i) \right]$$

That is, the proportion (p) of each species (i) is multiplied by the natural log (ln) of that proportion, and then the sum of those results is multiplied by –1. In a simple example, let's imagine two ponds, each of which has five fish species. In

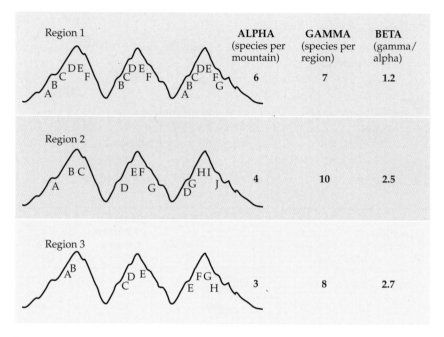

FIGURE 2.5 Biodiversity indexes for three regions, each consisting of three separate mountains. Each letter represents a population of a species; some species are found on only one mountain, while other species are found on two or three mountains. Alpha, gamma, and beta diversity values are shown for each region. If funds were available to protect only one region, Region 2 should be selected because it has the greatest gamma (total) diversity. However, if only one mountain could be protected, a mountain in Region 1 should be selected because these mountains have the highest alpha (local) diversity—that is, the greatest average number of species per mountain. Each mountain in Region 3 has a more distinct assemblage of species than the mountains in the other two regions, as shown by the higher beta diversity. If Region 3 were selected for protection, the relative priority of the individual mountains should then be judged based on how many unique species were found on each mountain.

Pond A, 60% of the individuals are orange carp and each of the remaining four species only represents 10% of the individuals. In Pond B there are also five fish species, but all the species have the same number of individuals, or 20% of the total. Using the Shannon diversity index, Pond B will have a greater diversity than Pond A (**FIGURE 2.6**). In some cases, one pond may even have a greater number of species but a lower diversity index than another pond if its community is dominated by one or just a few species. Note that, like the richness values explained previously, diversity measures of this type can be calculated at different scales and therefore are useful only as relative, rather than absolute, values. Furthermore, these quantitative definitions of diversity capture only part of the broad definition of biodiversity used by conservation biologists, and new ones continue to be developed (Iknayan et al. 2014). Although each has its limitations, they are useful for comparing regions and highlighting areas that have large numbers of native species requiring conservation protection.

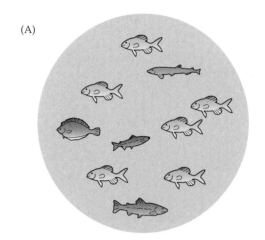

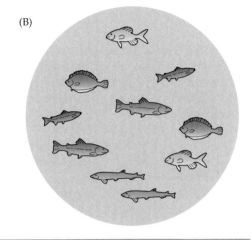

	Number in sample (*n*)	*n*/total = proportion in sample (*p*)	ln(*p*)	ln(*p*) × (*p*)
	6	0.60	−0.51	−0.31
	1	0.10	−2.30	−0.23
	1	0.10	−2.30	−0.23
	1	0.10	−2.30	−0.23
	1	0.10	−2.30	−0.23
Total	10 fish	1 (100%)		−1.23
				H = 1.23

	Number in sample (*n*)	*n*/total = proportion in sample (*p*)	ln(*p*)	ln(*p*) × (*p*)
	2	0.20	−1.61	−0.32
	2	0.20	−1.61	−0.32
	2	0.20	−1.61	−0.32
	2	0.20	−1.61	−0.32
	2	0.20	−1.61	−0.32
Total	10 fish	1 (100%)		−1.61
				H = 1.61

FIGURE 2.6 If each circle represents a random sample of fish from a pond and colors represent species, both have the same species richness: 5. However, the Pond A sample is dominated by a single species (6 orange fish out of a total of 10 fish, or 60% of the total), while each of the other 4 species has only 10% of the total. In contrast, the sample from Pond B has perfect evenness; that is, each of the 5 species has the same number of individuals, or 20% of the total. Therefore, we would consider Pond B to have greater species diversity. We can further quantify this by calculating *H*, a measure of diversity, as shown in each table, with Pond A having a diversity of 1.23 and Pond B having a diversity of 1.61.

2.2 Genetic Diversity

LEARNING OBJECTIVES

By the end of this section you should be able to:

2.2.1 Predict possible outcomes of novel situations in which there is limited or no genetic diversity.

2.2.2 Identify when molecular tools are appropriate for different conservation applications.

Conservation biology also concerns itself with the preservation of genetic diversity within a species. This level of diversity is important because it provides evolutionary flexibility: when environmental conditions change, a genetically diverse species is more likely to have traits that allow it to adapt. Rare species often have less genetic variation than widespread species and, consequently, are more vulnerable to extinction (Szczecin´ska et al. 2016) (see Chapter 6).

How does genetic diversity arise?

Genetic diversity arises because individuals have slightly different forms of their **genes**, the units of the chromosomes that specify the synthesis of specific proteins. These different forms of a gene are known as **alleles**, and their physical position on the chromosome is the gene's **locus** (plural is *loci*). The different alleles originally arise through **mutations**—changes that occur in the deoxyribonucleic acid (DNA) that constitutes an individual's chromosomes. These changes to the DNA can be minute or quite large, as occurs when entire segments of DNA move about the genome; these DNA sequences are called **transposable elements**. For example, it was discovered that transposable elements were responsible for a mutation that turned British peppered moths (*Biston betularia*) from light to dark (van't Hof et al. 2016). Most mutations do not result in new traits, but when they do, they contribute to genetic variation that could become important for a population's survival. When pollution during the Industrial Revolution killed lichens, thus turning previously light-barked trees black, those individual moths with the mutation giving them dark wings were better able to hide from predators (Cook et al. 2012). In more recent times, however, environmental regulations have led to cleaner air, and lighter-winged moths have become more abundant. Thus, genetic diversity arising from mutations has been important for the persistence of a species in the face of environmental change over time.

Although the peppered moth example shows how a single allele can make a profound difference for a species, most traits occur because of the combined effects of many genes. Thus, genetic variation can also increase when offspring receive unique combinations of genes and chromosomes from their parents via the **recombination** of genes that occurs during sexual reproduction. Genes are exchanged between chromosomes, and new combinations are created when chromosomes from two parents combine to form a genetically unique offspring. Although mutations provide the basic material for genetic variation, the random rearrangement of alleles in different combinations that characterizes sexually reproducing species dramatically increases the potential for genetic variation (**FIGURE 2.7**).

The total array of genes and alleles in a population is the **gene pool** of the population, while the particular combination of alleles that any individual possesses is its **genotype**. The **phenotype** of an individual represents the morphological, physiological, anatomical, and biochemical characteristics of that individual that result from the expression of its genotype in a particular environment. Examples of phenotypes include eye color and blood type, physical qualities that are determined predominantly by an individual's genotype.

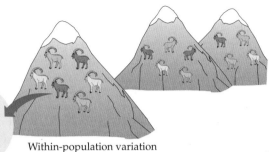

Two different
alleles of gene A

A   a

B   B

Individual genes
A and B

Within-individual variation

Chromosomes
(one from each parent)

Individual

Within-population variation

Among-population variation

FIGURE 2.7   There is genetic variation within individuals due to variation in the alleles of genes and variation between chromosomes. There is also genetic variation between individuals within populations and among separate populations. (After K. T. Scribner et al. 2006. In *Principles of Conservation Biology*, 3rd Ed., M. J. Groom et al. [Eds.], p. 377. Oxford University Press/Sinauer, Sunderland MA.)

The amount of genetic variation in a population is determined both by the number of **polymorphic genes**—genes that have more than one allele—and by the number of alleles for each of these genes. The existence of a polymorphic gene also means that some individuals in the population will be **heterozygous** for the gene; that is, they will receive a different allele of the gene from each parent. On the other hand, some individuals will be **homozygous**; they will receive the same allele from each parent. All these levels of genetic variation contribute to a population's (and therefore a species') ability to adapt to a changing environment. We will see examples in Chapter 6 of how a species' survival can be affected by its capacity to evolve.

Measuring genetic diversity

As you can see, genetic diversity can be quantified in several ways, including measuring rates of homozygosity, figuring number and/or abundance of alleles for a gene, and identifying combinations of genes that contribute to trait diversity within a population or a species. *Conservation genetics* refers to the use of such genomic information to address issues within conservation biology. Genetic tools can also allow us to identify new species and determine when hybridization has occurred. Other uses include but are not limited to the following:

- Describing or comparing the **genetic structure** (patterns of genotypes) of managed and wild populations
- Determining the capacity of a small or endangered population to respond to environmental change
- Observing population dynamics in the context of human impacts

- Detecting the presence of a cryptic species by sampling DNA in the environment
- Identifying species and origin of biological material as a means of controlling trade in endangered species

These uses of genomics in conservation will be discussed in greater detail in the chapters that follow; as you will see, this technology touches nearly every area of conservation biology today.

In most of the uses just mentioned, genotypes of a sample of individuals are determined, such as is done with leaf samples of individual rare plants (Zhou et al. 2018). Alternatively, the diversity of an entire population can be sampled at once, such as by genotyping the microorganisms in soil or water samples, referred to as **environmental DNA (eDNA) sampling** (Deiner et al. 2021; Delgado-Baquerizo et al. 2018). Environmental DNA sampling can also be used to detect the presence of specific species that are otherwise difficult to track, such as in testing water to detect rare amphibians (Goldberg et al. 2018). The use of genomics tools ranges from identifying DNA sequences themselves to looking at the diversity of proteins created by those sequences (called allozymes), among many other approaches. The genetic material used for such analyses may be taken from the nucleus of the cell or, alternatively, from organelles such as mitochondria or chloroplasts. These latter sources of DNA are inherited only from the egg, not the sperm, and thus can provide information about the matrilineal line. Genomics is a rapidly evolving field that is increasingly being used to advance the goals of conservation biology.

> Genetic variation within a species can allow the species to adapt to environmental change. New technologies allow scientists to measure multiple types of genetic diversity.

2.3 Ecosystem Diversity

LEARNING OBJECTIVES

By the end of this section you should be able to:

2.3.1 Draw, interpret, and analyze a food web, identifying trophic levels.

2.3.2 Apply the concept of "keystone" to specific species and resources.

Ecosystems are diverse, and this diversity is apparent even across a particular landscape. As we climb a mountain, for example, the structure of the vegetation and the kinds of plants and animals gradually change from those found in a tall forest to those found in a low, moss-filled forest to alpine meadow to cold, barren rock (see the chapter opening photo). As we move across the landscape, physical conditions (soil, temperature, precipitation, and so forth) change. One by one, the species present at our starting point drop out, and we encounter new species that were not found there. The landscape as a

whole is dynamic and changes in response to the overall environment and the types of human activities that are associated with it. For example, climate change is expected to significantly change distributions of species and thus the biodiversity found at any given location (Weiskopf et al. 2020). In this section, we will review some key ecological terms and processes that underlie biodiversity at different scales.

What are communities and ecosystems?

A **biological community** is defined as the species that occupy a particular locality and the interactions among those species. A biological community, together with its associated physical and chemical environment, is termed an ecosystem (**FIGURE 2.8**). Many characteristics of an ecosystem result from ongoing processes, including water cycles, nutrient cycles, and energy capture. These processes occur at geographic scales that range from square meters to hectares to regional scales involving tens of thousands of square kilometers (see Table 1.1 for definitions of these metric terms). For example, in a temperate forest, rain falls and is absorbed by the soil. Some of that rain evaporates from the surface, some percolates to groundwater reserves, and some is taken up by plants that use it in photosynthesis, converting atmospheric CO_2 into carbohydrates. These plants may then be eaten by animals, which convert the carbohydrates to energy through respiration, which releases CO_2 back into the environment. Other plants may decompose on the forest floor, releasing nutrients and providing energy for bacteria, fungi, and animals.

The physical environment, especially annual cycles of temperature and precipitation and the characteristics of the land surface, affects the structure and characteristics of a biological community and profoundly influences whether a site will support a forest, grassland, desert, or wetland. In aquatic ecosystems, physical characteristics such as water turbulence and clarity, as well as water chemistry, temperature, and depth, affect the characteristics of the associated **biota** (a region's flora and fauna). In turn, the biological community can alter the physical characteristics of an environment. For example, wind speeds are lower and humidity is higher inside a forest than in a nearby grassland. Marine communities such as kelp forests and coral reefs can affect the physical environment as well by buffering wave action.

Within a biological community, species play different roles and differ in what they require to survive. For example, many species of plants are pollinated only by certain types of insects and/or have their seeds dispersed by certain bird species (Carstensen et al. 2018). Similarly, animal species differ in their requirements, such as the types of food they eat and the types of resting and breeding places they prefer, collectively referred to as **habitat**. Even though a forest may be full of vigorously growing green plants, an insect species that feeds on only one rare and declining plant species may be unable to develop and reproduce because it cannot get the specific food that it requires. Any of these requirements may become a **limiting resource** when it restricts the size of a population.

> Within a community, each species has its own requirements for food, temperature, water, and other resources, any of which may limit its population size and its distribution.

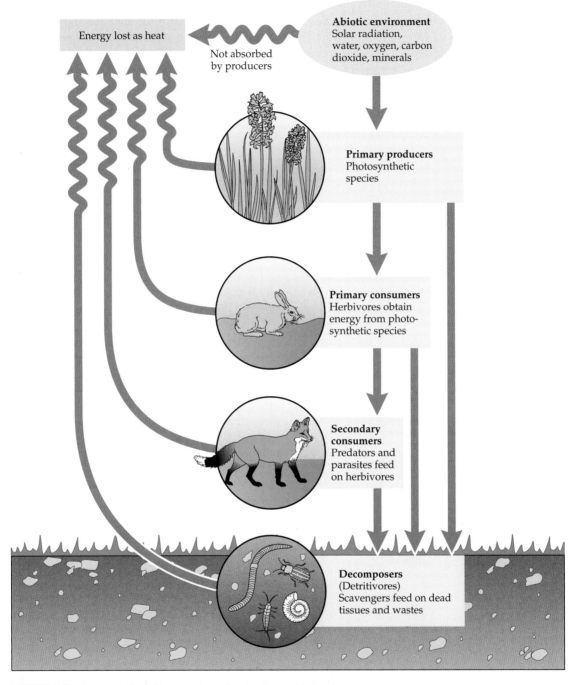

FIGURE 2.8 A model of a field ecosystem, showing its trophic levels and simplified energy pathways.

Species interactions within ecosystems

The composition of ecosystems is often affected by **competition** and **predation** (Sher and Molles 2021). **Predators** are animals that hunt and eat **prey**, which are the organisms that are eaten. Predation on plants is generally referred to as **herbivory**. Predators of all types may dramatically reduce the densities of certain prey species and even eliminate some species from particular habitats. Indeed, predators may indirectly increase the number of prey species in an ecosystem by keeping the density of each species so low that severe competition for resources does not occur.

In many ecosystems, predators keep the number of individuals of a particular prey species below the number that the resources of the ecosystem can support, a number termed the habitat's **carrying capacity**. If the predators (e.g., wolves; *Canis lupus*) are removed by hunting, poisoning, or some other human activity, the prey population (e.g., deer; *Odocoileus* spp.) may increase to carrying capacity, or it may increase beyond carrying capacity to a point at which crucial resources are overtaxed and the population crashes. In addition, the population size of a species may be controlled by other species that compete with it for the same resources; for example, the population size of terns that nest on a small island may decline or grow if a gull species that uses the same nesting sites becomes abundant or is eliminated from the site.

Community composition is also affected when two species benefit each other in a **mutualism**. Mutualistic species reach higher densities when they occur together than when only one of the species is present. One example of mutualism is the relationship between fruit-eating birds and plants with fleshy fruit containing seeds that are dispersed by birds (**FIGURE 2.9**). Another example is flower-pollinating insects and flowering plants. In some cases, these mutualistic relationships are **symbiotic**, and the species apparently cannot survive without each other. For example, certain symbiotic algae living inside coral animals are ejected following unusually high water temperatures in tropical areas, leading to the weakening and subsequent death of their associated coral species (see Figure 5.4).

Trophic levels

Biological communities can be organized into **trophic levels** that represent the different ways in which species obtain energy from the environment (see Figure 2.8). **Primary producers** make up the *first trophic level*. These organisms obtain their energy directly from the sun via photosynthesis. In terrestrial environments, higher plants, such as flowering plants, gymnosperms, and ferns, are responsible for most photosynthesis, while in aquatic environments, seaweeds, single-celled algae, and cyanobacteria (also called blue-green

FIGURE 2.9 This bohemian waxwing (*Bombycilla garrulus*) will disperse the fruit of a mountain ash tree (*Sorbus aucuparia*) by ingesting the fruit and then defecating the seed in some other location far from the parent tree. In this way, both species benefit.

algae) are the most important. All these species use solar energy to build the organic molecules they need to live and grow. Because less energy is transferred to each successive trophic level, the greatest biomass (living weight) in a terrestrial ecosystem is usually that of the plants.

The *second trophic level* contains the **herbivores**, which eat primary producers and are thus known as **primary consumers**. The intensity of grazing by herbivores often determines the relative abundance of plant species and even the amount of plant material present.

Carnivores are in the third and higher trophic levels. **Carnivores** are animals that obtain energy by eating other animals. At the third trophic level are **secondary consumers**, predators that eat herbivores (e.g., foxes that eat rabbits). At the fourth trophic level are **tertiary consumers**, predators that eat other predators (e.g., bass that eat frogs).

Some secondary and higher consumers combine direct predation with scavenging behavior, as lions do. Others, known as **omnivores**, include both animal and plant foods in their diets. Brown bears are a good example of this, consuming other mammals, insects, fish, and birds, as well as fruits, nuts, roots, and leaves (Coogan et al. 2018). In general, predators occur at lower densities than their prey, and populations at higher trophic levels contain fewer individuals than those at lower trophic levels (Carbone and Gittleman 2002). A single savanna can support many more zebras than lions.

Parasites and disease-causing organisms, **pathogens**, form an important subclass of predators. Parasites of animals, including mosquitoes, ticks, intestinal worms, and protozoans, as well as microscopic disease-causing organisms such as some bacteria and viruses, do not kill their hosts immediately, if ever. Plants can also be attacked by bacteria, viruses, and a variety of parasites that include fungi, other plants (such as mistletoe), nematode worms, and insects. The effects of parasites range from imperceptibly weakening their hosts to totally debilitating or killing them over time. The spread of parasites and disease from captive or domesticated species, such as dogs, to wild species, such as lions, is a major threat to many rare species (see Chapter 5), but we must also consider that parasites contribute to biological diversity and thus may be targets for conservation themselves.

Decomposers and **detritivores** feed on dead plant and animal tissues and wastes (detritus), breaking down complex tissues and organic molecules into the simple chemicals that are the building blocks of primary production (**FIGURE 2.10**). Decomposers release minerals such as nitrates and phosphates back into the soil and water, where they can be taken up again by plants and

Photo by Anna Sher

FIGURE 2.10 Fungi, such as this basidiomycete (mushroom) growing on a forest floor, break down dead material, thus cycling the nutrients such as nitrogen back into the soil, where they can be taken up again by plants. Most ecosystems would not function without detritivores such as these.

algae. Decomposers are usually much less conspicuous than herbivores and carnivores, but their role in the ecological community is vital. The most important decomposers are fungi and bacteria, but a wide range of other species play a role in breaking down organic materials. For example, vultures and other scavengers tear apart and feed on dead animals, dung beetles feed on and bury animal dung, and worms break down fallen leaves and other organic matter. Crabs, worms, molluscs, fish, and numerous other organisms eat detritus in aquatic environments. If decomposers were to die off, organic material would accumulate, and plant growth would decline greatly (Gessner et al. 2010).

Food chains and food webs

Although species can be organized into the general trophic levels just described, their actual requirements or feeding habits within those trophic levels may be quite restricted. For example, in salt marsh ecosystems, there are species of mirid bugs that only eat the eggs of a specific plant hopper species, which in turn only eats one species of cordgrass (Wimp et al. 2011). These specific feeding relationships are termed **food chains**. The more common situation in many biological communities, however, is for one species to feed on several other species at the lower trophic level, to compete for food with several species at its own trophic level, and, in turn, to be preyed on by several species at the higher trophic level. Consequently, a more accurate description of the organization of biological communities is a **food web**, in which species are linked together through complex feeding relationships (**FIGURE 2.11**). When considering food web dynamics, scientists are often interested in comparing the effects of higher trophic levels on lower trophic levels and vice versa, referred to as top-down versus bottom-up impacts (e.g., Vidal and Murphy 2018). Species at the same trophic level that use approximately the same environmental resources are considered to be a **guild** of competing species.

Humans can substantially alter the relationships in food webs (Alva-Basurto and Arias-Gonzalez 2014). For example, an increase in salt inputs to the Pecos River in New Mexico resulting from human alterations was associated with changes in aquatic food webs due to decreased diversity and shortened trophic chains (East et al. 2017).

GLOBAL CHANGE
CONNECTION

All these relationships between species influence how they evolve over time. The traits that determine how species look, function, and behave are shaped not only by the **abiotic** (nonliving) environment, but also by the **biotic** (living) environment. When a plant produces a toxin in its leaves, it exerts a selective pressure on the animals that might eat it; those individual insects that have a genetic capacity to tolerate the toxin will be more likely to survive and grow, thus increasing that trait in the insect population over generations. The same process of **evolution** occurs in the context of competition and mutualism, each species responding to every other species it interacts with. This is one of the reasons ecosystems can be so altered and often damaged when new species are introduced, species go extinct, or environmental conditions change. An impact on even one species is likely to reverberate throughout the entire community that has evolved with that species.

> Evolution is the process of species change, including adaptations to the biotic environment. Because organisms evolve in response to each other, disturbances to one species will usually affect other species.

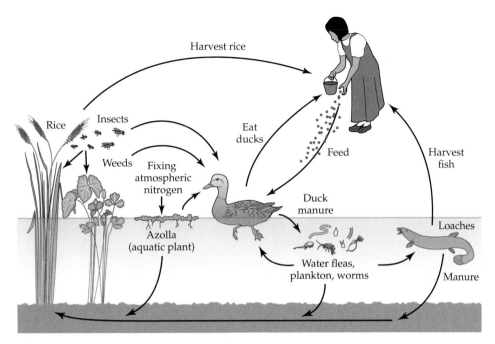

FIGURE 2.11 A simple food web in a traditional agricultural ecosystem. Photosynthetic plants are eaten by people, ducks, and insects. Insects and aquatic invertebrates are eaten by ducks and fish, which are then eaten by people.

Keystone species and resources

Within biological communities, certain species or guilds of species with similar ecological features may determine the ability of many other species to persist in the community (**FIGURE 2.12**). These **keystone species** affect the organization of the community to a far greater degree than we would predict if we considered only their numbers or biomass (Valls et al. 2015). Protecting keystone species is a priority for conservation efforts because loss of a keystone species or guild will lead to losses of numerous other species as well.

Top predators are often considered keystone species because they can markedly influence herbivore populations. The elimination of even a small number of individual predators, even though they constitute only a minute fraction of the community biomass, may result in dramatic changes in the vegetation and a great loss in biodiversity, sometimes called a **trophic cascade**. For example, in some places where gray wolves (*Canis lupus*) have been hunted to extinction by humans, deer (*Odocoileus virginianus*) populations have exploded (Beschta and Ripple 2016). The deer severely overgraze the habitat, eliminating many herb and shrub species. The loss of these plants, in turn, is detrimental to the deer and to other herbivores, including insects. The reduced plant cover may lead to soil erosion, also contributing to the loss of species that inhabit the soil. When wolves are restored to ecosystems, trophic relationships can sometimes be reestablished (Ripple and Beschta 2012).

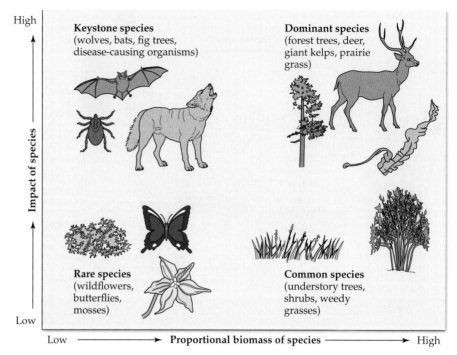

FIGURE 2.12 Keystone species determine the ability of large numbers of other species to persist within a biological community. Although keystone species make up only a small percentage of the total biomass, a community's composition would change radically if one of them were to disappear. Rare species have minimal biomass and seldom have significant effects on the community. Dominant species constitute a large percentage of the biomass and affect many other species in proportion to this large biomass. Some species, however, have a relatively low impact on the community organization despite being both common and heavy in biomass. (After M. E. Power et al. 1996. *BioScience* 46: 609–620.)

Species that extensively modify the physical environment through their activities, often termed **ecosystem engineers**, are also considered keystone species (Jones et al. 1996; Romero et al. 2015) (**FIGURE 2.13**). Losing keystone species can create a series of linked extinction events, known as an **extinction cascade**, resulting in a degraded ecosystem with much lower biodiversity at all trophic levels. This may already be happening in tropical forests where overharvesting has drastically reduced the populations of birds and mammals that act as predators, seed dispersers, and herbivores (Naniwadekar et al. 2015). While such a forest appears to be green and healthy at first glance, it is really an "empty forest" in which ecological processes have been irreversibly altered such that many plant and animal species will be eliminated over succeeding decades or centuries (Green et al. 2020; Redford 1992). In the marine environment, the loss of key structural species such as sea grasses and seaweeds can lead to the loss of specialized species that inhabit such communities, such as delicate sea dragons and sea horses (Hughes et al. 2009). If the few keystone species in a community being affected by human

FIGURE 2.13   The North American beaver (*Castor canadensis*) is considered an ecosystem engineer because the dams it constructs cause overbank flooding, thereby creating new wetland habitat for themselves and other species.

activities can be identified, they can sometimes be carefully protected or actively managed to increase their numbers. Increasingly in recent years, keystone species have been successfully reintroduced to ecosystems (Hale and Koprowski 2018).

Habitats may also contain **keystone resources**, often physical or structural, that occupy only a small area yet are crucial to many species in the ecosystem (Hunter et al. 2017). These resources can be important when restoring ecosystems and will be discussed further in Chapter 11. For example, deep pools in streams, springs, and ponds may be the only refuges for fish and other aquatic species during the dry season, when water levels drop. For terrestrial animals, these water sources may provide the only available drinking water for a considerable distance. Hollow tree trunks and tree holes are keystone resources as breeding sites for many bird and mammal species and may limit their population sizes (Cockle and Martin 2015). Protecting old, hollow trees as a keystone resource is a priority during certain logging activities (Lindenmayer 2017).

Ecosystem dynamics

An ecosystem in which the processes are functioning normally, whether or not there are human influences, is referred to as a **healthy ecosystem**. In many cases, ecosystems that have lost some of their species will remain healthy because there is often some redundancy in the roles performed by ecologically similar species. Ecosystems that can remain in the same state are referred to as **stable ecosystems**. These systems remain stable either because of lack of disturbance or because they have special features that allow them to remain stable in the face of disturbance. Such stability despite disturbance can result from one or both of two features: resistance and resilience. **Resistance** is the ability to maintain the same state even with ongoing disturbance; a river ecosystem that retained its major ecosystem processes after an oil spill would be considered resistant. **Resilience** is the ability to return to an original state quickly after disturbance has occurred; that would be the case if, following contamination by an oil spill and the deaths of many animals and plants, a river ecosystem soon returned to its original condition.

2.4 Biodiversity Worldwide

LEARNING OBJECTIVES

By the end of this section you should be able to:

2.4.1 Interpret a pie chart and a bar graph.

2.4.2 Contrast two different estimates for global diversity, making an argument for which is more likely.

2.4.3 For any given ecosystem type, predict whether it is likely to have low or high biodiversity, based on geologic, climatic, and/or biological features.

Developing a strategy for conserving biodiversity requires a firm grasp of how many species exist on Earth and how those species are distributed across the planet. The answers to both questions can be complex.

How many species exist worldwide?

At present, just over 2 million species have been described (**FIGURE 2.14**). However, the actual number of species on our planet remains unknown. Our knowledge of species numbers is imprecise because inconspicuous species have not received their proper share of taxonomic attention and so most species remain undescribed. For example, spiders, nematodes (microscopic worms), and fungi living in the soil and insects living in the tropical forest canopy are small and difficult to study. These poorly known groups could number in the hundreds of thousands, or even millions, of species, especially considering that each species or genus of animal and plant has many unique species of bacteria, protists, and fungi living on or inside it. Furthermore, genetic analysis is regularly turning up cryptic species. Thus, the actual proportions of different taxa may be quite different from what we currently observe, and scientists do not necessarily agree about these numbers (**FIGURE 2.15**). A study that included such considerations estimated that there could be between 1 billion and 6 billion species on Earth, with 78% being bacteria (Larsen et al. 2017). Viruses, too, might have been grossly underestimated (Carlson et al. 2019), as have insects. One well-cited study suggests that 80% of insects are yet to be discovered and that other terrestrial arthropods (such as spiders, centipedes, and millipedes) may be just as diverse (Stork 2018). If

(A)

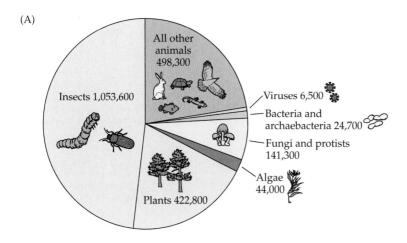

(B)

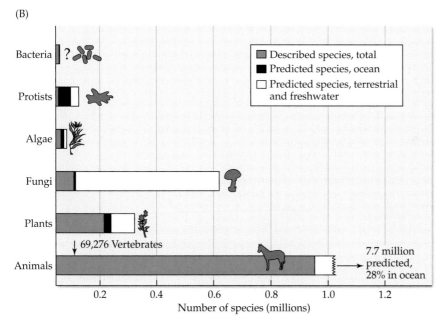

FIGURE 2.14   (A) Over 2 million species have been identified and described by scientists; the majority of these species are insects and plants. These represent only a fraction of the total species likely to exist. (B) Estimates of predicted total numbers according to Mora et al. (2011). The numbers of described species are indicated by the green portions of the bars; estimates for undescribed species are shown in black (marine species) and yellow (terrestrial and freshwater species). Note that because of the rate of species discovery, these 2011 numbers of known species are low compared with current numbers, and even estimates of total species have already been surpassed by counts of currently described species in some groups, such as plants. (A, data from IUCN 2021. *The IUCN Red List of Threatened Species.* Vrsn 2021-2, using https://www.iucnredlist.org/resources/summary-statistics; B, after C. Mora et al. 2011. *PLOS Biol* 9: e1001127. doi.org/10.1371/journal.pbio.1001127. © 2011 Mora et al. CC BY 4.0.)

(A)

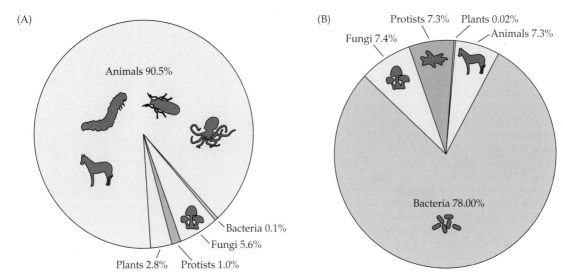

(B)

FIGURE 2.15 Proportions of the numbers of total, actual species may be very different from those of currently described species (see Figure 2.14) and is contentious among scientists. (A) Mora et al. 2011 predicted that animals, primarily insects, probably account for 90.5% of all species on Earth. (B) In contrast, a more recent and still controversial estimate (Larsen et al. 2017) takes into account the frequency of cryptic species and also assumes that every animal has at least 10 unique bacterial symbionts, making bacteria the group with the largest number of species and yielding a total number of species between 1 and 6 billion, more than a couple of orders of magnitude larger than the Mora et al. estimate of 8.7 million. (After B. B. Larsen et al. 2017. *Q Rev Biol* 92: 229–265. Republished with permission of University of Chicago Press.)

true, insects and terrestrial arthropods alone would have 30 million species. These dramatic departures from previous estimates underline the difficulty in making such projections.

Amazingly, as many as 20,000 new species are described each year. While certain groups of organisms, such as birds, mammals, and temperate flowering plants, are relatively well known, a small but steady number of new species in these groups are discovered each year (Joppa et al. 2011). Even among a group as well studied as primates, dozens of new monkey species have been found in Brazil, and dozens of new species of lemurs have been discovered in Madagascar—all since 1990. Every decade, 500–600 new species of amphibians are described. Most of the new animals discovered are insects and other invertebrates. For example, the Natural History Museum of London discovered 503 new species in 2020; although the doors were closed to the public due to the COVID-19 pandemic for most of the year, scientists were busy identifying specimens that had been previously collected in the field (Davis 2020). Most of these were insects, including 170 new beetle species.

Species surveys are a great opportunity to involve "citizen scientists," who can work with field biologists to collect and identify specimens. One example of this is the **bioblitz**, in which scientists and others

Many scientists are working to determine the number of species on Earth. The most widely accepted estimate is that there are 8 million to 10 million species, about half of them insects.

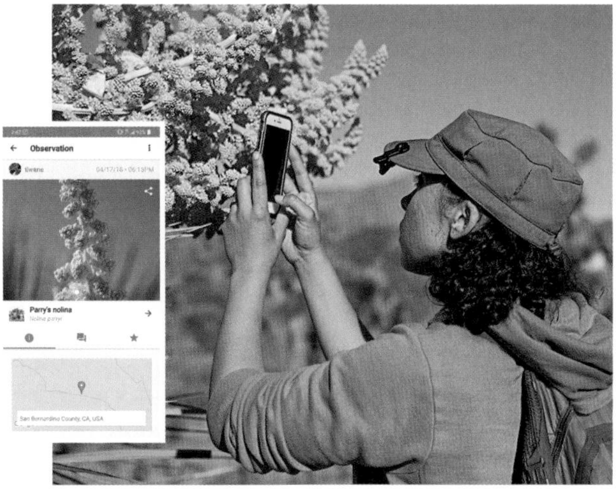

FIGURE 2.16 Here, iNaturalist is being used to photograph, date, and georeference a Parry's nolina plant (*Nolina parryi*; inset). The image will be identified through crowdsourcing and the information stored in a database for use by scientists and others. This particular specimen was originally identified as Bigelow's nolina (*Nolina bigelovii*) but was corrected by two different iNaturalist users, who were listed as "top identifiers of Parry's nolina."

perform an intensive biological survey of a designated area in a short time with the goal of documenting all living species there. Digital tools are constantly being developed to help laypeople collect quality data, such as the smartphone application and web platform iNaturalist, a joint initiative of the California Academy of Sciences and the National Geographic Society with partners across the globe (www.inaturalist.org). Citizen scientists can take photos of organisms they encounter, either as a part of a bioblitz or casually, and upload them to this crowdsourced identification resource, where they are georeferenced and dated (**FIGURE 2.16**). Most observations and photos are licensed for reuse (sometimes restricted to noncommercial use), including for scientific study. Observations with complete data and verified species identifications are shared with other databases, including the Global Biodiversity Information Facility (www.gbif.org). Since its release in 2008, iNaturalist had compiled more than 58 million records of 320,000 species as of February 2021. It is important to note that the value of these types of tools depends on the involvement of biologists and other experts who can validate the identifications.

In addition to new species, entire biological communities continue to be discovered, particularly in extremely remote and inaccessible localities. These communities often consist of inconspicuous species, such as bacteria, protists, and small invertebrates, that have escaped the attention of earlier taxonomists (**FIGURE 2.17**). Specialized exploration

FIGURE 2.17 A community of phytoplankton, including various species of diatoms and dinoflagellates. Microscopic organisms such as these are generally less understood than those that are more easily seen, and new species are frequently discovered.

techniques have aided in these discoveries, particularly in the deep sea and in the forest canopy. For example, each new deep-sea hydrothermal vent explored reveals dozens, if not hundreds, of species previously unknown to science. Drilling projects have shown that diverse bacterial communities exist 2.8 km deep in the Earth's crust at densities of up to 100 million bacteria per gram of solid rock (Brockman and Murray 2018). In the depths of the northwestern Pacific near Japan, 50 new species of nematodes were found on a single collecting trip (Fadeeva et al. 2015). The unique features of these discoveries not only expand our understanding of the biodiversity of our planet, but also raise questions regarding evolutionary and physiological processes. These organisms are also being actively investigated as sources of novel chemicals, as medicines, for their potential usefulness in degrading toxic chemicals, and for insight into whether life could exist on other planets.

The bacteria and archaebacteria (i.e., prokaryotes) in particular are very poorly known and are thus underrepresented in estimates of the total species on Earth, even as new advances in technology provide an ever more accurate count of species in this group. These single-celled organisms have an important role to play in ecosystem functioning, as suggested by their sheer abundance—analyses of DNA indicate that there may be as many as 100,000 prokaryotic cells in a milliliter of seawater (Zeng et al. 2014). Sender et al. (2016) have demonstrated that the human body is occupied by at least as many bacterial cells as human cells (in the order of 10^{13} for a 70 kg "reference man"). It has even been argued that the diversity and abundance of these species reflect the health of an individual (Rosen and Palm 2017). Only about 24,700 species of bacteria and archaebacteria are currently identified (Parks et al. 2020), because they are difficult to grow and identify. However, there could be millions of undescribed bacterial and archaebacterial species, especially if we consider that there are likely unique symbionts in every plant and animal species (see Figure 2.15B) (Larsen et al. 2017). One study even argued that there are trillions of species (Locey and Lennon 2016). However, a study designed to address the limitations of previous estimates, using genetic analysis of almost 500 samples from across the world, arrived at an estimate of 0.8–1.6 million prokaryotic "operational taxonomic units"—that is, what we might consider distinct species of bacteria and archaebacteria (Louca et al. 2019).

> DNA analyses suggest that many thousands of species of prokaryotes have yet to be described. One reason for this is that most species have multitudes of bacteria and archaebacteria inside of them, including species yet unknown to science.

Considering that between 15,000 and 18,000 new, multicellular species are described each year and that at least 3 million more are waiting to be identified, the task of describing the world's species will not be completed for over 180 years if continued at the present rate. This fact underlines the critical need for taxonomists trained to use the latest molecular technology and for web-based information sharing. Online museum and herbarium records, plus international databases such as the IUCN, the Global Biodiversity Information Facility, the Catalogue of Life (www.catalogueoflife.org), and the Encyclopedia of Life project (www.eol.org) will make species names and descriptions more widely and readily available for study.

Where is the world's biodiversity found?

The most species-rich environments appear to be tropical forests, coral reefs, the deep sea, large tropical lakes and river systems, and regions with Mediterranean climates.

TROPICAL FORESTS   Even though tropical forests occupy less than 7% of the world's land area, they contain more than half of its species, most of which are insects (Corlett and Primack 2010). Tropical forests also have many species of birds, mammals, amphibians, and plants. Among flowering plants, gymnosperms, and ferns, about 40% of the world's 275,000 species occur in tropical forest areas. Each of the rain forest areas in the Americas, Africa, Madagascar, Southeast Asia, New Guinea, Australia, and various tropical islands has a different biogeographic history, resulting in unique assemblages of species (see Figure 4.5). For example, lemurs are found only in Madagascar, and hummingbirds are found only in the Americas.

OCEANIC DIVERSITY   Approximately 71% of the Earth's surface is water, and 96.5% of this is ocean. In the oceans, diversity is spread over a much broader range of phyla and classes than in terrestrial ecosystems. These marine systems contain representatives of 31 of the 35 animal phyla that exist today; many of these 31 phyla exist only in the marine environment (Grassle 2001). The broad diversity in the ocean may be due to its great age, enormous water volume, degree of isolation of some seas by intervening landmasses, and other factors.

Within oceans, coral reef ecosystems are particularly diverse; although they occupy less than 0.1% of ocean surface area, they are home to one-third of marine fish species (Bowen et al. 2013; Fisher et al. 2015). The reefs' physical structure is created by the corals (invertebrate animals), which provide habitat for many other organisms. The photosynthetic algae that live mutualistically inside the corals provide them with abundant carbohydrates. One explanation for the richness of coral reefs is their high primary productivity: 2500 g of biomass (living matter) per square meter per year, in comparison with 125 $g/m^2/y$ in the open ocean. Extensive niche specialization among coral species and adaptations to varying levels of disturbance may also account for the high species richness found in coral reefs. The world's largest coral reef is Australia's Great Barrier Reef, with an area of 349,000 km^2, which contains over 400 species of corals, 1500 species of fish, 4000 species of molluscs, and 6 species of turtles. Coral reefs also support some 252 species of birds.

> Species diversity is greatest in the tropics, particularly in tropical forests and coral reefs.

LAKES AND RIVERS   Freshwater systems are some of the most productive on the planet, not only for the organisms that live in them, but for the diversity of life that is fed by them. Freshwater makes up only 0.01% of the Earth's water but supports over 7% of the world's biodiversity, including 45% of all fishes and 25% of all molluscs. Other major groups include insects, amphibians, plants, and mammals (IUCN 2019). The greatest biodiversity within lakes and rivers can be found in the tropics, particularly in the Amazon basin. Because of the difficulties in observing aquatic life and the abundance of cryptic species in this system, environmental DNA is an especially useful tool for measuring diversity in freshwater systems (Cilleros et al. 2019).

MEDITERRANEAN-TYPE COMMUNITIES   Great diversity is found among plant species in southwestern Australia, the Cape region of South Africa, California, central Chile, and the Mediterranean basin, all of which are characterized by a Mediterranean climate of moist winters and hot, dry summers. Of these regions, the Mediterranean basin is the largest in area (2.1 million km^2) and has the most plant species (22,500) (Conservation International and Caley 2008) and more than 17,000 marine species, representing 4%–18% of named marine species (UNEP 2021). The Cape Floristic Region of South Africa has an extraordinary concentration of unique plant species (9000) in a relatively small area (78,555 km^2) (**FIGURE 2.18**). The shrub and herb communities in these areas are rich in species, apparently because of their combination of considerable geologic age, complex site characteristics (such as topography and soils), and severe environmental conditions. The high frequency of fire in these areas may also favor rapid speciation and prevent the dominance of just a few species.

The distribution of species

At various spatial scales, there are concentrations of species in particular places, and there is a rough correspondence in the distribution of species richness between different groups of organisms (Domisch et al. 2015). For example, in North America, large-scale patterns of species richness are highly correlated for amphibians, birds, butterflies, mammals, reptiles, land snails, all vascular plants (including trees), and tiger beetles. A region with numerous species of one group will tend to have numerous species of the other groups as well (Ricketts et al. 1999). On a local scale, however, this relationship may break down; for example, amphibians may be most diverse in wet, shady habitats, whereas reptiles may be most diverse in dry, open habitats. At a global scale, each group of living organisms may reach its greatest species richness in a different part of the world because of historical circumstances or the suitability of the site to its needs.

Places with large concentrations of species often have a high percentage of **endemic species**—that is, species that occur there and nowhere else. The countries with the largest numbers of endemic mammals, which represent important targets for conservation efforts, are Indonesia (259), Australia (241), Brazil (183), Madagascar (187), and Mexico (158) (IUCN 2021). Geographically isolated countries and islands also tend to have high percentages of endemic species. For example, most of the species in Madagascar are endemic because it is a large, ancient island on which many unique species have evolved in isolation. Species-rich countries on

FIGURE 2.18   The Cape region of South Africa has evolved a unique ecosystem, collectively called the fynbos (meaning "fine-leaved plants" in Dutch), characterized by large numbers of plant species that are found nowhere else.

Photo by Anna Sher

FIGURE 2.19 There is a strong latitudinal pattern of global species richness, with the greatest diversity near the equator. Here, current highest species richness for terrestrial vertebrates is shown in red, with change toward the blue end of the spectrum reflecting decreasing diversity poleward. (After P. D. Mannion et al. 2014. *Trends Ecol Evol* 29: 42–50. doi.org/10.1016/j.tree.2013.09.012. © 2013 Elsevier Ltd. CC BY 3.0, with data from C. N. Jenkins et al. 2013. *Proc Natl Acad Sci USA* 110: E2602–E2610.)

the mainland, such as Tanzania, have comparatively fewer endemic species because they share many species with neighboring countries.

Almost all groups of organisms show an increase in species diversity toward the tropics. In a survey of 10 taxonomic groups, 78% of the world's species were found at tropical latitudes (Barlow et al. 2018) (**FIGURE 2.19**). For example, Thailand has 5393 species of animals, while France has less than half that number, even though both countries have roughly the same land area (**TABLE 2.1**). The

TABLE 2.1 **Number of Animal Species in Selected Tropical and Temperate Countries Paired for Comparable Size**

Tropical country	Area (1000 km²)	Number of animal species	Temperate country	Area (1000 km²)	Number of animal species
Brazil	8456	6256	Canada	9220	2196
DRC[a]	2268	4488	Argentina	2737	2750
Indonesia	1812	9508	Iran	1636	2117
Colombia	1039	6845	South Africa	1221	4733
Venezuela	882	4580	Chile	748	1756
Thailand	511	5393	France	550	2659
Philippines	298	5268	United Kingdom	242	1211
Rwanda	25	1139	Belgium	30	835

Source: Data from 2020 IUCN Red List, accessed in 2021, using https://www.iucnredlist.org/resources/summary-statistics.
[a]DRC = Democratic Republic of the Congo.

contrast is particularly striking for trees and other flowering plants; 10 hectares of forest in Amazonian Peru or Brazil might have 300 or more tree species, whereas an equivalent forest area in temperate Europe or the United States would probably contain 30 species or less. Even within a given continent, the number of species increases toward the equator.

For many groups of marine species, the greatest diversity of coastal species is also found in the tropics, with a particular richness of species in the western Pacific. For open-ocean species, the greatest diversity is found at midlatitudes. Temperature is the most important variable explaining these patterns (Tittensor et al. 2010).

Local variation in climate, sunlight and rainfall, topography, and geologic age also affects patterns of species richness (Zellweger et al. 2015). In terrestrial communities, species richness tends to increase with decreasing elevation, increasing solar radiation, and increasing precipitation; that is, hot, rainy lowland areas have the most species. Species richness can be greater where complex topography and great geologic age provide more environmental variation, which allows genetic isolation, local adaptation, and speciation to occur. Geologically complex areas can produce a variety of soil conditions with very sharp boundaries between them, leading to multiple communities and plant species adapted to one specific soil type or another.

GLOBAL CHANGE
CONNECTION

With better methods of exploration and investigation, we are now able to appreciate the great diversity of the living world. This is truly a golden age of biological exploration. Natural history societies and clubs that combine amateur and professional naturalists contribute to this effort. Yet with this knowledge of biodiversity come both the awareness of the damaging impact of human activity, which is diminishing biodiversity right before our eyes, and the responsibility to protect and restore that biodiversity that remains.

Summary

- Taxonomists use morphological and genetic information to describe and identify the world's species. Places vary in their species richness, the number of species found in a particular location.

- There is genetic variation among individuals within a species. Genetic variation allows species to adapt to a changing environment, including in agriculture. Genomic tools are increasingly important in conservation biology.

- Within an ecosystem, species play different roles and have varying requirements for survival. Certain keystone species are important in determining the ability of other species to persist in an ecosystem.

- It is estimated that there are 6 million to 10 million species, most of which are insects, by some estimates, or bacteria, by others. The majority of the world's species have still not been described and named; further work is needed to describe microorganisms such as bacteria.

- The greatest biological diversity is found in tropical regions, with particular concentrations of species in rain forests and coral reefs. The ocean may also have great species diversity but needs further exploration.

For Discussion

1. How many species of birds, trees, and insects can you identify in your neighborhood? How could you learn to identify more? Is it important to be able to identify species in the wild?

2. What are the factors promoting species richness? Why is biological diversity diminished in particular environments? Why aren't species able to overcome these limitations and undergo the process of speciation?

3. Conservation efforts usually target genetic variation, species diversity, biological communities, and ecosystems for protection. What are some other components of natural systems that need to be protected? What do you think is the most important component of biodiversity, and why do you believe it is most important?

Suggested Readings

Barlow, J., et al. 2018. The future of hyperdiverse tropical ecosystems. *Nature* 559: 517–526. The high diversity of the tropics makes it a special target for conservation efforts.

Díaz, S., and M. Cabido. 2001. *Vive la difference*: Plant functional diversity matters to ecosystem processes. *Trends in Ecology & Evolution* 16(11): 646–655. A widely cited, classic paper explaining why species traits, not just their taxonomy, are important.

Larsen, B. B., et al. 2017. Inordinate fondness multiplied and redistributed: The number of species on Earth and the new pie of life. *Quarterly Review of Biology* 92: 229–265. A dramatic reevaluation of the potential diversity of Earth, given how genetic analysis has revealed the commonness of cryptic species.

Magurran, A. E. 2013. *Measuring Biological Diversity: Frontiers in Measurement and Assessment*. Oxford University Press, Oxford, UK. A widely cited text that discusses both classic and emerging methods.

Mastretta-Yanes, A., et al. 2018. An initiative for the study and use of genetic diversity of domesticated plants and their wild relatives. *Frontiers in Plant Science* 9: 209. An example of a proposal to maintain plant genetic diversity, based in Mexico.

Methorst, J., et al. 2021. The importance of species diversity for human well-being in Europe. *Ecological Economics* 181: 106917. A compelling quantitative analysis of the impact of species diversity on life satisfaction and socioeconomic factors.

Ricklefs, R. E., and F. He. 2016. Region effects influence local tree species diversity. *Proceedings of the National Academy of Sciences USA* 113: 674–679. Regional species diversity results largely from geologic and geographic properties that affect evolution.

Sher, A. A., and M. Molles. 2021. *Ecology Concepts and Applications*, 9th ed. McGraw Hill Education, New York. A resource for students seeking to learn more about ecosystem structure and function, including the role that biodiversity plays.

Vidal, M. C., and S. M. Murphy. 2018. Bottom-up vs. top-down effects on terrestrial insect herbivores: A meta-analysis. *Ecology Letters* 21(1): 138–150. Trophic cascades mean that species feel impacts both from changes to their food sources and from their predators.

Ecotourism is one way that biodiversity can have value for local people and also fund conservation. The greater one-horned rhino in Asia is one such success story.

The Value of Biodiversity

3

The economics of biodiversity received international attention when a British trophy-hunting boss sent out an email in early 2021 inviting clients to come to South Africa to kill endangered animals he said had become "plentiful" during the COVID-19 pandemic (dailymail.co.uk February 14, 2021). Such hunting is big business. For example, several years earlier an American big-game hunter paid poachers $50,000 to kill a lion (CNN 2015). The 13-year-old male, affectionately named "Cecil the Lion," had been an important animal icon and tourism draw at the Hwange National Park in Matabeleland North, Zimbabwe; there was an immediate outcry from conservationists and animal lovers around the world when his death was announced. What Cecil and other wild organisms are "worth," both alive and dead, is a critical question in evaluating attitudes and implementing actions that affect biodiversity.

The 2020 *Living Planet Report*, published by the World Wildlife Fund (WWF) and the Zoological Society of London (ZSL), reported that there has been an overall decline of 68% in population sizes of vertebrate species between 1970 and 2020 (WWF 2020). This is a tragic increase from 60% in the 2018 report (WWF 2018). Ultimately, the trend of biodiversity decline will be reversed only if people believe that we are truly losing something of value. But what exactly are we losing? Why should anyone care if a species declines or even becomes extinct or an ecosystem is destroyed? The *Living Planet Report* states that of the world's annual GDP, US$44 trillion (that is, over half) is highly or moderately dependent on nature. But should species only be protected if they have a clear use for humans?

What are we willing to spend to protect species, and how will this spending be financed? Governments and communities throughout the world are coming to realize that biodiversity is extremely valuable—indeed, it is essential to human existence.

Most environmental degradation and species losses occur as accidental byproducts of human activities. Species are hunted to extinction. Sewage is released into rivers. Low-quality land is cleared for short-term cultivation. **Economics**, the study of the transfer of the production, distribution, and consumption of goods and services, can both help us understand the reasons why people treat the environment in what appears to be a shortsighted manner and provide tools to help protect environmental resources.

One of the most universally accepted tenets of modern economic thought centers on the "voluntary transaction": the idea that a monetary transaction takes place only when it is beneficial to both of the parties involved. For example, a baker who sells his loaves of bread for $50 will find few customers. Likewise, a customer who is willing to pay only 5 cents for a loaf will soon go hungry. A transaction between seller and buyer will occur only when a mutually agreeable price is set that benefits both parties: perhaps $5 for that loaf of bread. Adam Smith, the eighteenth-century philosopher whose ideas are the foundation of much modern economic thought, wrote, "It is not upon the benevolence of the butcher, the baker, or the brewer that we eat our daily bread, but upon his own self-interest" (Smith 1909). The sum of each individual acting in his or her self-interest results in a more prosperous society.

There are exceptions to Smith's principle, however, that apply directly to environmental issues. For example, Smith assumed that all the costs and benefits of free exchange are accepted and borne by the participants in the transaction. In some cases, however, associated costs (or sometimes benefits) befall individuals not directly involved in the exchange. These hidden costs and benefits are known, respectively, as negative and positive **externalities** (**FIGURE 3.1**) (Chen et al. 2017). Companies, individuals, or other stakeholders involved in production that results in ecological damage generally do not bear the full cost of their activities, but they gain substantial private economic benefits. The company that owns an electric power plant that burns coal and emits toxic fumes benefits from the sale of low-cost electricity, as does the consumer who buys this electricity. Yet the negative externalities of this transaction—decreased air quality and visibility, increased respiratory disease for people and animals, damage to plant life, and a polluted environment—are distributed throughout society. A current example is the accelerating environmental impact of shipping services, sparked by e-commerce generally and the introduction of "free" two-day shipping specifically. Although these are convenient, there are concerns about the recent dramatic increase in the use of cardboard (with its own production and waste externalities) and emissions from diesel-fueled delivery trucks and planes (Nguyen et al. 2018).

Even more important and more frequently overlooked is the negative externality of environmental damage done to **open-access resources**,

FIGURE 3.1 Smog in the Andes, looking eastward over Santiago, Chile. Air pollution is responsible for 3.3 million premature human deaths (and unknown numbers of wildlife deaths) worldwide per year, a devastating externality of the combined emissions from industry, transportation, and agriculture.

such as water, air, and soil, as a consequence of human economic activity. Open-access resources are those that are collectively owned by society at large—also called **common-property resources**—or are owned by no one. These resources are available for everyone to use and are essentially free. When there are no regulations on their use, then people, industries, and governments use and damage the resources without paying more than a minimal cost, or nothing at all. This situation, which has been referred to as the **tragedy of the commons** (a phrase popularized by Garett Hardin's 1968 essay of that name on human population control), means that the value of the open-access resource is gradually lost to all of society. The unregulated dumping of industrial sewage into a river as a byproduct of manufacturing is a common example. The externalities of this activity are degraded drinking water and an increase in disease, loss of opportunity to bathe and swim in the water, fewer fish that are safe to eat, and the loss of many species unable to survive in the polluted river. The factory owner gains free disposal of sewage, but the society pays the price in terms of lost products and services.

When externalities are not identified and managed, or there is inadequate regulation of common-property resources, certain economic activities make the society as a whole *less* prosperous, not more prosperous (**FIGURE 3.2**). When an economic system fails, and thus the balance of supply and demand for products or services is lost, economists call this **market failure**. Avoiding market failure is arguably the primary goal of government policies that regulate economic activities that can affect the environment.

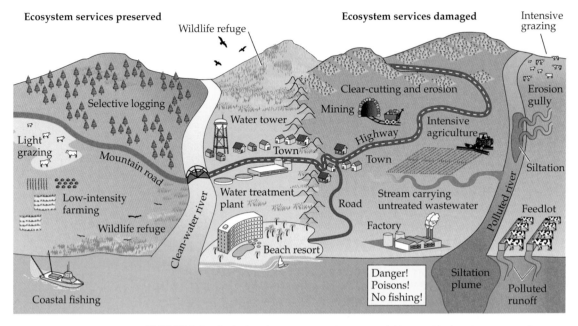

FIGURE 3.2 Agricultural ecosystems, forestry activities, and industries are usually valued by the products that they produce. In many cases, these activities have negative externalities in that they erode soil, degrade water quality, and harm aquatic life (right side of figure). Farming, forestry, and other human activities could also be valued on the basis of their public benefits, such as soil retention and maintaining water quality and fish populations, and their owners might receive subsidies for these benefits (left side of figure).

3.1 Ecological and Environmental Economics

LEARNING OBJECTIVES

By the end of this section you should be able to:

3.1.1 Conduct a general cost-benefit analysis of a situation in which natural resources are used.

3.1.2 Interpret a scatterplot, including summarizing a trend.

3.1.3 Determine the elements of a given ecosystem service, with intermediate services, benefits, and values.

A major problem for conservation biology is that natural resources have often been undervalued by modern society. Thus, the costs of environmental damage have been ignored, the depletion of natural resource stocks has been disregarded, and the future value of resources has been discounted (Lambin et al. 2018). Because the underlying causes of environmental damage are so often economic in nature, the solution must incorporate economic principles. In an effort to account for all costs of economic transactions, including environmental costs, two closely related research areas have evolved—environmental economics and ecological economics—that integrate economics, environmental science, ecology, and public policy and that include valuations of biodiversity in economic analyses (Tietenberg and Lewis 2016). **Environmental economics** is a subdiscipline of economics that places a value on components of the environment. Its modern form can be traced to a popular 1972 book, *The Limits to Growth*, by environmental scientist Donella Meadows and coauthors; they used system dynamics modeling to predict potential future balances of the human population, pollution, and agricultural production (Hanley et al. 2015). **Ecological economics**, which is more closely allied to conservation biology, seeks to integrate the thinking of ecologists and economists into a transdiscipline aimed at developing a sustainable world (Costanza et al. 2014) but also can be useful at a more local scale, such as for improving outcomes of restoration projects (Iftekhar et al. 2017). It is a field that seeks to create new valuation systems that integrate economic valuation with ecology, environmental science, sociology, and ethics, which will facilitate designing better public policies. The fundamental challenge facing conservation biologists is to ensure that all the costs of economic activity, as well as all the benefits, are understood and taken into account when decisions are made that will affect biodiversity (Smith et al. 2017).

GLOBAL CHANGE
CONNECTION

Arguments for the protection of biodiversity are often strengthened by evidence provided by ecological economics.

Cost-benefit analysis

Economic methods are now being used to review development projects and evaluate their potential environmental effects *before* the projects proceed. **Environmental impact assessments**, in particular, consider the present and future effects of projects on the environment. "The environment" is often broadly defined to include not only harvestable natural resources but also air and water quality; the livelihoods of local people, including quality of life; and biodiversity. In its most comprehensive form, **cost-benefit analysis** compares the benefits gained against the costs of a project or resource use (e.g., Becker et al. 2018). In practice, though, cost-benefit analyses are notoriously difficult to calculate accurately because benefits and costs change over time and are difficult to measure. Today, there is an increasing tendency by governments, conservation groups, and economists to apply the **precautionary principle**. That is, it may be better not to approve a project that has risk associated with it and to err on the side of doing no harm to the environment, rather than doing harm unintentionally or unexpectedly, as by building wind turbines where they could harm endangered birds (Braunisch et al. 2015). The precautionary principle is a key feature of many national and international policies and agreements regarding environmental management, even though its interpretation can be vague and variable (Garnett and Parsons 2017).

It would be highly beneficial to apply cost-benefit analysis to many of the basic industries and practices of modern society. Many environment-damaging economic activities appear to be profitable even when they are actually losing money, because governments subsidize the industries involved in them with tax breaks, direct payments or price supports, cheap fossil fuels, free water, and road networks—sometimes referred to as **perverse subsidies** (Jiang and Lin 2014). The elimination of such subsidies that are harmful to biodiversity is one of the explicit targets of the Convention on Biological Diversity (CBD) Decision X/2. Without these subsidies, many environmentally damaging or expensive activities—such as farming in areas with high labor, energy, and water costs; overfishing in the ocean; and inefficient and highly polluting energy use—would be reduced (Merckx and Pereira 2015).

Attempts have been made to include the loss of natural resources in calculations of gross domestic product (GDP) and other indexes of national production. The problem with GDP is that it measures economic activity in a country without accounting for all the costs of unsustainable activities (such as overfishing of coastal waters and poorly managed strip-mining), which cause the GDP to increase, even though they may be destructive to a country's long-term economic well-being. In actuality, the economic costs associated with environmental damage can be considerable, and they often offset the gains attained through agricultural and industrial development. For example, a recent meta-analysis of 139 studies found that electricity and transportation systems cost us over US$24.6 trillion annually in externalities (Sovacool et al. 2021).

> Unsustainable activities such as clear-cut logging, strip-mining, and overfishing may cause a country's apparent productivity to increase temporarily but are generally destructive to long-term economic well-being.

A system that accounts for natural resource depletion, pollution, and unequal income distribution in measures of national production is the Index of Sustainable Economic Welfare (ISEW), the updated version of which is called the Genuine Progress Indicator (GPI). This index incorporates personal consumption as the starting point, against which environmental and societal costs and benefits are considered (Fox and Erickson 2020). It includes factors such as the loss of farmlands, the loss of wetlands, the impact of air and water pollution, the number of people living in poverty, and climate change, among others. According to the GPI, the world economy reached a peak around 1978 and has been slowly declining since then, even though the standard GDP index showed a dramatic gain (Kubiszewski et al. 2013). In a recent analysis, climate change was found to be among the top nine factors that influenced GPI the most, finding if climate change decreased to zero, GPI would increase by 14% (Fox and Erickson 2020). The GPI suggests what conservation biologists have long feared: many modern economies are achieving their growth only through environmental degradation and the unsustainable consumption of natural resources. As these resources run out and as humans suffer the effects of pollution, the true economic situation will continue to deteriorate.

A third measure of national productivity is the Environmental Performance Index (EPI; epi.yale.edu), which uses 20 environmental indicators to rank countries according to the health of, and threats to, their ecosystems,

GLOBAL CHANGE
CONNECTION

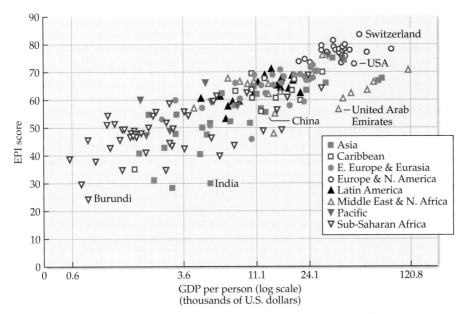

FIGURE 3.3 Wealthy countries with higher gross domestic product (GDP) per person tend to have higher scores on the Environmental Performance Index (EPI), as measured by various indicators: health and stress level of ecosystems, human vulnerability to environmental change, ability of the society and institutions to cope with environmental changes, and cooperation in international environmental initiatives. (After Z. A. Wendling et al. 2018. *2018 Environmental Performance Index*. Yale Center for Environmental Law & Policy: New Haven, CT, https://epi.yale.edu/.)

the vulnerability of their human population to adverse environmental conditions, the ability of their society to protect the environment, and their participation in global environmental protection efforts. In general, developing countries with a low GDP per person also have low EPI scores, including the large country of India, as **FIGURE 3.3** shows. China has a moderate GDP and EPI. Higher-income countries, such as the United States, Japan, and Germany, tend to have much higher EPI scores. There is a concern among many economists and businesspeople that a country that rigorously protects its environment, as shown by a high EPI, might not be competitive in the world economy, as measured by a competitiveness index that includes worker productivity and a country's ability to grow and prosper. But in fact, environmental sustainability is *not* linked to a country's economic competitiveness. Countries such as Switzerland have an economy that is both sustainable and competitive, whereas the United Arab Emirates is competitive but ranks poorly in sustainability. The rapidly growing economy of India is intermediate in competitiveness but ranks low in environmental sustainability.

GLOBAL CHANGE
CONNECTION

New measures of national productivity take environmental sustainability into account. These measures include both the benefits and the costs of human activities.

Financing conservation

Another important aspect of environmental economics is the cost of conservation. Especially when a species is already rare or endangered, it is not enough to simply do no harm. People must intervene to protect, manage, and otherwise support its health and survival. But such interventions can be expensive. For example, a reintroduction program for the endangered whooping crane in the Eastern US is projected to cost $1.3 million (USFWS 2020a). The effort increased this species from 16 individuals in 1941 to 483 as of 2018. Although the rescue of the whooping crane is a success story, there are many other less charismatic species that do not have adequate financial support. Thus, in addition to cost-benefit analysis, it is important to do **cost-effectiveness analysis** as well, in effect asking, Where do we get the most with our conservation dollar? (Martin et al. 2018).

We must also concern ourselves with where and how we get these funds. Conservation is financed in many ways, including by the sometimes controversial practice of hunting (Crosmary et al. 2015). Some countries, such as Namibia, depend on sales of expensive trophy licenses, even for the hunting of endangered species, to support conservation (Rust 2015) and as a source of jobs for local people (Angula et al. 2018). In the United States, most state wildlife conservation efforts are primarily funded by the sale of hunting licenses, tags, and stamps; hunters' excise taxes generated nearly $1 billion to support state conservation programs in 2019 (USFWS 2020b). This is considered a utilitarian approach to conservation: it assumes that species and their habitats will be conserved if people find them valuable, even if only to kill for sport. It has also been argued that management for hunting requires maintaining larger populations than are necessary for ecotourism, which is beneficial for conservation (discussed in more detail in the section "Financing Conservation") (Di Minin et al. 2016). Alternatively, others believe that using hunting as a foundation for conservation is morally wrong and will inevitably lead to poaching and decreased biodiversity (Selier et al. 2014). The death of charismatic species and even individuals (such as the lion mentioned in the beginning of this chapter) has fueled pressure to find alternatives.

Ultimately, only that which is perceived to have value will be saved, so the basis on which we assign this value is of utmost importance. It will also inevitably raise questions regarding moral and ethical values.

What are species worth?

There are many classification systems used to evaluate the benefits we receive from our natural environment (De Groot et al. 2010; Haines-Young and Potschin 2018). As yet there is no universally accepted framework for assigning value to biodiversity, but a variety of approaches have been proposed. Among the most useful is the framework used by McNeely et al. (1990) and Barbier et al. (1994), in which economic values are first divided into **use values** and **non-use values**. Use values of biodiversity are divided between **direct use values** (also known in other frameworks as **commodity values** and **private goods**) and **indirect use values**. Direct use values are assigned to products harvested by people, such as timber, seafood, and medicinal plants from the wild, while indirect use values are assigned to benefits provided by biodiversity that do not involve harvesting or destroying the resource. Indirect use values provide current benefits to people, such as recreation, education, scientific research, and scenic amenities, and include the benefits of ecosystem services, such as water quality, pollution control, natural pollination and pest control, ecosystem productivity, soil protection, and regulation of climate. **Option value** is determined by the prospect for possible future benefits for human society, such as new medicines, possible future food sources, and future genetic resources. **Existence value** is the non-use value that can be assigned to biodiversity—for example, economists can attempt to measure how much people are willing to pay to protect a species from going extinct or an ecosystem from being destroyed.

GLOBAL CHANGE
CONNECTION

Ecosystem services

The many and varied environmental benefits provided by biodiversity and ecosystems in general to humans are collectively referred to as **ecosystem services** (Ehrlich and Ehrlich 1982). There are currently several classifications of ecosystem services (e.g., MA 2005; TEEB and Kubar 2010). The Common International Classification of Ecosystem Services (CICES V5.1; Haines-Young and Potschin 2018) is today one of the most widely used and includes three categories:

1. *Provisioning services* are the material or energy outputs of an ecosystem, including food, medicine, and raw materials. The worth of biological outputs to humans will be discussed in detail in the section "Direct use values."

2. *Regulating services* are services provided by the ecosystem acting as regulator of the quality of the air, water, and soil. Forests provide many of these by regulating local climate, removing pollutants from the atmosphere, and holding soil, with their roots, that would otherwise blow or wash away. Other examples can be found in the section "Indirect use values."

3. *Cultural services* include inspiration for art, design, music, and other cultural expression, aesthetics, intellectual stimulation, and spiritual value. Examples of these can be found in the section "Amenity value," "Educational and scientific value," and Section 3.4.

These three types of services can be considered in terms of not only the final services themselves, but also the "supporting/intermediate" services that support them (Potschin-Young et al. 2017) (**FIGURE 3.4**). For example, a lake can provide habitat for fish (an intermediate service), with the fish caught by fisheries as the final service. Considering ecosystems services in this way helps us to identify the aspects at risk of degradation or devaluation. Next we will discuss ecosystem services in terms of the benefits that organisms (both living and dead) provide for humans and how this benefit is quantified as a value in financial terms.

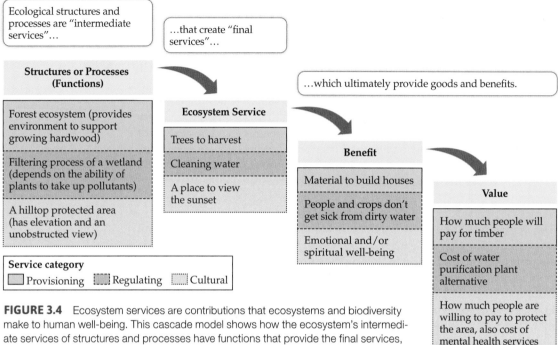

FIGURE 3.4   Ecosystem services are contributions that ecosystems and biodiversity make to human well-being. This cascade model shows how the ecosystem's intermediate services of structures and processes have functions that provide the final services, which in turn are valued for their goods and benefits. Here we can see an example of a forest, a wetland, and a hilltop for each of the three service categories, respectively: provisioning, regulating, and cultural services. In all cases, the "value" of these services is determined by the market, or how much people are willing to pay for it. Differentiating between these elements is useful for understanding what exactly should be protected and why. (After R. H. Haines-Young and M. Potschin. 2010. In *Ecosystem Ecology: A New Synthesis*, D. Raffaelli and C. Frid [Eds.]. pp. 110–139. BES Ecological Reviews Series. Cambridge University Press, Cambridge.)

3.2 Use Values

LEARNING OBJECTIVES

By the end of this section you should be able to:

3.2.1   Distinguish between different types of use values.

3.2.2   Provide original examples of productive use values and consumptive use values.

3.2.3   Determine how you would calculate a monetary value for any given use value.

Use value is the benefit of a commodity, such as a species. Figure 3.4 is a conceptual model that describes use values in terms of ecosystem services; here we will use another model to consider (and measure) use value: direct versus indirect.

Direct use values

Direct use values are those that fall under the "provisioning" category of ecosystem services. They can be readily quantified in economic terms, such as by examining import and export statistics. Direct use values are divided into **consumptive use value**, for goods that are consumed locally, and **productive use value**, for products that are sold in markets.

Consumptive use value

People living close to the land often derive a considerable proportion of the goods they require for their livelihood from the surrounding environment. These goods, such as fuelwood and wild meat, are consumed locally and are therefore assigned consumptive use value (**FIGURE 3.5**). These goods do not appear in a nation's GDP, because they are neither bought nor sold beyond the village or local region and do not appear in the national or international marketplace. However, if rural people are unable to obtain these products (as might occur following environmental degradation, overexploitation of natural resources, or even the creation of a protected reserve), their standard of living will decline, possibly to the point where they are forced to relocate. Consumptive use value can be assigned to a product by considering how much people would have to pay if they had to buy an equivalent product when their local source was no longer available. This valuation is sometimes referred to as a **replacement cost approach**.

> Consumptive use value can be calculated by considering how much people would have to pay to buy an equivalent product if their local source were no longer available.

Studies of traditional societies in the developing world show how extensively these people use their natural environment to supply themselves with fuelwood (see Figure 3.5A), meat, vegetables, fruit, medicine, rope and string, and building materials (Angelsen et al. 2014). In rural areas, nearly 100% of households depend on wood fuel, a resource that is rapidly diminishing in many areas (Waswa et al. 2020). About 80% of the world's population still relies principally on traditional medicines derived from plants and animals as their primary source of treatment (Ekor 2014, but see Oyebode et al. 2016).

One of the crucial requirements of rural people is protein, which they obtain by hunting and collecting wild animals for meat. In some places, this meat is called **bushmeat**. One study estimated bushmeat to be about 40% of peoples' diets in Botswana and about 80% in the Democratic Republic of the Congo (formerly Zaire; Powell et al. 2013). However, another study that interviewed over 2000 people across sub-Saharan Africa found that 62% of men and 72% of women said they would never eat it (Luiselli et al. 2018). Bushmeat extraction rates for Africa are undeniably unsustainable, perhaps by a factor of six. This wild meat includes not only birds, mammals, and fish, but also spiders, snails, caterpillars, and insects. In certain areas of Africa, because of overharvesting of larger animals, insects may constitute the majority of the dietary protein and supply critical vitamins.

In areas along coasts, rivers, and lakes, wild fish represent an important source of protein (see Figure 3.5B). Throughout the world, 91 million tons of fish were caught in 2016, with 79 million tons from the oceans and 12 million

GLOBAL CHANGE
CONNECTION

(A)

© Zamzam Images/Alamy Stock Photo

(B)

© Pulsar Imagens/Alamy Stock Photo

FIGURE 3.5 Examples of natural products with consumptive use value. (A) Women in India return to their village with loads of wood. Fuelwood is one of the most important natural products consumed by local people, particularly in Africa and southern Asia. (B) Along a river in India, fishermen catch small fish to eat. The value of these products can be estimated based on what these people would have to spend to purchase rather than harvest them.

tons from freshwater sources (FAO 2018a). Much of this catch is consumed locally. In coastal areas, fishing is often the most important source of employment, and seafood is the most widely consumed protein. Aquaculture, including fish, shrimp, and other species, is increasing rapidly throughout the world; much of the feed for these activities is fish meal and fish oil derived from wild-caught fish, but more recently plant ingredients derived from agricultural crops like soybean and rapeseed are becoming common (Øverland and Skrede 2017).

Although dependency on local natural products is primarily associated with the developing world, there are rural areas of the United States, Canada, Europe, and other developed countries where millions of people are dependent on fuelwood for heating, on wild game and seafood for their protein needs, and on intact ecosystems for clean drinking water and sewage treatment. Many of these people would be unable to survive in these locations if they had to pay for these necessities.

Productive use value

Resources that are harvested from the wild and sold in national or international commercial markets are assigned productive use value (**FIGURE 3.6**). In standard economics, these products are valued at the price paid at the first point of sale minus the costs incurred up to that point, whereas other methods value the resource at the final retail price of the products. For example, the bark and leaves from wild shrubs and trees of the common witch hazel (*Hamamelis virginiana* and related species) are used to make a variety of astringent herbal products, including aftershave lotions, insect-bite creams, and hemorrhoid preparations. The final retail price of the medicine, which includes the values of all inputs (labor, energy, other materials, transportation, and marketing, as well as witch hazel bark and leaves), is vastly greater than the purchase price of the witch hazel raw materials.

GLOBAL CHANGE
CONNECTION

The productive use value of natural resources is significant, even in industrial nations. It was once estimated that approximately 4.5% of the US GDP depends in some way on wild species (Prescott-Allen and Prescott-Allen 1986). This would translate to about $873 billion (out of a GDP of $19.4 trillion) for the year 2018. The percentage is far higher for developing countries that have less industry and a higher percentage of their population living in rural areas. For example, in the Democratic Republic of the Congo, nontimber, wild forest products contribute between 25% and 40% of the average household income (Ingram et al. 2012). The international trade in wildlife products reported to the Harmonized System (HS) administered by the World Customs Organization (WCO) in 2012 was $187 billion (Chan et al. 2015). However, it is difficult to accurately calculate the total value of wild-harvested products because of the unknown contribution of "invisible trades," due to low detection rates, underreporting, and nonreporting, especially of illegal products, such as endangered species (Phelps and Webb 2015).

A wide variety of natural resources are sold commercially and have enormous total market value. Their value can be considered the productive value of biodiversity.

(A)

(B)

FIGURE 3.6 Examples of productive use value. (A) A wide variety of marine animals are collected in the wild or produced by aquaculture and then sold as seafood, as shown by this market in South Korea. (B) The productive use value of the trees in this forest in Canada is simply calculated as their worth on the market.

The range of products obtained from the natural environment and sold in the marketplace is enormous: these products include fuelwood, construction timber, fish and shellfish, medicinal plants, wild fruits and vegetables, wild meat and skins, fibers, rattan (a vine used to make furniture and other household articles), honey, beeswax, natural dyes, seaweed, animal fodder, natural perfumes, and plant gums and resins. Additionally, there are large international industries associated with collecting tropical cacti, orchids, and other plants for the horticultural industry and birds, mammals, amphibians, and reptiles for zoos and private collections (see Chapter 5). The value of ornamental fishes in the aquarium trade is estimated at $340 million per year, with wild-caught fish representing about 10% of the total (Evers et al. 2019; Monticini 2010).

GLOBAL CHANGE
CONNECTION

FOREST PRODUCTS   Wood is one of the most significant products obtained from natural environments, with an export value of about $244 billion per year (FAO 2019). The total value of timber and other wood products is far greater—likely as high as $2.6 trillion per year—because most wood is used locally and is not exported. Non-wood products from forests, including bushmeat, fruits, gums and resins, rattan, and medicinal plants, also have a large productive use value. These non-wood products are sometimes erroneously called "minor forest products"; in reality, they are often very important economically and may rival the value of wood.

THE NATURAL PHARMACY   Effective drugs are needed to keep people healthy, and they represent an enormous industry, with worldwide sales projected to reach $1.2 trillion by 2024 (Evaluate Pharma 2018). The natural world is an important source of medicines currently in use, as well as possible future medicines. All 20 of the pharmaceutical products most frequently used in the United States are based on chemicals that were first identified in natural organisms. More than 25% of the prescriptions filled in the United States contain active ingredients derived directly from plants, and many of the most important antibiotics, including penicillin and tetracycline, are derived from fungi or microorganisms (Waterman et al. 2016).

Many modern medicines were first discovered in a wild species used in traditional medicine and then produced synthetically by chemists (Buenz et al. 2018; Van Ooij 2016). For example, the use of coca (*Erythroxylum coca*) by natives of the Andean highlands eventually led to the development of synthetic derivatives such as Novocain (procaine) and lidocaine, commonly used as local anesthetics in dentistry and surgery. The rose periwinkle (*Catharanthus roseus*) from Madagascar (**TABLE 3.1**) is the source of two potent drugs that have increased the rate of survival of childhood leukemia from 10% to 90%. Venomous animals such as rattlesnakes, bees, and cone snails have been especially rich sources of chemicals with valuable medical and biological applications. An enzyme derived from a heat-tolerant bacterium (*Thermus aquaticus*) collected from hot springs at Yellowstone National Park forms a key component in the polymerase chain reaction used to amplify

TABLE 3.1	Twenty Drugs from the Plant World First Discovered in Traditional Medical Practice		
Drug	Medical use	Plant source	Common name
Ajmaline	Treats heart arrhythmia	*Rauwolfia* spp.	Rauwolfia
Aspirin	Analgesic, anti-inflammatory	*Spiraea ulmaria*	Meadowsweet
Atropine	Dilates eyes during examination	*Atropa belladonna*	Belladonna
Caffeine	Stimulant	*Camellia sinensis*	Tea plant
Cocaine	Ophthalmic analgesic	*Erythroxylum coca*	Coca plant
Codeine	Analgesic, antitussive	*Papaver somniferum*	Opium poppy
Digitoxin	Cardiac stimulant	*Digitalis purpurea*	Foxglove
Ephedrine	Bronchodilator	*Ephedra sinica*	Ephedra plant
Ipecac	Emetic	*Cephaelis ipecacuanha*	Ipecac plant
Morphine	Analgesic	*Papaver somniferum*	Opium poppy
Pseudoephedrine	Decongestant	*Ephedra sinica*	Ephedra plant
Quinine	Antimalarial prophylactic	*Cinchona pubescens*	Cinchona
Reserpine	Treats hypertension	*Rauwolfia serpentina*	Rauwolfia
Sennoside A, B	Laxative	*Cassia angustifolia*	Senna
Scopolamine	Treats motion sickness	*Datura stramonium*	Thorn apple
THC	Antiemetic	*Cannabis sativa*	Marijuana
Toxiferine	Relaxes muscles during surgery	*Strychnos guianensis*	Strychnos plant
Tubocurarine	Muscle relaxant	*Chondrodendron tomentosum*	Curare
Vincristine	Treats pediatric leukemia	*Catharanthus roseus*	Rose periwinkle
Warfarin	Anticoagulant	*Melilotus* spp.	Sweet clover

Sources: M. J. Balick and P. A. Cox. 1996. *Plants, People, and Culture: The Science of Ethnobotany.* Sci Am Library: New York; E. Chivian and A. Bernstein. 2008. *Sustaining Life: How Human Health Depends on Biodiversity.* Oxford University Press: New York.

DNA in the biotechnology industry and in biological research (**FIGURE 3.7**). This enzyme is also used in the medical field to detect human diseases. The industries using this enzyme have generated hundreds of billions of dollars of value and employ hundreds of thousands of people. How many more such valuable species will be discovered in the years ahead—and how many will go extinct before they are discovered?

Indirect use values

Many organisms provide benefits to humans only when alive and in their natural ecosystem, such as plants growing on a hillside that capture carbon dioxide, provide summer shade, and prevent soil erosion. These resources have **nonconsumptive use value** because they are not consumed. These include both regulating and cultural ecosystem services (see Figure 3.4).

Indirect use values are nonconsumptive use values that can be assigned to all aspects of biodiversity that provide economic benefits without being harvested or destroyed. Because these benefits are not goods or services in the usual economic sense, they do not typically appear in the statistics of national economies, such as the GDP. They are often called **public goods** because they belong to society in general, without private ownership. However, these benefits may be crucial to the continued availability of the natural products on which economies depend. If natural ecosystems are not available to provide these benefits, substitute sources must be found—often at great expense—or local and even regional economies may face a decline in prosperity or even collapse.

Human societies are totally dependent on the free services that we obtain from natural ecosystems, because we could not pay to replace these ecosystems if they were permanently degraded or destroyed. Especially important in this regard are wetland ecosystems because of their role in water purification and nutrient recycling as well as their enormous importance in flood control (see the section "Water and soil protection").

Economists are actively improving calculations of the indirect use value of ecosystem services at regional and global levels (Bateman et al. 2013;

FIGURE 3.7 Specialized bacteria, such as these bright-orange bacteria at Prism Lake, grow abundantly in Yellowstone National Park's hot mineral springs and have contributed essential enzymes for the high-temperature reactions used in the biotechnology industry.

Photo by Richard Primack

Reyers et al. 2013). However, many ecological economists sharply disagree about how calculations of indirect use value should be done, or even whether they should be done at all (Peterson et al. 2010). One such calculation suggests that the annual value of ecosystem services worldwide is as much as $145 trillion (Costanza et al. 2014), exceeding the current $84.5 trillion annual value of the world's economy (O'Neill 2021). Using different approaches, other ecological economists have come up with lower estimates, but those estimates have still amounted to trillions of dollars a year. The disparity in these various estimates indicates that much more work needs to be done on this topic.

The great variety of environmental services that ecosystems provide can be assigned particular types of indirect use value. The following sections discuss some of the specific indirect use values derived from biodiversity. Later in the chapter, we will consider two other nonconsumptive ways of valuing biodiversity: option value (the value that biodiversity may have in the future) in Section 3.3 and existence value (the amount that people are willing to pay to protect biodiversity [or other environmental goods or services] even if they never expect to experience it) in Section 3.4.

Ecosystem productivity

All life on Earth is made possible by the energy of the sun, which is converted into usable energy through photosynthesis in plants and algae. Humans depend on the energy stored in plants for many direct uses, such as food, fuelwood, and fodder for animals. This plant material is also the starting point for innumerable food chains (see Chapter 2), from which people harvest many animal products. Humans appropriate approximately half of the productivity of the terrestrial environment to meet their needs for natural resources (MA 2005), and most of the remaining half performs services that have indirect use value to humans, including the production of oxygen (O_2) by plants through the process of photosynthesis.

The destruction of the vegetation in an area through overgrazing by domestic animals or overharvesting of timber will destroy the system's ability to perform these functions (**FIGURE 3.8**). Eventually, it will lead to losses of plant biodiversity and of the associated production of plant biomass, loss of the animals that live in that area, and losses of natural resources and ecosystem services for people.

Likewise, coastal estuaries are areas of rapid plant and algal growth that provide the starting point for food chains leading to commercial stocks of fish and shellfish. When these coastal areas are filled in for development, their value to society is lost. Even when degraded or damaged wetland ecosystems are rebuilt or restored—usually at great expense—they often do not function as well as they initially did and almost certainly do not contain their original species composition or species richness.

Scientists are actively investigating how the loss of species from biological communities affects ecosystem processes such as the total growth of plants, the ability of plants to absorb atmospheric CO_2, and the ability of

GLOBAL CHANGE
CONNECTION

> Ecosystems with reduced species diversity are less able to adapt to the altered conditions associated with rising carbon dioxide levels and global climate change.

Photo by Anna Sher

FIGURE 3.8 Although many grassland systems are adapted to grazing, overgrazing reduces ecosystem services and can lead to decreased ecosystem productivity and increased soil erosion.

(A)

(B)

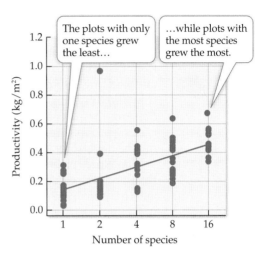

FIGURE 3.9 (A) Healthy prairie ecosystems are naturally diverse, with many species of grasses and forbs. (B) Varying numbers of grassland species were grown in experimental plots over seven years. The plots containing the most species had the greatest overall productivity (the amount of biomass produced by plants). (B, after C. Wagg et al. 2017. *Ecology* 98: 2952–2961. © 2017 by the Ecological Society of America.)

communities to adapt to global climate change (Handa et al. 2014; Liu et al. 2018). Many studies of natural and experimental grassland communities confirm that as species are lost, overall productivity declines and the community is less flexible in responding to environmental disturbances such as drought (Hautier et al. 2015; Wagg et al. 2017) (**FIGURE 3.9**). However, it is also true that in many locations, ecosystem services can be maintained if the loss of native species is balanced by their replacement by non-native species. Recently ecologists have even found a link between loss of biodiversity in mammal communities and the increased transmission of pathogens, such as Lyme disease (Kilpatrick et al. 2017).

GLOBAL CHANGE
CONNECTION

Water and soil protection

Biological communities are of vital importance in protecting watersheds, buffering ecosystems against extremes of flood and drought, and maintaining water quality. Plant foliage and dead leaves intercept the rain and reduce its impact on the soil. Plant roots and soil organisms aerate the soil, increasing its capacity to absorb water. This increased water-holding capacity reduces the flooding that would otherwise occur after heavy rains and allows a slow release of water for days and weeks after the rains have ceased.

When logging, farming, and other human activities disturb vegetation, the rates of soil erosion, and even occurrences of landslides, increase rapidly, decreasing the value of the land for some human activities. Climate change has been shown to be directly linked to the incidence of landslides through its effects on vegetation and the increased intensity of rainstorms (Paranunzio et al. 2018). Damage to the soil limits the ability of plant life to recover from disturbance and can render the land useless for agriculture. Erosion and flooding also contaminate drinking water supplies for humans in the communities along rivers, leading to an increase in human health problems. Soil erosion increases sediment loads entering the reservoirs behind dams, causing a loss of electrical output, and it creates sandbars and islands, reducing the navigability of rivers and ports.

Floods are currently the most common natural disaster in the world, killing thousands of people each year, and losses of wetland and floodplain ecosystems have contributed to these disasters. In the industrial nations of the world, wetlands protection has become a priority in order to prevent flooding of developed areas (**FIGURE 3.10**). A regional study of the northeastern United States found that wetlands prevented $625 million in direct damages from a single hurricane and provided an average of 16% reduction in annual flood losses (Narayan et al. 2017). The conversion of floodplain habitat to farmland along the Mississippi, Missouri, and Red Rivers in North America and the Rhine River in Europe is considered a major factor in the massive, damaging floods along those rivers in past years. It is estimated that the floodwater holding

© Patricia Hofmeester/Shutterstock

FIGURE 3.10 Wetlands perform many vital functions for humans. The trees, their root mats, and other vegetation act like a living sponge that traps and slowly releases rainwater and snowmelt, while creating resistance that slows water flows, thus preventing damage from floodwater. Aquatic plants also clean water by taking up excess nitrogen, phosphorus, heavy metals, and other substances that can be harmful to humans.

capacity of the wetlands along the Mississippi river is only 20% of what it once was (https://www.epa.gov/wetlands/why-are-wetlands-important). The year 2021 brought many dramatic examples of the consequences of such wetland loss, especially when combined with climate change; according to floodlist. com, there were 124 flood events across 20 countries, resulting in nearly 1,000 human fatalities in July alone.

Wetlands also perform important functions for filtering excess nutrients and toxins in water. An entire field of research and at least one scientific journal is dedicated to understanding and maximizing the capacity of wetland plants to clean water, typically for the intent of subsequent human use. For example, a recent meta analysis determined that a higher diversity of plants facilitated greater uptake of nitrogen (Brisson et al. 2020). The monetary value of temperate forests for provisioning water and treating waste has been estimated as $198 per hectare per year (FAO 2018b).

Most aquatic ecosystems such as swamps, lakes, rivers, floodplains, tidal marshes, mangroves, estuaries, the continental shelf, and the open ocean are capable of breaking down and immobilizing toxic pollutants, such as the heavy metals and pesticides that have been released into the environment by human activities (Ostroumov 2017). However, when such aquatic communities, especially the bacteria and other microorganisms they contain, are damaged by a combination of sewage overload and habitat destruction, alternative systems have to be developed. These contrived systems, such as waste treatment facilities and giant landfills, cost tens of billions of dollars. In regions that cannot afford to build such facilities, people's quality of life can be severely harmed.

GLOBAL CHANGE
CONNECTION

Wetland ecosystem services whose value is typically not accounted for in the current market system include waste treatment, water purification, and flood control—all of which are essential to healthy human societies.

Climate regulation

GLOBAL CHANGE
CONNECTION

Plant communities are important in moderating local, regional, and even global climate conditions. At the local level, trees provide shade and evaporate water from their leaf surfaces during photosynthesis, reducing the local temperature in hot weather. This cooling effect reduces the need for fans and air conditioners and increases people's comfort and work efficiency. Trees are also locally important because they act as windbreaks for agricultural fields, reducing soil erosion by wind and reducing heat loss from buildings in cold weather.

At the regional level, plants capture water that falls as rain and then transpire it back into the atmosphere, from which it can fall as rain again. The loss of vegetation from large forested regions such as the Amazon basin and western Africa may result in a reduction of average annual rainfall or greatly altered weather patterns over large areas.

In both terrestrial and aquatic environments, plant growth is tied to the carbon cycle. A reduction in plant life results in reduced uptake of CO_2, contributing to the rising CO_2 levels that lead to global warming (see Chapter 5) (Brienen et al. 2015; Schebek et al. 2018). Environmental economists also recognize the value of intact and restored forests in retaining carbon and absorbing atmospheric CO_2 (Bluffstone et al. 2017). Many countries and corporations are paying to protect and restore forests and other ecosystems as part of the worldwide effort to address global climate change by reducing CO_2 emissions (see Chapter 12).

GLOBAL CHANGE
CONNECTION

Species relationships and environmental monitors

Many of the species harvested by people for their direct consumptive use value depend on other wild species for their continued existence. For example, the wild game and fish harvested by people are dependent on wild insects and plants for their food. Thus, a decline in a wild species of little immediate value to humans may result in a corresponding decline in a harvested species that is economically important.

Crop plants benefit from wild insects, birds, and bats. Predatory insects such as praying mantises and ladybugs, as well as many bird and bat species, feed on pest insect species that attack crops, increasing crop yields and reducing the need to spray pesticides (Taylor et al. 2018). Insects, birds, and bats also act as pollinators for numerous crop species (**FIGURE 3.11**). About 150 species of crop plants in the United States require insect pollination of their flowers, which is often performed by a combination of wild insects and domesticated honeybees (Vanbergen and the Insect Pollinator Initiative 2013). Pollinators are necessary for the production of 75% of our crop species (Kremen 2018). The value of wild insect pollinators will increase in the near future if they take over the pollination role of honeybees, whose populations are declining in many places because of disease and pests. Many useful wild plant species depend on fruit-eating birds, primates, and other animals to act as seed dispersers. Where these animals have been overharvested, fruits remain uneaten, seeds are not dispersed,

> Relationships between species are often essential for preserving biodiversity and providing value to people. For example, many insects pollinate the crops on which people depend for food.

(A)

Photo by David McIntyre

(B)

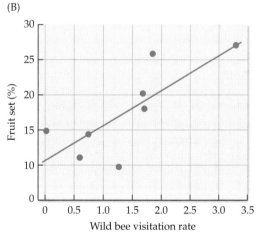

FIGURE 3.11 (A) Bees provide ecosystem services for humans by pollinating food crops such as this peach tree; without pollination, the tree will not produce fruit. The indirect economic value of these insects can be estimated by what it would cost to pay humans to hand-pollinate the trees—something that is sometimes necessary when bee populations decline. (B) Higher visitation rates by wild bees (the number of bees observed in 15-minute intervals foraging on 1000 flowers) increase the fruit set of sweet cherries (the percentage of flowers that develop into fruits). (B, after A. Holzschuh et al. 2012. *Biol Conserv* 153: 101–107. Copyright 2012. Reprinted with permission from Elsevier.)

and plant species head toward local extinction (Corlett 2017). It should be noted, however, that there is redundancy in guilds of similar species, and the services of one natural predator, pollinator, or seed disperser may be carried out equally well by another species.

One of the most economically significant relationships in ecosystems is the one that many forest trees and crop plants have with the soil organisms that provide them with essential nutrients (Wood and Bradford 2018). Fungi and bacteria break down dead plant and animal matter in the soil, which they use as their energy source (see Figure 2.10). In the process, the fungi and bacteria release mineral nutrients such as nitrogen into the soil. These nutrients are used by plants for further growth. The poor growth and dieback of many trees in certain areas of North America and Europe is attributable in part to the deleterious effects that acid rain and air pollution have on the soil fungi that help supply the trees with mineral nutrients and water (Lawrence et al. 2018).

GLOBAL CHANGE
CONNECTION

Amenity value

Ecosystems provide many recreational services for humans; for instance, they furnish a place to enjoy nonconsumptive activities such as hiking, photography, and birdwatching. This experience of nature not only is enjoyable, but also leads to improved health for the participants (Donovan et al. 2013). The monetary value of these activities, sometimes called their **amenity value**, can be considerable and can have a major impact on local economies. Even sportfishing and hunting, which in theory are consumptive uses, are in practice both consumptive and nonconsumptive because the food value of the animals caught by fishermen and hunters is often insignificant compared with the time and money spent on these activities.

Ecotourism is a special category of recreation that involves people visiting places and spending money wholly or in part to experience unusual biological communities (such as rain forests, African savannas, coral reefs, deserts, the Galápagos Islands, or the Everglades) and to view particular "flagship" species (such as elephants on safari trips) (**FIGURE 3.12**). Tourism, valued at $8.9 trillion per year, is among the world's largest industries; representing 10.3% of global GDP, it is much larger than even the automobile industry (3.5%) (WTTC 2019).

Ecotourism has traditionally been a key industry in many parts of the world, such as Kenya and Costa Rica. The Great Barrier Reef in Australia is estimated to be worth $56 billion and employs 36 times more people than the commercial fishing industry in Australia (ABC News 2017). Direct expenditures of shark-diving tourists in Australia have been estimated to be at least $25.5 million (Huveneers et al. 2017). In addition to international tourism, the rapidly growing middle classes in developing countries, such as China and India, are increasingly traveling within their own countries to visit national parks and nature reserves (Lyngdoh et al. 2017). In India and Nepal, ecotourism has been an important component of the remarkable comeback of the greater one-horned rhino (*Rhinoceros unicornis*) from the brink of extinction (Chaudhary and Gautam 2020; see chapter opener).

FIGURE 3.12 (A) Ecotourism can provide an economic justification for protecting biodiversity and can also provide benefits to people living nearby. (B) The diagram illustrates some of the main elements in a successful ecotourism program. (B, after R. W. Braithwaite. 2001. In *Encyclopedia of Biodiversity*, Vol. 5, S. A. Levin [Ed.], pp. 667–679. Academic Press, San Diego, CA. Copyright 2001. Republished with permission from Elsevier; permission conveyed through Copyright Clearance Center.)

(A)

© Tiger Tracker/Alamy Stock Photo

(B)

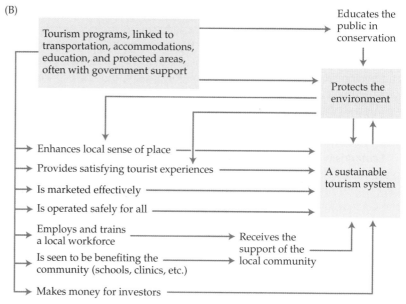

The revenue provided by ecotourism has the potential to provide one of the most immediate justifications for protecting biodiversity, particularly when ecotourism activities are integrated into overall management plans (Wabnitz et al. 2018). In integrated conservation and development projects (ICDPs), local communities develop accommodations, expertise in nature guiding, local handicraft outlets, and other sources of income; the income from ecotourism allows the local people to give up unsustainable or destructive hunting, fishing, or grazing practices (see Chapter 10). The local community benefits from the learning of new skills, employment opportunities, greater protection for the environment, and the development of additional community infrastructure such as schools, roads, medical clinics, and stores.

A danger of ecotourism is that the tourists themselves will unwittingly damage the sites they visit—by leaving trash behind, trampling wildflowers, breaking coral, or disrupting nesting bird colonies, for instance—thereby contributing to the degradation and disturbance of sensitive areas (Büscher and Fletcher 2017; Shutt et al. 2014). Tourists might also indirectly damage sites by creating a demand for fuelwood for heating and cooking, thus contributing to deforestation. In addition, the presence, affluence, and demands of tourists can transform traditional human societies in tourist areas by changing employment opportunities (Devine 2017). As local people increasingly enter a cash-based economy, their values, customs, and relationship to nature may be lost along the way. A final potential danger of this industry is that ecotourist facilities may provide a sanitized fantasy experience rather than helping visitors understand the serious social and environmental problems that endanger biodiversity.

> The rapidly developing ecotourism industry can provide income to protect biodiversity, but possible costs must be weighed along with benefits.

Fortunately, there are many intentional efforts to capitalize on the benefits of ecotourism while minimizing its costs; for example, increasingly environmental education is being incorporated into ecotourism, and ecotourism companies often contribute funds to mitigate the negative effects of their industry (**FIGURE 3.13**). In some cases, visitors may even participate in conservation activities as a part of their experience. There is a growing body of research documenting the benefits of ecotourism for conservation, both direct (funds raised, volunteer hours) and indirect (e.g., education) (Wardle et al. 2018).

Educational and scientific value

Many books, television programs, movies, and websites produced for educational and entertainment purposes are based on nature themes (Tietge 2018). These natural history materials are continually incorporated into school curricula and are worth billions of dollars per year. To take one example, movies with penguins as main characters or themes have had world revenues estimated at around

FIGURE 3.13 Conservation education can be incorporated into tourism. At each stop on this zipline route in Queenstown, New Zealand, Aashray Lal and fellow guides use interpretive panels to teach guests from around the world about local ecology and the importance of preserving ecosystems. The zipline company is also able to market itself as eco-friendly by making significant financial contributions to conservation.

Photo by Anna Sher

$1.6 billion. They represent a nonconsumptive use value of biodiversity because they use nature only as intellectual content. A considerable number of professional scientists, as well as highly motivated amateurs, are engaged in making ecological observations and preparing educational materials for such films. While these scientific activities provide economic benefits to the communities surrounding field stations, their real value lies in their ability to increase human knowledge, enhance education, and enrich the human experience.

Multiple uses of a single resource: A case study

Horseshoe crabs (*Limulus polyphemus*) provide an example of the diverse values that can be provided by just one species. They are usually noticed only as clumsy creatures that seem to move with difficulty in shallow seawater. In the United States, commercial fishermen harvest these animals in large quantities for use as cheap fishing bait. In recent years, however, ecologists have realized that horseshoe crab eggs and juveniles are extremely important as a food source for shorebirds and coastal fish, which have a major role in local tourism related to birdwatching and sportfishing. Without horseshoe crabs, shorebird and sport fish populations decline in abundance (WHSRN 2015). Additionally, the blood of horseshoe crabs is collected to make limulus amoebocyte lysate (LAL), a highly valuable chemical used to detect bacterial contamination in injection-administered medications and vaccines (Krisfalusi-Gannon et al. 2018). This chemical cannot be manufactured synthetically; horseshoe crabs are its only source. Without this natural source of LAL, our ability to determine the purity of injected medicines would be compromised.

Currently, commercial fishing and sportfishing interests, environmental groups, birdwatching groups, and the biomedical industry are competing for control of horseshoe crabs along United States coastlines. Each group can make a good argument for its own right to use or protect horseshoe crabs. Hopefully, the final result will be a working compromise that allows the crabs a place in a functioning ecosystem and still provides for the needs of people living in the area and elsewhere in the world.

3.3 The Long-Term View: Option Value

LEARNING OBJECTIVES

By the end of this section you should be able to:

3.3.1 Produce novel examples of option values.

3.3.2 Explain why option value is an important justification for conservation.

The potential of biodiversity to provide an economic benefit to human society at some point in the future is its option value. As the needs of society change, so must the methods of satisfying those needs, and such methods often lie in previously untapped animal or plant species. For example, the continued genetic improvement of cultivated plants is necessary not only for increased yield, but also to guard against pesticide-resistant insects and more virulent strains of fungi, viruses, and bacteria (Grossi-de-Sá et al. 2017).

We are continually searching the biological communities of the world for new plants, animals, fungi, and microorganisms that can be used to fight human diseases or to provide some other economic value, an activity referred to as **bioprospecting** (Borsetto and Wellington 2017). These searches are generally carried out by government research institutes and pharmaceutical companies, sometimes in collaboration with universities. For example, the Center for Botanicals and Metabolic Syndrome (CBMS; cbms.rutgers.edu) was established in Tajikistan in 2014 by scientists from Rutgers University with a grant from the US National Institutes of Health (NIH), in part to investigate new potential medicines, particularly for treating diabetes (**FIGURE 3.14**). The NIH also funds the US National Cancer Institute's Natural Products Branch, which has been carrying out a program to test extracts of thousands of wild species for their effectiveness in controlling cancer cells. Another approach has been to target plants and other natural products used in traditional medicine for screening, often in collaboration with local healers and rural villagers. Programs such as these provide financial incentives for countries to protect their natural resources and the biodiversity knowledge of their Indigenous inhabitants.

> Sometimes a specific species or ecosystem can provide a variety of goods and services to human society. Compromises are often needed to balance competing uses.

FIGURE 3.14 Herbal medicines sold at a marketplace in Mary, Turkmenistan. Molecular biologists are currently testing the properties of plants used for traditional cures in a search for cures to diseases such as diabetes.

The search for valuable natural products is wide-ranging: in addition to ethnobotanists looking for new cures for diseases, entomologists search for insects that can be used as biological control agents, microbiologists search for bacteria that can assist in biochemical manufacturing processes, and so on. The growing biotechnology industry is finding new ways to reduce pollution, to develop better industrial processes, and to fight diseases threatening human health. The gene-splicing techniques of molecular biology are allowing unique, valuable genes found in one species to be transferred to another species. Both newly discovered and well-known species are often found to have exactly those properties needed to address some significant human problem. If biodiversity is reduced, the ability of scientists to locate and utilize a broad range of species will also be reduced.

A question currently being debated among conservation biologists, governments, environmental economists, and corporations is, Who owns the commercial development rights to the world's biodiversity? The answer to this is complex, especially in the modern era. An excellent example is provided by the immunosuppressant drug cyclosporine, which occurs naturally in the fungus *Tolypocladium inflatum*. The Swiss company Sandoz (which later merged with another company to become Novartis) developed cyclosporine into a family of drugs with sales of nearly $2 billion per year (Verified Market Research 2021). Cyclosporine is a key medicine used following transplant surgery to prevent organ rejection. The fungus was found in a sample of soil that was collected in Norway, without permission, by a Sandoz biologist on vacation and was later screened for biological activity. Norway has not yet received any payment for the use of this fungus in drug production. Similarly, the US National Park Service has not received any payments for the bacteria collected from hot springs at Yellowstone National Park that led to breakthroughs in DNA technology.

Such past and present unauthorized bioprospecting for commercial purposes is now often termed **biopiracy**. To help mitigate this problem, 129 countries have signed what is known as the Nagoya Protocol, which dictates that any company using "genetic resources" from a participating nation must negotiate an agreement to ensure that any profits or benefits are shared. Many developing countries have also passed laws that require permits for collecting biological material for research and commercial purposes, imposing criminal penalties and fines for the violation of such laws. Both developing and developed countries now frequently demand a share in the commercial activities derived from the biodiversity contained within their borders, and rightly so. Local people in developing countries who possess knowledge of species, protect them, and show them to scientists should also share in the profits from any use of them. Writing treaties and developing procedures to guarantee participation in this process continue to be major diplomatic challenges.

> Given the enormous value of biodiversity, the question of who should benefit financially from it is important and sometimes complex.

While most species may have little or no direct economic value and little option value, a small proportion may have enormous potential to supply medical treatments, to support a new industry, or to prevent the collapse of a major agricultural crop. For example, wild arabica coffee is globally endangered, and populations are projected to shrink by 50% or more due to climate change alone; this is important genetic stock to support an industry with an export value of over $13 billion per year (Moat et al. 2018). Other species or sets of species may provide other kinds of future values even if they don't provide them now. If just one of these species goes extinct before it is discovered, it could be a tremendous loss to the global economy, even if the majority of the world's species are preserved. As Aldo Leopold commented in *Round River* (1953):

GLOBAL CHANGE
CONNECTION

> If the biota, in the course of aeons, has built something we like but do not understand, then who but a fool would discard seemingly useless parts? To keep every cog and wheel is the first precaution of intelligent tinkering.

The diversity of the world's species can be compared to a manual on how to keep the Earth running effectively. The loss of a species is like tearing a page out of the manual. If we ever need the information from that page to save ourselves and the Earth's other species, the information will have been irretrievably lost.

3.4 Existence Value

LEARNING OBJECTIVES

By the end of this section you should be able to:

3.4.1 Identify cases in which the primary justification for conservation is existence value.

3.4.2 Create a diagram to determine the total economic value of a given ecosystem.

Many people throughout the world care about wildlife, plants, and entire ecosystems and want to see them protected. Their concern may be associated with a desire to someday visit the habitat of a unique species and see it in the wild; alternatively, concerned individuals may not expect, need, or even desire to see a species personally or experience the habitat in which it lives. For this reason, existence value is considered a non-use value: people value the resource without any intention to use it now or in the future. In economic terms, existence value is the amount that people are willing to pay to prevent species from going extinct, habitats from being destroyed, and genetic variation from being lost (Martín-López et al. 2014). A related idea is **beneficiary value**, or **bequest value**: how much people are willing to pay to protect something of value for their own children and descendants, or for future generations.

Particular species—the so-called charismatic megafauna, such as pandas, whales, lions, and many birds—elicit strong responses in people, who

> People, governments, and organizations annually contribute large sums of money to ensure the continuing existence of certain species and ecosystems.

FIGURE 3.15 Whale watching can make a significant contribution to a local economy, and participants often later contribute money to organizations promoting whale conservation. Here, people greet a California gray whale (*Eschrichtius robustus*) in Magdalena Bay, on the Pacific side of the Baja Peninsula. Such meetings can enrich human lives.

sometimes also want to protect them from going extinct (**FIGURE 3.15**). Special groups have been formed to appreciate and protect butterflies and other insects, wildflowers, and fungi. People place value on wildlife in a direct way by joining and contributing billions of dollars to organizations that protect species. In the United States, billions of dollars are contributed each year to conservation and environmental organizations, with The Nature Conservancy ($1230 million in 2020), the WWF ($276 million), Ducks Unlimited ($230 million), and the Sierra Club ($137 million) high on the list. Citizens also show their concern by directing their governments to spend money on conservation programs and to purchase land for habitat. For example, the US government spent millions of dollars to protect a single rare species, the brown pelican (*Pelecanus occidentalis*), which was protected initially under the US Endangered Species Act and is now considered recovered.

Existence value can also be attached to ecosystems, such as temperate old-growth forests, tropical rain forests, coral reefs, and prairie remnants, and to areas of scenic beauty. Growing numbers of people and organizations contribute large sums of money annually to ensure the continuing existence of these habitats, and the environment is increasingly an important issue during national elections. Further, people want environmental education included in public school curricula (e.g., see the EPA's Environmental Education website, www.epa.gov/education).

In summary, ecological economics has helped to draw attention to the wide range of goods and services provided by biodiversity. That attention has enabled scientists to account for environmental impacts that were previously left out of the equation. When analyses of large-scale development projects have finally been completed, some projects that initially appeared to be successful have been seen to actually be running at an economic loss. For example, to evaluate the success of an irrigation project using water diverted from a tropical wetland ecosystem, the short-term benefits (improved crop yields) must be weighed against the environmental costs. **FIGURE 3.16** shows the total economic value of a tropical wetland ecosystem, including its use value, option value, and existence value. When the wetland ecosystem is damaged by the removal of water, the ecosystem's ability to provide the goods and services shown in the figure is curtailed, their value greatly diminishes, and the economic success of the project is called into question. It is only by incorporating the value of the wetland into this equation that an accurate view of the total project can be gained.

Total Economic Value of a Tropical Wetland Ecosystem

Use Values

Existence Value
Protection of biological
 diversity
Maintaining culture of
 local people
Continuing ecological and
 evolutionary processes

Direct Use Values
Fish and meat
Fuelwood
Timber and other
 building materials
Medicinal plants
Edible wild fruits
 and plants
Animal fodder

Indirect Use Values
Flood control
Soil fertility
Pollution control
Drinking water
Transportation
Recreation and tourism
 (e.g., birdwatching)
Education
Biological services
 (e.g., pest control, pollination)

Option Value
Future products:
 Medicines
 Genetic resources
 Biological insights
 Food sources
 Building supplies
 Water supplies

FIGURE 3.16 Evaluation of the success of a development project must incorporate the full range of its environmental impacts. This figure shows the total economic value of a tropical wetland ecosystem, including direct and indirect use value, option value, and existence value. A development project such as an irrigation project lowers the value of the wetland ecosystem when water is removed for crop irrigation. When that lowered value is taken into account, the irrigation project may represent an economic loss. (After M. J. Groom et al. 2006. *Principles of Conservation Biology*, 3rd ed. Oxford University Press/Sinauer, Sunderland MA.; based on data in L. Emerton. 1999. *Evaluating Eden Series Discussion Paper No. 5*, IIED, London.)

3.5 Environmental Ethics

In most modern societies, people attempt to protect biodiversity, environmental quality, and human well-being through regulations, incentives, fines, environmental monitoring, and assessments. A complementary approach is to change the fundamental values of our materialistic society. **Environmental ethics**, a vigorous and growing discipline within philosophy, articulates the ethical value of the natural world (Stenmark 2017; Traer 2018). As a corollary, it challenges the materialistic values that tend to dominate modern societies. If contemporary societies deemphasized the pursuit of wealth and instead focused on furthering genuine human well-being, the preservation of the natural environment and the maintenance of biodiversity would probably become honored practices rather than occasional afterthoughts (Muiños et al. 2015).

Ethical values of biodiversity

Environmental ethics provides virtues and values that make sense to people today. At a time when there are unprecedented threats to the environment, ethical arguments can and do convince people to conserve biodiversity. Ethical arguments are also important because although economic arguments by themselves provide a basis for valuing some species and ecosystems, economic valuation can also provide grounds for extinguishing species, or for saving one species and not another (Rolston 2012). According to conventional economic thinking, a species with low population numbers, an unattractive appearance, no immediate use to people, and no relationship to any species of economic importance will be given a low value. Halting profitable developments or making costly attempts to preserve these species may not have any obvious economic justification. In fact, in many circumstances, economic cost-benefit analyses will support the destruction of endangered species that stand in the way of "progress."

> Ethical arguments can complement economic and biological arguments for protecting biodiversity. Such ethical arguments are readily understood by many people.

Despite any economic justification, however, many people would make a case against species extinctions on ethical grounds, arguing that the conscious destruction of a natural species is morally wrong, even if it is economically profitable. Similar arguments can be advanced for protecting unique ecosystems and genetic variation. Ethical arguments for preserving biodiversity resonate with people because they appeal to our nobler instincts or to belief in a divine creation (Melin 2016).

The following arguments, based on the intrinsic value of species and on our duties to other people, are important to conservation biology because they provide a rationale for protecting all species, including those with no obvious economic value.

EACH SPECIES HAS A RIGHT TO EXIST

All species represent unique biological solutions to the problem of survival. For this reason, the survival of each species must be respected, regardless of its importance to humans. This statement is true whether the species is large or small, simple or complex, ancient or recently evolved; whether it is economically important or of little immediate economic value to humans; and whether it is loved or hated by humans. Each species has value for its own sake—an **intrinsic value** unrelated to human needs or desires (Smith 2016). This argument suggests not only that we have no right to destroy any species, but also that we have a moral responsibility to actively protect species from going extinct as the result of our activities, as articulated by the deep ecology movement described later in this section. This argument also recognizes that humans are part of the larger biotic community and reminds us that we are not the center of the universe (**FIGURE 3.17**).

> An argument can be made that people have a responsibility to protect species and other aspects of biodiversity because of their intrinsic value, not because of human needs.

GLOBAL CHANGE
CONNECTION

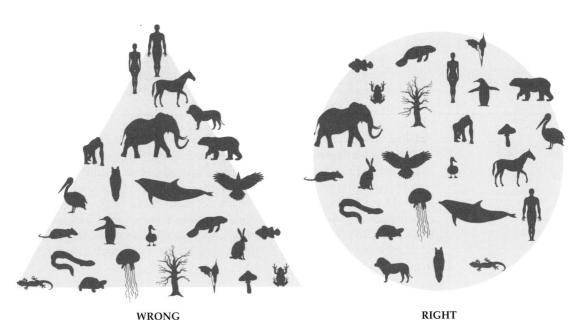

WRONG

RIGHT

FIGURE 3.17 According to certain principles of environmental ethics, it is wrong for humans to act as if they are at the top of the living world and have the right to exploit and damage other species; rather, there is a moral imperative for humans to behave as if all species have an equal right to exist. (Silhouettes © glyph/Shutterstock, © ntnt/Shutterstock, and © theromb/Shutterstock.)

ALL SPECIES ARE INTERDEPENDENT   Species interact in complex ways in natural communities. The loss of one species may have far-reaching consequences for other members of the community (as described in Chapter 2). Other species may become extinct in response, or the entire ecosystem may become destabilized as the result of cascades of species extinctions. For these reasons, if we value some parts of nature, we should protect all of nature (Leopold 1949). We are obligated to conserve the system as a whole because that is the appropriate survival unit. For example, parasites (arguably the least charismatic of all creatures) are the most abundant form of life, and they face possible extinction when hosts are threatened (Thompson et al. 2018). Given the important evolutionary relationships with their hosts, the loss of parasites can have dangerous cascading impacts throughout an ecosystem because of their role in density regulation of hosts, directly or by altering host morphology or behavior.

> All species and ecosystems are interdependent, and so all parts of nature should be protected. It is in the long-term survival interest of people to protect all of biodiversity.

PEOPLE HAVE A RESPONSIBILITY TO ACT AS STEWARDS OF THE EARTH   Many religious adherents find it wrong to destroy species because they are God's creations (see Chapter 1). If God created the world, then presumably the species God created have value. Within the Jewish, Islamic, and Christian traditions, human responsibility for protecting animal species is explicitly described in the Bible as part of the covenant with God. For example, Muslim clerics in Malaysia recently joined Indonesia in a fatwa (a religious edict) against hunting endangered species. Asrorun Niam Sholeh of the Indonesian Ulema Council stated, "As Muslims, we have a duty to maintain the ecological balance" (Yi 2016). Other major religions, including Hinduism and Buddhism, strongly support the preservation of nonhuman life. As Mahatma Gandhi (1869–1948) taught, "The good man is the friend of all living things."

PEOPLE HAVE A RESPONSIBILITY TO FUTURE GENERATIONS   If in our daily living we degrade the natural resources of the Earth and cause species to become extinct, future generations will pay the price in terms of a lower standard of living and quality of life (Gardiner et al. 2010). As species are lost and wild lands developed, children are deprived of one of the most exciting experiences in growing up: the wonder of seeing "new" animals and plants in the wild. To remind us to act more responsibly, we might imagine that we are borrowing the Earth from future generations who expect to get it back in good condition.

RESPECT FOR HUMAN LIFE AND HUMAN DIVERSITY IS COMPATIBLE WITH A RESPECT FOR BIODIVERSITY   Some people worry that recognizing intrinsic value in nature requires taking resources and opportunities away from human beings, but respect for and protection of biodiversity can be linked to greater opportunities and better health for people (Kilpatrick et al. 2017). Some of the most exciting developments in conservation biology involve supporting the economic development of disadvantaged rural people in ways that are linked to the protection of biodiversity. In developed countries, the **environmental justice** movement seeks to empower poor and politically weak people, who are often members of minority groups, to protect their own environments; in the process, their well-being and the protection of biodiversity are enhanced (Robinson 2011).

PEOPLE BENEFIT FROM AESTHETIC AND RECREATIONAL ENJOYMENT OF BIODIVERSITY Throughout history, poets, writers, painters, and musicians of all cultures have drawn inspiration from wild nature (Thoreau 1854; see Chapter 1). Nearly everyone enjoys wildlife and landscapes aesthetically, and this joy increases the quality of our lives (**FIGURE 3.18**). Nature-related activities are important in childhood development (Luck et al. 2011) and may even improve human health (Donovan et al. 2013).

Paul Shepard (1925–1996) argued that civilization has made people out of sync with our environment and that we need this contact for our emotional and psychological well-being. A related idea, called "nature deficit disorder," was popularized by author Richard Louv (2005) to describe the wide variety of problems people suffer when deprived of interactions with the natural world. The presence and abundance of trees in urban settings has frequently been associated with a lower incidence of crime (e.g., Gilstad-Hayden et al. 2015). Recreational activities such as hiking, canoeing, and mountain climbing are physically, intellectually, and emotionally satisfying. People spend tens of billions of dollars annually in these pursuits, proof enough that they value

Photo by Anna Sher

FIGURE 3.18 The enjoyment people receive from being in nature is an important justification for the preservation of biodiversity. Here, visitors can use a boardwalk to experience the scenery at the base of Aoraki/Mount Cook in New Zealand, without danger of trampling sensitive species.

them highly. A loss of biodiversity diminishes this experience. What if there were no more migratory birds or no more meadows filled with wildflowers and butterflies? Would we still enjoy nature as much?

PEOPLE BENEFIT FROM THE KNOWLEDGE THE NATURAL WORLD PROVIDES

Three of the central mysteries in the world of science are (1) how life originated, (2) how the diversity of life interacts to form complex ecosystems, and (3) how humans evolved. Thousands of biologists are working on these questions and are coming ever closer to the answers. New techniques of molecular biology allow greater insight into the relationships of living species as well as some extinct species, which are known to us only from fossils. When species become extinct and ecosystems are damaged, however, important clues are lost, and the mysteries become harder to solve. For example, if *Homo sapiens'* closest living relatives—chimpanzees, bonobos, gorillas, and orangutans—disappear from the wild, we will lose important clues to human physical and social evolution.

GLOBAL CHANGE
CONNECTION

Deep ecology

One well-developed environmental philosophy that supports environmental activism is known as **deep ecology** (Naess 2008; Witoszek and Mueller 2017). Deep ecology builds on the basic premise of biocentric equality, which expresses "the intuition … that all things in the biosphere have an equal right to live and blossom and to reach their own individual forms of unfolding" (Devall and Sessions 1985). Humans have a right to live and thrive, as do the other organisms with whom we share the planet (see Figure 3.17). Deep ecologists oppose what they see as the dominant worldview, which places human concerns above all and views human happiness in materialistic terms. Deep ecologists see acceptance of the intrinsic value of nature less as a limitation than as an opportunity to live better lives.

Deep ecology is an environmental philosophy that advocates placing greater value on protecting biodiversity through changes in personal attitude, lifestyle, and even societies.

Because present human activities are destroying the Earth's biodiversity, our existing political, economic, technological, and ideological structures must change. The changes called for by deep ecology entail enhancing the quality of life for all people and emphasizing improvements in environmental quality, aesthetics, culture, and spirituality rather than higher levels of material consumption. The philosophy of deep ecology includes an obligation to work to implement needed programs through political activism and a commitment to personal lifestyle changes, in the process transforming the institutions in which we work, study, pray, and shop (Bearzi 2009). The perspectives of the deep ecology movement and other relative ethical and religious arguments allow professional biologists, ecologists, and all concerned people (such as you?) to escape from their narrow, everyday concerns and act and live "as if nature mattered" (Naess 1989). These ideas of deep ecology are now merging with "science-based visions of the relationship between humans and nature" and ecological economics to create a more unified set of arguments for the protection of biodiversity (Witoszek and Mueller 2017; Akamani 2020).

Summary

- Ecological economics is developing methods for valuing biodiversity and, in the process, providing arguments for its protection. Benefits humans derive from nature are called ecosystem services. These services can be described and quantified in terms of their use value.

- Direct use values are assigned to products harvested from the wild, such as timber, fuelwood, fish, wild animals, edible plants, and medicinal plants. Direct use values can be further divided into consumptive use values (for products that are used locally) and productive use values (for products harvested in the wild and later sold in markets).

- Indirect use values can be assigned to aspects of biodiversity that provide economic benefits to people but are not harvested during their use. Non-consumptive use values include ecosystem productivity, protection of soil and water resources, positive interactions of wild species with commercial crops, and regulation of climate. Biodiversity also provides value to recreation, education, and ecotourism activities.

- The option value of biodiversity is its potential to provide future benefits to human society, such as new medicines, industrial products, and crops. Biodiversity also has existence value, which is the amount of money people and their governments are willing to pay to protect species and ecosystems without any plans for their direct or indirect use.

- Environmental ethics appeals to religious and secular value systems to justify preserving biodiversity. The most central ethical argument asserts that people must protect species and other aspects of biodiversity because they have intrinsic value, unrelated to human needs. Further, biodiversity must be protected because human well-being is linked to a healthy and intact environment.

For Discussion

1. Explain why Adam Smith's basic principles of economics are not enough to protect biodiversity, and identify some situations that are vulnerable to market failure.

2. Find a recent large development project in your area, such as a dam, office park, shopping mall, highway, or housing development, and learn all you can about it. Estimate the costs and benefits of this project in terms of biological diversity, economic prosperity, and human health. Who pays the costs and who receives the benefits? Consider other projects carried out in the past and determine their impact on the surrounding ecosystem and human community.

3. Consider the natural resources that people use near where you live. Can you place an economic value on those resources? If you can't think of any products harvested directly, consider the many regulating ecosystem services such as flood control, freshwater provisioning, and soil retention.

4. Imagine that the only known population of a dragonfly species will be destroyed unless money can be raised to purchase the pond where it lives and the surrounding land. How could you assign a monetary value to this species?

5. Do living creatures, species, biological communities, and physical entities, such as rivers, lakes, and mountains, have rights? Can we treat them any way we please? Where should we draw the line of moral responsibility?

Suggested Readings

Chan, K. M., et al. 2016. Opinion: Why protect nature? Rethinking values and the environment. *Proceedings of the National Academy of Sciences USA* 113: 1462–1465. Considering the value of nature in terms of relational values (instrumental) as distinct from those that are of the object itself (intrinsic).

Fox, M. J. V., and J. D. Erickson. 2020. Design and meaning of the genuine progress indicator: A statistical analysis of the US fifty-state model. *Ecological Economics* 167: 106441. An analysis of which factors that make up the GPI have the greatest impact.

Godet, L., and V. Devictor. 2018. What conservation does. *Trends in Ecology & Evolution* 33: 720–730. Evidence from the conservation biology literature that the discipline is having positive and significant impacts on the protection of species.

Hardin, G. 1968. The tragedy of the commons. *Science* 162: 1243–1248. This often-cited work suggests that population control is the only solution to the overuse of common-property resources.

Kilpatrick, A. M., et al. 2017. Conservation of biodiversity as a strategy for improving human health and well-being. *Philosophical Transactions of the Royal Society B* 372: 20160131. https://doi.org/10.1098/rstb.2016.0131. An investigation of the relationship between biodiversity and the transmission of pathogens such as Lyme disease.

Kremen, C. 2018. The value of pollinator species diversity. *Science* 359: 741–742. We depend on wild pollinators such as bees for 75% of our crop species.

Sovacool, B. K., J. Kim, and M. Yang. 2021. The hidden costs of energy and mobility: A global meta-analysis and research synthesis of electricity and transport externalities. *Energy Research & Social Science* 72: 101885. A quantitative look at the impact of electricity and transportation on the environment and human health.

Wardle, C., et al. 2018. Ecotourism's contributions to conservation: Analysing patterns in published studies. *Journal of Ecotourism*, https://doi.org/10.1080/14724049.2018.1424173. An overview of the literature on the importance of ecotourism.

Witoszek, N., and M. L. Mueller. 2017. Deep ecology. *Worldviews: Global Religions, Culture, and Ecology* 21: 209–217. A modern framing of an eco-philosophy.

Habitat degradation and loss, particularly from agriculture, is a leading cause of biodiversity loss. Satellite images like this one of land being burned for agriculture, taken near the borders of Bolivia, Paraguay, and Brazil, are increasingly used to document environmental changes that impact biodiversity.

Threats to Biodiversity: Habitat Change

Maintaining a healthy environment means preserving all of its components in good condition—ecosystems, biological communities, species, populations, and genetic variation. If species, ecosystems, and populations are adapted to local environmental conditions, why are they being lost? Why don't they tend to persist in the same place over time? Why can't they adapt to a changing environment? These questions have a single, simple answer: massive anthropogenic disturbances (that is, disturbances caused by human activities) have altered, degraded, and destroyed the landscape on a vast scale, destroying populations, species, and even whole ecosystems.

GLOBAL CHANGE
CONNECTION

There are seven major threats to biodiversity: habitat destruction, habitat fragmentation, habitat degradation (including pollution), global climate change, the overexploitation of species for human use, the invasion of nonnative species, and the spread of disease (**FIGURE 4.1**). These direct threats are the consequence of the indirect threat of increasing human population and consumption (see Figure 1.2). Most threatened species face at least two of these threats, which may interact to speed their way toward extinction and hinder efforts to protect them (Tilman et al. 2017). Moreover, multiple threats may interact additively, or even synergistically, such that their combined impact on a species or an ecosystem is greater than the sum of their individual effects.

GLOBAL CHANGE
CONNECTION

In this chapter, we will explore the nature of the expansion of human populations and how it can lead to reducing, fragmenting, polluting, and otherwise degrading habitat. As with most threats to biodiversity, not all habitat loss is human caused, but the rate and magnitude can be attributed to our activities. Typically, these threats develop so rapidly, and on such a large scale, that species cannot adapt genetically to the changes or disperse to a more hospitable location. It is the role of conservation biology to document these habitat changes and measure their impact so that priorities can be set for action.

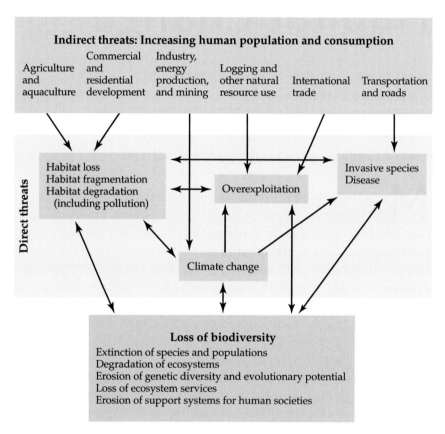

FIGURE 4.1 The major direct threats to biodiversity (yellow boxes) are the result of human activities. Although individual human activities (green box) are associated with particular threats, all can contribute to each threat; for example, while agriculture is primarily associated with loss of habitat and pollution, it is also a vector for invasive plants and contributes to climate change through use of fossil fuels. Furthermore, these direct threats can interact synergistically to speed up the loss of biodiversity. (After M. J. Groom et al. 2006. *Principles of Conservation Biology*, 3rd Ed. Oxford University Press/Sinauer Associates: Sunderland, MA, with updates from IUCN 2018. Threats Classification Scheme. Version 3.2. https://www.iucnredlist.org/resources/threat-classification-scheme.)

4.1 Human Population Growth and Its Impact

LEARNING OBJECTIVES

By the end of this section you should be able to:

4.1.1 Interpret Figure 4.1, linking increasing human population to loss of biodiversity.

4.1.2 Use the equation $I = PAT$ and data shown in Figure 4.3 to determine why the geographically small country of Japan has a greater total impact on world resources than many larger countries like Mongolia.

Up until about 300 years ago, the rate of human population growth had been relatively slow, with the birthrate only slightly exceeding the mortality rate. The greatest destruction of ecosystems has occurred since the Industrial Revolution, during which time the human population exploded from 1 billion in 1850 to 7.7 billion in 2018. The human population could reach a maximum of 9–10 billion by the middle of the twenty-first century (see Figure 1.1). Humans have increased in such numbers because birthrates have remained high while mortality rates have declined—a result of both modern medical achievements (specifically, the control of disease) and more reliable food supplies, caused in large part by the advent of nitrogen fertilizer. Population growth has slowed in the industrialized countries of the world, as well as in some developing countries in Asia and Latin America, but it is still high in other areas, particularly in tropical Africa.

Humans and their activities dominate ecosystems worldwide (**FIGURE 4.2**). People use large amounts of natural resources, such as fuelwood, timber, wild meat, and wild plants, and convert vast areas of natural habitat into agricultural and residential lands. Agricultural systems and other human activities now occupy one-fourth of the Earth's land surface (Krausmann et al. 2013). All else being equal, more people equal greater human impact, more land clearing for agriculture, and less biodiversity (Clark and Tilman 2017; Turcotte et al. 2017). For example, nitrogen pollution is greatest in rivers flowing through landscapes with high human population densities, and rates of deforestation are greatest in countries with the highest rates of human population growth. A study of African nations found that population density and economic activity were the best predictors of environmental degradation (Bradshaw and Di Minin 2019). Therefore, some scientists have argued strongly that controlling the size of the human population is the key to protecting biodiversity (Crist et al. 2017; Ehrlich et al. 2012; Tilman et al. 2017).

Contradicting the view that slowing human population growth is the only answer to protecting biodiversity, many have pointed to the fact that we have seen time and again that technology has improved our ability to meet human needs (McGurn 2018) with the added benefit of reducing population growth rates (Henderson and

GLOBAL CHANGE
CONNECTION

The major threats to biodiversity—habitat destruction, habitat fragmentation, pollution, global climate change, overexploitation of resources, invasive species, and the spread of disease—are all rooted in the expanding human population.

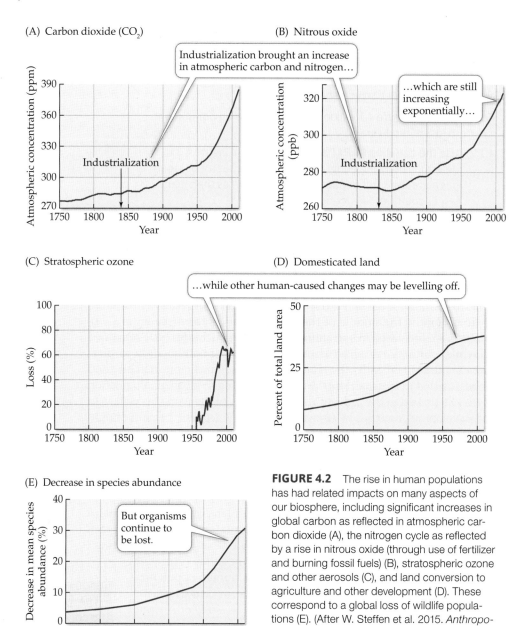

FIGURE 4.2 The rise in human populations has had related impacts on many aspects of our biosphere, including significant increases in global carbon as reflected in atmospheric carbon dioxide (A), the nitrogen cycle as reflected by a rise in nitrous oxide (through use of fertilizer and burning fossil fuels) (B), stratospheric ozone and other aerosols (C), and land conversion to agriculture and other development (D). These correspond to a global loss of wildlife populations (E). (After W. Steffen et al. 2015. *Anthropocene Rev* 2: 81–98; data from misc. sources.)

Loreau 2019). Furthermore, focusing attention on population growth (which is higher in developing nations) rather than consumption (which is higher in more industrialized nations) is highly problematic from a human rights standpoint (Kaplan 2021). As we will see in this chapter, consumption on the individual level varies widely across the globe and is arguably a more significant problem than population growth.

Healthy ecosystems can persist close to areas with high population densities, even large cities, as long as human activities are regulated by local customs or government officials. The sacred groves of trees that are preserved next to villages in Africa, India, and China are examples of locally managed biological communities. When this regulation breaks down during war, political unrest, or other periods of social instability, the result is usually a scramble to collect and sell resources that had been used sustainably for generations. The higher the human population density, and the larger the city, the more closely human activities must be regulated, because the potential for both destruction and conservation is greater (Adams and Klobodu 2017).

People in industrialized countries (and the wealthy minority in developing countries) consume a disproportionately large share of the world's energy, minerals, wood products, and food (Mills Busa 2013) and therefore have disproportionate effects on the environment. The impact (I) of any human population on the environment is roughly captured by the formula $I = PAT$ where P is the number of people, A is the average income, and T is the level of technology (Elrich and Holdren 1971; Vivanco et al. 2016). It is important to recognize that the impact of a population is often felt over a great distance; for example, citizens of Germany, Canada, and Japan affect the environment in other countries through their use of foods, luxury goods, and other materials produced elsewhere. The increasing interconnectedness of resource and labor markets is termed **globalization**. The fish eaten quietly at home in Washington, DC, may have come from Alaskan waters, where its capture may have contributed to the population decline of sea lions, seals, and sea otters; the chocolate cake and coffee consumed at the end of a meal in Italy or France were made with cacao and coffee beans that might have grown in plantations carved out of rain forests in western Africa, Indonesia, or Brazil. It has been estimated that 40% of all primary production in biodiversity hotspots is being harvested for human use each year, with 20% driven by consumption in other countries (Weinzettel et al. 2018). Globally, 20% of all primary production is appropriated by the conversion of natural ecosystems into cropland and pastureland (Weinzettel et al. 2019). Residents of industrialized countries also affect other countries through the production of waste, including the greenhouse gases responsible for climate change.

The per capita influence a group of people has on both the surrounding environment and locations across the globe has been called the **ecological footprint** (Mancini et al. 2018; Wackernagel and Rees 1996) (**FIGURE 4.3**). The ecological footprint per person is high in developed countries such as the United States and Canada and relatively low in developing countries such as India and Guatemala.

> The enormous consumption of resources in an increasingly globalized world is not sustainable in the long term.

A modern city in a developed country typically has an ecological footprint that is hundreds of times its area. For example, the city of Toronto, Canada, occupies an area of 630 km², but each of its citizens requires the environmental services of 8 ha (0.08 km²) to provide food, water, and waste disposal sites (Global Footprint Network 2018). With a population of 2.4 million people, Toronto has an ecological footprint of 185,000 km², an area equal to the state of New Jersey or the country

(A)

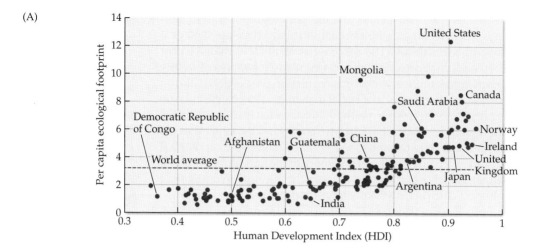

(B)

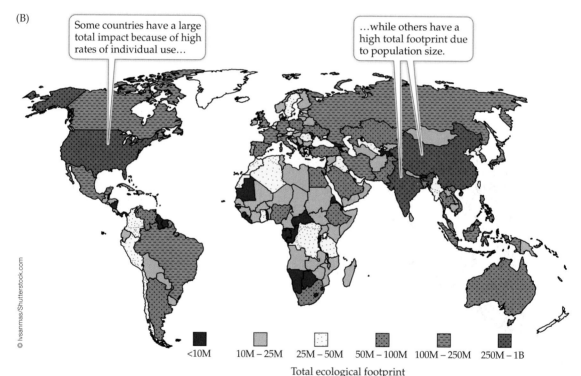

FIGURE 4.3 An ecological footprint for a nation is arrived at by calculations that estimate the number of global hectares needed to support an average citizen of that nation. Calculations shown here include resources used and carbon dioxide emitted. (A) When plotted against the Human Development Index (HDI), which reflects living standards, ecological footprints graphically illustrate the disproportionate use of natural resources by people in developed nations. (B) However, the total impacts (measured in millions of hectares, "M") of developing countries such as China, with 1.4 billion people, are also large because of their large populations. (A, based on data from United Nations Development Programme/Human Development Reports/Data (1990–2017) and Global Footprint Network/National Footprint Accounts 2018 [GFN/NFA 2018]. B, from GFN/NFA 2018, www.footprintnetwork.org/resources/data. CC BY-SA 4.0.)

of Syria and more than 300 times its own area. This excessive consumption of resources is not sustainable in the long term. Unfortunately, this pattern of consumption is now being adopted by the expanding middle class in the developing world, including the large, rapidly developing countries of China and India, increasing the probability of massive environmental disruption (Fan et al. 2017; Franco et al. 2017). China, in particular, has emerged as a rapidly growing industrial powerhouse that not only exports manufactured goods but also imports resources from around the world; China is now the world's greatest energy consumer (see Figure 1.2). The affluent citizens of developed countries must confront their excessive consumption of resources and reevaluate their lifestyles while at the same time offering aid to curb population growth, protect biodiversity, and assist industries in the developing world to grow in a responsible way.

GLOBAL CHANGE
CONNECTION

However, it has also been argued that we must look beyond the dichotomy of "rich versus poor" countries and instead focus on the rise of consumption across the globe, seeking ways to promote individual development that will in turn address population growth (Crist et al. 2017). In particular, when there is more gender equity and women have better access to education, families typically have fewer children (Marphatia et al. 2020, Scott et al. 2021). Furthermore, economic development can have a positive effect on biodiversity because wealthier countries can better afford to establish and maintain their national parks and other natural areas. Developed countries with a gross domestic product (GDP) of $10,000 per capita per year or greater are able to fund environmental projects such as species management, ecosystem services, climate change impact assessments, invasive species control, and pollution detection, using funding from government, universities, or private sources (McKinley et al. 2017). Developing countries cannot afford this funding and typically rely on heavy agriculture or deforestation to increase their economies, which increases environmental degradation (Alvarado and Toledo 2017).

4.2 Habitat Destruction

LEARNING OBJECTIVES

By the end of this section you should be able to:

4.2.1 Summarize causes of habitat destruction in tropical regions.

4.2.2 Interpret bar and column graphs.

4.2.3 Compare the threat of "shifting cultivation" with "commodity-driven" reasons for deforestation.

4.2.4 Contrast the threats suffered by other ecosystems with those of tropical forests.

The primary cause of the reduction in biodiversity, including variation at the genetic, species, and ecosystem levels, is the habitat loss that inevitably results

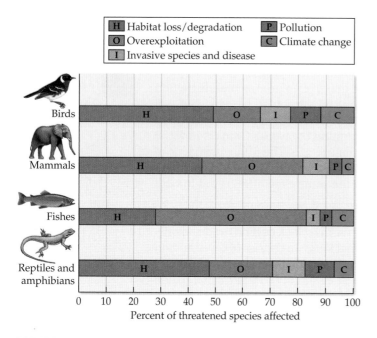

H	Habitat loss/degradation	P Pollution
O	Overexploitation	C Climate change
I	Invasive species and disease	

FIGURE 4.4 The indirect threats of habitat loss, degradation, and fragmentation are the greatest threats to vertebrates, followed by the direct mortality threats of overexploitation and invasive species. Groups of species face different threats: birds are more threatened by habitat loss, whereas fish are more threatened by overexploitation. Many species face multiple threats. (After WWF. 2018. *Living Planet Report - 2018: Aiming Higher*. M. Grooten and R. E. A. Almond [Eds]. WWF, Gland, Switzerland.)

from the expansion of human populations and activities (**FIGURE 4.4**). Thus, the most important means of protecting biodiversity is habitat preservation. *Habitat loss* does not necessarily mean wholesale habitat destruction; habitat fragmentation and habitat damage associated with pollution can also mean that the habitat is effectively "lost" to species that cannot tolerate these changes, even though, to the casual onlooker, the habitat appears intact. In many areas of the world, particularly on islands and in locations where the human population density is high, most of the original habitat has been destroyed. By one assessment, only 25% of the land on Earth has not been modified in some way by people (WWF 2018); 22.4% is used specifically for agriculture (FAO 2011). Because the world's human population, as well as its standard of living, will continue to increase at least until 2050, some say that the world's farmers will need to increase agricultural output by up to twice the current level by 2050 while also drastically reducing ecological impacts (Pikaar et al. 2018; Rockström et al. 2017). Others say that this is unnecessary if we address the socioeconomic problems that prevent access to food and promote waste and unnecessary excess (Berners-Lee et al. 2018; Kc et al. 2018) and if we also take advantage of scientific and technological advances (Chávez-Dulanto et al. 2020). If models are correct and global human populations will stop growing in the latter half of the century (Chapter 1), such concerns may be alleviated so long as there is not a proportional increase in individual resource consumption. Clearly, it is important that the standard of living improve for the most disadvantaged, which means that the wealthiest countries must be especially careful not to increase their ecological footprint. Protection of biodiversity is directly tied to our modern challenge of meeting global human needs.

GLOBAL CHANGE CONNECTION

Habitat disturbance has been particularly severe throughout Europe; in southern and eastern Asia, including India, Thailand, China, and Japan; in southeastern and southwestern Australia; on the southeastern and northern coasts of South America; in the Caribbean; and in central and eastern North America, as well as parts of Mexico (**FIGURE 4.5**). In many of these regions, more than 75%

The main threat to biodiversity is habitat destruction.

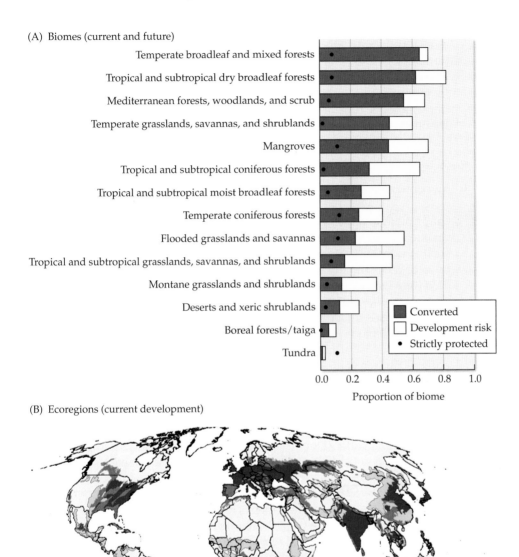

(A) Biomes (current and future)

(B) Ecoregions (current development)

FIGURE 4.5 Many of the world's major biomes have already had a large proportion of their area converted to human uses. (A) Current and potential future conversion. (B) Current conversion. (After J. R. Oakleaf et al. 2015. *PLOS ONE* 10(10): e0138334. © 2015 Oakleaf et al. CC BY 4.0, creativecommons.org/licenses/by/4.0.)

of the natural habitats have been significantly disturbed or removed (Pandit et al. 2018). Only 20% of the land area in Europe remains unmodified by human activities, and in some regions of Europe, the percentage is even lower (European Environment Agency. In Germany or the United Kingdom, for example, one can hardly find any habitat that has not been modified by humans at one time or another. Across Europe, 86% of land changes are attributable to land use by humans (Song et al. 2018).

The principal human activities that threaten the habitats of endangered species, in order of decreasing importance, are overexploitation (72% of IUCN listed species affected), agriculture (62%), urban development (35%), and invasion by nonnative species and disease (26%) (Maxwell et al. 2016). The remaining threats affect lower percentages of species but are hardly insignificant: pollution (22%), system modification such as dams and changes to fire regime (21%), climate change (19%), and other types of human disturbance such as war (14%). Most species that were assessed had two or more threats; of all of the plant and vertebrate species that have gone extinct since 1500, 75% were harmed by overexploitation or agricultural activity or both.

GLOBAL CHANGE
CONNECTION

As a result of farming, logging, and other human activities, very little **frontier forest**—intact blocks of undisturbed forest large enough to support all aspects of biodiversity—remains in many countries; the global decline in frontier forest is estimated at approximately 0.5% per year during the past decade (Hansen et al. 2013). Between 2001 and 2019, the world lost 386 million ha of tree cover; that's a 9.7% decrease, equivalent to 105 gigatons of CO_2 emissions since 2000 (Global Forest Watch 2020). In 2017, in response to demand by the public and other stakeholders, 447 companies made commitments to curb deforestation linked to the production of palm, soy, timber and pulp, and cattle (Donofrio et al. 2017); changes such as these will be necessary but not sufficient alone to stop losses of biodiversity.

Tropical rain forests

The destruction of tropical rain forests has come to be synonymous with the rapid loss of species. The tropics occur between approximately 23.5° north and south of the equator, and *rain forests* (sometimes written *rainforests*) are those characterized by a closed, evergreen canopy often more than 25 m tall with abundant, woody vines and generally more than 2000 mm rain per year. Tropical rain forests occupy 7% of the Earth's land surface, but they are estimated to contain over 50% of its species (Corlett and Primack 2010) (see Chapter 2). Simulations of the impact of deforestation on tropical regions predicted a 41% reduction in biodiversity relative to undisturbed forest, with species compositions shifting to favor globally "weedy" species such as rats (Alroy 2017). Deforestation of tropical forests, particularly in the form of fire, also contributes to climate change and is responsible for 10% of total human carbon emissions (Bebber and Butt 2017). These inputs combined with the loss of tree biomass mean that tropical forests are no longer a sink for atmospheric carbon (Baccini et al. 2017). Loss of tree cover globally from fires has increased in recent years, nearly equaling losses from agriculture and forestry (Global Forest Watch 2020).

GLOBAL CHANGE
CONNECTION

Many tropical rain forest species are important to local economies and have the potential for greater use by the entire world population (see Chapter 3). Tropical rain forests also have local significance as home to numerous Indigenous cultures, regional importance in protecting watersheds and moderating climate, and global importance as sinks to absorb some of the excess CO_2 that is produced by the burning of fossil fuels.

GLOBAL CHANGE
CONNECTION

The world's largest rain forest is the Amazon rain forest, covering approximately 40% of South America, with nearly two-thirds in Brazil. For many years, Brazil led the world in deforestation, but rates decreased between 2004 and 2015 due to government policies and enforcement (**FIGURE 4.6**). Meanwhile, Southeast Asia had become the leader, with 80 million ha of rain forest lost during this same period (Estoque et al. 2019). Unfortunately, changes in political leadership in Brazil led to record high levels of deforestation in that country in 2016, while deforestation levels remained high in Southeast Asia. It is reported that rates have accelerated even more in 2020 in Brazil under the country's current administration (BBC 2020).

Scientists have used high-resolution Google Earth images to create predictive models of forest disturbances across the globe (Curtis et al. 2018). From

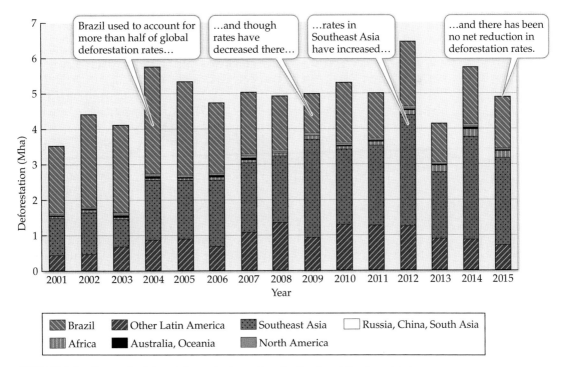

FIGURE 4.6 Change in deforestation extent over time for Brazil and Southeast Asia as shown in million hectares (Mha). Political leadership and global demand has shifted amount of deforestation in these two regions, which is currently very high. (After P. G. Curtis et al. 2018. *Science* 361: 1108–1111. doi.org/10.1126/science.aau3445. Reprinted with permission from AAAS and P. Curtis.)

these models, scientists have determined that globally most deforestation is "commodity-driven," defined as the long-term permanent conversion to agriculture, mining, or energy infrastructure (27 ± 5%). This is followed closely by forestry (26 ± 4%), shifting cultivation (24 ± 3%), and wildfire (23 ± 4%), with urbanization accounting for only 0.6 ± 0.3%. Deforestation in only 1% of the area analyzed was attributable to other causes, including all other natural disturbances (such as insect outbreaks or storms) put together. In the tropics, losses are primarily commodity-driven clearance by large landowners and commercial interests to create pasture for cattle ranching or to plant cash crops, such as oil palms, soybeans, and rubber trees (Austin et al. 2017) (**FIGURE 4.7**). These are grown in large-scale, often pesticide-laden plantations in which large areas are maintained under a uniform crop cover (Mukul and Herbohn 2016). Commercial agriculture displaces poor farmers and justifies the expansion of roads. In addition, large areas of rain forest are damaged during commercial logging operations, most of which are poorly managed selective logging. These logged forests are prone to widespread fires due to the large numbers of branches and dead trees on the ground. In many cases, logging operations precede conversion of tropical forest land to agriculture and ranching.

FIGURE 4.7 Complex and diverse tropical forests give way to an African tea plantation, a sea of green that supports almost no biodiversity—not only because it is a monoculture, but also because of the liberal use of pesticides. Conversion of tropical forests to agriculture is considered the most common cause of loss of these ecosystems. This plantation and others like it were initially established to be buffer zones around protected remnant forests; however, over time, the plantations facilitated further encroachment and conversion by the growing local human population. In this region of western Kenya, almost 60% of the forest cover was lost between 1984 and 2009, with 90% of this loss attributable to agriculture (Cordeiro et al. 2015).

Less prevalent but still significant are tropical forest losses from **shifting cultivation**, a kind of subsistence farming sometimes referred to as slash-and-burn, or swidden, agriculture, in which trees are cut down and then burned (see Chapter opening photo). The cleared patches are farmed for two or three seasons, after which soil fertility has usually diminished and soils have eroded to the point where crop production is so low that the patches are then abandoned and a new area is cleared (Coomes and Miltner 2017). Although these patches may recover with time and are less detrimental than commercial agriculture, studies show that shifting cultivation has a negative effect on both plants and animals (Mukul and Herbohn 2016). This small-scale cultivation of crops is often done by poor farmers who are forced onto remote forest lands by poverty (or who sometimes are moved there by government-sponsored resettlement programs) (Ravikumar et al. 2017). Rain forests are also destroyed by fuelwood production, mostly to supply local villagers with wood for cooking fires (see Figure 3.5). More than 2 billion people cook their food with firewood, so their impact is significant.

The destruction of tropical rain forests is frequently caused by demand in industrialized countries for cheap agricultural products, such as rubber, palm oil, cocoa, soybeans, orange juice, and beef, and for low-cost wood products. At present, most consumers in industrialized countries are not aware of how their choices affect land use. Many people would be surprised to learn how widely palm oil is used in processed food and consumer products, not to mention where it comes from. It is also true that increasing proportions of these agricultural and wood products are consumed within the countries that produce them or exported to rapidly industrializing countries, such as China and India. If the current rates of loss continue, there will be little tropical forest left after the year 2050, except in the relatively small national parks and remote, rugged, or infertile areas of the Amazon basin, Congo River basin, and New Guinea (Walker et al. 2019). The move to establish large new parks in many tropical countries is cause for some hope; however, these parks will need to be well funded and managed to be effective in preserving biodiversity, as described in Chapter 9. There are also more than 50 certification standards that help consumers choose sustainable forest products, the largest being the Programme for the Endorsement of Forest Certification (PEFC) and the Forest Stewardship Council (FSC).

GLOBAL CHANGE
CONNECTION

Other threatened habitats

The plight of the tropical rain forests is perhaps the most widely publicized case of habitat destruction, but many other habitats are also in grave danger.

TROPICAL DECIDUOUS FORESTS   A month of less than 100 mm rain is considered dry in tropical forests, and trees that experience drought regularly for long periods adapt by becoming deciduous (i.e., dropping their leaves). The land occupied by tropical deciduous forests is more suitable for agriculture and cattle ranching than is the land occupied by tropical rain forests. These forests are also easier than rain forests to clear and burn. Moderate rainfall in the range of 250–2000 mm per year allows mineral nutrients to be retained in the soil, from which they can be taken up by plants. Consequently, human population density is five times greater in deciduous forest areas of Central America than in adjacent rain forest areas. Researchers using historical maps plus forest cover data from Landsat satellites have determined that Madagascar, home to endangered lemurs, an endemic group of primates, has lost 44% of forest cover, mostly deciduous forest, since 1953, with 37% since 1973 (Vieilledent et al. 2018) (see Figure 6.11). By one estimate, suitable habitat is expected to decline by 81% by 2070 if deforestation continues at the current rate (Morelli et al. 2020).

GLOBAL CHANGE
CONNECTION

GRASSLANDS Temperate grassland, sometimes also called prairie, is another habitat type that has been almost completely destroyed by human activities. It is relatively easy to convert large areas of grassland to farmland or cattle ranches. As of 2016, 2.5 million acres are converted to cropland annually (McCarthy 2017). The remaining area of grassland is fragmented and widely scattered across the landscape. Increasing prices for agricultural products are currently driving the conversion of even marginal grasslands to farmland (Wright and Wimberly 2013). Worldwide, only 4.6% of temperate grasslands are protected (Carbutt et al. 2017) (see Chapter 9).

Between 1800 and 1950, as much as 98% of North America's tallgrass prairie was converted to farmland.

FRESHWATER HABITATS Wetlands and other freshwater aquatic habitats are critical habitats for fish, aquatic invertebrates, amphibians, and birds. They are also a resource for flood control, water filtration, and power production (as described in Chapter 3; see Figure 3.10). Freshwater systems are often filled in or drained for development or are otherwise altered by dams, channelization of watercourses, and chemical pollution (Mitsch and Gosselink 2015). Over half of the wetland ecosystems that existed in the early twentieth century have been lost in North America, Europe, Australia, and China (Moreno-Mateos et al. 2012). In the United States, according to the National Rivers and Streams Assessment, 36% of streams are in poor condition due to the impacts of urbanization, agriculture, ranching, and mining, but it is important to note that this is an improvement from the 43% reported a decade earlier (EPA 2016). Degradation means that their ecosystem functions and services are lost, including their ability to serve as dispersal routes for aquatic animals and plants (see Chapter 3). Destruction of wetlands and streams has been equally severe in other parts of the industrialized world, such as Europe and Japan. About 56% of wetlands in Europe have been lost, with higher rates of degradation starting in the twentieth century (Davidson 2014). Only 2 of Japan's 30,000 rivers can be considered wild, without dams or some other major modification.

GLOBAL CHANGE
CONNECTION

In the last few decades, major threats to wetlands and other aquatic environments in developing countries have included massive development projects involving drainage, irrigation, and dams, organized by governments and often financed by international aid agencies. The Three Gorges Dam on the Yangtze River of China is one example (Yang et al. 2014). The dam is the largest hydroelectric power plant in the world, generating much-needed clean and renewable energy. However, the dam and reservoir have displaced more than 1 million people, destroyed untold numbers of ecosystems and archaeological sites, and altered the river and delta systems, with unknown ecological consequences. The economic benefits of such projects are important, but the rights of local people and the value of ecosystems are often not adequately considered.

MARINE COASTAL AREAS   Human populations are increasingly concentrated in marine coastal areas. Some 20% of marine coastal areas have already been degraded, filled in, or highly modified by human activity, despite their importance in the harvesting of fish, shellfish, seaweeds, and other marine products (see Figure 3.6A). Coastal wetlands are also threatened by pollution, dredging, sedimentation, destructive fishing practices, invasive species, and now rising temperatures (see Chapter 5). Human impacts on these habitats are less well studied than in the terrestrial environment, but they are probably equally severe, especially in shallow coastal areas. Two coastal habitats of special note are mangroves and coral reefs.

GLOBAL CHANGE
CONNECTION

MANGROVES   Mangrove forests are among the most important wetland communities in tropical areas, and it is estimated that the world has lost over half of its mangrove forests (Thomas et al. 2017) (**FIGURE 4.8**). Composed of species that are among the few woody plants able to tolerate salt water, mangrove forests occupy coastal areas with saline or brackish water, typically where there are muddy bottoms. Such habitats, like salt marshes in the temperate zone, are extremely important breeding grounds and feeding

GLOBAL CHANGE
CONNECTION

© Tim Laman/Minden Pictures

FIGURE 4.8   Mangrove forests, like this one in Malaysia, have high biological diversity and provide many ecosystem services. They are threatened by overharvesting and clearing for agriculture and aquaculture.

areas for shrimp and fish. In Australia, for example, two-thirds of the species caught by commercial fishermen depend to some degree on the mangrove ecosystem. Despite their great economic value and their utility for protecting coastal areas from storms and tsunamis, mangroves are often cleared for rice cultivation and commercial shrimp and prawn hatcheries, particularly in Southeast Asia (Bryan-Brown et al. 2020). Mangroves have also been severely degraded by overcollection of wood for fuel, construction poles, and timber throughout the region; it has been argued that there are too few incentives for local users and managers to stop this overexploitation (Máñez et al. 2014). Today, almost 40% of mangrove-dependent animal species are considered to be at high risk of extinction (Daru et al. 2013). Mangroves are also important for sequestering atmospheric carbon; in 2018 Apple announced that it would protect and restore a 27,000-acre mangrove forest in Colombia to help reduce the technology company's ecological footprint (Campbell 2018).

CORAL REEFS Tropical coral reefs contain an estimated one-third of the ocean's fish species in only 0.2% of its surface area (Hughes et al. 2017) (**FIGURE 4.9**). The most severe destruction is taking place in the Philippines, where a staggering 90% of the reefs are dead or dying. In China, coral reefs have declined by 80% over the past 30 years (Hughes et al. 2013). The effects of global climate change, including ocean warming, ocean acidification, and the increasing intensity of tropical storms fueled by warmer surface waters, are playing a major role in the rapid degradation of coral reefs (Comeau et al. 2015; see Chapter 5). Other culprits are pollution, which either kills the corals directly or allows excessive growth of algae; sedimentation resulting from deforestation, agriculture, and coastal development; overharvesting of fish, clams, and other animals; and finally, stunning with dynamite or cyanide to collect the few remaining living creatures for food and the aquarium trade. Worst of all is the combination of local and global impacts, which makes it difficult or impossible for coral reefs to bounce back from

GLOBAL CHANGE
CONNECTION

FIGURE 4.9 Coral reefs such as this one support complex ecosystems with very high biodiversity.

© iStock.com/prasit chansarekorn

natural disturbances, such as hurricanes, to which they may otherwise be relatively resilient. Further losses of coral reefs are expected within the next 40 years, especially in tropical East Asia, the areas around Madagascar and East Africa, and throughout the Caribbean (see Figure 5.4).

Over the last 15 years, scientists have discovered extensive reefs of corals living in cold water at depths of 300 m or more, many of which are in the temperate zone of the North Atlantic. These coral reefs are rich in species, many of which are new to science. Yet just as these communities are being explored for the first time, they are being destroyed by trawlers, which drag nets across the seafloor to catch fish. The trawlers destroy the very coral reefs that protect and provide food for young fish. The damage to these cold-water reefs by careless harvesting is costing the industry its resource base in the long run.

Desertification

Many ecosystems in seasonally dry climates are degraded by human activities into human-made deserts, a process known as *desertification* (Geist 2017; Lavauden 1927). These dryland ecosystems include grasslands, scrub, and tropical deciduous forests as well as temperate shrublands, such as those found in the Mediterranean region, southwestern Australia, South Africa, central Chile, and California. Naturally occurring dry areas cover about 41% of the world's land area and are home to about 1 billion people. Approximately 10%–20% of these drylands are at least moderately degraded, and more than 25% of the productive capacity of their plant growth has been lost (Sayre et al. 2013). These areas may initially support agriculture, but their repeated cultivation, especially during dry and windy years, often leads to soil erosion and loss of water-holding capacity in the soil. The land may also be chronically overgrazed by domesticated livestock, and woody plants may be cut down for fuel (**FIGURE 4.10**). Frequent fires during long dry periods often damage the remaining vegetation. The result is the progressive and largely irreversible degradation of the ecosystem and the loss of soil cover.

GLOBAL CHANGE
CONNECTION

© Georg Gerster/Science Source

FIGURE 4.10 Desertification is the process of land degradation that leads to loss of productivity and ecosystem function. These degraded lands are in contrast with desert biomes, which may support rich and complex ecosystems.

Ultimately, formerly productive farmland and pastures take on the appearance of a desert. Desertification has been ongoing for thousands of years in the Mediterranean region and was known even to ancient Greek observers.

Worldwide, 9 million km^2 of arid lands have been converted to human-made deserts. These areas are not functional desert ecosystems but wastelands, lacking the flora and fauna characteristic of natural deserts. The process of desertification is most severe in the Sahel region of Africa, just south of the Sahara, where most of the native large mammal species are threatened with extinction. The human dimension of the problem is illustrated by the fact that the Sahel region is estimated to have 2.5 times more people (150 million currently) than the land can sustainably support (Çonkar 2020). Further desertification appears to be almost inevitable, especially given the higher temperatures and lower rainfall associated with predictions of future climate change (Benjaminsen and Hiernaux 2019; Guo et al. 2017).

In such areas, the solution will be programs involving the implementation of sustainable agricultural practices, the elimination of poverty, increased political stability, and control of the human population. The United Nations Convention to Combat Desertification is one source of such programs, committing to restore the productivity of degraded lands through a variety of initiatives. A major project of this international group, in cooperation with the International Union for Conservation of Nature (IUCN) and others, is to work with nations across the globe to set land degradation neutrality (LDN) targets; as of 2019, 120 countries had officially joined in this effort (United Nations Convention to Combat Desertification 2021).

4.3 Habitat Fragmentation

LEARNING OBJECTIVES

By the end of this section you should be able to:

4.3.1 Predict the effects of habitat fragmentation on a given species.

4.3.2 Calculate the amount of edge caused by fragmentation of a shape of known area.

4.3.3 Interpret the results of a meta-analysis.

In addition to being destroyed outright, habitats that formerly occupied wide, unbroken areas are often divided into pieces by roads, fields, towns, and a broad range of other human constructs. **Habitat fragmentation** is the process whereby a large, continuous area of habitat is both reduced in area and divided into two or more fragments. When habitat is destroyed, a patchwork of habitat fragments may be left behind. These fragments are often isolated from one another by a highly modified or degraded landscape, and their edges experience altered conditions, referred to as **edge effects**, such as increased wind, fire, species invasion, or predation (e.g., Didham 2017). The fragments are often on the least desirable land for human uses, such as steep slopes, poor soils, and inaccessible areas.

Fragmentation almost always occurs during a severe reduction in habitat area, but it can also occur when habitat area is reduced to only a minor degree by roads, railroads, canals, power lines, fences, oil pipelines, fire lanes, or other barriers to the free movement of species. In a recent study of mangrove deforestation, it was found that in some regions such as Cambodia and the southern Caribbean, there was little overall habitat loss, but a great deal of fragmentation (Bryan-Brown et al. 2020). In many ways, habitat fragments resemble islands of original habitat in an inhospitable, human-dominated landscape. Habitat fragmentation is a serious threat to biodiversity, as species are often unable to survive under the altered conditions found in fragments.

GLOBAL CHANGE
CONNECTION

Habitat fragments differ from the original habitat in three important ways:

1. Fragments have a greater amount of edge per area of habitat (and thus a greater exposure to edge effects).
2. The center of a habitat fragment is closer to an edge.
3. When a formerly continuous habitat hosting a large population is divided into fragments, each fragment hosts a smaller population.

Threats posed by habitat fragmentation

The effects of habitat fragmentation have been studied extensively by scientists, in some cases by creating natural experiments that manipulate degree of habitat isolation, habitat size, and proportion of edge. One such large-scale, long-term experiment is the Biological Dynamics of Forest Fragmentation Project in Brazil (**FIGURE 4.11**), established to address the debate regarding

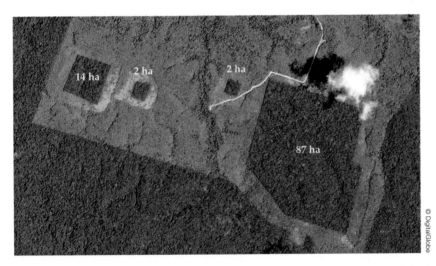

FIGURE 4.11 The Biological Dynamics of Forest Fragmentation (BDFF) study in Brazil is the largest and longest-running experiment of tropical forest fragmentation; forest fragments of different sizes were created in 1980 when the land was cleared for cattle ranches and farms. Shown in this aerial photograph are forest fragments of 2 ha, 14 ha, and 87 ha. These fragments were surveyed for species composition, microclimate, and other ecosystem characteristics before the forest was cleared and have been regularly surveyed over time. In addition, forest blocks of equivalent size inside intact forest have been surveyed as controls. In the photograph, note that the fragments vary in their distance from the intact forest. Long-term fragmentation studies established since this one have focused more on the effect of fragment isolation (see Figure 4.12).

the biological effects of fragment size (explored more in Chapter 9). Since then, other experiments have been established in the United States, Australia, Canada, the United Kingdom, France, and Borneo, each producing dozens of published papers. When there is a large body of findings from different research projects, one way that scientists can evaluate the results is by using a **meta-analysis**, a statistical tool that makes it possible to combine data from multiple sources. The magnitude of an impact of interest (such as the effect that reducing fragment size has on community) is called the **effect size**. A meta-analysis on the results of these long-term experiments, which covered five continents, several biomes, and 35 years, found that fragmentation reduced species diversity by 13%–75% (Haddad et al. 2015). Studies consistently found negative impacts on community composition and ecosystem processes such as succession and nutrient retention (**FIGURE 4.12A**). Furthermore, species losses generally increased over time (**FIGURE 4.12B**). More recent analyses have suggested that habitat fragmentation does not

(A)

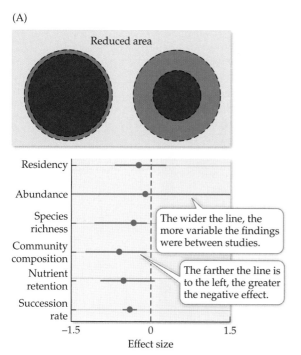

(B)

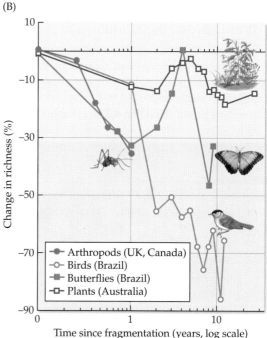

FIGURE 4.12 Fragmentation by roads, walls, or other human-made barriers has a variety of impacts on ecosystems. Here we see the summarized results of 76 field experiments spanning 35 years of research done across the globe to investigate the impact of increased isolation. (A) The results of a meta-analysis of all the studies on the effects of reduced habitat area caused by fragmentation. The position of the dots reflects the average effect size, or magnitude of the observed impact, with the error bar reflecting the variability of the results among studies. The

greatest negative effects of isolation from fragmentation were on community composition. Species abundances varied widely, as some (especially invasive) species increased with fragmentation. (B) This research also showed the decline in several species groups over time in these fragments. The location of the study is shown in parentheses. (After N. M. Haddad et al. 2015. *Sci Adv* 1: e1500052. doi.org/10.1126/sciadv.1500052. © 2015 The Authors. CC BY 4.0, creativecommons.org/licenses/by/4.0.)

always harm biodiversity, particularly when it is experimentally separated from the effects of habitat destruction (Fahrig 2017; Fahrig et al. 2019). However, others have argued that we must be careful when quantifying "good" versus "bad" effects, particularly when biodiversity is maintained through weedy and invasive species at the expense of endemics (Miller-Rushing et al. 2019). Furthermore, fragmentation can lead to loss; a recent study found that the smaller a tropical forest fragment, the greater the risk of habitat loss, likely because of easier access (Hansen et al. 2020). What follows are ways in which habitat fragmentation can directly affect biological systems.

LIMITS TO DISPERSAL AND COLONIZATION   Fragmentation limits a species' potential for dispersal and colonization by creating barriers to normal movements (Rossetti et al. 2017). In an undisturbed environment, seeds, spores, and animals move passively or actively across the landscape. Over time, populations of a species may build up and go extinct on a local scale as the species disperses from one suitable site to another and the biological community undergoes succession. At a landscape level, a series of populations exhibiting this pattern of extinction and recolonization is sometimes referred to as a *metapopulation* (see Figure 7.12).

When a habitat is fragmented, the potential for dispersal and colonization is often reduced. Many bird, mammal, and insect species of the forest interior will not cross even short stretches of open ground (Rossetti et al. 2017). If they do venture into the open, they may find predators waiting on the forest edge to catch and eat them. Agricultural fields 100 m wide may represent impassable barriers to the dispersal of many invertebrate species. Roads, too, may be significant barriers to animal movement (see Chapter 9). Many species avoid crossing roads, which represent an environment totally different from the habitat they are leaving. For animals that do attempt to cross roads, motor vehicles are a major source of mortality, as has been observed in the endangered Florida panther (Kroll 2015). To deal with such problems, highway officials are building animal underpasses, overpasses, and other improvements to minimize animal mortality (see Figure 9.12). For example, in 2018 Washington State built a massive wildlife overpass, 20 m wide and 11 m tall, intended to facilitate elk migration and reduce the number of vehicle collisions with deer (Valdez 2018).

As species go extinct within individual fragments through natural successional and metapopulation processes, new species will be unable to arrive because of barriers to colonization, and the number of species present in the habitat fragment will decline over time. Extinction will be most rapid and severe in small habitat fragments.

> The barriers that fragment a habitat reduce the ability of animals to forage, find mates, disperse, and colonize new locations. Habitat fragmentation often creates small subpopulations that are vulnerable to local extinction.

RESTRICTED ACCESS TO FOOD AND MATES   Many animal species need to move freely across the landscape, either as individuals or in social groups, to feed on widely scattered resources. For example, elephants in central Africa travel nearly 3000 km per year and have home ranges of over 700 km^2 on average (Mills et al. 2018). A study by wildlife biologists determined that the building of a continuous wall between the United States and Mexico to prevent people from crossing illegally into the United States would threaten 117 terrestrial and freshwater animal species by cutting off 50% or more of their ranges (Peters et al. 2018).

A given resource may be needed for only a few weeks each year, or even only once in a few years, but when a habitat is fragmented, species confined to a single habitat fragment may be unable to migrate over their normal home range in search of that scarce resource. Gibbons and other primates, for example, typically remain in forests and forage widely for fruits. Finding scattered trees with abundant fruit crops may be crucial during episodes of fruit scarcity (see Figure 7.5). Clearings and roads that break up the forest canopy may prevent these primates from reaching nearby fruiting trees because they are unable or unwilling to descend to the ground and cross the intervening open landscape. Fences may prevent the natural migration of large grazing animals such as wildebeest and bison, forcing them to overgraze unsuitable habitat, which eventually leads to starvation and further degradation of the habitat (Jakes et al. 2018).

Barriers to dispersal can also restrict the ability of widely scattered species to find mates, leading to a loss of reproductive potential for many animal species. Plants may have reduced seed production if butterflies and bees are less able to migrate among habitat fragments to pollinate flowers (see Figure 3.11).

CREATION OF SMALLER POPULATIONS   Habitat fragmentation may precipitate population decline and extinction by dividing an existing widespread population into two or more subpopulations in restricted areas. These smaller populations are then more vulnerable to inbreeding depression, genetic drift, and other problems associated with small population size (see Chapter 6). While a large area may support a single large population, it is possible that none of the smaller subpopulations in the fragments will be sufficiently large to persist, even if the total area is the same. Connecting the fragments with properly designed movement corridors may be the key to maintaining populations.

INTERSPECIES INTERACTIONS   Habitat fragmentation increases the vulnerability of the fragments to invasion by nonnative and native pest species (Aguirre-Acosta et al. 2014). Road edges themselves may represent dispersal routes for invasive species. The forest edge represents a high-energy, high-nutrient, disturbed environment in which many pest species of plants and animals can increase in number and then disperse into the interior of the fragment.

Omnivorous animals may increase in population size along forest edges, where they can eat foods, including eggs and nestlings of birds, from both undisturbed and disturbed habitats. In the coniferous forests of Finland, predation on nests was found to be higher in edges of clear-cut areas than in either the forest interior or forest corridors, and sometimes higher than in the clear-cut area (Huhta and Jokimäki 2015). The invading predators include not only native omnivores—such as raccoons, skunks, and blue jays

GLOBAL CHANGE
CONNECTION

in North America—but also introduced domesticated cats. In a study done in the southeastern United States, successful hunting cats were found to capture an average of 2.4 prey items during seven days of roaming, with Carolina anoles (*Anolis carolinensis*) being the most common prey species (Loyd et al. 2013). In rural Poland, an examination of both scat and stomach contents and prey brought home by cats found small mammals to be the most common prey, followed by birds (Krauze-Gryz et al. 2012).

The combination of habitat fragmentation, increased nest predation, and destruction of tropical wintering habitats is probably responsible for the dramatic decline of certain migratory songbird species of North America, such as the cerulean warbler (*Dendroica cerulea*), particularly in the eastern half of the United States (With 2015). Populations of deer and other herbivores can also build up in edge areas, where plant growth is lush, eventually overgrazing the vegetation and selectively eliminating certain rare and endangered plant species for several kilometers into the forest interior.

Habitat fragmentation also puts wild populations of animals in closer proximity to domesticated animals. Diseases of domesticated animals can then spread more readily to wild species, which often have no immunity to them (see Chapter 5). There is also the potential for diseases to spread from wild species to people, once the level of contact increases, as happened with COVID-19. Fragmented forest habitats characteristic of suburban development often have high densities of white-footed mice (*Peromyscus leucopus*) and black-legged ticks (*Ixodes scapularis*) with high rates of infection with Lyme disease, which leads to a corresponding increase in Lyme disease in people living in those areas (see Figure 5.17) (Ostfeld et al. 2018).

Edge effects

Habitat fragmentation greatly increases the amount of edge relative to the amount of interior habitat. A simple example will illustrate the characteristics of edges and the problems they can cause. Consider a square conservation reserve 1000 m (1 km) on each side (**FIGURE 4.13**). The total area of the reserve is 1 km² (100 ha). The perimeter (or edge) of the reserve totals 4000 m. A point in the middle of the reserve is 500 m from the nearest perimeter. If the principal edge effect for birds in the reserve is predation by domesticated cats and introduced rats, which forage 100 m into the forest from the perimeter of the reserve and prevent forest birds from successfully raising their young, then only the reserve's interior—64 ha—is available to the birds for breeding. If the reserve is divided into four fragments by a road and a railroad, each of which is 495 m in area, the nesting habitat is further reduced to 8.7 ha in each fragment, for a total of 34.8 ha. Even though the road and railroad remove only 2% of the reserve area, they reduce the habitat available to the birds by about half. The implications of edge effects can be seen in the decreased ability of birds to live and breed in small forest fragments compared with larger blocks of forest habitat. Currently, 70% of remaining forest across the globe is within 1 km of the forest's edge, with significant implications for forest-dependent species (Haddad et al. 2015).

> Habitat fragmentation increases edge effects—changes in light, humidity, temperature, and wind that may be less favorable for many species living there.

GLOBAL CHANGE
CONNECTION

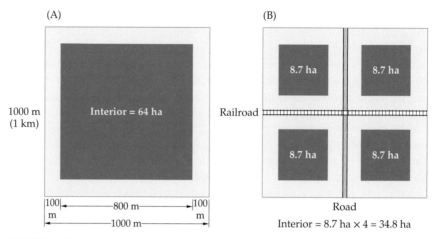

(A)

1000 m
(1 km)

Interior = 64 ha

|100 m ←————800 m————→|100 m
|←————————1000 m————————→|

(B)

Railroad

8.7 ha   8.7 ha

8.7 ha   8.7 ha

Road

Interior = 8.7 ha × 4 = 34.8 ha

FIGURE 4.13 A hypothetical example shows how habitat area is reduced by fragmentation and edge effects. (A) A 1 km² protected area. Assuming that edge effects (gray) penetrate 100 m into the reserve, approximately 64 ha are available as usable habitat for nesting birds. (B) The bisection of the reserve by a road and a railroad, although taking up little actual area, extends the edge effects so that almost half the breeding habitat is destroyed. Edge effects are proportionately greater when forest fragments are irregular in shape, as is usually the case.

The microenvironment at a forest fragment edge is different from that in the forest interior. Some of the more important differences are greater fluctuations in levels of light, temperature, humidity, and wind (Reynoso and Williams-Linera 2017). These edge effects are most evident up to 100 m inside the forest, although certain effects are detectable up to 400 m from the forest edge. Because so many plant and animal species are precisely adapted to certain levels of temperature, humidity, and light, changes in those conditions eliminate many species from forest fragments. Shade-tolerant wildflower species of temperate forests, late-successional tree species of tropical forests, and humidity-sensitive animals such as amphibians are often rapidly eliminated by habitat fragmentation. A study that used data on 1673 vertebrate species collected across seven major biogeographic regions found that 85% of the species were affected by proximity to a forest edge (Pfeifer et al. 2017). Furthermore, endangered species across taxa were much more likely to require forest interior habitat, having the highest abundance 200–400 m away from edges. On the other hand, certain species benefit from the higher light levels and disturbance on the edge and actually reach higher densities on the edge of fragments (Fahrig 2017; Fahrig et al. 2019).

When a forest is fragmented, increased wind, lower humidity, and higher temperatures make fires more likely (Numata et al. 2017). Fires may spread into habitat fragments from nearby agricultural fields that are being burned regularly, as in sugarcane harvesting, or from the irregular activities of farmers practicing slash-and-burn agriculture (Zhao et al. 2021). Forest fragments may be particularly susceptible to fire damage when wood has accumulated at the edge of the forest where trees have died or have been blown down by the wind.

4.4 Environmental Degradation and Pollution

Even when a habitat is unaffected by overt destruction or fragmentation, the ecosystems and species in that habitat can be profoundly affected by human activities. Ecosystems can be damaged and species driven to extinction by external factors that do not obviously change the structure of the dominant plants or other features in the community. Sometimes this damage and its cause(s) are quite obvious. For example, keeping too many cattle in a grassland community gradually damages it, often eliminating many native species and favoring invasive species that can tolerate grazing and trampling. On the other hand, out of sight from the public, fishing trawlers drag across as much as 80% of the ocean shelf in some European seas (Amoroso et al. 2018). The trawling destroys delicate creatures such as anemones and sponges, reduces species diversity and biomass, and alters community structure (Lauria et al. 2017).

> Pollution of the air, water, and soil by chemicals, wastes, and the byproducts of energy production destroys habitats in insidious ways.

Other types of environmental damage are not visually apparent even though they occur all around us, every day, in nearly every part of the world. The most subtle and universal form of environmental degradation is pollution, commonly caused by pesticides, herbicides, sewage, fertilizers from agricultural fields, industrial chemicals and wastes, emissions from factories and automobiles, and sediment deposits from eroded hillsides. The general effects of pollution on water quality, air quality, and even the global climate are cause for great concern, not only because of their threats to biodiversity but also because of their effects on human health and agriculture (Khaniabadi et al. 2017; Kim et al. 2017). The subtle, unseen forms of pollution are probably the most threatening—primarily because they are so insidious.

GLOBAL CHANGE
CONNECTION

Pesticide pollution

The dangers of pesticides were brought to the world's attention in 1962 by Rachel Carson's influential book *Silent Spring* (see Figure 1.4). Carson described a process known as **biomagnification**, through which dichloro-diphenyl-trichloroethane (DDT) and other organochlorine pesticides become concentrated as they ascend the food chain (Köhler and Triebskorn 2013) (**FIGURE 4.14**). These pesticides, used on crop plants to kill insects and sprayed on water bodies to kill

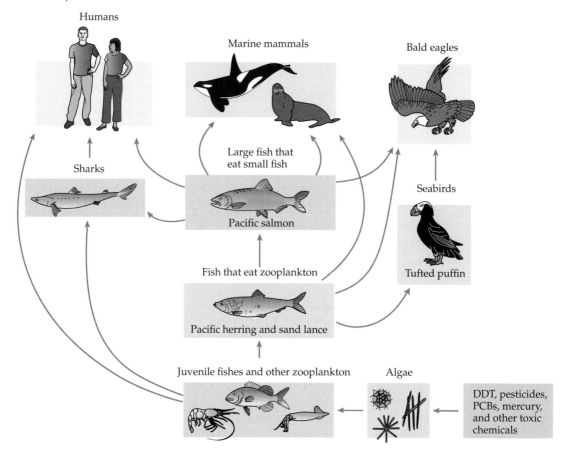

FIGURE 4.14 Toxic chemicals become successively concentrated at higher levels in the food chain, leading to health problems for humans, marine mammals, seabirds, and raptors. (After P. S. Ross and L. S. Birnbaum. 2003. *Hum Ecol Risk Assess* 9(1): 303–324. Taylor & Francis Ltd, http://www.tandfonline.com.)

mosquito larvae, were harming wildlife populations, especially birds, such as hawks and eagles, that eat large amounts of insects, fish, or other animals exposed to DDT and its byproducts.

Recognition of this situation in the 1970s led many industrialized countries to ban the use of DDT and other chemically related pesticides. The ban eventually allowed the partial recovery of many bird populations, including the peregrine falcon, which now has stable populations and is categorized as a species of "least concern" by the IUCN (**FIGURE 4.15**). Nevertheless, massive use of pesticides—even DDT—persists because of

FIGURE 4.15 Peregrine falcons are now breeding in many areas across North America. Their population declined dramatically when DDT use began in the 1940s and then recovered after DDT was banned in 1972 (Hoffman and Smith 2003).

Courtesy of the US Fish and Wildlife Service

their benefits to people. For example, DDT is still highly effective in controlling mosquitoes, which, through the malaria they spread, are still a significant cause of death in tropical regions. These benefits must be weighed against harm not only to endangered animal species but also to people, particularly the workers who handle these chemicals in the field and the consumers of the agricultural products, such as crops and even chicken eggs, exposed to these chemicals (Bouwman et al. 2015). Plants, animals, and people living far from where the chemicals are applied can still be harmed; surprisingly high concentrations of pesticides are found in the tissues of polar bears in northern Norway and Russia, where they have a harmful effect on bear health (Elliott and Elliott 2013).

GLOBAL CHANGE
CONNECTION

Water pollution

Water pollution has negative consequences for all species: it destroys important food sources and contaminates drinking water with chemicals that can cause immediate and long-term harm to the health of humans and other species that come into contact with the polluted water (Landrigan et al. 2018). In the broader picture, water pollution often severely damages aquatic ecosystems. Rivers, lakes, and oceans are sometimes used as open sewers for industrial wastes and residential sewage. And higher densities of people almost always mean greater levels of water pollution. Pesticides, herbicides, petroleum products, heavy metals (such as mercury, lead, and zinc), detergents, toxic chemicals such as polychlorinated biphenyls (PCBs), and industrial wastes directly kill organisms such as insect larvae, fish, amphibians, and even marine mammals living in aquatic environments. An increasing source of pollution in coastal areas is the discharge of nutrients and chemicals from shrimp and salmon farms. Medicines used by people or given to domesticated animals can enter the aquatic environment through sewage, either because waste treatment plants cannot remove them or because they leak into wells (Schaider et al. 2014). According to the IUCN Freshwater Fish Specialist Group (2020), antidepressants and birth control pills are now major threats, and some researchers have raised the concern about the impact of COVID-19 treatment on environmental pollution levels (Horn et al. 2020). Biologically active chemicals, especially hormones, can have an adverse effect on the physiology, behavior, and reproduction of fish and other animals that ingest them (Guedes-Alonso et al. 2017).

GLOBAL CHANGE
CONNECTION

In contrast to wastes in the terrestrial environment, which have primarily local effects, toxic wastes in aquatic environments diffuse over a wide area (**FIGURE 4.16**). Toxic chemicals, even at very low concentrations in the water, can be lethal to aquatic organisms through biomagnification. Many aquatic environments are naturally low in essential minerals, such as nitrates and phosphates, and aquatic plant and animal species have adapted to their natural absence by developing the ability to process large volumes of water and concentrate these minerals. When these species process polluted water, they concentrate toxic chemicals along with the essential minerals, and the toxins may eventually poison them. Species that feed on these aquatic species ingest the toxic chemicals they have concentrated. One of the most serious consequences for humans is the accumulation of mercury and other toxins

GLOBAL CHANGE
CONNECTION

FIGURE 4.16 This 80-mile-long spill of toxic materials from the Gold King Mine, which extended through three states and two Native American reservations in 2015, prompted a state of emergency in the southwestern United States. The extent of the harm done to wildlife and people by the arsenic, lead, mercury, and cadmium released is not fully known. According to a report from the US Environmental Protection Agency (EPA), human error, in the form of a botched cleanup effort, was to blame for the 3-million-gallon spill (Denver Post 2016). As of 2019, the EPA had spent $2.9 million to clean up the spill. This photograph is of the Animas River at Bakers Bridge near Durango, Colorado, taken August 6, 2015.

© Whit Richardson/Alamy Stock Photo

by long-lived predatory fishes, such as swordfish and sharks, and the toxic effects on the nervous systems of people who eat these types of fish frequently (Eagles-Smith et al. 2018).

GLOBAL CHANGE
CONNECTION

In 2015 a new aquatic pollutant was discovered: **microplastics**, plastic particles between 1 μm and 5 mm in diameter that arise from a variety of human sources and are mostly created when plastic debris in the ocean is broken down by wave action and sunlight (**FIGURE 4.17A**). Microplastics are now ubiquitous in the world's oceans and are ingested by marine life, particularly filter feeders (Zhang et al. 2020) (**FIGURE 4.17B**). These particles can physically block nutrient absorption in filter feeders and can even cause mechanical damage to their digestive tracts. However, possibly more dangerous to marine animals are the high levels of pollutants, some toxic, that can be present in or on the particles (Germanov et al. 2018). Bioaccumulation of these toxins over decades in long-lived filter feeders such as some types of rays, filter-feeding sharks, and various marine mammals, such as baleen whales (*Balaenoptera physalus*), poses a significant risk for the animals themselves and for those who eat fish (Sharma and Chatterjee 2017).

GLOBAL CHANGE
CONNECTION

Even essential minerals that are beneficial to plant and animal life can become harmful pollutants at high concentrations. Anthropogenic releases of human sewage, agricultural fertilizers, detergents, and industrial wastes often add large amounts of nitrates and phosphates to aquatic systems, initiating the process of **eutrophication**. Humans release as much nitrate into the environment as is produced by all natural processes, and the anthropogenic release of nitrogen is expected to keep increasing as the human population continues to increase. Furthermore, changes in precipitation patterns as a result of global climate change have been predicted to increase rates of eutrophication through increased runoff (Sinha et al. 2017). Even small amounts of these nutrients can stimulate growth, and high concentrations often result in thick "blooms" of algae at the surfaces of ponds and lakes. These algal blooms may be so dense that they outcompete other plankton

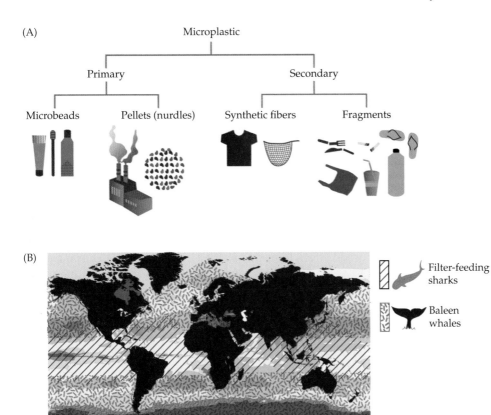

FIGURE 4.17 (A) Microplastics are plastic particles from 1 μm to 5 mm in diameter, arising from a variety of human-made sources. The primary sources include microbeads, from personal care products (scrubs and toothpastes), and industrial particles called nurdles, used to manufacture plastic items. Secondary sources of microplastics arise from the breakdown of larger plastic waste. Many of these plastics contain toxic substances or have toxins adhering to them. (B) Overlaying the results of a survey of microplastic concentrations with distributions of the dominant filter-feeding megafauna suggests the magnitude of risk of ingestion. Areas of highest microplastic concentrations overlap with distributions of both filter-feeding sharks (*Rhincodon typus*, lined overlay) and baleen whales (*Balaenoptera physalus*, dotted overlay). (After E. S. Germanov et al. 2018. *Trends Ecol Evol* 33: 227–232. © 2018. Reprinted with permission from Elsevier. Data from E. van Sebille et al. 2015. *Environ Res Lett* 10: 124006 and IUCN Red List of Threatened Species 2018-2.)

species and shade bottom-dwelling plant species. As the algal mat becomes thicker, its lower layers sink to the bottom and die. The bacteria and fungi that decompose the dying algae multiply in response to this added sustenance and consequently absorb all the oxygen in the water. Without oxygen, much of the remaining animal life dies off, sometimes visibly in the form of masses of dead fish floating on the water's surface. The result is a greatly impoverished and simplified community, a *dead zone* consisting of only those species that are tolerant of polluted water and low oxygen levels.

This process of eutrophication can also affect entire marine systems, particularly coastal areas and bodies of water in confined areas, such as the Gulf of Mexico, the Mediterranean, the North and Baltic Seas in Europe, and the enclosed seas of Japan, which have large anthropogenic inputs of nutrients (Huang et al. 2017; Ménesguen and Lacroix 2018). In warm tropical waters, eutrophication favors algae, which grow over coral reefs and completely change the reef community (Díaz-Ortega and Hernández-Delgado 2014).

The key to stopping eutrophication and its negative effects is to reduce the release of excess nutrients through improved sewage treatment and better farming practices, including reduced applications of fertilizer and establishment of buffer zones between fields and waterways.

Sediments eroded from logged or farmed hillsides can also harm aquatic ecosystems. The sediment covers submerged plant leaves and other green surfaces with a muddy film that reduces light availability and can block O_2 and CO_2 exchanges, thus decreasing the rate of photosynthesis. Increasing water turbidity (a decrease in water clarity) reduces the depth at which photosynthesis can occur and may prevent animal species from seeing, feeding, and living in the water. Sediment loads are particularly harmful to the many coral species that require crystal clear waters to survive.

Air pollution

GLOBAL CHANGE
CONNECTION

In the past, people assumed that the atmosphere was so vast that materials they released into the air would be widely dispersed and their effects would be minimal. But today, several types of air pollution are so widespread that they have damaged whole ecosystems. These same pollutants also have severe effects on human health, demonstrating again the common interests shared by people and nature (see Figure 3.1).

ACID RAIN   **Acid rain** is produced when industries such as smelting operations and coal- and oil-fired power plants release huge quantities of nitrogen oxides and sulfur oxides into the air, where those chemicals combine with moisture in the atmosphere to produce nitric and sulfuric acids. These acids are incorporated into cloud systems and dramatically lower the pH (the standard measure of acidity) of rainwater, leading to the weakening and death of trees over wide areas. Acid rain also lowers the pH of soil moisture and water bodies and increases the concentration of toxic metals such as aluminum in the soil and water.

Increased acidity alone damages many plant and animal species; as the acidity of water bodies increases, many fish either fail to spawn or die outright. Increased acidity, along with both aquatic and terrestrial pollution, is a contributing factor to the dramatic decline of many amphibian populations throughout the world (Alton and Franklin 2017). Most amphibian species depend on bodies of water for at least part of their life cycle, and a decline

Acid rain and other effects of air pollution are increasing rapidly in Asian countries as they industrialize. Acid rain is particularly harmful to freshwater species.

in water pH causes a corresponding increase in the mortality of eggs and young animals. Acidity also inhibits the microbial process of decomposition, lowering the rate of mineral recycling and ecosystem productivity. Many ponds and lakes in industrialized countries have lost large portions of their animal communities as a result of acid rain. These damaged water bodies are often in supposedly pristine areas hundreds of kilometers from major sources of urban and industrial pollution, such as the North American Rocky Mountains and Scandinavia. In developing countries such as China, acid rain is becoming a greater problem as the country powers its rapid industrial development with fuels that are high in sulfur. As a result of international agreements to reduce industrial discharge of sulfur and nitrogen oxide into the atmosphere, levels of acidification have decreased in parts of Europe and the United States, and in some cases amphibian populations have recovered (Dolmen et al. 2018).

GLOBAL CHANGE
CONNECTION

OZONE PRODUCTION AND NITROGEN DEPOSITION Automobiles, power plants, and industrial activities release hydrocarbons and nitrogen oxides as waste products. In the presence of sunlight, these chemicals react with the atmosphere to produce ozone and other secondary chemicals, collectively called *photochemical smog*. Although ozone in the upper atmosphere is important in filtering out ultraviolet radiation, high concentrations of ozone at ground level damage plant tissues and make them brittle, harming biological communities and reducing agricultural productivity. Ozone and smog are detrimental to people and other animals when inhaled, so both people and biological communities benefit from air pollution controls. Environmental regulations have been successful in halting the increase in global ozone (see Figure 4.2C), evidence of the potential to curb other pollutants.

Unfortunately, nitrogen levels are continuing to increase exponentially (see Figure 4.2B). When airborne nitrogen compounds are deposited in terrestrial environments by rain and dust, ecosystems throughout the world are damaged and altered by potentially toxic concentrations of nitrogen, and many species are unable to survive in the altered conditions (Nijssen et al. 2017). In particular, the combination of nitrogen deposition and acid rain is responsible for a decline in the density of soil fungi that form beneficial relationships with trees (see Chapter 3).

GLOBAL CHANGE
CONNECTION

TOXIC METALS Leaded gasoline (still used in many developing countries, despite its clear danger to human health), coal burned for heat and power, mining and smelting operations, and other industrial activities can release large quantities of lead, zinc, mercury, and other toxic metals into the water and atmosphere (Wade 2013). The releases are often the consequence of normal operations but can be particularly severe as the consequence of natural disaster or negligence (see Figure 4.16). These elements are directly poisonous to plant and animal life and can cause permanent injury to children. The effects of these toxic metals are particularly evident in areas surrounding large smelting operations, where there may be measurable negative effects on life for miles around.

Enforcement of local and national policies and regulations may sometimes reduce levels of air pollution; eliminating lead from gasoline is one such example. Concentrations of air pollutants are declining in certain areas of North America and Europe, but they continue to rise in many other areas of the world that are undergoing industrialization. Because air and water pollutants do not observe political boundaries, international agreements addressing these issues are important. As we will see in Chapter 12, lowering levels of air and water pollution is necessary for the health of both human populations and biodiversity.

GLOBAL CHANGE
CONNECTION

4.5   A Concluding Remark and Reason for Hope

This chapter has described the sources and consequences of destruction and alteration of habitats. Given that habitat loss due to destruction, fragmentation, and degradation is the single greatest threat to biodiversity, it should be the priority for conservation efforts. Fortunately, there are many positive actions being taken by governments and individuals across the globe to address the loss of habitat. Among them are extensive *reforestation*, that is, the planting of trees, which counteracts the effect of forest habitat loss, prevents erosion, and is important for absorbing atmospheric carbon. For example, Vietnam's prime minister has called for the planting of 1 billion trees nationwide (Taterski 2021). In Brazil, 4.2 million ha have regrown since 2000, and globally an area the size of France has been reforested over the same period (WWF 2021). Awareness of the impacts of habitat loss is increasing, and with it, meaningful commitments to reduce its effects. When these and other threats to biodiversity are well understood, protection and recovery efforts have the best chance for success.

Summary

- The major threats to biodiversity are habitat destruction, fragmentation, degradation (which includes pollution), climate change, overexploitation, invasive species, and disease. All of these threats result from the use of the world's natural resources by an increasing human population.

- Habitat destruction particularly threatens rain forests, wetlands, coral reefs, and other species-rich communities.

- Habitat fragmentation is the process whereby a large, continuous area of habitat is both reduced and divided into two or more fragments. Habitat fragmentation can lead to the rapid loss of some of the remaining species because it creates barriers to the normal processes of dispersal, colonization, and foraging. Particular fragments may contain altered environmental conditions that make them less suitable for the original species. Habitat fragmentation can also facilitate habitat loss.

- Environmental pollution eliminates many species from ecosystems even where the structure of the community is not obviously disturbed. Environmental pollution results in pesticide biomagnification; contamination of water with industrial wastes, sewage, and fertilizers; and air pollution resulting in acid rain, excess nitrogen deposition, photochemical smog, and high ozone levels.

For Discussion

1. Human population growth is often blamed for the loss of biological diversity. Is this valid? What other factors are responsible, and how do we weigh their relative importance? Is it possible to find a balance between providing for increasing numbers of people and protecting biodiversity?

2. Consider the most damaged and the most pristine habitats near where you live or go to school. Why have some habitats been preserved and others fragmented and degraded?

3. Learn about one endangered species in detail. What is the full range of immediate threats to this species? How do these immediate threats connect to larger social, economic, political, and legal issues?

Suggested Readings

Crist, E., C. Mora, and R. Engelman. 2017. The interaction of human population, food production, and biodiversity protection. *Science* 356(6335): 260–264. A comprehensive look at the importance of population control for conservation.

Hansen, M. C., et al. 2020. The fate of tropical forest fragments. *Science Advances* 6(11): eaax8574. A look at the relationship between fragmentation and habitat loss and the implications for forest management.

Miller-Rushing, A. J., et al. 2019. How does habitat fragmentation affect biodiversity? A controversial question at the core of conservation biology. *Biological Conservation* 232: 271–273. A thoughtful review of the debate over the impact of habitat fragmentation.

Newbold, T., et al. 2020. Tropical and Mediterranean biodiversity is disproportionately sensitive to land-use and climate change. *Nature Ecology & Evolution* 4(12): 1630–1638. An analysis comparing biomes found that both the types of land use and the effect of climate had disproportionate effects on species in tropical and Mediterranean ecosystems.

Walker, R. T., et al. 2019. Avoiding Amazonian catastrophes: Prospects for conservation in the 21st century. *One Earth* 1(2): 202–215. Projections of Amazon forest cover by 2050 under different scenarios.

Weinzettel, J., D. Vačkář, and H. Medková. 2018. Human footprint in biodiversity hotspots. *Frontiers in Ecology and the Environment* 16(8): 447–452. Advanced quantitative methods were used to determine the impact of consumption and supply chains within highly diverse regions across the globe.

Ocean biodiversity has many threats. Among them, overexploitation is a primary concern for most marine endangered species; however, policies to limit harvesting have been successful in preventing several species from going extinct, including humpback whales (*Megaptera novaeangliae*). Climate change also directly impacts ocean life, including by reducing sea ice, upon which many marine mammals depend.

Climate Change and Other Threats to Biodiversity

5

I n Chapter 4, we reviewed the impacts of three of the seven major threats to biodiversity, all related to direct changes in habitat availability or quality. While loss of habitat accounts for more species loss than any other factor, we will not stop losses of biodiversity without addressing the remaining four threats: climate change, overexploitation, invasive species, and disease. These issues are typically complex, often involving cascades of impacts or indirect effects. They also frequently have an international component, facilitated by processes occurring at continental and even global scales.

Excessive fishing, hunting, or collecting is the second leading threat for most taxa, and it even exceeds habitat loss for marine organisms (see Figure 4.4). Some of this overexploitation is driven by the demand for protein by growing numbers of people on Earth, but much of it is driven by increasing demand for exotic pets, ornamental and medicinal plants, and unique biologically based products, such as elephant ivory and sea turtle jewelry. Similarly, global climate change is not simply a function of human population growth but also is due to the lifestyles we choose. In this chapter we will learn about the nature of these remaining threats, including ways and instances in which they can be (and have been) successfully addressed.

GLOBAL CHANGE
CONNECTION

5.1 Global Climate Change

LEARNING OBJECTIVES

By the end of this section you should be able to:

5.1.1 Draw a diagram to explain the greenhouse effect.

5.1.2 Summarize the evidence supporting the human connection to climate change.

5.1.3 Identify novel examples of the impacts of climate change on living systems.

5.1.4 Make predictions about the impact of climate change on the world's aquatic and coastal ecosystems.

There has long been a scientific consensus that climate change (1) is a problem and (2) is caused by human activity (Cook et al. 2016). The Intergovernmental Panel on Climate Change (IPCC), part of the United Nations Environment Programme and including 195 countries, made the following conclusion after considering all available data:

> *It is unequivocal that human influence has warmed the atmosphere, ocean, and land. Widespread and rapid changes in the atmosphere, ocean, cryosphere and biosphere have occurred.* (IPCC 2018)

However, climate change has increasingly been politicized (Chinn et al. 2020). As a result, a great deal of misinformation has been circulated in public and social media around this issue (Treen et al. 2020). The resulting misunderstanding threatens our capacity as a society to address the threat climate change poses. This section will elucidate the science of climate change and its affect on biological organisms.

Carbon dioxide, methane, and other trace gases in the atmosphere allow light energy to pass through and warm the surface of the Earth. However, these gases, as well as water vapor (in the form of clouds), are able to trap the energy radiating from the Earth as heat, slowing the rate at which heat leaves the Earth's surface and radiates back into space. These gases are called **greenhouse gases** because they function much like the glass in a greenhouse, which is transparent to sunlight but traps energy inside the greenhouse once light is transformed into heat. The similar warming effect of Earth's atmospheric gases is called the **greenhouse effect**. We can imagine these gases as "blankets" over the Earth's surface: the denser the concentration of gases, the more heat is trapped near the Earth, and the higher the planet's surface temperature.

The greenhouse effect allows life to flourish on Earth—without it, the temperature at the Earth's surface would fall dramatically. Today, however, as a result of human activities, concentrations of greenhouse gases are increasing so much that they are significantly affecting the Earth's climate (IPCC 2018). The term *global warming* is used to describe the rise in temperatures

resulting from the increase in greenhouse gases, and *global climate change* refers to the complete set of climate characteristics that are changing now and will continue to change in the future because of this increase, including patterns of precipitation and wind.

The relationship between carbon and temperature

How do we know that humans are contributing to global climate change, and what do conservation biologists need to do to address the problems created by climate change? The IPCC was organized by the United Nations in 1988 to address the first of these questions by evaluating the available research. Thousands of scientists and others contribute to the work of the IPCC, volunteering their time and expertise to provide and assess current data. Due to the overwhelming body of evidence, there is now broad scientific agreement on these points:

- The increased levels of greenhouse gases have been caused by human activities.
- These gases have affected the world's climate and ecosystems.
- These effects will likely increase in the future (IPCC 2018).

One source of the data that has led to these conclusions has been the analysis of gases trapped in ice in the Arctic and Antarctic; these were laid down over geologic time and thus allow us to determine climate conditions hundreds of thousands of years ago (**FIGURE 5.1A**). The first of these gas samples was analyzed by Soviet and French scientists in the 1980s from a 2083 m core of ice taken near the Antarctic station of Vostok. This and ice cores from other locations have since been analyzed by other teams of scientists, including the European Project for Ice Coring in Antarctica (EPICA), whose data are presented in **FIGURE 5.1B**. Together, these data have revealed three important findings:

1. Carbon dioxide (CO_2) concentrations have varied significantly over time.
2. Temperature fluctuations have followed CO_2 levels; when the amount of atmospheric CO_2 went up, the temperature increased shortly after, in a consistent pattern.
3. For hundreds of thousands of years, the level of atmospheric CO_2 was never higher than 300 parts per million by volume (ppmv), while the current levels are greater than 415 ppmv.

After Vostok, other Antarctic ice cores extended the measurements closer to the current day. The most recent of these, collected at Siple Station, revealed that atmospheric CO_2 began rising exponentially in the mid-1800s, corresponding to the Industrial Revolution. This indirect evidence that human emissions were contributing to the rise in CO_2 has been followed by more direct evidence. The annual global burning of fossil fuels (in the form of oil, coal, or natural gas) has been calculated to produce 36 billion tonnes (almost 40 billion US tons) of CO_2 emissions, while the annual increase in atmospheric CO_2 is about 2.5 ppmv; thus, there is ample industrial carbon produced by human activities to account for the atmospheric increase (Richie and Rosner 2020). Although

GLOBAL CHANGE
CONNECTION

(A)

FIGURE 5.1 (A) By analyzing air bubbles trapped in ice cores, scientists can determine the climate that existed even hundreds of thousands of years ago. (B) Ice cores sampled in Vostok, Antarctica, revealed that temperature has fluctuated as much as 11°C over the past 800,000 years and that these fluctuations closely mirrored carbon dioxide measurements. More recent data are collected from direct measurements. (B, after D. Lüthi et al. 2008. *Nature* 453[7193]: 379–382. https://doi.org/10.1038/nature06949, and J. Jouzel et al. 2007. *Science* 317 [5839]: 793–796. Reprinted with permission from AAAS. https://doi.org/10.1126/science.1141038.)

(B)

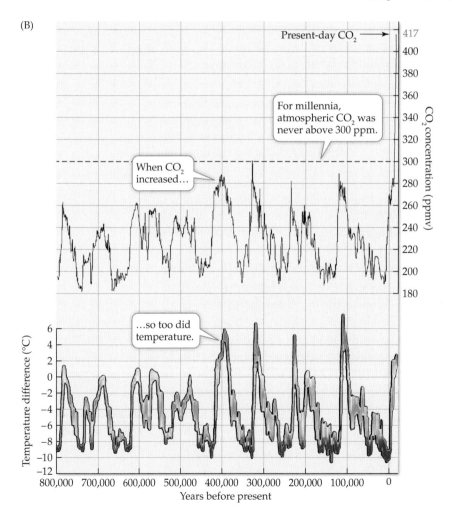

there was a dramatic dip during the peak of the COVID-19 pandemic in 2020 (Tollefson 2020), emissions appear to have quickly returned to pre-pandemic levels (Zheng et al. 2020). Extensive clearing of tropical forests also contributes to the rise in CO_2. It is important to note that the current rate of increase is about 100 times faster than the increases observed after the last ice age when the Earth also began to warm. Further, we now have direct measurements of temperature and atmospheric CO_2; if we plot industrial carbon production against these, we can see that all three are rising at approximately the same rate (**FIGURE 5.2**), suggesting that human burning of fossil fuels is directly responsible for rising CO_2 levels, and indirectly responsible for rising temperatures.

However, the most compelling evidence for a link between humans and increased atmospheric carbon may be the measurement of the proportion of atmospheric CO_2 attributable to industrial sources. Fossil fuels come from carbon sources buried for millions of years and as a consequence have a different carbon isotope signature than other sources (see Chapter 1 for another example of isotope analysis). Trees use the carbon from atmospheric CO_2 to form wood. Thus, by measuring the carbon isotope ratios in tree rings, which can be easily dated, scientists have been able to measure an increasing proportion of carbon from fossil fuels beginning at about 1850 (Keeling et al. 1989).

GLOBAL CHANGE
CONNECTION

There is a broad consensus among scientists that increased atmospheric concentrations of carbon dioxide and other greenhouse gases produced by human activities have already resulted in warmer temperatures and will continue to affect the Earth's climate in the coming decades.

(A) Carbon dioxide

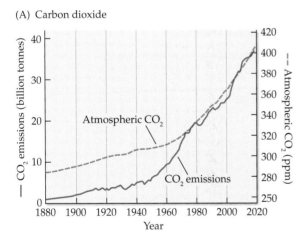

(B) Temperature

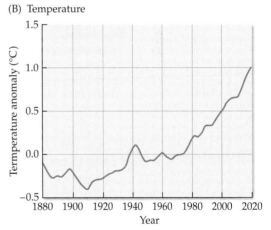

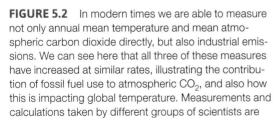

FIGURE 5.2 In modern times we are able to measure not only annual mean temperature and mean atmospheric carbon dioxide directly, but also industrial emissions. We can see here that all three of these measures have increased at similar rates, illustrating the contribution of fossil fuel use to atmospheric CO_2, and also how this is impacting global temperature. Measurements and calculations taken by different groups of scientists are consistent (Tollefson 2018). (A, after NOAA Climate.gov [https://www.climate.gov/media/12990], adapted from original by Howard Diamond [NOAA ARL]. Atmospheric CO2 data from NOAA and ETHZ. CO2 emissions data from Our World in Data [https://ourworldindata.org/co2-and-other-greenhouse-gas-emissions] and the Global Carbon Project. B, after https://climate.nasa.gov/vital-signs/global-temperature.)

In its 2018 report, the IPPC's extensive review of the evidence supported the conclusion that human activities have caused global surface temperatures to increase between 0.8°C and 1.2°C since preindustrial times; the temperature increase is expected to reach 1.5°C between 2030 and 2052 if the current rate continues (IPCC 2018). In fact, July 2021 was the hottest month in the Earth's recorded history (Lindsey 2021). Over the 136 years for which we have precise temperature records, all 18 of the warmest years have occurred since 1998.

Cascading effects of increased temperature

The impacts of global warming are varied and multifaceted (**TABLE 5.1**), in part due to the fact that temperature increases are not uniform across the globe. Temperatures at high northern latitudes, such as in Siberia, Alaska, and Canada, have increased more than in other regions (**FIGURE 5.3**). It appears that this pattern may be due to increased concentrations of **anthropogenic** (from human activity) aerosols in these regions. The aerosols act in a way that is similar to the action of greenhouse gases, and they work in conjunction to trap heat (Lembo et al. 2018). As a result of increasing temperatures, climate regions in the northern and southern temperate zones will be shifted toward the poles. This warming is expected to create a snowball effect in which thawing tundra and melting permafrost (the layer of soil that typically stays frozen throughout the year) will further increase the rate of CO_2 release through respiration by soil microbes, although more research is needed to understand its extent (Schuur et al. 2020).

TABLE 5.1	Some Effects of Global Warming

1. INCREASED TEMPERATURES AND INCIDENCE OF HEAT WAVES
Example: 2020 (tied with 2016) was the hottest year worldwide in the historical record so far.

2. MELTING GLACIERS AND POLAR ICE
Example: Arctic Ocean summer ice has declined in area by 15% over the past 25 years. September 2020 had the second-lowest extent of sea ice recorded. Since 1850, glaciers in the European Alps have disappeared from more than 30%–40% of their former range.

3. RISING SEA LEVEL
Example: Since 1938, one-third of the coastal marshes in a wildlife refuge in the Chesapeake Bay have been submerged by rising seawater.

4. EARLIER SPRING ACTIVITY
Example: One-third of English birds are now laying eggs earlier in the year, and two-thirds of temperate plant species are flowering earlier than they did several decades ago

5. SHIFTS IN SPECIES RANGES
Example: Two-thirds of European butterfly species studied are now found 35–250 km farther north than recorded several decades ago.

6. POPULATION DECLINES
Example: Adélie penguin populations have declined over the past 25 years as their Antarctic sea ice habitat melts away.

Sources: Union of Concerned Scientists (www.ucsusa.org), NASA, and NOAA.

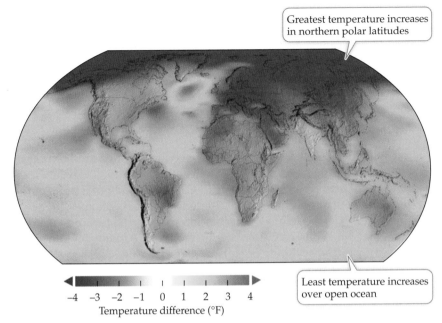

Greatest temperature increases in northern polar latitudes

Least temperature increases over open ocean

−4 −3 −2 −1 0 1 2 3 4
Temperature difference (°F)

FIGURE 5.3 Global temperature differences from average, for 2020. Here, temperature changes are reported in terms of difference (anomaly) from the average annual temperature from 1951 to 1980. Note that temperature changes differ across the surface of the Earth, with the greatest temperature increases at northern latitudes. At publication of this edition, the latest data were for 2020, which tied with 2016 as the hottest year on record. (From NASA/GISS, Scientific Visualization Studio. 2020. https://climate.nasa.gov/vital-signs/global-temperature.)

Uneven distributions of elevated temperatures across the globe due to a variety of factors will mean that impacts of climate change will vary by region. For example, rainfall has already started to increase on a global scale and will continue to do so, while some places are expected to be drier than before. The effects of climate change will also increase the probability of extreme weather events, including hurricanes, floods, snowstorms, and even regional drought. For example, increasing temperatures are associated with rising mortality rates of trees in Israel (Klein et al. 2019). Coastal areas will be especially impacted, as the combination of storms and rising sea levels will cause increased destruction of cities and other human settlements and will severely damage coastal vegetation, including beaches and coral reefs.

The effects of global climate change on temperature and rainfall are also expected to have particularly dramatic effects on tropical ecosystems. Many tropical species and biological communities appear to have narrow tolerances for temperature and rainfall, so even small changes in the climate could have major effects on species composition, cycles of plant reproduction, patterns of migration, and other effects (Seddon et al. 2016). Major contractions in the

GLOBAL CHANGE
CONNECTION

As rainfall patterns change and most regions become warmer, many terrestrial plant and animal species may not be able to adapt quickly enough to survive. Climate change may also have huge impacts on marine ecosystems and coastal areas occupied by people.

area of rain forest are quite likely, as are species shifts. By evaluating records from 106 long-term studies of lowland Amazonian forests, researchers have observed significant species shifts among new trees over the past 30 years, with drought-adapted species becoming more abundant while wet-adapted species are disappearing (Esquivel-Muelbert et al. 2019). Cool-adapted species that live atop tropical mountains are also likely to be highly vulnerable to increasing temperatures; as bands of vegetation move higher on mountains, the species at the top will have nowhere to go and thus face extinction (Campos-Cerqueira et al. 2017).

In tropical deciduous forests, savannas, and areas with a Mediterranean climate, these warmer and drier conditions will result in an elevated incidence of fire. The increase of wildfires in California and other parts of the western United States over the last two decades, damaging homes, property, and protected forest, could be an indication of what the future may bring (Abatzoglou and Williams 2016). The increase in fire frequency can affect biodiversity directly, by killing organisms and changing habitat availability, and also indirectly by changing soil chemistry. A study of 48 sites across the globe determined that frequent burns in savanna and forests decreased both carbon and nitrogen in surface soils, which can affect primary productivity and the rest of the community that depends on plant growth (Pellegrini et al. 2018). Thus, we can see how one change, increase in temperature, can lead to a cascade of impacts for an ecosystem.

Ocean acidification, warming, and rising sea level

Evidence indicates that ocean water temperatures have increased over the last 50 years by about 0.75°C (NOAA 2021). As a consequence, certain marine species are expanding their ranges farther from the equator, and coral reefs and other marine habitats are threatened by rising seawater temperatures (Henson et al. 2017). Zooplankton are declining in some areas because of warmer seawater temperatures, with dire consequences for the marine animals that feed on them (Benedetti et al. 2018).

Abnormally high water temperatures in oceans also cause corals to sicken and expel the symbiotic algae that would normally live inside their bodies and provide them with essential carbohydrates (**FIGURE 5.4A**). These "bleached" corals then often die from starvation. The coral bleaching event of June 2014–April 2017 was the worst on record, with more than 70% of the Earth's seas experiencing conditions associated with mortality (**FIGURE 5.4B**). This was the third global bleaching event in recorded history, the others taking place in 2010 and 1998; all three events coincided with periods of abnormally high ocean temperatures. Although fluctuations in ocean temperature are normal, the geographic pattern and intensity of these recent warming events have been very unusual. Although reefs in some places have been able to recover (Gilmour et al. 2013), many have not; for example, half of the Great Barrier Reef is now dead as a result of the most recent bleaching event. Satellite tracking of the world's ocean temperatures by NOAA's Coral Reef Watch program provides managers with information to help struggling reefs (see Figure 5.4B).

(A)

(B)

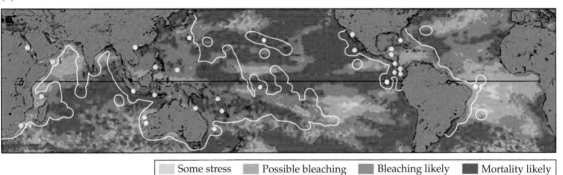

| Some stress | Possible bleaching | Bleaching likely | Mortality likely |

FIGURE 5.4 High ocean temperatures cause corals to expel the symbiotic algae that provide them with food. (A) The 2015 coral bleaching event had dramatic effects on this reef in American Samoa. The photo on the left was taken in December 2014, the one on the right in February 2015. The photos were taken by the XL Catlin Seaview Survey, a group documenting bleaching across the globe. (B) Temperature increases that threaten coral reefs can be tracked using satellite data. Here, we can see that for the period June 2014–April 2017, more than 70% of the Earth's seas experienced conditions that usually cause coral mortality. Furthermore, most of the seas of American Samoa reached this point at least twice during this period, making it the worst coral bleaching event on record. Major coral reef regions are outlined here, with World Heritage sites (i.e., reefs of particular biological or cultural significance) indicated with dots. Corals begin to starve after bleaching, but some may recover if conditions improve. The NOAA Coral Reef Watch program uses surface temperature data collected by satellite to determine where managers should focus efforts. Global warming is the leading cause of coral bleaching. (After NOAA *Coral Reef Watch 2017* and L. Burke et al. 2011. *Reefs at Risk Revisited*. World Resources Institute, Washington, DC. CC BY 4.0.)

Increased anthropogenic CO_2 in the atmosphere has caused another problem: as much as a third of the CO_2 is absorbed by our oceans, causing a reduction in their pH. In World Wildlife Fund's *Living Planet Report* (WWF 2020), ocean acidification is highlighted as one of the critical ways that human activities are threatening the biosphere. It is estimated that the rate at which ocean acidification is now occurring has not been seen in at least 300 million years (Hönisch et al. 2012). Ocean acidification is associated with many problems, most notably the decrease in calcium carbonate saturation states that affect shell-forming marine animals and corals. Some marine species will be able to adapt, but others will not; the last great acidification event 55 million years ago was marked by mass extinctions. Mussels, clams, urchins, and starfish all have difficulty forming shells in low pH, which also causes corals to be weaker and more vulnerable to erosion. However, other species grow better in acidic conditions, such as sea grasses and some types of algae. Given that all of these species create habitat for fish and other marine life, entire ecosystems will likely be affected by ocean acidification. Some models suggest that species diversity is likely to decrease in coral reefs, mussel beds, and some macroalgal habitats, while diversity may increase in sea grass and other macroalgal habitats (Sunday et al. 2017).

Beyond its impacts on biodiversity, ocean acidification is expected to negatively affect fisheries and food security for millions of people. One study of the impacts of ocean acidification in the United Kingdom predicted a loss in fish and shellfish biomass of between 10% and 60% by 2050, with resulting significant loss in revenue and jobs (Fernandes et al. 2017).

Increased temperatures also mean that sea levels around the world are rising. This is due to the combination of melting glaciers and the thermal expansion of water in the ocean; the volume of water increases as its temperature rises. Significant melting of glaciers and sea ice has been extensively documented, especially at the poles (**FIGURE 5.5**). The perimeter of the Arctic ice has visibly shrunk in recent years, as observed by satellite images; the minimum extent of sea ice recorded in 2020 was second only to the lowest extent, measured in 2012 (Scott 2020). Loss of sea ice equates to loss of habitat for several animals, including harp seals, polar bears, and walrus (**FIGURE 5.6**). Although losses in the north are more pronounced, new and more sophisticated methods of measuring ice melt have discovered that losses in Antarctica may be greater and more rapid that previously supposed (Rignot et al. 2019). By considering aspects of the ice such as topography and mass rather that simple areal extent, scientists have determined that the melting of ice in Antarctica has accelerated by 280% since 1979. This has contributed to a sea level rise of 3.1 ± 0.3 mm yearly since 1993 (Dangendorf et al. 2019). In total, sea level has already risen by about 20 cm (8 in.) over the last 100 years (IPCC 2018). An increase in snowfall in Antarctica could partially offset ice mass loss (Seroussi et al. 2020), but more rapid and dramatic mass loss is also possible given that deep glaciers with thick marine-terminating fronts are prone to instabilities (Bell et al. 2020).

It is predicted that by 2100 there will be a global sea level rise (relative to 1986–2005) of 0.26 to 0.77 m with an increase of 1.5°C (IPCC 2018). In regions

(A)

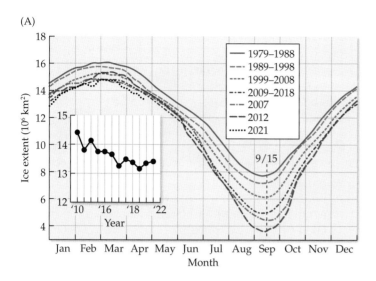

(B)

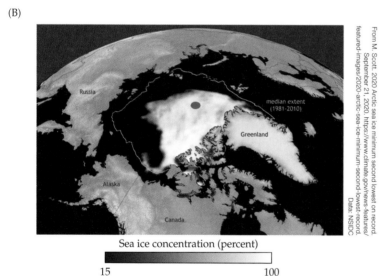

Sea ice concentration (percent)

15   100

FIGURE 5.5 (A) All climate models predict that the greatest warming will take place in the northern polar regions. The graph shows satellite-data-derived measurements of the extent of arctic sea ice cover by month, for selected years from 1979 to April 2021. The inset shows how the ice extent has decreased over time for the month of April. (B) The satellite-derived map shows the Arctic sea ice extent (in white) in summer of 2020 as compared with the median modern ice edge between 1981 and 2010 (yellow outline). The red dot marks the North Pole. (A, after J. C. Comiso et al. *Current state of sea ice cover*. Accessed April 2021. https://earth.gsfc.nasa.gov/cryo/data/current-state-sea-ice-cover.)

From M. Scott. 2020 Arctic sea ice minimum second lowest on record. September 21, 2020. https://www.climate.gov/news-features/featured-images/2020-arctic-sea-ice-minimum-second-lowest-record. Data: NSIDC

with low elevational gradients, this seemingly modest rise in sea level could affect areas thousands of meters inland; massive flooding of coastal cities and communities will certainly occur. Much of the current land area of low-lying countries such as Bangladesh could be underwater within 100 years. Somewhere between 25% and 80% of coastal wetlands could be also altered by rising sea level. Many coastal cities, such as Miami and New York, will have to build expensive seawalls or be flooded by the rising waters. If the temperature rise is greater than expected, which is a distinct possibility, the impacts will be worse; an increase of 2.0°C will mean an additional sea level rise of 0.1 m, which would mean that up to 10 million more people would be exposed to the related risk of coastal flooding (IPCC 2018). A slower rate of rise

GLOBAL CHANGE CONNECTION

FIGURE 5.6   Polar ice caps are melting at alarming rates, as these walrus crowded onto an ice floe in the Bering Sea off Alaska seem to attest.

Photograph by Budd Christman, courtesy of NOAA

would provide greater opportunities for people to adapt to the changing conditions by building seawalls or by moving infrastructure farther from the coast and would give wildlife an opportunity to migrate or possibly evolve. Whether the rate will be fast or slow will depend on our response to this crisis.

Shifting species ranges and other impacts

GLOBAL CHANGE CONNECTION

Global climate change has the potential to radically restructure ecosystems by changing the ranges of many species. For example, marine species are expected to shift their ranges farther from the poles in response to warming temperatures (Molinos et al. 2016). Climate modeling of habitat suitability predicts that the boreal forests of Canada are likely to become fragmented, subjecting moose, the boreal chickadee, caribou, and other species to dangerously smaller ranges and making them vulnerable to invasion (Murray et al. 2017) (see Chapter 4). Many changes have already been documented, particularly at high latitudes and elevations. The distributions of many bird and plant species have moved poleward and also to higher elevations, and reproduction is occurring earlier in the spring (Mainwaring et al. 2017). For example, Rocky Mountain wildflower species are blooming as much as a month earlier now than a century ago (Munson and Sher 2015), which puts them at potential risk of frost damage and may disconnect them from their pollinators (Pardee et al. 2018). Alpine mammals are moving into new ranges (**FIGURE 5.7**). Migrating birds have been observed spending longer times at their summer breeding grounds (Bussière et al. 2015). Some species will be able to persist and even thrive with these changes, while others are likely to face extinction. As mentioned earlier, alpine species have a limit to how far "up" they can migrate before they run out of habitat. Furthermore, some species are capable of rapid evolution in response to climate change, while for others the change is too fast for them to adapt.

It is likely that, as the climate changes, many existing protected areas will no longer preserve the rare and endangered species that currently live in them (Davies et al. 2017). We need to establish new conservation areas now

(A)

Alan Schmierer, Public Domain

(B)

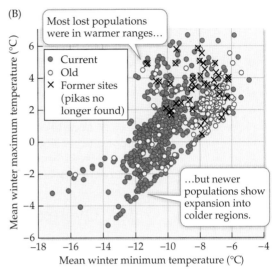

Most lost populations were in warmer ranges…

Mean winter maximum temperature (°C)

- Current
- Old
- × Former sites (pikas no longer found)

…but newer populations show expansion into colder regions.

Mean winter minimum temperature (°C)

FIGURE 5.7 (A) American pikas (*Ochotona princeps*, a relative of the rabbit) live at high altitudes and so are the subject of concern in the face of warming climates. (B) A survey of 2387 sites throughout the Great Basin was compared with historical records of pikas ("old sites"), including where they were known to be extirpated. All extirpated sites were in warmer, drier regions. Although this may appear to be bad news in a warming climate, these data also show that the species appears to be adapting to a wider range of tolerated climates, possibly indicating an evolutionary trend that could help the species adapt. (B, after C. I. Millar et al. 2018. *Arct Antarct Alp Res* 50: e1436296. https://doi.org/10.1080/15230430.2018.1436296. CC BY 4.0. https://creativecommons.org/licenses/by/4.0.)

to protect sites that will be suitable for these species in the future, such as sites with large elevational and latitudinal gradients. Potential future migration routes, such as north–south river valleys, also need to be identified and established now (Nuñez et al. 2013). If species are in danger of going extinct in the wild because of global climate change, the last remaining individuals may have to be maintained in captivity (see Chapter 8). Another strategy that we need to consider is to move isolated populations of rare and endangered species to new localities at higher elevations and closer to the poles, where they can survive and thrive. This approach has been termed *assisted colonization*. There is some debate within the conservation community about whether assisted colonization represents a valid strategy or whether it is too problematic because of the potential for transplanted species to become invasive species in their new ranges. Even if global climate change is not as severe as predicted, establishing new protected areas can only help to protect biodiversity.

The work to slow climate change is still primarily motivated by concern for its impact on humans (Nelson et al. 2013). Global climate change will have an enormous effect on human populations due to the rising sea level and increased hurricane impacts, wildfires, drought stress, and desertification. In many areas of the world, crop yields have already declined and will decline even more because of less favorable growing conditions, especially high temperatures and drought (Ray et al. 2019). Crop yields are predicted to

GLOBAL CHANGE
CONNECTION

decline by an average of 8% in tropical areas of Africa and South Asia, where chronic, severe hunger is already an enormous problem, and by 30% or more in certain populous countries such as Brazil and Indonesia. Rice production in China may decrease by 13.5% by 2060 (Chen et al. 2020). The disadvantaged people of the world will be least able to adjust to these changes and will suffer the consequences disproportionately. However, all countries of the world will be affected, and it is time for people and their governments to recognize the urgent need to address the problem of global climate change. Because the implications of global climate change are so far-reaching, biological communities, ecosystem functions, and climate must be carefully monitored over the coming decades to prevent disaster.

5.2 Overexploitation

LEARNING OBJECTIVES

By the end of this section you should be able to:

5.2.1 Identify at least two reasons why a poor economy can lead to overexploitation.

5.2.2 Consider whether the reported numbers for internationally traded organisms are likely to be over- or underestimates.

5.2.3 Analyze arguments for and against commercial whaling.

Although climate change receives more press, the threat posed by overexploitation is second to habitat loss, affecting 6241 (72%) of the 8688 threatened or endangered species listed by the **International Union for Conservation of Nature (IUCN)** (Maxwell 2016). As human populations have increased, our methods of harvesting have become dramatically more efficient, and our impact on the environment has escalated (Dearing et al. 2006; Ellis et al. 2010). In many areas, this has led to an almost complete depletion of large animals from many biological communities, leading to the creation of "empty forest syndrome" (Green et al. 2020). Even in preindustrial societies, intense exploitation, particularly for meat, led to the decline and extinction of local species of birds, mammals, and reptiles (Cazzolla Gatti 2016).

GLOBAL CHANGE
CONNECTION

In modern times, technological advances have meant that, even in the developing world, guns are used instead of blowpipes, spears, or arrows for hunting in tropical forests and savannas, further increasing human impact. Small-scale local fishermen now have outboard motors on their boats, which allow them to harvest wider areas more rapidly. Meanwhile, powerful motorized fishing boats and enormous factory ships harvest fish from the world's oceans to support the global market (**FIGURE 5.8**).

Today's vast human population and improved technology have resulted in unsustainable harvest levels of many biological resources.

Some traditional societies imposed restrictions on themselves to prevent overexploitation of jointly owned common property or natural resources. For example, the rights to specific harvesting

© Images & Stories/Alamy Stock Photo

FIGURE 5.8 Intensive harvesting has reached crisis levels in many of the world's fisheries. These bluefin tuna (*Thunnus thynnus*) are being transferred from a fishing trawler to a factory ship, aboard which huge quantities of fish are efficiently processed for human consumption. Such efficiency can result in massive overfishing.

territories were rigidly controlled, and hunting and harvesting in certain areas were banned. There were often prohibitions against harvesting female, juvenile, and undersized animals. Certain seasons of the year and times of the day were closed for harvesting. Certain efficient methods of harvesting were not allowed. In fact, these restrictions, which allowed some traditional societies to exploit communal resources on a long-term, sustainable basis, are almost identical to some of the fishing restrictions regulators have imposed on or proposed for many fisheries in industrialized nations.

Such self-imposed restrictions on using common-property resources are often less effective today. In much of the world, resources are exploited opportunistically without restraint. In economic terms, a regulated common-property resource sometimes becomes an open-access resource and available to everyone without regulation. The lack of restraint applies to both ends of the economic scale—the poor and hungry as well as the rich and greedy. If a market exists for a product, local people will search their environment to find and sell it. Sometimes poor residents of an area will sell the rights to a resource, such as a forest or mining area, for cash. In rural areas, the traditional controls that regulate the extraction of natural products have generally weakened. Whole villages are mobilized to systematically remove every usable animal and plant from an area of forest. Where there has been substantial human migration, civil unrest, or war, controls of any type may no longer exist. In countries beset with civil conflict, such as Somalia, Cambodia, the former Yugoslavia, the Democratic Republic of the Congo, and Afghanistan, firearms have come into the hands of rural people. The breakdown of food distribution networks in countries such as these leaves the resources of the natural environment vulnerable to whoever can exploit them (Eklund and Thompson 2017).

GLOBAL CHANGE
CONNECTION

In some areas, populations of large primates, such as gorillas and chimpanzees, as well as ungulates and other mammals may be reduced by 54% to more than 90% by hunting and trapping. Populations of certain species may be eliminated altogether, especially those that occur within a few kilometers of a road (Estrada et al. 2017). In many places, hunters are extracting animals at a rate six or more times greater than the resource base can sustain. The result is a forest with a mostly intact plant community that is lacking its animal community (Galetti and Dirzo 2013). Without large animals, many plant species lack effective seed dispersal and decline in abundance, thus triggering a cascade of impacts.

The decline in terrestrial animal populations caused by intensive hunting for food, which has been termed the *bushmeat crisis*, is a major concern for wildlife officials and conservation biologists, especially throughout Africa (van Velden et al. 2018). Furthermore, eating primate bushmeat increases the possibility that new diseases will be transmitted to human populations; both the human immunodeficiency virus (HIV) and the Ebola virus, which cause human diseases, have been linked to such practices (Calmy et al. 2015). Solutions involve restricting the sale and transport of bushmeat, restricting the sale of firearms and ammunition, closing roads following logging, extending legal protection to key endangered species, establishing protected reserves where hunting is not allowed, and most importantly, providing alternative protein sources to reduce the demand for bushmeat.

International wildlife trade

GLOBAL CHANGE
CONNECTION

The legal and illegal trade in wildlife is responsible for the decline of many species. Most species of rhino are endangered due primarily to the international market for rhino horn, which is believed by some to have medicinal properties despite lack of supporting research. For example, the northern white rhinoceros (*Ceratotherium simum cottoni*) has only two females left (Hildebrant et al. 2021). Fortunately, there are several factors that are contributing to the conservation of this and other species of rhinos, including dramatically decreased rates of poaching in 2020 during the COVID-19 pandemic. Even before that, numbers of poached rhinos appeared to be decreasing in many countries, including South Africa (UNODC 2020, PoachingFacts 2021). This is likely at least in part due to the fact that as of 2019, nearly all of the top international auction houses (such as Christie's) no longer support the trade in rhino horn. However, international demand continues to contribute to the overharvesting of many other species, including butterflies by insect collectors; of parrots by aviculturists; of orchids, cacti, and other plants by horticulturists; of marine molluscs by shell collectors (see Figure 6.6); and of tropical fishes by aquarium hobbyists.

FIGURE 5.9 Orchids are the most extensively traded taxonomic group on the CITES watch list. They are extremely popular ornamental plants, the showiest of which originate in the tropics. The trade of all species of orchids is tracked by international agreement.

The most powerful mechanism to protect species threatened by international trade has been the Convention on International Trade in Endangered Species (CITES), an agreement among 183 "parties" (182 countries plus the European Union) to track and limit the trade of certain species to protect them from becoming extinct (see Chapter 12; www.cites.org). Over 38,700 species that are tracked are listed in what is called the Appendices. Among these, the largest group legally traded, in terms of volume, is plants, and the most traded plant type are orchids, with more than 35 million imported around the world annually (CITES 2021) (**FIGURE 5.9**). Per the terms of CITES, all species of orchids are tracked, and legal trade is required to benefit the species in some way. Harvesting tropical orchids from the wild has historically involved felling trees to collect plants growing in the branches, thus potentially harming hundreds if not thousands of other organisms. Fortunately, 99% of the orchids traded today are cultivated. However, other taxonomic groups aren't so easily propagated by people, so wild collection continues.

Although most international trade in wildlife is legal and therefore can be regulated, $7 billion to $23 billion per year is not, making wildlife fourth in value among illegally traded items, after humans, arms, and drugs (Nellemann et al. 2016). When number of seizures is used as a metric, mammals are the largest group illegally traded, followed by reptiles, corals, and plants (WWF 2020) (**FIGURE 5.10A**). However, if we consider these in terms of value, plants, specifically trees, are the greatest proportion (**FIGURE 5.10B**). A broad category of tropical trees referred to as "rosewood" accounts for nearly 32%; this category encompasses hundreds of species. According to the 2020 United Nations World Wildlife Crime Report 2020, produced by the UN Office of Drugs and Crime (UNODC), most of the illegal trade of rosewood originates in Africa to supply demand for furniture, with some 82% imported to China (UNODC 2020).

Many say that the pangolin is now the most heavily trafficked animal in the world (Wildlife Justice Commission 2019, Heighton and Gaubert 2021), even if elephant parts represent a larger number of seizures at national borders (UNDOC 2020). There are eight species of pangolin (Manidae, the only family in the order Pholidota, Mammalia); as a result of overexploitation, all eight are listed in CITES Appendix I, and the IUCN lists all as either Vulnerable, Endangered, or Critically Endangered. According to the UNODC,

FIGURE 5.10 Major groups illegally traded, worldwide. (A) Share of all seizure incidents in the World WISE database, by taxonomic category, aggregated 1999–2018. (B) Share of total seizures, by type (aggregated on the basis of standard value), 2014–2018. (After UNODC 2020. *World Wildlife Crime Report 2020: Trafficking in Protected Species.* United Nations Office on Drugs and Crime. Data: UNODC World WISE Database.)

(A) Share of wildlife seizure incidents

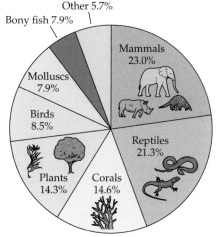

(B) Share of wildlife seizures by market value

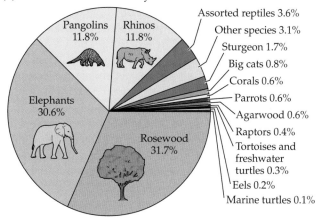

the trade in pangolins has increased by an order of magnitude since 2014 (**FIGURE 5.11**). The pangolin trade is driven by demand for use in traditional Chinese medicine (TCM) and other traditional medicines, but pangolins are also occasionally killed for their meat or for ornamental use (Baiyewu et al., 2018). Pangolin trade came to the public's attention in 2020 due to its connection with COVID-19; the theory was that the new coronavirus originated in bats, but was spread to humans via pangolins (Raza et al. 2021). Interactions between animals and humans are now the source of as many as 75% of new diseases (CDC 2021a) (see Section 5.4). In response to the pandemic, in June 2020 the Chinese government increased protections for their native pangolin species and made the use of pangolin scales for medicinal purposes illegal (WWF 2021). Although genetic analysis has since shed doubt on the theory that pangolins were the vector for COVID-19 (Runwall 2021), even

GLOBAL CHANGE
CONNECTION

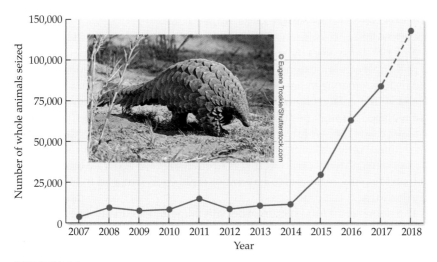

FIGURE 5.11 The ant-eating pangolin (eight species in the family Manidae) has become the most illegally trafficked animal in the world, with number of individuals seized at borders increasing dramatically beginning in 2014. The number for 2018 is estimated. (After UNODC 2021. *Wildlife Crime: Pangolin Scales, 2020.* United Nations Office on Drugs and Crime. Data: World WISE.)

a plausible connection between illegal wildlife trade and an international pandemic is a poignant example of the far-reaching importance of the work of biological conservation.

It is important to note that most countries are involved with the international black market in wildlife, with no single country accounting for more than 9% of all seizures (World Wildlife Report 2020). This is in part because illegal trade in wildlife links poor local people, corrupt customs officials, rogue dealers, and criminal gangs with wealthy buyers, often in other countries, who do not question the sources from which they buy. This trade has many of the same characteristics, the same practices, and sometimes the same players as the illegal trade in drugs and weapons, and it is extremely widespread and highly profitable. Confronting those who perpetuate such illegal activities has become a major and dangerous job for international law enforcement agencies.

Illegal harvesting of wildlife is often facilitated through mislabeling or use of ambiguous common names. However, scientific tools are increasingly being employed to assist with tracking. **DNA barcoding**, one of these tools, is a method of species identification using a designated section of a gene or genes. A study used this technology to determine whether endangered and/or prohibited species of sharks were being sold by fishmongers and in fish-and-chips takeaways in England, and it found that a species that is globally threatened and critically endangered in the northeast Atlantic, the spiny dogfish (*Squalus acanthias*), was the most commonly used fish in almost all takeaway products (Hobbs et al. 2019). Thus, this is yet another example of the application of genetic tools in conservation biology.

Commercial harvesting

Governments and industries often claim that they can avoid the overharvesting of wild species by applying modern scientific management. As part of this approach, an extensive body of literature has developed in wildlife and fisheries management and in forestry to describe the *maximum sustainable yield* (the greatest amount of a resource that can be harvested each year and replaced through population growth without detriment to the population). In many real-world situations, however, industry representatives and government officials managing commercial harvesting operations may lack the key biological information that is needed to make accurate calculations. Not surprisingly, attempts to harvest at high levels can lead to abrupt species declines.

The demand for food has an especially high impact on our oceans; 10%–12% of the world's people rely on fisheries and aquaculture for their livelihood, and 4.3 billion people rely on fish as a protein source (WWF 2018). Humans have caught almost 6 billion tons of seafood since 1950, including both fish and invertebrates. However, catch from the wild has been decreasing steadily since a peak in 1996, in large part because of the rapid increase in yield from aquaculture.

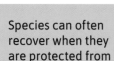

GLOBAL CHANGE
CONNECTION

For many marine species, direct exploitation is less important than the indirect effects of commercial fishing (Burgess et al. 2013). Many marine vertebrates and invertebrates are caught incidentally during fishing operations; most of these organisms, referred to as *bycatch*, are killed or injured in the process. One-third of a million tonnes of bycatch is dumped back into the sea yearly by US fisheries alone (Savoca et al. 2020). The declines of skates, rays, sharks, turtles, and seabirds continue to be linked to their wholesale death as bycatch; a recent study found that over 90 endangered species were threatened this way (Roberson et al. 2020). The huge number of sea turtles and dolphins killed by commercial fishing boats as bycatch resulted in a massive public outcry and led to the development of improved nets to reduce these accidental catches (see Chapter 1, "The interdisciplinary approach: A case study with sea turtles," for one success story). The development of improved nets and hooks, as well as other methods to reduce bycatch, is an active area of current fisheries research. Fortunately, there is good evidence that these approaches are working and bycatch is decreasing (Savoca et al. 2020).

> Species can often recover when they are protected from overexploitation.

One of the most heated debates over the harvesting of wild marine species has involved the hunting of whales. The debate is due in part to the strong emotional attachment to whales that many people in Western countries have. Several international agreements govern whaling, the most important of which is the International Whaling Commision (IWC), established in 1946 in response to the rapid declines of whales, to identify species that had been hunted to dangerously low levels. The IWC finally banned all commercial whaling in 1986. Many species have likely been saved from extinction as a result, including the gray whale (*Eschrichtius robustus*) and the humpback whale (*Megaptera novaeangliae*), which appear to have recovered (see chapter opening photo) (**TABLE 5.2**). However, certain species remain at densities far

TABLE 5.2	Worldwide Populations of Whale Species Harvested by Humans			
Species	Numbers prior to whaling[a]	Present numbers	Primary diet	IUCN status[b]
BALEEN WHALES				
Blue	200,000	5,000–15,000	Plankton	Endangered
Bowhead	56,000	10,000	Plankton	Least concern
Fin	475,000	100,000	Plankton	Vulnerable
Gray (Pacific stock)	23,000	15,000–22,000	Crustaceans	Least concer
Humpback	150,000	84,000	Plankton, fish	Least concern
Minke	140,000	200,000	Plankton, fish	Least concern
Northern right	Unknown	450	Plankton	Endangered
Sei	250,000	50,000	Plankton, fish	Endangered
Southern right	100,000	10,000	Plankton	Least concern
TOOTHED WHALES				
Beluga	Unknown	136,000	Fish, squid, crustaceans	Least concern
Narwhal	Unknown	123,000	Fish, squid, crustaceans	Least concern
Sperm	1,100,000	300,000–450,000	Fish, squid	Vulnerable

Sources: IUCN Red List of Threatened Species 2018, NOAA Fisheries 2019.

[a]Pre-exploitation population numbers are highly speculative. Genetic evidence suggests the populations may have been greater.

[b]Status is determined by a combination of numbers, threats, and trends. For example, numbers of southern right whales are low but are increasing.

below their original numbers despite many years of protection. Those species include the blue whale (*Balaenoptera musculus*) and the northern right whale (*Eubalaena glacialis*), even though they have had hunting banned by the IWC since 1967 and 1935, respectively.

The slow recovery of some whale species may be due to continued hunting, both legal and illegal. Despite the blanket IWC ban in 1986, Japan continued whaling for so-called scientific research. In 2019 Japan withdrew from the International Whaling Commission and resumed commercial whaling officially after a 30-year hiatus (BBC 2021). Countries such as Japan, Norway, and Iceland argue that whaling is a part of their culture and as such should continue, albeit sustainably. Japan primarily hunts the minke whale (*Balaenoptera* spp.) and the Bryde's whale (*Balaenoptera brydei*), which are not currently endangered. Most other maritime countries maintain that no whaling should be permitted.

As we will see in Chapter 10, projects linking the conservation of biodiversity to local economic development represent one possible approach to protect and manage the remaining individuals in an overharvested species. In some cases, this linkage may be made possible by acknowledging the sustainable harvesting of a natural resource with a special certification that allows producers to receive a higher price for their product. Certified timber products and seafoods are already entering the market, but it remains to be seen whether they will have a significant positive effect on biodiversity, particularly among increasingly affluent consumers in China and other Asian countries. It is also unknown whether the regulations associated with certification will be enforced in practice and will restrain the ongoing threats (Christian et al. 2013).

GLOBAL CHANGE CONNECTION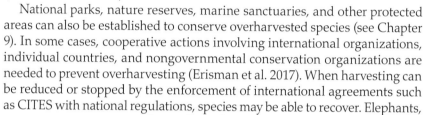

National parks, nature reserves, marine sanctuaries, and other protected areas can also be established to conserve overharvested species (see Chapter 9). In some cases, cooperative actions involving international organizations, individual countries, and nongovernmental conservation organizations are needed to prevent overharvesting (Erisman et al. 2017). When harvesting can be reduced or stopped by the enforcement of international agreements such as CITES with national regulations, species may be able to recover. Elephants, sea otters, sea turtles, seals, and certain whale species provide hopeful examples of species that have recovered—once overexploitation was stopped—in certain places in the world (Magera et al. 2013). A group that has seen marked improvement is birds; once the European Union banned trade in wild-caught birds in 2005, the number of birds traded decreased dramatically, from approximately 1.3 million per year to less than 200,000 (Reino et al. 2017). Historically, 66% of bird imports worldwide were by five EU countries.

> Certifying timber, seafood, and other products as sustainable may be a way to prevent overharvesting.

5.3 Invasive Species

LEARNING OBJECTIVES

By the end of this section you should be able to:

5.3.1 Provide examples for each of the sources of invasive species.

5.3.2 Contrast the risk and impacts of invasive species in Cuba versus Florida.

5.3.3 Evaluate the risk to ecosystems of genetically modified organisms (GMOs).

Species that become established and proliferate in new (i.e., nonhistorical) ranges where they cause environmental harm are considered **invasive**. The threat posed by invasive species is third behind habitat loss/degradation and overexploitation. Not unlike chemical pollution and other risk factors, invasive species have negative effects on ecosystems because the native organisms have not evolved adaptations to deal with the new conditions they impose. In most cases, the changes caused by an invasive species are novel because it has only recently been introduced to that ecosystem. However, native species can also become problematic, usually in response to human alterations to their ecosystem, such as the availability of human food waste supporting great increases in corvids (crows and ravens), which impact the rest of the ecosystem through their predation on other birds' nests (Carey et al. 2012). Across the world, fragmentation of forests, suburban development, and easy access to garbage have all allowed the numbers of coyotes, red foxes, and certain gull species in the US, to increase and sometimes negatively impact the ecosystem.

Invasive species and disease are most commonly listed as threats to individual species of reptiles, amphibians, and mammals, although birds and plants are also frequently affected, particularly on islands (WWF 2018). The effects of invasive species have been estimated to cost countries from 1.4% to 12% of their gross domestic product (GDP), amounting to billions of dollars per year (Marbuah et al. 2014; Pysêk et al. 2020). Aquatic invasive species alone have cost the global economy US$345 billion (Cuthbert et al. 2021).

GLOBAL CHANGE
CONNECTION

Species invasions have occurred by a variety of means:

- *European colonization*   European settlers arriving at new colonies released hundreds of European bird and mammal species into places like New Zealand, Australia, North America, and South Africa to make the countryside seem familiar and to provide game for hunting. Numerous species of fish (e.g., trout, bass, carp) have been widely released to provide food and recreation. In some cases, food sources introduced to support game species have disrupted food webs and led to declines in populations of game species.

- *Agriculture, horticulture, aquaculture*   Large numbers of plant species have been introduced and grown as crops, ornamentals, pasture grasses, or soil stabilizers. Many of these species have escaped from cultivation and have become established in local ecosystems. As aquaculture develops and the ornamental fish industry expands, there is a constant danger of more plant and animal species escaping and becoming invasive in marine and freshwater environments (Xu et al. 2014).

- *Accidental transport*   Species are often transported unintentionally. For example, weed seeds are accidentally harvested with commercial seeds and sown in new localities; rats, snakes, and insects stow away aboard ships and airplanes; and disease-causing microbes, parasitic organisms, and insects travel along with their host species, particularly in the leaves and roots of plants and the soil of potted plants (Hulme et al. 2018). Around 70% of the nonnative forest pest insects in the United States arrived on imported living plants. Seeds, insects, and microorganisms on shoes, clothing, and luggage can be transported across the world in a few days by people traveling by plane. Ships frequently carry organisms in their ballast tanks, which can hold up to 150,000 tons of water, releasing vast numbers of bacteria, viruses, algae, invertebrates, and small fish into new locations. Governments have developed regulations to reduce the transport of species in ballast water, such as requiring ships to exchange their ballast water 320 km (200 miles) offshore in deep water before approaching a port.

- *Biological control* When a nonnative species becomes invasive, one solution is to release an animal species from its original range that will consume the pest and hopefully control its numbers. In the past, this was done without much thought given to impacts on other species, leading to cases in which a biological control agent itself has become invasive, attacking native species along with (or instead of) the intended target species. For example, stoats (*Mustela erminea*, a type of weasel) were released in New Zealand to help control European rabbits but then became a significant threat to ground-dwelling birds. To minimize the risk of such effects, species being considered as biological control agents are now extensively tested before release to determine whether they will restrict their feeding to the intended target species.

Threats posed by invasive species

The negative effects of invasive species can be both direct and indirect. Abundant research has documented these effects for many types of organisms and has shown how effects may cascade through an ecosystem. When invasive plant species dominate a community, the diversity and abundance of native plant species and the insects that feed on them may show a corresponding decline. Evidence also indicates that invasive plants can even change the community of soil microbes (Ramirez et al. 2019). Perhaps because of their position in the trophic cascade, invasive plants and algae are consistently found to have some of the greatest negative impacts on biodiversity, followed by terrestrial arthropods and vertebrates (Mollot et al. 2017). Mechanisms whereby invasive species cause harm include the following:

GLOBAL CHANGE
CONNECTION

- *Competition for resources* Through their dramatic expansion, invasive organisms may use up or block access to resources needed by native organisms, including food, space, light, and even mates. Nonnative plants can compete with native plants, reducing their growth or excluding them altogether (**FIGURE 5.12**). For example, in the United Kingdom, invasive pontic rhododendron (*Rhododendron ponticum*) grows into thick bushes that smother other plant species and help spread sudden oak death (a plant disease); the leaves are also toxic to many native animals. Invasive plants can even compete for pollinators or overwhelm native plants with nonnative pollen (Beans and Roach 2015).

- *Predation and parasitism* Extinction risk from invasion is often associated with predation. Introduced rats are a primary management concern on many islands, as are snakes and other types of predators. American bullfrogs (*Lithobates catesbeianus*) are voracious predators and invasive in many parts of the world; a study in

(A)

(B)

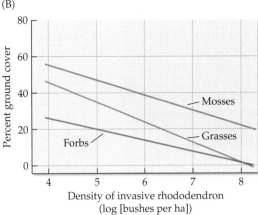

FIGURE 5.12 Some invasive plants competitively exclude nearly all other species. (A) Pontic rhododendron (*Rhododendron ponticum*) is native to southern Europe and southwest Asia but has become invasive in the United Kingdom, particularly in the oak woodland of western Scotland, an ecosystem with high conservation importance. (B) Researchers found that with increasing density of rhododendron, grasses, forbs (broad-leaved plants), and mosses all decreased. (B, after J. E. Maclean et al. 2018. *J Appl Ecol* 55(2): 874–884. © 2017 The Authors. *Journal of Applied Ecology.* © 2017 British Ecological Society.)

Argentina found them to consume 40 different prey taxa, including several species of native frogs and crustaceans (Quiroga et al. 2015).

- *Changes in ecosystem processes* Both cheatgrass (*Bromus tectorum*) and tamarisk (*Tamarix* spp.), two of the most pervasive weeds in the western United States, indirectly kill native plants by providing fuel that promotes wildfires (Drus et al. 2013). Soil microbial communities are also different under tamarisk trees, lacking important symbionts for native plants (Meinhardt and Gehring 2013). The presence of domesticated cats and dogs has been shown to affect a wide range of ecological processes, including bird migration patterns, seed dispersal, and breeding and parental behaviors (Terraube and Bretagnolle 2018).

- *Alteration of abiotic conditions* Some invasive plants have been found to alter soil properties, including moisture, acidity, and enzymes, which can facilitate more invasions or otherwise harm native species (Kuebbing et al. 2014). Both invasive plants and animals can alter the structural environment (Gutiérrez and Bernstein 2014): animals can do this through behaviors such as burrowing, and plants because of their particular growth forms, which may differ in shape or density from those of natives. Such changes to the environment may then affect ecosystem processes.

Invasive species may have more than one of these effects on an ecosystem at once, particularly those species that are in the middle of the food chain, as insects and other invertebrates usually are. Insects introduced both

Invasive species may displace native species through competition for limiting resources, prey on native species to the point of extinction, or alter the habitat so that native species can no longer persist.

GLOBAL CHANGE
CONNECTION

deliberately, such as European honeybees (*Apis mellifera*), and accidentally, such as fire ants (*Solenopsis invicta*) and gypsy moths (*Lymantria dispar*), can build up huge populations, both competing with native animals and preying on native plants and animals.

Invasive species on oceanic islands

The isolation of oceanic island habitats encourages the evolution of a unique assemblage of endemic species (see Chapter 6), but it also leaves those species particularly vulnerable to depredations by invading species. Many plants that grow on islands with few herbivores do not produce the bad-tasting, tough vegetative tissue that discourages herbivores, nor do they have the ability to resprout rapidly following damage. Some island birds that have evolved in the absence of predators have lost the power of flight and simply build their nests on the ground. Such species often succumb rapidly when invasive herbivores or predators are introduced. Mammals and other vertebrates introduced to islands often prey efficiently on endemic animal species and have grazed some native plant species to extinction (Russell et al. 2017). It has been estimated that 41%–75% of endangered or threatened species of birds, mammals, and aquatic organisms found on islands could be saved from extinction by eradication of invasive mammals, especially rodents (Spatz et al. 2017). Moreover, island species often have no natural immunity to mainland diseases. For example, when nonnative domesticated birds (e.g., chickens, ducks) are brought to islands by people, they frequently carry pathogens or parasites that, though relatively harmless to the carriers, devastate the native populations (e.g., wild birds).

The introduction of just one nonnative species to an island may cause the local extinction of numerous native species. The brown tree snake (*Boiga irregularis*), accidentally introduced onto a number of Pacific islands, is devastating endemic bird populations. The snake eats eggs, nestlings, and adult birds. On Guam alone, the brown tree snake has driven 10 of 13 forest bird species to extinction (Perry and Vice 2009). The government spends $3.4 million per year on attempts to control the brown tree snake population, including aerial drops of dead mice injected with acetaminophen (a poison for snakes); after decades of effort, their work is finally showing positive results (Siers et al. 2018, Joshua 2020).

A quarter of the native bird species of New Zealand have gone extinct since the arrival of humans and the resulting introduction of many invasive species to which the endemic birds (and other native organisms) were not adapted (Russell et al. 2015). New Zealand's national symbol, the kiwi (*Apteryx* spp.), is an endemic flightless bird that did not evolve with mammalian predators such as stoats (mentioned previously), domestic dogs (*Canis lupus familiaris*), and common brushtail possums (*Trichosurus vulpecula*). It is estimated that New Zealand is losing 2% of the kiwi population each year to these predators. Possums were introduced to establish a fur trade but now represent a particularly significant threat to many native species (**FIGURE 5.13A**). They prey on birds' eggs and chicks and destroy nesting habitat, while also competing with

FIGURE 5.13 (A) The common brushtail possum was introduced from Australia to New Zealand in 1837 and has since become a significant threat to native species. The possums are omnivores that prey on bird eggs and invertebrates and compete with native species for food. (B) A new method of control is being employed that instantly kills the possum when, attracted by a scent bait, it inserts its head looking for food. The benefit of this type of trap is that it is more humane than poison plus is self-resetting, able to discharge several times.

(A)

Greg Schechter/CC BY 2.0

(B)

Courtesy of Anna Sher

native birds and reptiles for leaves, berries, and nectar. Possums also eat endangered invertebrates such as the weta (a group of 70 endemic, large cricket-like insects). The Predator Free 2050 program of the New Zealand Department of Conservation *Te Papa Atawhai* was launched with the goal of extirpating all invasive mammals from the country, beginning with its smallest islands. Efforts have already been successful in increasing populations of birds, including the iconic kiwi, in some areas.

Control methods are being developed to be increasingly humane, practical, and cost-effective (**FIGURE 5.13B**). For example, in the Galápagos Islands, drones are being used to distribute pesticides to remote areas, and in New Zealand new traps are being used that reset automatically and spare caught animals from unnecessary suffering.

Invasive species in aquatic habitats

The diversity and abundance of both freshwater and saltwater species is frequently found to decline in association with aquatic invaders, which include plants, fishes, and invertebrates (Gallardo et al. 2016). Freshwater ecosystems are somewhat similar to oceanic islands in that they are isolated habitats and are thus at increased risk of invasion. There has been a long history of introductions of nonnative commercial and sport fishes into lakes, such as the introduction of the Nile perch (*Lates niloticus*) into Lake Victoria in East Africa, which was followed by the extinction of numerous endemic cichlid fishes. Often the introduced nonnative fishes are larger and more aggressive than the native fish fauna, and they may eventually drive the local fishes to extinction. But once these invasive species are removed from aquatic habitats, the native species are sometimes able to recover (Love et al. 2018).

Aggressive aquatic invaders also include plants and invertebrates. One of the most alarming invasions in North America was the arrival in 1988 of the zebra mussel (*Dreissena polymorpha*) in the Great Lakes. This small, striped native of the Caspian Sea apparently was a stowaway in the ballast tanks of a European cargo ship. Within two years, zebra mussels had reached almost unbelievable densities of 700,000 individuals per square meter in parts of Lake Erie, encrusting every hard surface and choking out native mussel species in the process (**FIGURE 5.14A**). As they continue to spread throughout the waters of the United States, these nonnative molluscs are causing

GLOBAL CHANGE
CONNECTION

enormous economic damage, estimated at $3.5 billion per year to fisheries, dams, power plants, water treatment facilities, and boats and devastating the aquatic communities of the Great Lakes (JOCI 2017). Both the zebra mussel and, more recently, the quagga mussel (*Dreissena rostriformis bugensis*) have become significant problems not only in the United States but in Europe as well (**FIGURE 5.14B**).

One-third of the worst invasive species in aquatic environments are aquarium and other ornamental species. Notable examples include water hyacinth (*Eichhornia*

(A)

(B)

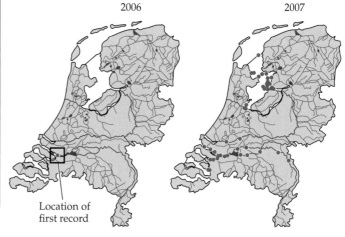

Location of first record

FIGURE 5.14 Invasive mollusks are a significant problem in many lakes. (A) For example, zebra mussels (*Dreissena polymorpha*) spread very quickly, covering native species and clogging pipes. They are dispersed in part by attaching themselves to boats. (B) These maps show the rapid spread of a similar species, the Eastern European quagga mussel (*Dreissena rostriformis bugensis*) in the Netherlands since it was first detected in the Rhine-Meuse estuary in 2006. In Europe, the quagga mussel has traveled between 23 and 383 km/y (mean 120 ± 53.8 km/y). (B, after J. Matthews et al. 2014. *Biol Invasions* 16: 23. https://doi.org/10.1007/s10530-013-0498-8.)

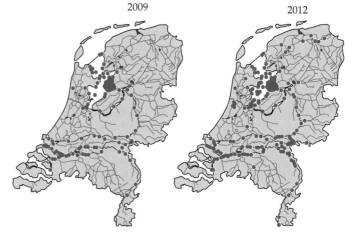

crassipes), Eurasian water chestnut (*Trapa natans*), and other ornamental plants that create huge floating mats (**FIGURE 5.15**). These mats deprive submerged plants of light; use up oxygen in the water, thereby killing fish and turtles; are barriers to animals that fish; and even act as vectors of disease by creating habitat for mosquitoes that carry malaria and for a snail that hosts a flatworm that causes disease in humans. Another aquatic invader, *Caulerpa taxifolia*, is a species of marine algae that has blanketed the seafloor, covering native algae, sea grass, and sessile animals in the Mediterranean Sea. Both water hyacinth and *Caulerpa* are listed by the IUCN as among the world's top 100 worst invasive species.

The ability of species to become invasive

The great majority of introduced species do not survive outside of their native ranges, and of those that do, only a

FIGURE 5.15 Every year volunteers remove Eurasian water chestnut, an invasive floating plant, from the Mystic River in northeastern Massachusetts. Floating invasive plants like this one create thick mats that exclude native aquatic plants that provide habitat and food for native fish and other animals. They are also bad for animals that require open water to forage, such as eagles. This effort has been successful in removing thousands of "baskets" of the weed annually since 2010, with good results.

small fraction (perhaps less than 1%) are capable of increasing, spreading, and becoming invasive in their new locations. Why are certain species able to invade and dominate new habitats and displace native species so easily? These species may be better suited to taking advantage of disturbed conditions than native species (Vitousek et al. 2017), particularly if the disturbance is new to that system and alters the way physical or chemical resources are made available. Human activity that causes disturbances may create unusual environmental conditions, such as higher or lower mineral nutrient levels, increased or decreased incidence of fire, or enhanced or lowered light availability, to which nonnative species are sometimes better adapted than native species. In fact, the highest concentrations of invasive species are often found in those habitats that have been most altered by human activity. Many of the threats to biodiversity mentioned in this and Chapter 4 can cause or exacerbate invasion; for example, ocean acidification was found to increase predation by invasive snails on native oysters by 48% (Sanford et al. 2014). When habitats are altered by global climate change, they become even more vulnerable to invasion (Merow et al. 2017).

GLOBAL CHANGE
CONNECTION

Another explanation for why some introduced species become invasive is the **predator release hypothesis**, which attributes their rapid proliferation in their new habitat to the absence of the specialized natural predators and parasites that would otherwise control the invaders' population growth (Davis 2009). In Australia, for example, introduced rabbits spread uncontrollably, grazing native plants to the point of extinction, because there were no effective checks on their numbers. Australian control efforts have focused in part on introducing specific diseases as a biological control for rabbit populations.

In one of the key generalizations of this field, the species that are most likely to become invasive and have significant effects in a new location are those species that have already been shown to do so somewhere else or, similarly, have close relatives that are invasive. However, unique problems arise when an introduced species instead has close relatives in the native biota. When invasive species hybridize with the native species and varieties, unique genotypes may be eliminated from local populations, and taxonomic boundaries may become obscured (see Chapter 2)—a process called *genetic swamping*. This appears to be the fate of native trout species when confronted by introduced species. In the southwestern United States, the Apache trout (*Oncorhynchus apache*) has had its range reduced by habitat destruction and competition with introduced species. The species has also hybridized extensively with rainbow trout (*O. mykiss*), an introduced sportfish, blurring its identity as a distinct species.

Control of invasive species

Invasive species often thrive where human activities have changed the environment.

While the effects of habitat degradation, fragmentation, and pollution can potentially be corrected and reversed in a matter of years or decades as long as the original species are present, well-established invasive species may be impossible to remove from ecosystems. They may have built up such large numbers and become so widely dispersed and so thoroughly integrated into an ecosystem that eliminating them may be extraordinarily difficult and expensive (Kopf et al. 2017).

The threats posed by invasive species are so severe that reducing the rate of their introduction has become a priority for conservation efforts in many areas. Actions have included passing and enforcing of laws, including customs restrictions prohibiting the transport and introduction of nonnative species. In the United States, the National Invasive Species Council, a body that helps shape legislation, reports directly to the president. The formation of the European Union facilitated the enactment of broad legislation; control of alien invasive species is Target 5 of the EU 2020 Biodiversity Strategy. Down under, the Australian Weeds Strategy 2017 to 2027 provides a national framework for a country that has had a long history of invasive species problems and consequently is at the forefront of prevention and control. Conservation biologists play a key role in advising these efforts. Often the work of these groups is to require restrictions and inspections related to the movement of soil, wood, plants, animals, and/or other items across international borders and even through checkpoints within countries.

A variety of strategies for controlling and eliminating invasive species includes three general approaches to their removal: use of pesticides (chemical poisons), mechanical culling (physical removal, such as pulling weeds and trapping or shooting animals), and biological control (use of one organism to control another). Each of these approaches, used alone or in combination with the others, has both benefits and limitations. In many cases, use of pesticides is the most effective means of controlling undesirable species, but the chemicals can be expensive and may hurt nontarget, desirable species. Mechanical approaches usually have the fewest nontarget effects but are the most labor-intensive. Biological control, such as use of a specialist insect or pathogen, can be the cheapest approach to managing invaders and is usually the gentlest on the ecosystem as a whole, but biological controls will not usually eradicate the species in question and may take time to show any effect.

> Governments must act to prevent the introduction of new invasive species, to monitor the arrival and spread of invasives, and to eradicate new populations of invasives.

An extensive public education program is often necessary so that people are aware of why invasive species need to be removed or killed, especially when they are charismatic mammals such as cats, dogs, and rabbits (Crowley et al. 2017). An emphasis on "compassion for the ecosystem" is one approach that has been suggested to help explain why lethal controls are sometimes warranted (Russell et al. 2016).

Habitat manipulation, changing agricultural practices, and the use of crop strains that are more resistant to pests are other methods frequently employed to prevent invasions or lessen their effects. Using a combination of these techniques for the long-term management of pests is called **integrated pest management**.

Currently, vast sums are spent controlling widespread outbreaks of invasives, but inexpensive, prompt control and eradication efforts at the time of first sighting can stop a species from getting established in the first place (Epanchin-Niell 2017). Training citizens and protected-areas staff to monitor vulnerable habitats for the appearance of known invasive species and then promptly implementing intensive control efforts can also be an effective way to stop the establishment and early spread of a potential new invader. Effective prevention may require a cooperative effort on the part of multiple levels of government and private landowners.

Even though the effects of invasive species are generally considered to be negative, they may provide some benefits as well, especially when the habitat is so degraded that the original species are unlikely to reestablish themselves, even once the invasive species are removed. Invasive plant species can sometimes stabilize eroding lands, provide nectar for native insects, and supply nesting sites for birds and mammals (e.g., Bateman et al. 2013; Shackelford et al. 2013). In such situations, the trade-offs need to be evaluated to determine whether the potential benefits of removal will outweigh the overall costs. Increasingly, biologists are recognizing the value of accepting "novel ecosystems" (see Chapter 11) in which a mixture of native and nonnative species is best suited to the new conditions created by human activities (Hobbs et al. 2013).

GMOs and conservation biology

A special topic of concern for conservation biologists is the increasing use of **genetically modified organisms** (**GMOs**) in agriculture, forestry, aquaculture, and toxic waste cleanup. GMOs are organisms whose genetic codes have been changed through recombinant DNA technology; scientists have added genes from different ("source") species. Such gene transfers can be done not only across species but across taxonomic domains, as when a bacterial gene that produces a chemical toxic to insects is transferred into a crop species such as corn. Enormous amounts of cropland—especially in the United States,

Argentina, China, and Canada—have already been planted with GMOs, mainly soybeans, corn (maize), cotton, and oilseed rape (canola). Genetically modified animals are under development, with salmon and pigs showing commercial potential. In 2017 the first GMO salmon was sold by an American company to Canadian consumers (Waltz 2017).

Humans have been genetically modifying domesticated crop and animal species since the dawn of civilization by means of selective breeding, hybridization, and other forms of artificial selection (Chaurasia et al. 2020). However, many species being investigated as potential sources of transferable genes, including viruses, bacteria, insects, fungi, and shellfish, have not previously been used in breeding programs, and this has led to concerns about the risks of GMOs, especially in Europe (Rose et al. 2020). The primary risk cited is that genetically modified crop species will hybridize with related species, leading to invasion by new, aggressive weeds and diseases (Bagla 2010). For example, the development of crops that are made resistant to particular viruses by encoding virus DNA into the plants could lead to new viral diseases. Further, there is concern that eating food from GMO crops may result in dangers to human health, such as new allergic reactions or the development of antibiotic-resistant gut flora (Phillips 2008). For all these reasons, we might apply the precautionary principle (see Chapter 3), limiting use of GMOs until there are strong assurances of their safety (Karalis et al. 2020). However, others argue that this is a misapplication of the precautionary principle, particularly in the face of dire human needs that could be addressed using GMOs.

GMOs have the potential to increase crop production to feed a growing human population, to produce new and cheaper medicines, and to reduce the use of pesticides on agricultural fields and the pesticide pollution associated with such use (Gbashi et al. 2021; Tsatsakis et al. 2017). However, a primary driver of the development of GMOs has been economic. For example, one hugely popular GMO—the Roundup Ready soybean—has been genetically engineered by the manufacturer of the herbicide glyphosate, commonly sold as the weed killer Roundup, so that the crop can be treated with more—not less—Roundup. There is no consensus on whether Roundup itself is dangerous for humans. However, the negative impacts of commercial herbicides on wild species are well documented (e.g., Lanzarin et al. 2020; Santovito et al. 2020), making their increased use the most likely negative outcome for this type of genetic engineering. Crops are also known to hybridize with closely related species, and there is evidence of engineered genes "escaping" in the case of corn (Castro Galvan et al. 2019). However, a combination of factors has decreased the profitability of using GMO crop varieties, thus decreasing their use (and consequent risk) in recent years (Teferra 2021).

In summary, the benefits of GMOs need to be examined and weighed against both known and unknown potential risks. As of the publication of this edition, there is no scientific consensus on risks, to either the environment or human health, other than that more research is needed (Karalis et al. 2020). The best approach involves proceeding cautiously, investigating GMOs thoroughly before commercial releases are authorized, and monitoring environmental and health effects after release.

5.4 Disease

LEARNING OBJECTIVES

By the end of this section you should be able to:

5.4.1 Name specific diseases that impact threatened species.

5.4.2 Discuss how disease interacts with other threats.

The increased transmission of disease as a result of human activities is a major threat to many endangered species and ecosystems. Pathogens (disease-causing organisms) such as bacteria, viruses, fungi, and protists can have major effects on vulnerable species and even the structure of entire ecosystems. Human activities can lead to increased populations of many pathogens, leading to outbreaks of both animal and human disease. The movement of viruses from wildlife or domesticated animals to humans is called **zoonosis**. For example, coronaviruses that originated in animals are responsible for at least seven different human diseases (CDC 2021b). In addition, interaction with humans and their domesticated animals exposes wild animals to dangerous, new diseases (Turcotte et al. 2017) (**FIGURE 5.16**).

Disease may be the single greatest threat to some rare species. The decline of numerous frog populations in visually pristine montane habitats across the

GLOBAL CHANGE
CONNECTION

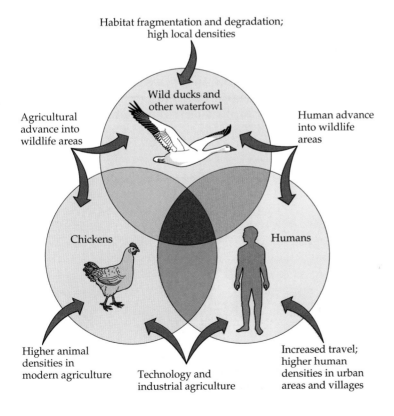

FIGURE 5.16 Infectious diseases (such as rabies, Lyme disease, influenza, bird flu, hantavirus, and canine distemper) spread among wildlife populations, domesticated animals, and humans as a result of increasing population densities and the advance of agriculture and human settlements into wild areas. This diagram illustrates the infection and transmission routes of bird flu, which is caused by a virus that infects wild waterfowl, chickens, and humans. The shaded areas of overlap indicate that the disease can be shared among the three groups. Single arrows indicate factors contributing to higher rates of infection; double arrows indicate factors contributing to the spread of disease among the three groups. (After P. Daszak et al. 2000. *Science* 287: 443–449.)

world is apparently due to the introduction of a nonnative fungal disease. Millions of sea stars ("starfish") of at least 20 species mysteriously died off along the west coast of North America from Mexico to Alaska from sea star wasting disease, caused by a virus eventually discovered through genetic analysis. The last population of black-footed ferrets (*Mustela nigripes*) known to exist on its own in the wild was destroyed by the canine distemper virus in 1987, though a few healthy individuals were caught for a captive breeding program (see Figure 8.2A). One of the main challenges of managing the captive breeding program has been protecting the captive ferrets from canine distemper, human viruses, and other diseases; this is being done through rigorous quarantine measures and subdivision of the captive colony into geographically separate groups. Fortunately, these conservation efforts have been successful; although still endangered, black-footed ferrets have wild populations once more.

> Increased incidence of infectious disease threatens wild and domesticated species as well as humans. Transfer of disease between species is a subject of special concern.

White-nose syndrome (WNS) is a fungal disease that is currently killing millions of bats across the eastern United States. In some caves, 90% of the bats have died. The disease is characterized by a powdery white fuzz on a bat's snout and other membranous areas (**FIGURE 5.17**). Bats die when the fungus causes skin irritation that wakes them early from hibernation in midwinter, when its diet of flying insects is not available, causing them to starve to death. Discovered in one cave in New York State in 2006, the disease has spread rapidly across the region, probably during bat migration. It is possible that cave explorers or bat researchers accidentally introduced the fungus to the United States as a contaminant on their clothes, boots, or equipment following a visit to a European bat cave. Currently, the only effective way to prevent its spread to new colonies is to close bat caves to all human visitors except for scientists who sterilize their clothes and equipment before entering, but scientists are actively looking for a cure. Comparative genomics has been used to discover that the fungus responsible for the disease is highly sensitive to ultraviolet light, pointing to a potential treatment for WNS (Palmer et al. 2018).

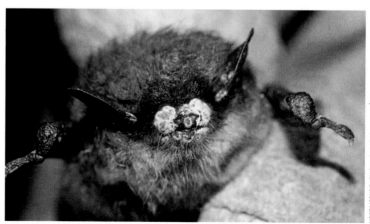

FIGURE 5.17 A little brown bat (*Myotis lucifugus*) infected with white-nose syndrome is being examined by a researcher. This fungal disease, which first appeared in the United States in 2006, affects a variety of bat species and has killed up to 90% of the bats in many caves.

Photograph courtesy of Ryan Von Linden/New York Department of Environmental Conservation

Three basic principles of epidemiology have obvious practical implications for limiting disease in the management of endangered species. First, a high rate of contact between host and pathogen or parasite is one factor that encourages the spread of disease. In general, as host population density increases, so does risk of disease. In addition, a high density of the infective stages of a parasite in the environment of the host population can lead to increased incidence of disease. In natural situations, the rate of infection is typically reduced when animals migrate away from their droppings, saliva, old skin, dead conspecifics, and other sources of infection. However, in unnaturally confined situations, such as habitat fragments, zoos, or even nature reserves, the animals remain in contact with these potential sources of infection, and disease transmission increases. Furthermore, at higher densities, animals have abnormally frequent contact with one another, and once one animal becomes infected, the parasite can rapidly spread throughout the entire population.

Second, indirect effects of habitat destruction can increase an organism's susceptibility to disease. When a host population is crowded into a smaller area because of habitat destruction, its members often face lowered habitat quality and food availability, lowered nutritional status, and less resistance to infection in addition to higher contact rates (Zylberberg et al. 2013). Furthermore, crowding can lead to social stress within a population, which also lowers the animals' resistance to disease. Pollution may also make individuals more susceptible to infection by pathogens, particularly in aquatic or extreme environments (Schmeller et al. 2018). It has even been observed that biodiversity regulates disease by diluting the number of suitable host species or by constraining the sizes of host populations through predation and competition (Morand 2018).

GLOBAL CHANGE
CONNECTION

Third, in many conservation areas, zoos, national parks, exotic food markets, and newly cleared agricultural areas, species come into contact with other species that they would rarely or never encounter in the wild—including humans and domesticated animals—so zoonosis infections such as rabies, influenza, distemper, hantavirus, and bird flu can spread from one species to another. Infectious diseases can spread between wildlife populations, domesticated animals, and humans as a result of increasing human population densities and the advance of agriculture and human settlements into wildlife areas. Several coronaviruses have arisen this way over at least the last 100 years, including not only COVID-19 but also SARS and MERS. HIV and the deadly Ebola virus also both appear to have spread from wildlife populations to humans and to domesticated animals. Such examples are likely to become more common as a result of anthropogenic changes to the environment, the increase in international travel, and globalization of the economy.

> Steps must be taken to prevent the spread of disease in captive animals and to ensure that new diseases are not accidentally introduced into wild populations.

GLOBAL CHANGE
CONNECTION

In zoos, colonies of animals are often caged together in small areas, and similar species are often housed close to one another. Consequently, if one animal becomes infected, the pathogen can spread rapidly to other animals and to related species. Once they are infected with an exotic disease, captive animals cannot be returned to the wild without threatening the entire wild population. Furthermore, a species that is fairly resistant to a disease can act as a reservoir for the disease, which can then infect populations of susceptible species. For example, apparently healthy African elephants can transmit a fatal herpesvirus to related Asian elephants when they are kept together in zoos. Diseases can spread very rapidly between captive species, especially when kept in crowded conditions.

Finally, because disease organisms are expected to respond to their environment as any species does, we must consider the impacts of global climate change. Such impacts can be direct—changing conditions may favor the reproduction of the disease agent itself, as when a warmer environment allows bacteria to thrive. The impacts can also be indirect—new conditions can weaken the host or its ability to fight disease. For example, warming and acidification of our oceans can increase incidence of disease in fish and other marine poikilotherms by making them more susceptible (Burge and Hershberger 2020). Furthermore, relationships between parasites and hosts that had reached evolutionary homeostasis may be disrupted, either by the death of parasites or alternatively by increases in their numbers or impact, causing disease (Byers 2021). We are already observing some of these effects in agricultural systems (Burdon and Zhan 2020).

GLOBAL CHANGE
CONNECTION

5.5 A Concluding Remark

This chapter has described threats to biodiversity that often operate at a continental or even global scale. This means that the solutions frequently must include international cooperation, discussed further in Chapter 12. Imperiled species often face multiple threats, operating at different scales. Threats can also be synergistic or one can cause another, such as climate change creating habitat fragmentation, or pollution making an organism more susceptible to disease. Furthermore, we can see how humans can also be harmed if such dangers are ignored and species are not protected. It falls upon the field of conservation biology to investigate the scale, impact, and interactions of these threats and to propose solutions for the benefit of all.

GLOBAL CHANGE
CONNECTION

In some cases, however, the science is not enough, particularly when an issue has been politicized. While most of the threats discussed in this chapter have experienced this to some degree, climate change in particular faces a rising resistance. Politization of an issue typically occurs when significant economic interests and power are at stake; thus, it is no surprise that there is an organized effort to push back on international agreements that limit use of fossil fuels to address climate change (Davenport and Lipton 2017). In an attempt to understand why people can so readily reject the science of climate change, psychologists, philosophers, sociologists, and economists have found that there can be innate responses to uncomfortable information that contribute to the problem (Haltinner and Sarathchandra 2018; Kerr and Wilson 2021; Linde 2020; Rekker 2021). This is a dangerous trend, as climate change ultimately affects all species, including humans. As will be discussed in more detail in Chapter 13, it is therefore also the role of conservation biologists to help educate the public and policymakers about the science behind the issues.

Summary

- Global climate change, including warmer temperatures and changing precipitation patterns, is already occurring because of the large amounts of carbon dioxide and other greenhouse gases produced by the burning of fossil fuels and deforestation. Predicted temperature increases may be so rapid in coming decades that many species will be unable to adjust their ranges and will become extinct.

- Overexploitation is driving many species to extinction and can consequently undermine entire ecosystems. Overexploitation is the result of increasingly efficient methods of harvesting and marketing, increasing demand for products, and increased access to remote areas.

- Humans have deliberately and accidentally moved thousands of species to new regions of the world. Some of these nonnative species have become invasive, greatly increasing their numbers at the expense of native species.

- Levels of disease often increase when animals are confined to nature reserves, zoos, or habitat fragments and cannot disperse over wide areas. In zoos and botanical gardens, diseases sometimes spread between related species of animals and plants. Diseases may also spread between domesticated species and wild species, and even between humans and both wild and domesticated animals.

- Species may be threatened by a combination of factors, all of which must be addressed in a comprehensive conservation plan.

For Discussion

1. Find a statement from the popular press that questions whether we should be concerned about global warming. What evidence would you use to critique their statement? Write a response using scientific information.

2. Research what is currently being done to protect a particular species threatened with overexploitation. Does it seem to be working? Why or why not? What else might be effective?

3. Learn about an invasive species or disease affecting biodiversity in your area. What is the extent of the threat? How is it managed?

Suggested Readings

Andersson, A. A., et al. 2021. CITES and beyond: Illuminating 20 years of global, legal wildlife trade. *Global Ecology and Conservation* 26: e01455. An analysis of CITES data demonstrating global patterns of trade in plants and animals.

González, E., et al. 2020. Combined effects of biological control of an invasive shrub and fluvial processes on riparian vegetation dynamics. *Biological Invasions* 22(7): 2339–2356. A look at the ecology of the invasion of riverside ecosystems, as well as the impact of the use of a biocontrol insect, using a data set from 7 years of monitoring.

Heighton, S. P., and P. Gaubert. 2021. A timely systematic review on pangolin research, commercialization, and popularization to identify knowledge gaps and produce conservation guidelines. *Biological Conservation* 256: 109042. What is known about the animal that is the most illegally traded and also possibly linked with the origin of COVID-19.

IPCC (Intergovernmental Panel on Climate Change). 2018. *Global Warming of 1.5°C.* https://www.ipcc.ch/sr15/. An IPCC special report on the impacts of global warming of 1.5°C above preindustrial levels and related global greenhouse gas emission pathways, in the context of strengthening the global response to the threat of climate change, sustainable development, and efforts to eradicate poverty.

LeDee, O. E., et al. 2021. Preparing wildlife for climate change: How far have we come? *Journal of Wildlife Management* 85(1): 7–16. A synthesis and analysis of 2306 management recommendations from 509 published papers on the impact of climate change.

NOAA (National Oceanic and Atmospheric Administration). 2021. Climate at a glance: Global time series. Retrieved from www.ncdc.noaa.gov/cag/global/time-series/globe/ocean/ytd/12/1880-2021. An online database and visualization tool that can be used to explore changes in temperature on Earth.

Pyšek, P., et al. 2020. Scientists' warning on invasive alien species. *Biological Reviews* 95(6): 1511–1534. A sober review of the dangers of invasive species by some of the leading voices in the field.

Zheng, B., et al. 2020. Satellite-based estimates of decline and rebound in China's CO_2 emissions during COVID-19 pandemic. *Science Advances* 6(49): eabd4998. An illustration of the relationship between human activity and carbon emissions.

The splendid poison frog (*Oophaga speciosa*) was endemic to tropical forests in Panama but was last seen in 1992. Like many frogs, the splendid poison frog was primarily threatened by the chitrid fungus. The species was declared extinct in 2020 by the IUCN.

Extinction Risk

6

I n the fall of 2021, 22 different species, including the ivory-billed woodpecker, were officially declared extinct by the US Fish and Wildlife Service (CNN 2021). These species had been receiving protections under federal law, but most were already perilously close to extinction when efforts to save them began. Why losses such as these are cause for concern and how we can identify species to protect before it is too late are the topics of this chapter.

The diversity of species found on the Earth has been increasing since life originated. The increase has not been steady; rather, it has been characterized by periods of high rates of speciation followed by periods of minimal change and episodes of mass extinction. Five episodes of mass extinction are apparent in the fossil record, occurring at intervals ranging from 60 million to 155 million years (**FIGURE 6.1**). These episodes, which ended the last five geological periods, could be called natural mass extinctions. The most famous is the extinction of the dinosaurs during the Late Cretaceous, 65 million years ago, after which mammals achieved dominance in terrestrial communities.

The most massive extinction took place 250 million years ago between the Permian and Triassic Periods, when about 95% of all marine animal species and half of all animal families are estimated to have gone extinct. Two papers published recently shed light on the likely causes (Beradelli and Niemczyk 2021). This research follows a large body of work that uses detection and analyses of elements in rocks and fossils to reveal the cascade of events that created the "Great Dying," as it is sometimes called. An eruption of volcanoes at that time produced a lava bed kilometers thick and as big as half of the United States; the emissions from this massive lava bed dramatically increased temperatures (Kaiho et al. 2021). This increase in temperature led to a cascade of changes that included widespread fire, erosion, and eutrophication.

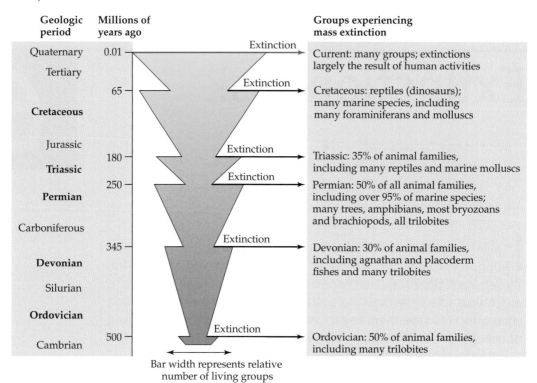

FIGURE 6.1 Relative numbers of animal families over geologic time. Although the total number of species groups on Earth has increased over the eons, a large percentage of these groups disappeared during each of five episodes of natural mass extinction (named in boldface at left). The most dramatic period of loss occurred about 250 million years ago, at the end of the Permian Period. A sixth mass extinction episode (red arrow at top of figure) began during the present geologic period and will continue for decades to come.

> The current rate of species loss is unprecedented, unique, and irreversible. Ninety-nine percent of current extinctions are caused in some way by human activities.

Marine shells from the Permian show that these gases, including CO_2, were also absorbed into the ocean, causing a dramatic increase in acidity (Jurikova et al. 2020). Most species could not tolerate the changes in environment; it took about 80–100 million years of evolution for Earth's biota to regain the number of families lost during the Permian extinction.

The linkages between greenhouse gas emissions and both increased temperatures and ocean acidification are the same that we see today (see Chapter 5). Although the levels of gases and temperatures reached were ultimately higher during the Permian-Triassic extinction episode, research reveals that the rate of change today is currently much greater: our industrial emissions are 10–20 times higher than the natural emissions that occurred at the end of the Permian (Jurikova et al. 2020). As we saw in Chapters 4 and 5, climate change is only one of several human-caused threats species face (see Figure 4.4). Thus, it is not surprising that we are witnessing so many extinctions today.

Extinction is as much a natural process as speciation is. One species may outcompete another for a vital resource, or predators may drive prey species to extinction. If extinction is a natural process, however, why is the current loss

GLOBAL CHANGE
CONNECTION

of species of such concern? The answer lies in the relative rates of extinction and speciation as well as in the causes of extinction. Speciation is typically a slow process, occurring through the gradual accumulation of mutations and shifts in allele frequencies over thousands, if not millions, of years. As long as the rate of speciation equals or exceeds the rate of extinction, biodiversity will remain constant or increase. In past geologic periods, the loss of existing species was eventually balanced and then exceeded by the evolution of new species. However, current extinction rates are much greater than speciation rates. Furthermore, even if an individual species is able to adapt to an environmental disruption, its ecosystem may still be at risk. A recent analysis of evolution in response to global warming found that even with rapid evolution, food webs will be disrupted because of the interdependence of organisms in an ecosystem (Yacine et al. 2021). With any anthropogenic impact, changes in one species can cause a trophic cascade of impacts, thus increasing the threat overall.

GLOBAL CHANGE
CONNECTION

One million species are now in danger of extinction, according to a recent report from the UN Intergovernmental Science-Policy Platform on Biodiversity and Ecosystem Services (IPBES), the most comprehensive biodiversity assessment to date (IPBES 2019). We are currently in the midst of a **sixth extinction episode**, caused by human activities rather than a natural disaster (Ceballos et al. 2015). Some geologists are referring to this current period of the Earth's history as the Anthropocene, from the Greek word for "human," but other scientists say that designating a new period is premature, as we have not yet arrived at a new stable state (WWF 2018).

Extinction is a top concern of conservation biology because the number of threatened species is so large. About 14% of the world's remaining bird species are currently threatened with extinction (IUCN 2021). **TABLE 6.1** shows certain animal groups for which the danger is particularly severe, including turtles, manatees, rhinoceroses, and penguins. The danger is even greater for other large groups: 26% of mammal species and 41% of amphibian species are threatened (IUCN 2021). Many plant species are also at risk, including 41% of gymnosperms. Most groups of plants, fungi, fishes, and insects are not well known, and for these groups the extinction risk cannot be accurately determined. But for those that have been adequately assessed, we can see that extinction risk is increasing over time (**FIGURE 6.2**). Corals are especially threatened, due to global warming (see Chapter 5).

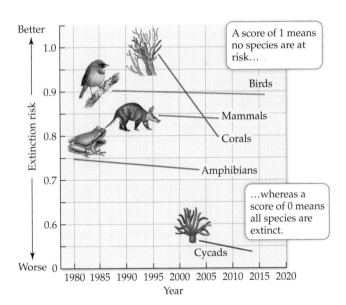

FIGURE 6.2 Using the IUCN database of known species, we can track extinction risk over time. It is currently available only for these five taxonomic groups, as these are the ones in which all species have been assessed at least twice. From these data, we can see that status of each of these major groups is declining, with some faster than others, especially corals. (After IUCN 2021. The IUCN Red List index. https://www.iucnredlist.org/assessment/red-list-index.)

TABLE 6.1	Numbers of Species Threatened with Extinction in Major Groups of Animals and Plants			
Group	Approximate number of species[a]	Percentage of known species evaluated	Number of species threatened with extinction in 2021	Percentage of species threatened with extinction[b]
VERTEBRATES				
Fishes	35,797	61%	3,210	Insufficient coverage
Amphibians	8,309	87%	2,442	41%
Reptiles	11,341	75%	1,458	Insufficient coverage
Birds	11,158	100%	1,481	14%
PLANTS				
Flowering plants	369,000	14%	20,883	Insufficient coverage
Gymnosperms (includes conifers, ginkgo, and cycads)	1,113	91%	403	41%
Ferns and related species	11,800	6%	266	Insufficient coverage
Mosses	21,925	1.3%	165	Insufficient coverage

Source: IUCN 2021. The IUCN Red List of Threatened Species. Version 2021. Retrieved May 2, 2021 from http://www.iucnredlist.org.
[a]Note that these numbers do not necessarily agree with other estimates, given the dynamic and complex nature of taxonomy (see Chapter 2).
[b]Estimate is calculated only for those species with sufficient data as of March 2021. For example, flowering plants have been only 14% evaluated; thus we cannot know what percentage of all flowering plant species is actually threatened.

6.1 The Meaning of "Extinct"

LEARNING OBJECTIVES

By the end of this section you should be able to:

6.1.1 Contrast the different types of extinction.

6.1.2 Provide evidence for why conservation of species on islands is a high priority.

The word *extinct* has many nuances, and its meaning can vary somewhat depending on the context:

- A species is considered **extinct** when, after a thorough search, no member of the species is found alive anywhere in the world; the dodo bird and the Tasmanian tiger-wolf (*Thylacinus cynocephalus*), for example, are extinct. More recently, the splendid poison frog was also declared extinct (see chapter opening photo).

- If individuals of a species remain alive only in captivity or in other human-controlled situations, the species is said to be **extinct in the wild**. The St. Helena ebony tree (*Trochetiopsis ebenus*) is nearly extinct in the wild, although it grows well under cultivation. The Spix's macaw, the blue parrot featured in the animated movies *Rio* and *Rio 2*, can be found in homes and zoos but was declared extinct in the wild in 2018.

- A species is **locally extinct**, or **extirpated**, when it is no longer found in a specific area it once inhabited but is still found elsewhere in the wild. For example, due to competition with invasive species, the native bee *Perdita meconis* is now extinct in Utah but still exists in other US states (Portman et al. 2018). This is in contrast to a species that no longer lives anywhere in the world, sometimes referred to as being **globally extinct**.

- Some conservation biologists speak of a species as being **functionally extinct** (or **ecologically extinct**) if it persists at such reduced numbers that its effects on the other species in its community are negligible. The baiji or Chinese river dolphin (*Lipotes vexillifer*) once found in the Yangtze River, for example, has not officially been designated as extinct, but no individuals could be found during an intensive survey in 2006 (Fisher and Blomberg 2012), and there have been only unconfirmed photographs of possible sightings since (**FIGURE 6.3**). Only 13 individuals were counted between 1997 and 1999; even at those numbers this species would no longer be playing a significant role in the river ecosystem.

© Mark Carwardine/Minden Pictures

FIGURE 6.3 The baiji or Chinese river dolphin (*Lipotes vexillifer*), nicknamed "the goddess of Yangtze," has had no confirmed sightings since 2006. An unconfirmed sighting in 2018 may mean that it is still extant, but functionally extinct. It was reduced to low numbers after industrialization, which brought hydroelectricity, increased transportation, and fishing. It is believed that the baiji has particularly suffered as the result of bycatch.

In addition to the various nuances of the term *extinct*, conservation biologists work with a variety of categories, including *endangered*, *vulnerable*, and *threatened*, that more specifically describe the status of species. These categories are discussed in detail in Chapter 7.

The current, human-caused mass extinction

GLOBAL CHANGE
CONNECTION

The global diversity of species reached an all-time high in the present geologic period. Many groups of organisms—such as insects, vertebrates, and flowering plants—reached their greatest diversity about 30,000 years ago. Since that time, however, species richness has slowly decreased as one species—*Homo sapiens*—has asserted its dominance. In our need to consume natural resources, humans have increasingly altered terrestrial and aquatic environments at the expense of other species.

The first noticeable effects of human activity on extinction rates can be seen in the elimination of large mammals from Australia and North and South America at the time humans first colonized these continents tens of thousands of years ago. After the arrival of these newcomers, approximately 70% of the *megafauna*—mammals weighing more than 44 kg (100 pounds)—became extinct, including the woolly mammoth (*Mammuthus primigenius*) and saber-toothed tiger (*Smilodon*) (Grayson 2016). Many of these extinctions were probably caused directly by hunting and indirectly by the burning and clearing of forests and grasslands and the introduction of invasive species and new diseases. For example, deliberate burning of savannas, presumably to encourage plant growth for browsing wildlife and thereby to improve hunting, has been occurring for 50,000 years in Africa. One analysis that compared estimated human populations with animal populations in North America found evidence that the extinctions of two out of five taxonomic groups were likely caused by humans (Broughton and Weitzel 2018). In South America, a recent study found the extinction of megafauna corresponded to the development of improved design of spear points (Prates and Perez 2021).

Island species have had higher rates of extinction than mainland species. Freshwater species are more vulnerable to extinction than marine species.

Extinction rates during the last 2000 years are best known for terrestrial vertebrates, especially birds and mammals, because these species are conspicuous and are therefore well studied. Extinction rates for the other 99% of the world's species are just rough guesses at present because scientists have not systematically searched the remote sites where many of these species occur, to determine whether they still exist. In fact, some species presumed to be extinct may actually be **extant** (still living), and vice versa. For example, the South Island takahē (*Porphyrio hochstetteri*), a goose-sized flightless bird in New Zealand, was presumed extinct for 40 years before it was dramatically rediscovered in 1948 (**FIGURE 6.4**). Intensive recovery efforts, including captive breeding (see Chapter 8) and closely monitored protected areas, have been successful in bringing the numbers from 280 in 2018 to 400 in 2021; this is an increase of nearly 32% in just three years! Its primary threat is introduced stoats (see Chapter 5).

It can sometimes be difficult to determine whether a species is truly extinct. A rare stick insect, *Dryococelus australis*, was driven to extinction on Lord Howe Island in Australia after the arrival of ship rats in the early 1900s. Eventually some similar-looking insects were found 20 km away, but they were not confirmed as the same species until 2017, when scientists used mitochondrial DNA to match them to museum specimens of *D. australis*, demonstrating that the species had experienced only a local extinction (Mikheyev et al. 2017).

The best available evidence so far indicates that 900 species have been documented to have gone extinct in the last 500 years, with 160 species being declared extinct in just the last decade (IUCN 2021), including the splendid poison frog (*Oophaga speciosa*) (see chapter opening photo). We also know that the majority of human-caused extinctions have occurred in the last 150 years.

FIGURE 6.4 The South Island takahē (*Porphyrio hochstetteri*) was believed to be extinct for at least 40 years but was rediscovered as extant. Captive breeding programs and protections from invasive predators have increased its numbers and range in New Zealand.

One trend to note is that the earliest extinctions documented by humans were on islands. In fact, 90% of known bird extinctions since 1500 have been on islands, even though greater than 80% of bird species occur on continents (Butchart et al. 2018). In contrast, extinctions of birds in mainland areas were first observed about 1800, but they have been increasing since then (**FIGURE 6.5**). In the future, mainland species will account for an increasing proportion of extinctions. Studying island extinctions has taught us much about population dynamics that is applicable to mainland conservation efforts.

Fortunately, the rate of species loss appears to have declined since 1950, in part due to deliberate efforts to protect rare species in danger of going extinct. These numbers can be misleading, however, because of the current practice of not declaring a species extinct until decades after any individuals of the species have been found. In coming years, numerous species will be declared to have gone extinct during the past half century. In the last decade, a number of species have not been found despite intensive searches and have subsequently been declared extinct, including the Monteverde golden toad of Costa Rica (*Bufo periglenes*) and the Alaotra grebe (*Tachybaptus rufolavatus*) of Madagascar. Many species not yet listed as extinct—and some species that have never been documented at all—have been reduced to such low numbers by human activities that their ability to persist is uncertain.

Many species today are represented only by scattered populations with reduced numbers of individuals. Although these isolated populations could persist for years or decades, the ultimate fate of such species is extinction.

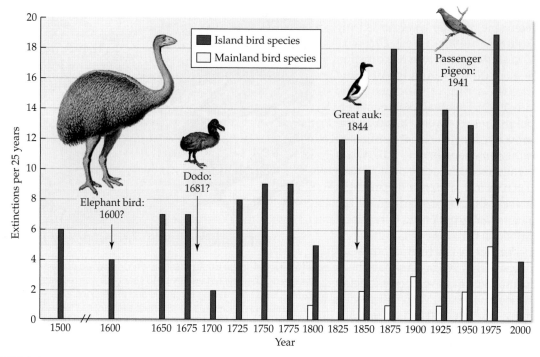

FIGURE 6.5 Rates of extinctions of bird species during 25-year intervals since 1500 on islands (dark blue) and on continents (light blue). Initially, extinctions were almost exclusively of island species, but extinctions of mainland species have increased since 1800. Shown are dates of famous bird extinctions, including the elephant bird of Madagascar (family Aepyornithidae), the dodo of Mauritius Island, the great auk of the Scottish Isles, and the passenger pigeon of North America. (After S. H. M. Butchart et al. 2006. *Bull Brit Orn Club* 126A: 7–24, with data updates from BirdLife International 2017. *We have lost over 150 bird species since 1500.* Downloaded from http://www.birdlife. org on 3/25/2019; Great Auk © Nicolas Primola/Shutterstock.com; Passenger Pigeon Tim Hough/CC BY-SA 3.0.)

Remaining individuals of species that are doomed to extinction have been called "zombie species" (Woodward 2019). There are many species in this category in the remaining fragments of species-rich tropical forests. The presumed eventual loss of species following habitat destruction and fragmentation is called the extinction debt (Tilman et al. 1994). Using data from the Argentine Dry Chaco, a deforestation hotspot, scientists calculated a delay of 10–25 years between land use change and species richness loss (Semper-Pascual et al. 2018). This means that there is a possibility to save such species from extinction, if action is taken before this window closes.

Local extinctions

In addition to the global extinctions that are a primary focus of conservation biology, many species are experiencing a series of local extinctions, or extirpations, across their range. Where habitats are degraded and destroyed, populations of plants and animals go extinct. For example, an analysis of 64 orchid species in the Czech Republic found that most had been locally extirpated from many of their sites at some point over the 120 years ana-lyzed; the number of occupied sites declined by 8%–92% (Štípková and Kindlmann 2021).

Concord, Massachusetts, was first assessed for wildflower species in the 1850s by the famous naturalist and philosopher Henry David Thoreau. Twen-ty-seven percent of the native species seen by Thoreau and other nineteenth-century Concord botanists could not be found when the area was surveyed 150 years later (Miller-Rushing and Primack 2008; Primack et al. 2009). A further 36% of the species now persist in only one or two populations and are therefore vulnerable to extinction. In some cases, only a few individual plants remain of species that were formerly common. Certain groups, such as orchids and lilies, have shown particularly severe losses. A combination of forest succession, invasive species, air and water pollution, grazing by deer, habitat destruction and fragmentation, and now climate change has contributed to these species' losses in Concord.

GLOBAL CHANGE
CONNECTION

> Species-rich tropical rain forests are being lost at a rate of 1% a year, a rate believed to result in the destruction of more than 13,500 biological populations each day. Population losses eventually result in species extinctions.

Local extinctions serve as important biological warning signs that something is wrong in the environment. Action is needed to prevent further local extinctions as well as global extinctions. The loss of local populations not only represents a loss of biodiversity but also diminishes the value of an area for nature enjoyment, scientific research, and the provision of crucial materials to local people in subsistence economies.

Extinction rates in aquatic environments

The oceans were once considered so enormous that it seemed unlikely that marine species could go extinct; many people still share this viewpoint. How-ever, as marine coastal waters become more polluted and species are harvested more intensely, even the vast oceans will not provide safety from extinction (McCauley et al. 2015). Many species of whales and large fishes have declined by 90% or more because of overharvesting and other human activities and are in danger of extinction. Many marine mammals are top predators, and their loss could have major effects on marine communities. Some marine species are the sole species of their genus, family, or even order, so the extinction of even a few of them could represent a serious loss to global biodiversity.

GLOBAL CHANGE
CONNECTION

Among vertebrates, the highest rates of extinction in the twentieth century have been recorded for freshwater fishes (WWF 2018). The modern extinction rate for North American freshwater fishes is conservatively estimated to be 877 times greater than the background extinction rate (Burkhead 2012). It is estimated that 30% of freshwater fish are threatened (Tickner et al. 2020). The fishes of California are particularly vulnerable because of the state's scarcity of water and intense development: in a recent study, it was determined that among 130 California freshwater fish evaluated, 99 (76%) were threatened to some extent, with 32 of these endangered (Leidy and Mole 2021). Large numbers of fishes and aquatic invertebrates, such as molluscs, are in danger of extinction because of dams, pollution, irrigation projects, overharvesting, invasive species, disease, and general habitat damage (**FIGURE 6.6**).

GLOBAL CHANGE
CONNECTION

(A)

Courtesy of NOAA

(B)

© Marc Tielemans/Alamy Stock Photo

FIGURE 6.6 (A) The endangered white abalone (*Haliotis sorenseni*), a type of mollusc, in its natural habitat on the coast of California. Several abalone species have reached dangerously low numbers due to overharvesting by humans, both for their muscular foot, which is a delicacy, and for their beautiful shells, from which jewelry is made. (B) Shown here is an antique abalone pin in the shape of the now-extinct huia bird from New Zealand. Abalones are also threatened by withering syndrome, a disease caused by the bacterium *Candidatus xenohaliotis californiensis*, for which a cure has yet to be found. High temperatures, such as those associated with climate change, have a synergistic effect on the disease. Abalone are also affected by ocean acidification (Swezey et al. 2020).

6.2 Measuring Extinction

LEARNING OBJECTIVES

By the end of this section you should be able to:

6.2.1 Calculate the number of species for an island with known constants, using the species-area formula.

6.2.2 Use the island biogeography model to determine the relative species richness of an island given its size and distance from the mainland.

6.2.3 Calculate extinction rate based on habitat loss rate.

Quantifying rates of extinction and likelihood of extinction is critical for understanding the magnitude of the problem and how best to address it. Calculating extinction rates requires information about many facets of organisms and their environments, including the extent to which humans are playing a role.

Background extinction rates

To better understand how calamitous the present extinction rates are, we can compare them with the natural extinction rates that would prevail regardless of human activity. What is the natural rate of extinction in the absence of human influence? Natural **background extinction rates** can be estimated by looking at the fossil record. On average, an individual species lasts about 1 million–10 million years before it goes extinct or evolves into a new species (Mace et al. 2005; Pimm and Jenkins 2005). Since there are perhaps 10 million species on the Earth today, we can predict that 1–10 of the world's species would be lost per year as a result of a natural extinction rate of 0.0001%–0.00001% per year. These estimates are derived from studies of wide-ranging marine animals, so they may be lower than natural extinction rates for species with narrower distributions, which are more vulnerable to habitat disturbance; however, they do appear to be applicable to terrestrial mammals. The current observed rate of extinction for birds and mammals is 1% per century (or 0.01% per year), which is 100–1000 times greater than would be predicted based on background rates of extinction. Putting it another way, about 100 species of birds and mammals were observed to go extinct between 1850 and 1950, but the natural rate of extinction would have predicted that, at most, only 1 species would have gone extinct. Therefore, the other 99 extinctions can be attributed to the effects of human activities.

Extinction rate predictions and the island biogeography model

Studies of island communities have led to general rules on the distribution of biodiversity, synthesized as the island biogeography model[1] by MacArthur and Wilson (1967). The **island biogeography model** was built to explain the species-area relationship: islands with large areas have more species than do islands with smaller areas (**FIGURE 6.7**). This rule makes intuitive sense because a large island will tend to have a greater variety of local environments and community types than a small island. In addition, large islands allow a larger number of separate populations per species and larger numbers of individuals in each population, increasing the likelihood of speciation and decreasing the probability of local extinction of newly evolved as well as recently arrived species.

The **species-area relationship** can be summarized by the empirical formula

$$S = CA^z$$

where S is the number of species on an island, A is the area of the island, and C and z are constants. The exponent z determines the slope of the relationship between species and area and C can be considered as the number of species when A=1. The values for C and z depend on the types of islands being compared (tropical vs. temperate, dry vs. wet, etc.) and the types of species involved (birds vs. reptiles, etc.). z values are typically about 0.25, with a range from 0.15 to 0.35. Island species with restricted ranges, such as reptiles and amphibians, tend to have z values near 0.35, while widespread mainland species tend to have z values closer to 0.15. Values of C tend to be high in

---

[1]A scientific model is a physical, conceptual, or mathematical representation of something observed in nature that can help us understand it and in some cases make predictions. Models are used to illustrate and explore specific aspects of real objects or phenomena and so are necessarily simplifications of reality.

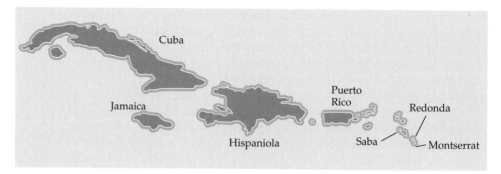

FIGURE 6.7 The number of species on an island can be predicted from the island's area. In this figure, the number of species of reptiles and amphibians is shown for seven islands in the West Indies. The numbers of species on large islands such as Cuba and Hispaniola far exceed those on the tiny islands of Saba and Redonda. (After R. H. MacArthur and E. O. Wilson. 1967. *The Theory of Island Biogeography.* Princeton University Press, Princeton, NJ.)

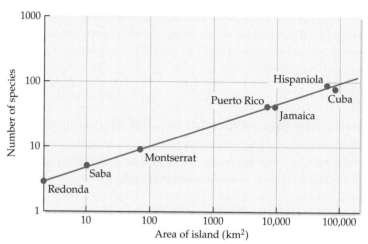

groups, such as insects, that contain many species, and they tend to be lower in groups, such as birds, that contain relatively few species (Connor and McCoy 2001).

Imagine the simplest situation, in which $C = 1$ and $z = 0.25$, for raptorial birds on a hypothetical archipelago:

$$S = (1)A^{0.25}$$

The formula predicts that islands of 10, 100, 1000, and 10,000 km^2 in area would have 2, 3, 6, and 10 species, respectively. It is important to note that a tenfold increase in island area does not result in a tenfold increase in the number of species; with this equation, each tenfold increase in island area increases the number of species by a factor of approximately 2.

The island biogeography model has since been empirically validated many times; for numerous groups of plants and animals, it has been found to describe the observed species richness reasonably well, explaining about half of the variation in numbers of species. For example, actual data from three Caribbean islands conform to this relationship: with increasing area, Nevis (93 km^2), Puerto Rico (8959 km^2), and Cuba (114,524 km^2) have 2, 10, and 57 species of

The island biogeography model can be used to predict how many species will go extinct due to habitat loss. The model can also be used to predict how many species will remain in protected areas of different sizes.

Anolis lizards, respectively; with a *C* of 0.5 and a *z* of 0.35, the islands would be predicted to have 2, 12, and 30 species, respectively. Note that *C* and *z* can differ based on taxa and by location. It was recently discovered that plant species on the Madeira and Canary Islands in the Atlantic Ocean, to the west of Africa and Europe, had a *z* = 0.30, whereas the nearby Azores Islands had much lower diversity overall and a *z* = 0.15 (Price et al. 2018).

In their classic text, MacArthur and Wilson (1967) also hypothesized that the number of species occurring on an island represents a dynamic equilibrium between colonization by (and evolution of) new species and extinctions of existing species. Starting with an unoccupied island, the number of species will increase over time, since more species will be arriving (or evolving) than will be going extinct, until the rates of extinction and immigration are balanced (**FIGURE 6.8**). Species establishment rates will be higher for large islands than for small islands because large islands represent larger targets

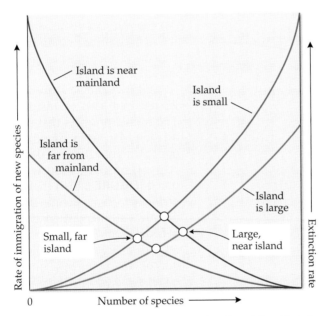

FIGURE 6.8 The island biogeography model describes the relationship between the rates of colonization and extinction in islands. The immigration rates (blue and red curves) on unoccupied islands are initially high, as species with good dispersal abilities rapidly take advantage of the available open habitats. The immigration rates slow as the number of species increases and sites become occupied. The extinction rates (green and gold curves) increase with the number of species on the island; the more species on an island, the greater the likelihood that a species will go extinct at any time interval. Colonization rates will be highest for islands near a mainland population source, since species can disperse over shorter distances more easily than longer ones. Extinction rates are highest on small islands, where both population sizes and habitat diversity are low. The number of species present on an island reaches equilibrium when the colonization rate equals the extinction rate (circles). The equilibrium number of species is greatest on large islands near the mainland and lowest on small islands far from the mainland. (After R. H. MacArthur and E. O. Wilson. 1963. *Evolution* 17: 373–387. © 1963 Society for the Study of Evolution.)

for dispersing organisms and are more likely to have open habitat suitable for colonization. Extinction rates will be lower on large islands than on small islands because large islands will have greater habitat diversity and greater numbers of populations. Furthermore, rates of immigration of new species will be higher for islands near the mainland than for islands farther away, since mainland species will be able to disperse to near islands more easily than to distant islands. The model predicts that for any group of organisms, such as birds or orchids, the number of species found on a large island near a continent will be greater than that on a small island far from a continent.

Extinction rates and habitat loss

The island biogeography model led to the development of a new field of study: **insular biogeography**. This is a subdiscipline of biogeography devoted to exploring species diversity in isolated natural communities. As we will see, the species-area relationship applies not only to oceanic islands, but also to mountaintops, desert oases, and any habitat fragmented by deforestation or other human disturbance. Therefore, it has become an important theoretical framework for protected areas, to be discussed in Chapter 9.

A reserve can be viewed as a "habitat island" in a "sea" of unsuitable habitat. The model predicts that when 50% of an island (or habitat island) is destroyed, approximately 10% of the species occurring on the island will be eliminated (**FIGURE 6.9**). If these species are **endemic** to that island—that is, if they occur there and nowhere else—they will become extinct. When 90% of the habitat is destroyed, 50% of the species will be lost; and when 99% of the habitat is gone, about 75% of the original species will be lost.

Predictions of extinction rates based on habitat loss vary considerably because each species-area relationship is unique. Using the conservative estimate that 1% of the world's rain forests are being destroyed each year, Wilson (1989) estimated that 0.2%–0.3% of all rain forest species will be lost per year. Assuming a total of 5 million rain forest species worldwide, that would result in a loss of 10,000 to 15,000 species per year, or 34 species per day. This estimate predicts that species extinctions by 2050 will be

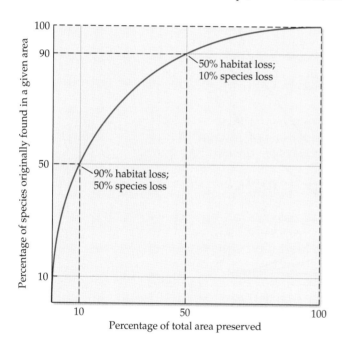

FIGURE 6.9 According to the island biogeography model, the number of species present in an area increases asymptotically—that is, it rises sharply and then levels off, as shown by the red curve in this example. The shape of the curve differs from region to region and among different species groups, but this model gives a general indication of the relationship between habitat loss and species loss. Here, if the area of habitat is reduced by 50%, then 10% of the species in the group will be expected to disappear; if the habitat is reduced by 90%, half the species will be lost. Stated in another way, a system of protected areas covering 10% of a country could be expected to include 50% of the country's species.

up to 35% in tropical Africa, 20% in tropical Asia, 15% in the Neotropics, and 8%–10% elsewhere. Extinction rates might in fact be higher because the highest rates of deforestation are occurring in countries with large concentrations of rare species and because large forest areas that remain are increasingly being fragmented by roads and development projects (Arima et al. 2016).

And yet, these grim predictions of high extinction rates do not seem to be coming true in the best-studied groups. For example, 1230 species of birds had been identified by 1900. However, only 13 bird species (1.1%) have since gone extinct (Pimm et al. 2014). In another example, only 17 plant taxa are known to have gone globally extinct in California, with another 30 that have gone locally extinct, together representing only 0.53% of California plant species (Rejmánek 2018). Similar extinction rates have been found in other Mediterranean climate countries in Europe (Croatia, France, Greece, Italy, Portugal, Slovenia, Spain), western Australia, and the Cape region of South Africa. The percentage is even lower in Chile. Given habitat losses in these regions, extinction rates would be expected to be much higher.

Extinction rates may be lower than predicted in these cases because threatened species in these taxa or locations are being given extra protection, and perhaps because hotspots particularly rich in endemic species are being targeted for conservation. It is also important to note that given how many species are yet to be discovered, we are surely losing species before we even know they exist. Lower extinction rates may also be observed because species that are committed to extinction are able to persist at low numbers for several generations. Regardless of which estimate gives the most reasonable prediction, all estimates indicate that tens of thousands—if not hundreds of thousands—of species are headed for extinction within the next 50 years (Pimm and Joppa 2015; Wagler 2021).

The time required for a given species to go extinct following a reduction in area or fragmentation of its range is a vital question in conservation biology, and the island biogeography model makes no prediction about how long it will take. Small populations of some species may persist for decades or even centuries in habitat fragments, even though they are committed to extinction (Ridding et al. 2021). Of the species that will eventually be lost, the best estimates predict that half will be lost in 50 years from a 1000 ha fragment, while half will be lost in 100 years from a 10,000 ha fragment (Brooks et al. 1999). In situations in which there is widespread habitat destruction followed by recovery, such as in New England and Puerto Rico over the last several centuries, species may be able to survive in small numbers in isolated fragments and then reoccupy adjacent recovering habitat. Even though 98% of the forests of eastern North America were cut down, the clearing took place in a patchwork fashion over hundreds of years, so forest always covered half of the area, providing refuges for mobile animal species such as birds.

This ability of species to persist for several generations and several decades in habitat fragments may also be why there have not been more observed species losses. This conclusion has two important implications for conservation. First, many species will go extinct in coming decades as their populations continue to decline in these fragmented habitats. And second,

GLOBAL CHANGE
CONNECTION

GLOBAL CHANGE
CONNECTION

the persistence of species in fragmented habitats provides a narrow window in which conservation actions have the potential to rescue declining species from extinction, as described in later chapters.

6.3 Vulnerability to Extinction

LEARNING OBJECTIVES

By the end of this section you should be able to:

6.3.1 List risk factors for extinction.

6.3.2 Determine whether a species has a high risk of extinction.

Populations (and ultimately species) that are declining in number are likely to go extinct unless the cause of decline is identified and corrected. As Charles Darwin pointed out more than 150 years ago in *On the Origin of Species* (1859):

> *To admit that species generally become rare before they become extinct, to feel no surprise at the rarity of the species, and yet to marvel greatly when the species ceases to exist, is much the same as to admit that sickness in the individual is the forerunner of death—to feel no surprise at sickness, but when the sick man dies, to wonder and to suspect that he died of some deed of violence.*

Past decline has been found to be the number one predictor of future decline (Di Marco et al. 2015). However, once a population is actively declining, it may be too late to prevent its extinction (see "The extinction vortex" later in this chapter). Thus, conservation biologists have sought to determine what features of species or populations might be predictive of future decline or extinction. Some groups of species clearly are more vulnerable to extinction than others. In Europe, a higher proportion of reptile species are threatened with extinction than plants, birds, mammals, or fishes (Dullinger et al. 2013). Ecologists have observed that across taxonomic lines, there are features that increase a species' risk of extinction, and through statistical modeling, they have identified those features that are most predictive of extinction (Purvis et al. 2000):

- *Narrow geographic range* This feature, which defines endemic species, has been found to be the most predictive of extinction or population decline (Di Marco et al. 2015; Ribeiro et al. 2016). This conclusion is intuitive: if the whole range is affected by human activity or a natural disaster, the species may become extinct. Bird species on oceanic islands and fish species confined to a single lake or watershed are good examples of species with limited ranges that are especially vulnerable to extinction. For example, many tropical bird species with narrow ranges will face increasing threats of extinction due to climate change in the coming decades (Gibson-Reinemer et al. 2015). Furthermore, a narrow range often (but not always) encompasses only one or a few populations or a small population size.

- *Only one or a few populations*   Any one population of a species may become extinct as a result of chance factors, such as earthquake, fire, an outbreak of disease, or human activity. Species with many populations are less vulnerable to extinction than are species with only one or a few populations. This feature is linked to the previous feature because species with few populations also tend to have narrow geographic ranges (**FIGURE 6.10**). Thus these "rare" species are at higher risk of extinction.

- *Small population size*   As we will see in Section 6.4, small populations are more likely to go locally extinct than large populations because of their greater vulnerability to loss of genetic diversity and to demographic and environmental variation. Species that characteristically have small population sizes, such as large predators and extreme specialists, are more likely to become extinct than species that typically have large populations. At the extreme are species whose numbers have declined to just a few individuals. A special category is species with a widely fluctuating population size in which the population is sometimes small.

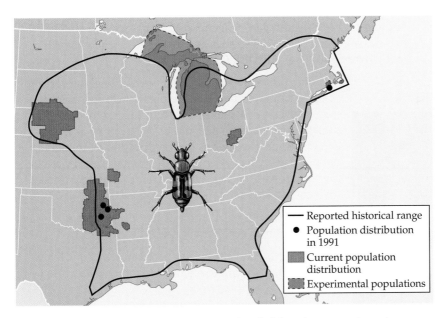

FIGURE 6.10   Although the American burying beetle (*Nicrophorus americanus*) was once widespread in the eastern and central United States (outlined in black), its range is now greatly reduced, which makes it vulnerable to extinction. When its recovery plan was published, it was found in the wild in only two separate areas in the United States: three populations in Oklahoma and one on Block Island in Rhode Island. There is now an experimental population in the Great Lakes region, in addition to seven other locations across 10 states. Intensive efforts have successfully increased its range and size of populations, such that in 2020 the US Fish and Wildlife Service changed its classification from endangered to threatened. (After ECOS, USFWS Environmental Conservation Online System, https://ecos. fws.gov/ecp0/profile/speciesProfile?spcode=I028; map © pingebat/Shutterstock.com.)

- *Island habitat*   As mentioned previously, the highest species extinction rates during historic times have occurred on islands. This is not surprising, given that species on islands often have limited range areas, small population sizes, and small numbers of populations and are more likely to be endemic. Of the terrestrial animal and plant species known to have gone extinct from 1600 to the present, almost half were island species, even though islands represent only a tiny fraction of the Earth's land surface. Island species usually have evolved and undergone speciation with a limited number of competitors, predators, and pathogens, which makes them particularly vulnerable to these threats if they are introduced. Species extinction rates peak soon after humans occupy an island and then decline after the most vulnerable species are eliminated (Wood et al. 2017). The loss of those species can result in an *extinction cascade* as coadapted species are affected, such as plants losing their pollinators or seed dispersers. In general, the longer an island has been occupied by people, the greater its percentage of extinct and endangered biota. In Madagascar, 75% of the 2000 tree species are endemic, with 782 threatened with extinction, according to an assessment completed by the IUCN (IUCN/SSC 2019). As another example, lemurs are endemic to Madagascar, and the IUCN has classified 103 of the 107 species as threatened or endangered (Schwitzer et al. 2014) (**FIGURE 6.11**).

GLOBAL CHANGE CONNECTION

- *Hunting or harvesting by people*   Overharvesting can rapidly reduce the population size of a species (see Chapter 5). If hunting and harvesting are not regulated, either by law or by local customs, the species can be driven to extinction. Utility has often been the prelude to extinction, as was the case with the dodo, now a symbol of modern extinctions (see Figure 6.5).

Some additional features of species or populations have been found to put them at risk:

- *Large home range*   A species in which individuals or social groups need to forage over wide areas is prone to die off when part of its range is damaged or fragmented by human activity.

© iStock.com/TiggyMorse

FIGURE 6.11   More than 100 species of lemurs are all endemic to Madagascar, a large island off the eastern coast of Africa. Their endemism, the fact that they live on an island, their small population sizes and numbers of populations, and even the fact that they are primates (a group with many endangered species) put them at higher anticipated risk of extinction. Shown here is a Coquerel's sifaka (*Propithecus coquereli*) and its young.

- *Large body size*   Large animals tend to have large individual ranges and low reproductive rates, they tend to require more food, and they tend to be hunted by humans. Top carnivores, especially, are often killed by humans because they compete with humans for wild game, sometimes damage livestock, and are hunted for sport. Larger animals are generally more vulnerable to the effects of habitat fragmentation (Riverá Ortíz et al. 2014). Within groups of species, the largest species is often the most prone to extinction; that is, the largest carnivore, the largest lemur, and the largest whale will go extinct first (Atwood et al. 2020).

- *Slow reproduction*   In some analyses, a low number of offspring in each reproductive event was one of the most predictive features of a species becoming endangered (Purvis et al. 2000). This characteristic often overlaps with large body size: elephants, whales, giraffes, rhinos, and gorillas all usually have only one or two young at a time. Other traits associated with slow reproduction are slow growth rates, late sexual maturity, long gestation periods, and long intervals between reproductive events. These traits mean that a species cannot compensate easily for increased mortality rates.

- *Limited dispersal ability*   Species unable to adapt physiologically, genetically, or behaviorally to changing environments must either migrate to more suitable habitat or face extinction. The rapid pace of anthropogenic changes often precludes evolutionary adaptation, leaving migration as a species' only alternative. Terrestrial species that are unable to cross roads, farmland, or disturbed habitats are doomed to extinction as their original habitat deteriorates. Dispersal is important in the aquatic environment as well, where dams, point sources of pollution, channelization, and sedimentation can limit movement. Limited dispersal ability may explain in part why freshwater fauna such as mussels and snails are more likely to be extinct or threatened with extinction than are dragonfly species, which are strong fliers and can move to new aquatic locations.

- *Seasonal migration*   Species that migrate seasonally depend on two or more distinct habitat types. If either one of those habitat types is damaged, the species may be unable to persist. The billion songbirds of 120 species that migrate each year between the northern United States and the Neotropics depend on suitable habitat in both locations to survive and breed. In addition, if barriers to dispersal are created by roads, fences, or dams between the needed habitats, a species may be unable to complete its life cycle. Salmon species that are blocked by dams from swimming up rivers and spawning are striking examples of this problem (O'Connor et al. 2015) (**FIGURE 6.12**).

- *Little genetic variation*   Genetic variation within a population can sometimes allow a species to adapt to a changing environment (Forsman 2014) (see "Loss of genetic diversity" later in this chapter). Species with little or no genetic variation may have a greater

FIGURE 6.12 Aggregations of salmon (*Oncorhynchus* spp.) migrate upstream to spawning pools. Impediments to their movement (such as dams) and other human activities (such as harvesting) that threaten these aggregations severely threaten populations.

tendency to become extinct when a new disease, a new predator, or some other factor alters their environment. However, one study determined that for mammals, genetic variation, while a contributing factor, was not as important as other, ecological factors, such as body size, for extinction risk (Polishchuk et al. 2015).

- *Specialized niche requirements* Once a habitat has been altered, it may no longer be suitable for specialized species. For example, wetland plants that require very specific and regular changes in water level may be rapidly eliminated when human activity affects the hydrology of an area (Quan et al. 2018). As another example, there are species of mites that feed only on the feathers of a single bird species. If the bird species goes extinct, so do its associated feather mite species.

- *Low tolerance for disturbance* Many species are adapted to stable, pristine environments where disturbance is minimal, such as old stands of tropical rain forests or the interiors of rich temperate deciduous forests. When these forests are logged, grazed, burned, or otherwise altered, these species are unable to tolerate the changed microclimate conditions (more light, less moisture, greater temperature variation) or an influx of invasive species.

- *Permanent or temporary aggregations* Species that group together in specific places are highly vulnerable to local extinction (see Figure 6.12). Herds of bison, flocks of passenger pigeons, and schools of spawning ocean fish all represent aggregations that have been exploited and overharvested by people. Many species of social animals may be unable to persist when population size or density falls below a certain number, because they may be unable to forage, find mates, or defend themselves, a situation termed the **Allee effect** (Allee 1931; Nagel et al. 2021).

- *No prior contact with people*   Species that have experienced prior human disturbance and persisted have a lower current extinction risk than species encountering people—along with the nonnative species associated with them—for the first time. Island species are often examples of this (see Figures 6.4 and 6.5).

- *Close relatives that are recently extinct or threatened with extinction* Some groups of species, such as primates, cranes, sea turtles, and orchids, are particularly vulnerable to extinction. The characteristics that make certain species in these groups vulnerable are often shared by related species.

These characteristics of extinction-prone species are not independent but tend to group together. For example, many orchid species have specialized habitat requirements, have specialized relationships with pollinators, and are overharvested by collectors; all those characteristics lead to small, declining populations and eventually to extinction. A high percentage of seabirds are also in danger of extinction because they have low reproductive rates; they form dense breeding aggregations, often in small areas, where their eggs and nestlings are prone to attack by introduced predators; they are killed by oil pollution and as bycatch during commercial fishing operations; and their eggs are overharvested by people (Munilla et al. 2007).

It also should be noted that several of these traits are associated with species that may be especially prone to the impacts of climate change. For example, mountaintop species are by definition isolated and typically have small populations. As the climate warms, habitat constricts even further as cold-adapted species move up in latitude; the top of the range cannot extend beyond the mountain top (see Chapter 5). The insular nature of mountaintops means that species there are also often limited in dispersal ability (Urban 2018).

GLOBAL CHANGE
CONNECTION

By using these features to identify extinction-prone species, conservation biologists can anticipate the need for managing their populations.

6.4 Problems of Small Populations

LEARNING OBJECTIVES

By the end of this section you should be able to:

6.4.1   Explain why the occasional arrival of new individuals is important for the survival of a population.

6.4.2   Calculate loss in heterozygosity due to genetic drift using effective population size.

6.4.3   Compare inbreeding depression with outbreeding depression and determine which is the greater risk for most rare species.

6.4.4   Distinguish between environmental and demographic stochasticity.

6.4.5   Calculate and define effective population size.

As mentioned in Section 6.3, species with small populations are in increased danger of going extinct. Small populations are subject to rapid decline in numbers and local extinction for three main reasons:

1. Loss of genetic diversity and related problems of inbreeding depression and genetic drift

2. Demographic fluctuations due to random variations in birth and death rates

3. Environmental fluctuations due to variation in predation, competition, disease, and food supply as well as natural catastrophes that occur at irregular intervals, such as fires, floods, storms, or droughts

We will now examine in detail each of these causes for decline in small populations.

Loss of genetic diversity

As described in Chapter 2, a population's ability to adapt to a changing environment depends on genetic variation, which occurs as a result of different individuals having different alleles of the same gene. Individuals with certain alleles or combinations of alleles may have just the characteristics needed to survive and reproduce under new conditions (Frankham 2015). Within a population, the frequency of a given allele can range from common to very rare. New alleles arise in a population either by random mutations or through the migration of individuals from other populations. Small populations often have very low genetic diversity, which compromises their ability to respond to environmental changes.

In small populations, allele frequencies may change significantly from one generation to the next simply because of chance—based on which individuals happen to survive to sexual maturity, mate, and leave offspring. This random process of allele frequency change, known as **genetic drift**, is a separate process from changes in allele frequency caused by natural selection. When an allele occurs at a low frequency in the gene pool of a small population, it has a significant probability of being lost in each generation (**FIGURE 6.13**). For example, if a rare allele occurs in 5% of all the gene copies present in a population of 1000 individuals, then 100 copies of the allele are present (1000 individuals × 2 gene copies per individual × 0.05 allele frequency), and the allele will probably remain in the population for many generations. However, in a population of 10 individuals, only 1 copy of the allele is present (10 individuals × 2 gene copies per individual × 0.05 allele frequency), and it is possible that the allele will be lost from the population by chance in the next generation. A problem related to the loss of alleles is the reduction of heterozygosity—that is, the proportion of individuals with two different alleles of a particular gene. Considering the general case of an isolated population in which there are two alleles of each gene in the gene pool, Wright (1931) proposed a formula to express the proportion of the original heterozygosity

Because of genetic drift, small populations lose genetic variation more rapidly than large populations. Some small populations may lack any genetic variation.

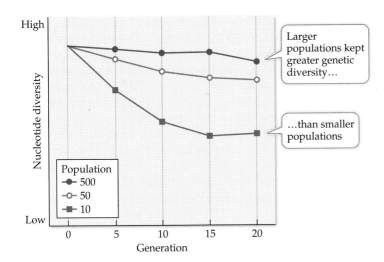

FIGURE 6.13 The rate at which genetic variation is lost through genetic drift varies with population size. This graph shows how nucleotide diversity of three populations changed over time in the model organism *Drosophila melanogaster*. Two DNA sequences of genetic variation were monitored for 20 generations in populations with various effective population sizes (N_e). After 20 generations, a population with an N_e of 500 had kept nearly all its nucleotide diversity, whereas a population of 10 had lost more than 50%. (After M. Shou et al. 2017. *Mol Ecol* 26: 6510–6523. © 2017 John Wiley & Sons Ltd.)

(H_0, which in this case is 1) remaining after each generation (H_1). The formula includes the **effective population size (N_e)**—the size of the population as estimated by the number of its breeding individuals[2]:

$$H_1 = [1 - 1/(2N_e)] \, H_0$$

According to this equation, a population of 50 breeding individuals would retain 99% of its original heterozygosity after 1 generation:

$$H_1 = (1 - 1/100) \, 1 = 1.00 - 0.01 = 0.99$$

The proportion of the original heterozygosity remaining after t generations (H_t/H_0) decreases over time, and the decrease is greater for smaller populations[3]:

$$H_t/H_0 = (H_1)^t$$

For our population of 50 individuals, then, the remaining heterozygosity would be 98% after 2 generations (0.99×0.99), 97% after 3 generations ($0.99 \times 0.99 \times 0.99$), and 90% after 10 generations (0.99^{10}). However, a population of 10 individuals would retain only 95% of its original heterozygosity after 1 generation, 90% after 2 generations, 86% after 3 generations, and 60% after 10 generations. Loss of genetic diversity, including heterozygosity, has been directly linked to extinction risk in some cases (Yoder et al. 2018); however, it has been recently argued that genome-wide diversity (called "neutral" genetic diversity) is not a good indicator of whether a species will necessarily have alleles that will protect it from extinction (Teixeira and Huber 2021).

---

[2]See "Factors that determine effective population size," later in this chapter.

[3]This is a simplification of the following equation: $H_t/H_0 = [1 - 1/(2N_e)]^t$.

This formula demonstrates that significant losses of genetic variation can occur in isolated small populations. Such small populations are often found on islands and in fragmented landscapes (Vranckx et al. 2012). However, the amount of genetic variation within a small population can increase over time through two means: regular mutation of genes and migration of even a few individuals from other populations. Mutation rates found in nature vary between 1 in 10,000 and 1 in 1 million per gene per generation; mutations may therefore make up for genetic drift in large populations and, to a lesser extent, in small populations. However, mutations alone are not sufficient to counter genetic drift in populations of 100 individuals or fewer. Fortunately, even a low frequency of movement of individuals between populations minimizes the loss of genetic variation associated with small population size (Weiser et al. 2013). As we will see in an example with lions, if even 1 or 2 immigrants arrive each generation in an isolated population of about 100 individuals, the effect of genetic drift will be greatly reduced. With 4–10 immigrants arriving per generation, the effects of genetic drift are negligible. In addition, genetic variation that increases fitness will tend to be retained longer in a population, even when there is genetic drift.

Field data confirm that a lower effective population size leads to a more rapid loss of genetic diversity from a population (**FIGURE 6.14**). An analysis of 15 different vertebrate species found a strong relationship between N_e and inbreeding coefficient—a measure of heterozygosity (Feng et al. 2019). Almost all species with populations over 10 million had over 60% heterozygosity, in contrast to most species with fewer than 100,000 individuals, which had less than 60% heterozygosity.

Low genetic diversity puts rare species at risk, but there are ways to mitigate this. Due to low genetic variation following major population declines, populations of the Tasmanian devil (*Sarcophilus harrisii*), a carnivorous marsupial mammal of Australia, are particularly susceptible to a contagious cancer; international breeding programs have been established recently to develop a more diverse metapopulation (Hogg et al. 2017). In addition to such breeding programs, wildlife biologists can move animals between populations, in what is known as assisted immigration, to artificially increase **gene flow**, the movement of genetic information from one population to another (see Chapter 8). Using a variety of such methods, conservation biologists hope to maintain and increase genetic diversity in rare species.

Consequences of reduced genetic diversity

Small populations subjected to genetic drift are susceptible to a number of deleterious genetic effects, such as inbreeding depression, outbreeding depression, and loss of evolutionary flexibility. These effects may contribute to a decline in population size, leading to an even greater loss of genetic diversity, a loss of fitness, and a greater probability of extinction.

INBREEDING DEPRESSION Matings between close relatives, such as between parents and their offspring, between siblings, or between cousins, and self-fertilization in hermaphroditic species may result in **inbreeding depression**, a condition that occurs when an individual receives two identical

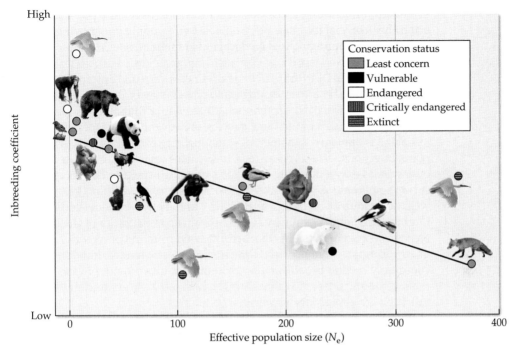

FIGURE 6.14   Smaller populations have a higher risk of inbreeding. Here we see the results from 14 different studies where inbreeding coefficients were determined for populations of species with known effective population size (N_e). We can see a strong negative relationship; as populations increase, inbreeding decreases. Four populations of the crested ibis (*Nipponia nippon*) were analyzed, three of which were from museum specimens of extinct populations. It is notable that although high inbreeding coefficients can be dangerous for species, there is not an obvious relationship between conservation status and degree of inbreeding, reflecting the fact that endangerment is complex and multifaceted. (After S. Feng et al. 2019. *Curr Biol* 29(2): 340–349. Copyright 2018. Reprinted with permission from Elsevier.)

copies of a defective allele (one from each of its parents or both from its self-fertilizing parent). Inbreeding depression is characterized by higher mortality of offspring, fewer offspring, or offspring that are weak or sterile or have low mating success. These factors result in even fewer individuals in the next generation, leading to more pronounced inbreeding depression.

Evidence for the existence of inbreeding depression comes from studies of human populations (in which there are records of marriages between close relatives for many generations), captive and wild animal populations, and cultivated plants (Frankham et al. 2014). In a wide range of captive mammal populations, matings among close relatives resulted, on average, in offspring with a 33% higher mortality rate than in non-inbred animals. This lower fitness resulting from inbreeding is sometimes referred to as a "cost of inbreeding." Inbreeding depression can be a severe problem in small captive populations in zoos, in domestic livestock breeding programs, and in wild populations with small effective population sizes (see Figure 6.14). For example, genetic analysis of lions from two protected areas in Nigeria revealed evidence of inbreeding and a lack of gene flow between the populations. To avoid the potential for inbreeding depression, trapping and moving lions as a conservation approach was recommended (Tende et al. 2014).

OUTBREEDING DEPRESSION   Individuals of different species rarely mate in the wild; there are strong ecological, behavioral, physiological, and morphological isolating mechanisms that ensure that mating occurs only between individuals of the same species. However, when a species is rare or its habitat is damaged, outbreeding—mating between individuals of different populations or species—may occur. Individuals unable to find mates of their own species may mate with individuals of related species. The resulting off-spring sometimes exhibit outbreeding depression, a condition that results in weakness, sterility, or lack of adaptability to the environment (Waller 2015). Outbreeding depression may be caused by incompatibility of the chromosomes and enzyme systems that are inherited from the different parents. To use an example from artificial selection, domestic horses and donkeys are commonly bred to produce mules. Although mules are not physically weak (on the contrary, they are quite strong, which is why humans find them useful), they are almost always sterile.

Outbreeding depression can also result from matings between different subspecies, or even matings between divergent populations of the same species. Such matings might occur in a captive breeding program or when individuals from different populations are kept together in captivity. In such cases, the offspring of parents with such different genotypes are unlikely to have the precise mixture of genes that allows individuals to survive and reproduce successfully in a particular set of local conditions. However, many other studies of animals have failed to demonstrate outbreeding depression; some have even shown that some hybrids are more vigorous than their parent species (Frankham 2016), a condition known as hybrid vigor. **Genetic rescue** is the practice of intentional hybridization of a rare species to keep it from extinction, although successful examples of this are rare (Todesco et al. 2016). Thus, outbreeding depression may be less of a concern for animals than inbreeding depression, the negative effects of which are well documented.

Outbreeding depression may be considerably more significant in plants, in which the arrival of pollen on the receptive stigma of a flower is to some degree a matter of the chance movement of pollen by wind or animal vectors. A rare plant species growing near a closely related common species may be overwhelmed by the pollen of the common species and fail to produce seeds (Willi et al. 2007). Even when hybrids are produced by matings between a common and a rare species, the genetic identity of the rare species is lost as its small gene pool is mixed into the much larger gene pool of the common species. The seriousness of this threat is illustrated by the fact that more than 90% of California's threatened and endangered plants occur in close proximity to other species in the same genus with which the rare plants could possibly hybridize. Such losses of identity can also take place in gardens when individuals from different parts of a species' range are grown next to one another. In an extreme example, some species of plants do best when self-fertilized, the most severe form of inbreeding; such species may show outbreeding depression when pollen from other individuals of even the same population is introduced (**FIGURE 6.15**).

LOSS OF EVOLUTIONARY FLEXIBILITY Genetic diversity is extremely important to a species' long-term survival. Rare alleles and unusual combinations of alleles that are harmless (or even slightly harmful) but confer no immediate advantage on the few individuals that carry them may turn out to be uniquely suited for a future set of environmental conditions. If such alleles and combinations do become advantageous, their frequency in the population will increase rapidly through natural selection because the individuals that carry them will be those most likely to survive and reproduce successfully, passing on the formerly rare alleles to their offspring.

Loss of genetic diversity in a small population may limit its ability to respond to new conditions and long-term changes in the environment, such as pollution, new diseases, or global climate change (Habel and Schmitt 2018). According to the fundamental theorem of natural selection, the rate of evolutionary change in a population is directly related to the amount of genetic variation in that population. A small population is less likely than a large population to possess the genetic variation necessary for adaptation to environmental changes and is therefore more likely to go extinct. In many plant populations, for example, a few individuals have alleles that promote tolerance for high concentrations of toxic metals such as zinc and lead, even when these metals are not present. If toxic metals become abundant in the environment because of pollution, individuals with these alleles will be better able to adapt to them and will grow, survive, and reproduce better than typical individuals; consequently, the frequency of these alleles in the

GLOBAL CHANGE
CONNECTION

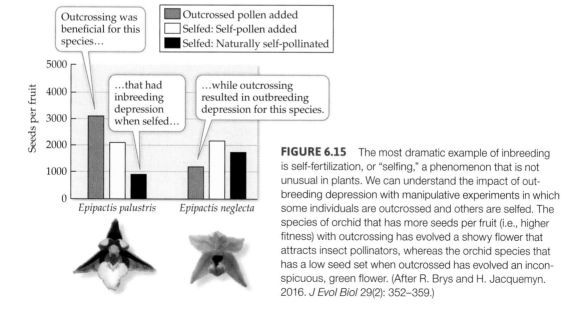

FIGURE 6.15 The most dramatic example of inbreeding is self-fertilization, or "selfing," a phenomenon that is not unusual in plants. We can understand the impact of outbreeding depression with manipulative experiments in which some individuals are outcrossed and others are selfed. The species of orchid that has more seeds per fruit (i.e., higher fitness) with outcrossing has evolved a showy flower that attracts insect pollinators, whereas the orchid species that has a low seed set when outcrossed has evolved an inconspicuous, green flower. (After R. Brys and H. Jacquemyn. 2016. *J Evol Biol* 29(2): 352–359.)

population will increase dramatically. However, if the population has become small and the genotypes for metal tolerance have been lost, the population could go extinct.

Factors that determine effective population size

Earlier in this section, we mentioned the effective population size, which is the size of a population as estimated by the number of its breeding individuals. The effective population size is lower than the total population size because many individuals do not reproduce, due to factors such as inability to find a mate, being too old or too young to mate, poor health, sterility, malnutrition, and small body size. Many of these factors are initiated or aggravated by habitat degradation and fragmentation. Furthermore, many plant, fungus, bacteria, and protist species have seeds, spores, or other structures that remain dormant in the soil unless stable conditions for germination appear. These individuals could be counted as members of the population though they are obviously not part of the breeding population.

The effective population size, N_e, will be much smaller than the total population size, N, when there is a small proportion of individuals reproducing, great variation in reproductive output, an unequal sex ratio, or wide fluctuations in population size.

Because of these factors, the effective size of a population (N_e) is often substantially smaller than its actual size (N). Because the rate of loss of genetic diversity depends on effective population size, loss of genetic diversity can be quite severe even in a large population. For example, consider a population of 1000 alligators consisting of 990 immature animals and only 10 mature breeding animals: 5 males and 5 females. The effective size of this population is 10, not 1000. In a population of a rare oak species, there might be 20 mature trees, 500 saplings, and 2000 seedlings, resulting in a population size of 2520 but an effective population size of only 20.

In addition, the effective population size is often even lower than the actual number of breeding individuals because of an unequal sex ratio, variation in reproductive output, or large annual changes in population size.

UNEQUAL SEX RATIO   A population may consist of unequal numbers of males and females due to chance, selective mortality, or the harvesting of one sex by people. If, for example, a population of a goose species that is monogamous (in which one male and one female form a long-lasting pair bond) consists of 20 males and 6 females, then only 12 individuals—6 males and 6 females—will be mating. In this case, the effective population size is 12, not 26. In other animal species, social systems may prevent many individuals from mating even though they are physiologically capable of doing so. Among elephant seals, many ungulates, and many primates, for example, a single dominant male usually mates with a large number of females and prevents other males from mating with them (**FIGURE 6.16**), whereas among African wild dogs and hyenas, the dominant female in the pack often bears all of the pups.

The effect of unequal numbers of breeding males and females on N_e can be described by this formula:

$$N_e = [4(N_f N_m)]/(N_f + N_m)$$

FIGURE 6.16 Two large male elk compete to mate with the many smaller females surrounding them. The effective population size is reduced because only one male is providing genetic input to many females.

where N_f and N_m are the numbers of adult breeding females and adult breeding males, respectively, in the population. In general, as the sex ratio of breeding individuals becomes increasingly unequal, the ratio of the effective population size to the number of breeding individuals (N_e/N) goes down. This occurs because only a few individuals of one sex are making a disproportionately large contribution to the genetic makeup of the next generation. In many fish and reptile species, sex expression is affected by temperature. As global climate change increases water and air temperatures in many places, the sex ratios of these species may become skewed, lowering their effective population sizes. In Switzerland, for example, grayling (*Thymallus thymallus*) populations that used to have highly variable sex ratios centered around 65% males before 1990 now consistently have 80%–90% males. The effective population size for this fish species will be far lower than the number of individuals in the population.

VARIATION IN REPRODUCTIVE OUTPUT   In many species, the number of offspring varies substantially among individuals. This variation is particularly pronounced in highly fecund species, such as plants and fishes, in which many or even most individuals produce a few offspring while others produce huge numbers. Unequal production of offspring leads to a substantial reduction in N_e because a few individuals in the present generation will be disproportionately represented in the gene pool of the next generation. In general, the greater the variation in reproductive output, the more the effective population size is lowered.

POPULATION FLUCTUATIONS AND BOTTLENECKS   In some species, population size varies dramatically from generation to generation. Particularly striking examples of this phenomenon are butterflies, annual plants, and amphibians. In a population with extreme size fluctuations, the effective population size is much nearer the lowest than the highest number of individuals, and it tends to be determined by the years in which the population has the smallest numbers. The effective size of a fluctuating population can be calculated over a period of t years using the number of individuals (N) breeding in any one year:

$$N_e = t/(1/N_1 + 1/N_2 + \ldots + 1/N_t)$$

Consider a butterfly population, monitored for five years, that has 10, 20, 100, 20, and 10 breeding individuals in those successive five years. In this case,

$$N_e = 5/(1/10 + 1/20 + 1/ + 1/20 + 1/10) = 5/(31/100) = 5(100/31) = 16.1$$

The effective population size over the course of five years is above the lowest population size (10) but well below the maximum (100) and the average (32) population sizes.

As these calculations suggest, a single year of drastically reduced population numbers will substantially lower the value of N_e. This principle applies to a phenomenon known as a **population bottleneck**, which occurs when a population is greatly reduced in size. Such a population will lose rare alleles if no individuals possessing those alleles survive and reproduce. With fewer alleles present and a decline in heterozygosity, the overall fitness of the individuals in the population may decline. Because of the relationship between genetic diversity and population size, molecular tools can be used to detect population bottlenecks in the evolutionary past. For example, according to genetic analysis, the Iberian lynx (*Lynx pardinus*) of Spain and Portugal has experienced three severe bottlenecks as a species that had led to its becoming the most endangered cat on the planet (Abascal et al. 2016). The current effective population is less than 200 individuals but is growing in response to conservation efforts.

A special category of population bottleneck, known as the **founder effect**, occurs when a few individuals leave one population and establish a new population. The new population often has less genetic diversity than the larger, original population. For example, the wolf population in Sweden was established by five individuals (Laikre et al. 2013). Similarly, if a population is fragmented by human activities, each of the resulting small subpopulations may lose genetic variation and go extinct. Such is the fate of many fish populations fragmented by dams (Hasegawa 2017). Bottlenecks can also occur when captive populations are established using relatively few individuals.

The lions (*Panthera leo*) of Ngorongoro Crater in Tanzania provide a well-studied example of a population bottleneck. The lion population in the crater consisted of 60–75 individuals until an outbreak of biting flies in 1962 reduced the adult population to 9 females and 1 male (Munson et al. 2008) (**FIGURE 6.17**). Two years later, 7 additional males immigrated to the crater, and no new immigration was observed for the next 49 years. The dramatic reduction in population size, the isolation of the population, and variation in reproductive success among individuals apparently created a population bottleneck, leading to inbreeding depression. In comparison with the large Serengeti lion population nearby, the crater lions showed reduced genetic diversity, high levels of sperm abnormalities, reduced reproductive rates, increased cub mortality, and higher rates of disease. After reaching a peak of

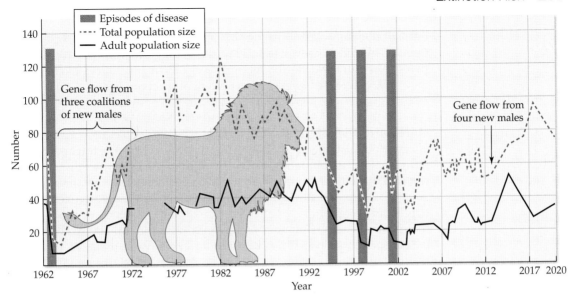

FIGURE 6.17 The Ngorongoro Crater lion population consisted of about 90 individuals in 1961 before crashing from disease in 1962, resulting in a population bottleneck and losing genetic diversity. However, gene flow from the arrival of new males facilitated recovery of the population, which peaked at 125 individuals in 1983 before collapsing again to 34 individuals (fewer than 20 of which were adults). A lack of census data for certain years is the cause of gaps in the lines. The four blue bars represent episodes of disease outbreak. The arrival of new males has likely contributed to an increase in numbers again; there were more total lions in 2019 than had been seen since the late 1980s. (After L. Munson et al. 2008. *PLOS ONE* 3: e2545. doi.org/10.1371/journal.pone.0002545. CC BY 4.0, creativecommons.org/licenses/by/4.0, with additional data from C. Packer, pers. comm.; lion silhouette © Shutterstock.com/Fairys_Shape Logo.)

125 animals in the 1980s, the population declined again. By 2003, following an outbreak of canine distemper virus that had spread from domestic dogs kept by people living just outside the crater area, the population had dropped to 34 lions. For the next decade the population size stayed between 50 and 70 individuals. However, the immigration of 4 groups of males to the area in the following decade facilitated a dramatic recovery in numbers and presumably genetic variation. This was supported by the fact that the population grew in 2019 to 97 lions, the largest since the late 1980s (updates from C. Packer 2019, personal communication). This unusual peak was likely facilitated by the ability of females to synchronize estrus when new males arrive. There are currently 65–75, a consequence of expected rates of cub mortality by competing adult males, and lower birthrate due to lionesses being busy raising juveniles (KopeLion 2021). Time will tell whether the impact of the new lions will persist.

Population bottlenecks do not always lead to greatly reduced heterozygosity, however. If a population expands rapidly in size after a temporary bottleneck, heterozygosity may be restored even though the number of alleles present is severely reduced. An example of this phenomenon is the high level of heterozygosity found in the greater one-horned rhinoceros (*Rhinoceros unicornis*) in Chitwan National Park, Nepal, even after the population passed through a bottleneck. Population size declined from 800 individuals to fewer than 100 individuals; fewer than 30 were breeding. With an effective population size of 30 individuals for one generation, the population would have lost only 1.7% of its heterozygosity. As a result of strict protection of the species by park guards, the population recovered to 645 individuals by 2015.

MANAGING FOR GENETIC VARIATION These examples demonstrate that effective population size is often substantially less than the total number of individuals in a population. A review of a wide range of wildlife studies revealed that effective population size averaged only 11% of total population size; that is, a population of 300 animals, seemingly enough to maintain the population, might have an effective size of only 33, which would indicate that it was in serious danger of extinction (Frankham 2005). For highly fecund species, such as fishes, seaweeds, and many invertebrates, the effective population size may be less than 1% of the total population size (Frankham et al. 2009). Consequently, management aimed toward simply maintaining large populations may not prevent the loss of genetic variation unless the effective population size is also large. In the case of captive populations of rare and endangered species, genetic variation may be effectively maintained by controlling breeding, perhaps by subdividing the population, periodically removing dominant males to allow subdominant males the opportunity to mate, and periodically transporting a few selected individuals among subpopulations.

The impact of stochasticity

Random variation, or **stochasticity**, in the biological or physical environment can cause variation in the population size of a species. For example, the population of an endangered butterfly species might be affected by fluctuations in the abundance of its food plants or its predators. Weather might also strongly influence the butterfly population: in an average year, the weather may be warm enough for caterpillars to feed and grow, whereas a cold year might cause many caterpillars to become inactive and consequently starve. There will also be random differences in reproduction and survival between individuals, causing variability in birth and death rates. Together, these sources of stochasticity will disproportionately affect smaller populations.

Random fluctuations in birth and death rates, disruption of social behavior by lowered population density, and environmental stochasticity all contribute to further decrease in the size of populations, often leading to local extinction. Intensive management is often required to prevent small populations from declining further in size and going extinct.

DEMOGRAPHIC STOCHASTICITY In an ideal, stable environment, a population would increase until it reached the carrying capacity (K) of the environment, at which point the average birthrate (b) per individual would equal the average death rate (d) and there would be no net change in population size. In any real population, however, individuals do not usually produce the average number of offspring: they may leave no offspring, somewhat fewer than the average, or more than the average. For example, in an ideal, stable population of giant pandas (*Ailuropoda melanoleuca*), each female would produce an average of two surviving offspring in her lifetime, but field studies show that rates of reproduction among individual females vary widely around that number. However, as long as the population size is large, the average birthrate provides an accurate description of the population. Similarly, the average death rate in a population can be determined only by examining large numbers of individuals because some individuals die young and others live a relatively long time. But as long as the population size is large, the average death rate provides an accurate and relatively stable description of the population.

Population size may fluctuate over time because of changes in the environment or other factors without ever approaching a stable value. Random fluctuations upward in population size are eventually bounded by the carrying capacity of the environment, after which the population may fluctuate downward again. Variation in population size due to random variation in reproduction and mortality rates is known as **demographic variation** or **demographic stochasticity**. If population size fluctuates downward in any one year because of a higher-than-average number of deaths or a lower-than-average number of births, the resulting smaller population will be even more susceptible to the negative effects of demographic fluctuations in subsequent years. For example, males of the Iberian lynx recently began mating at much younger ages, a demographically stochastic event, which increased birthrates but may have contributed to inbreeding between full siblings (Lucena-Perez et al. 2018).

Consequently, once a population decreases because of habitat destruction and fragmentation, demographic variation becomes important, and the population has a higher probability of declining more and even going extinct due to chance alone from low reproduction and/or high mortality (Haddad et al. 2015). Species with highly variable birth and death rates, such as the Iberian lynx, may be particularly susceptible to population extinction due to demographic stochasticity. The chance of extinction is also greater in species that have low birthrates, such as elephants, because these species take longer to recover from chance reductions in population size.

GLOBAL CHANGE
CONNECTION

When populations drop below a critical number, deviations from an equal sex ratio may also occur, leading to a declining birthrate and a further decrease in population size. This scenario is illustrated by the now-extinct dusky seaside sparrow (*Ammodramus maritimus nigrescens*); the last five individuals were males, so there was no opportunity to establish a captive breeding program.

Many small populations are demographically unstable because social interactions (especially mating) can be disrupted once population density falls below a certain level. As mentioned in Section 6.3, this interaction among population size, population density, population growth rate, and behavior is referred to as the Allee effect (Allee 1931). Perhaps the most significant aspect of the Allee effect for small populations involves reproductive behavior: many species that live in widely dispersed populations, such as bears, spiders, and tigers, have difficulty finding mates once the population density drops below a certain point. For example, the Antarctic fur seal (Arctocephalus gazella) was found to have higher pup survival when densities were higher (Nagel et al. 2021) (**FIGURE 6.18**). Even among plant species, as population size and density decrease, the distance between individual plants increases. Pollinating animals may fail to visit isolated, scattered plants, resulting in insufficient transfer

FIGURE 6.18 As an example of the Allee effect, Antarctic fur seals (*Arctocephalus gazella*) were found to have higher pup survival rates when populations were larger, despite increased competition for food (Nagel et al. 2021). This seal species is one of many that are struggling due to premature sea ice melt; the Allee effect thus exacerbates the negative impact of climate change.

of compatible pollen and a subsequent decline in seed production. In such cases, the birthrate will decline, population density will become lower yet, problems such as unequal sex ratio will worsen, and birthrates will drop even more. Once the birthrate falls to zero, extinction is guaranteed. Detecting and anticipating Allee effects are necessary for the management and recovery of endangered species.

ENVIRONMENTAL STOCHASTICITY AND CATAS-TROPHES Random variation in the biological and physical environment, known as **environmental stochasticity**, can also cause variation in the size of a population, and thus it disproportionately threatens small populations. For example, a population of an endangered rabbit species might be affected by fluctuations in the population of a deer species that eats the same types of plants, in the population of a fox species that feeds on the rabbits, and in the populations of parasites and disease-causing organisms that infect and weaken the rabbits. Variation in the physical environment might also strongly influence the rabbit population: rainfall during an average year might encourage plant growth and allow the rabbit population to increase, while dry years might limit plant growth and cause rabbits to starve. Environmental stochasticity affects all individuals in the population, unlike demographic stochasticity, which causes variation among individuals within the population.

Natural catastrophes that occur at unpredictable intervals, such as droughts, storms, earthquakes, and fires, along with cyclical die-offs in the surrounding biological community, can cause dramatic fluctuations in population levels. Natural catastrophes can kill part of a population or even eliminate an entire population from an area or can benefit a species by creating new habitat (Maslo et al. 2019). Even though the probability of a natural catastrophe in any one year is low, over the course of decades and centuries, natural catastrophes—including extended periods of unseasonable weather, excessive or insufficient rainfall, and events such as hurricanes and earthquakes—have a high likelihood of occurring.

Modeling efforts by Menges (1992) and others have shown that random environmental variation is generally more important than random demographic variation in increasing the probability of extinction in populations of small to moderate size. Environmental variation can substantially increase the risk of extinction even in populations that showed positive population growth under the assumption of a stable environment.

Imagine a rabbit population of 100 individuals in which the average birthrate is 0.2 and an average of 20 rabbits are eaten each year by foxes. On average, the population will maintain its numbers at exactly 100 individuals, with 20 rabbits born each year and 20 rabbits eaten each year. However, if there are 3 successive years in which the foxes eat 40 rabbits per year, the population size will decline to 80 rabbits, 56 rabbits, and 27 rabbits in years 1, 2, and 3, respectively. If there are then 3 years of no fox predation, the rabbit population will increase to 32, 38, and 46 individuals in years 4, 5, and 6, respectively. Even though the same average rate of predation (20 rabbits per year) occurred over this 6-year period, variation in year-to-year predation rates caused the rabbit population size to decline by more than 50%. At a population size of 46 individuals, the rabbit population will probably go extinct within the next 5–10 years if it is subjected to the average rate of 20 rabbits eaten by foxes per year. As we will see in Chapter 7, both environmental and demographic stochasticity can be included in models to determine more accurate estimates of extinction risk for a given population.

The extinction vortex

The smaller a population becomes, the more vulnerable it is to the combined effects of low genetic diversity, demographic variation, and environmental stochasticity that tend to lower reproduction, increase mortality rates, and so reduce its size even more, driving the population to extinction (**FIGURE 6.19**). This tendency of small populations to decline toward extinction has been likened to a vortex, a whirling mass of gas or liquid spiraling inward: the closer an object gets to the center of the vortex, the faster it moves. At the center of an **extinction vortex** is oblivion: the local extinction of the species. Once caught in such a vortex, it is difficult for a species to resist the pull toward extinction

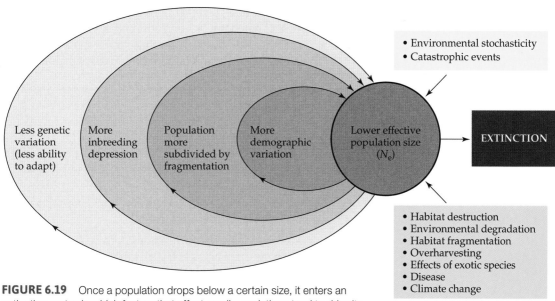

FIGURE 6.19 Once a population drops below a certain size, it enters an extinction vortex in which factors that affect small populations tend to drive its size progressively lower. This downward spiral often leads to the local extinction of species. (After Gilpin and Soulé. 1986. Minimum viable populations: Processes of species extinction. In M. E. Soulé [Ed.], *Conservation Biology: The Science of Scarcity and Diversity*, pp. 19–34. Oxford University Press/Sinauer Associates, Sunderland, MA.)

(Nabutanyi and Wittmann 2021). Small populations often require a careful program of population and habitat management, as described in later chapters, to increase their growth rates and allow them to escape from the extinction vortex.

Summary

- Many species are currently on the brink of extinction due to increasing human activity; this is particularly well documented in the well-studied groups of birds, mammals, and amphibians.

- Island species have had a higher rate of extinction than mainland species, due to small population sizes and less previous exposure to humans and mainland species. Among aquatic species, freshwater species apparently have a higher extinction rate than marine species.

- Current rates of extinction are between 100 and 1000 times greater than background extinction levels. More than 99% of modern species extinctions are attributable to human activity.

- The island biogeography model is used to predict the numbers of species that will persist in new protected areas and the numbers that will go extinct elsewhere due to habitat destruction and other human activities.

- Those species that are most vulnerable to extinction have particular features, including a narrow range, one or only a few populations, small population size, declining population size, and economic value to humans that leads to overexploitation.

- Small populations are vulnerable to further declines in size and eventual extinction due to genetic, demographic, and environmental factors, including those that are stochastic (randomly occurring). Intensive management of small populations may be required to prevent their extinction. Effective population size measures the number of breeding individuals, which is likely smaller than the total population and is affected by sex ratios, age structure, and other issues.

For Discussion

1. Explain what is meant by calling the present geologic period the "Anthropocene."

2. Why should conservation biologists—or anyone else—care if a species goes locally extinct if it is still found somewhere else?

3. Consider a species that has gone extinct in the last two centuries. Did it fall into one of the categories of extinction vulnerability listed in this chapter? What other biological or ecological traits might it have had that contributed to its extinction? Did it have one dominant feature that was predictive of extinction, or a combination of such features?

4. Find out about a species that is currently endangered in the wild. How might this species be affected by the problems of small populations? Address genetic, physiological, behavioral, and ecological aspects as appropriate.

Suggested Readings

Bilgmann, K., et al. 2021. Low effective population size in the genetically bottle-necked Australian sea lion is insufficient to maintain genetic variation. *Animal Conservation*. https://doi.org/10.1111/acv.12688. A case study of the use of N_e to model future loss of genetic diversity in two populations of sea lions.

Butchart, S. H., et al. 2018. Which bird species have gone extinct? A novel quantitative classification approach. *Biological Conservation* 227: 9–18. An attempt to predict which species are likely to be lost or are already gone.

Nagel, R., et al. 2021. Evidence for an Allee effect in a declining fur seal population. *Proceedings of the Royal Society B* 288(1947): 20202882. A surprising discovery about the importance of population size in the Antarctic fur seal, a species already suffering from the effects of global warming.

Nomoto, H. A., and J. M. Alexander. 2021. Drivers of local extinction risk in alpine plants under warming climate. *Ecology Letters* 24: 1157–1166. https://doi.org/10.1111/ele.13727. An exploration of extinction debt in the Swiss Alps.

Prates, L., and S. I. Perez. 2021. Late Pleistocene South American megafaunal extinctions associated with rise of Fishtail points and human population. *Nature Communications* 12: 1–11. New evidence of the impact of humans as extinction agents.

Ripple, W. J., et al. 2019. Are we eating the world's megafauna to extinction? *Conservation Letters* 12:e12627. https://doi.org/10.1111/conl.12627. Overview of extinction risk in large animals and the role of body size.

Teixeira, J. C., and C. D. Huber. 2021. The inflated significance of neutral genetic diversity in conservation genetics. *Proceedings of the National Academy of Sciences USA* 118(10): e2015096118. https://doi.org/10.1073/pnas.2015096118. A controversial challenge of the long-held assumption that all genetic diversity protects species from extinction.

Urban, M. C. 2018. Escalator to extinction. *Proceedings of the National Academy of Sciences USA* 115(47): 11871–11873. A review of the literature on the effect of climate change on local extinctions on mountaintops.

David Thandi releases a hooded vulture that has been carefully tagged as a part of a binational research project in Botswana. Vultures are threatened across the globe, but data collected on them will assist conservation efforts.

Conserving Populations and Species

7

How do we know that a species is in need of help? At any point in time, a population of any species can naturally be stable, increasing, decreasing, or fluctuating in number. In general, widespread human disturbance destabilizes populations of many native species, often sending them into sharp decline. But how can this disturbance be measured, and what actions should be taken to prevent or reverse it? How can conservation biologists determine whether a specific plan to manage a rare or endangered species has a good chance of succeeding?

We also need to know which species to protect. Conservation biologists and park managers do not have enough time and money to protect every species. Similarly, not all species need to be protected; many species have numerous large populations that are stable in size and cover a large area. Efforts need to be directed to identifying species in need of protection, using the most recent quantitative approaches that include both field data and modeling. This chapter discusses approaches for monitoring, managing, prioritizing, and protecting species and their populations in their natural environments, and how this information is used to establish legal protection for endangered species.

7.1 Applied Population Biology

By the end of this section you should be able to:

7.1.1 Create a list of questions to guide data collection for any given species.

7.1.2 Find existing data on a species or system within the literature.

7.1.3 Design a study to collect new data on a species.

To effectively protect and manage a rare or endangered species, it is vital to have a firm grasp of its ecology, its distinctive characteristics (sometimes called its **natural history**), and the status of its populations, particularly the dynamic processes that affect population size and distribution (i.e., its **population biology**). With more information about a rare species, land managers can more effectively maintain it, identify factors that place it at risk of extinction, and develop alternative management options.

Scientists who study the natural history and population biology of a species have historically pursued this knowledge for its own sake, but today these fields have very real and important applications in population-level conservation efforts for rare and endangered species.

Several types of natural history and population biology information are important to conservation biology:

- *Environment* What are the habitat types where the species is found, and how much area is there of each? How variable is the environment in time and space? How frequently is the environment affected by disturbances, and to what magnitude? How have human activities affected the environment?

- *Distribution* Where is the species found? Are individuals clustered together, distributed at random, or spaced out regularly? Do individuals of this species move and migrate among habitats or to different geographical areas over short or long time periods? How efficient is the species at colonizing new habitats? How have human activities affected the distribution of the species or its ability to move among habitats?

- *Biotic interactions* What types of food and other resources does the species need, and how does it obtain them? What other species compete with it for these resources? What predators, parasites, or diseases affect its population size? With what mutualists (pollinators, dispersers, etc.) does it interact? Do juvenile stages disperse by themselves, or are they dispersed by other species? How have human activities altered the relationships among species in the ecosystem?

- *Morphology* What does the species look like? What are the shape, size, color, surface texture, and function of its parts? How do the shapes of its body parts relate to their functions and help the species to survive in its environment? What are the characteristics that allow this species to be distinguished from species that are similar in appearance?

- *Physiology* What food, water, minerals, and other necessities does an individual need to survive, grow, and reproduce? How efficient is an individual at using its resources? How vulnerable is the species to extremes of climate, such as heat, cold, wind, and precipitation? What are the requirements during reproduction? Is the organism currently experiencing unusual stress?

- *Phenology* What is the timing of events for this species? When does the species come out of dormancy, reproduce, migrate, or other aspects of its life history that have evolved to occur at specific times? What are the environmental or developmental triggers for these? Have they been affected by global change?

- *Demography* What is the current effective population size, and what was it in the past? Is the number of individuals stable, increasing, or decreasing? Does the population have a mixture of adults and juveniles, indicating that new individuals are being recruited? At what age do individuals begin to reproduce? What is the sex ratio?

- *Behavior* How do the actions of an individual allow it to survive in its environment? How do individuals in a population mate and produce offspring? Do individuals of a species interact cooperatively or competitively? How do individuals find food? At what time of day or year is the species most visible for monitoring?

- *Genetics* How much variation occurs in morphological, physiological, and behavioral characteristics? How is the variation spread across the species range? How much of this variation is genetically controlled? What percentage of the genes is variable? How many alleles does the population have for each variable gene? Are there genetic adaptations to local sites? Is there gene flow between populations? What is the rate of inbreeding?

- *Interactions with humans* How do human activities affect the species? What human activities are harmful or beneficial to the species? Do people harvest or use this species in any way? What do local people know about this species?

Methods for studying populations have developed largely from the study of land plants and animals (**FIGURE 7.1**). Small organisms such as protists, bacteria, and fungi have not been investigated in comparable detail, and species that inhabit soil, freshwater, and marine habitats are also in need of greater study. In this section we will examine how conservation biologists undertake their studies of populations, while we keep in mind that the methods need to be modified for each species.

Knowledge of the natural history and population biology of a species is crucial to its protection, but urgent management decisions often must be made before all this information is available or while it is still being gathered.

FIGURE 7.1 Monitoring populations requires specialized techniques suited to each species. (A) Monitoring trees in a tropical forest often involves labelling and measuring girth as an estimate of total biomass. (B) Collecting very small animals that live in the sediments of streams is often done with "kick nets." (C) Plant ecologists can monitor rare plant populations by establishing transects, along which individuals are mapped and measured.

Researching existing information

When gathering information, it is important to remember that other people may have already investigated an ecosystem or studied the same (or a related) species. All science builds upon past work, so any exploration of a species or ecosystem will begin with a thorough search for existing information. In some cases, interviews with experts can be used to gather their collective knowledge about endangered species and ecosystems (Zepelini et al. 2017). In most cases, however, scientists will access existing information through published works, either in print or online. This is important because scientific literature may contain records of previous population sizes and distributions that can be compared with the current status of the species. Many papers with new insights in conservation biology are published using existing data, including those that employ meta-analysis (see Chapter 4) or more qualitative literature syntheses. The following is a list of types of existing information.

PEER-REVIEWED LITERATURE Academic publications, including journal articles and textbooks, are typically **peer-reviewed**, meaning that the content has been evaluated by others within that field ("peers") before it is approved for publication and distribution. Thus, there is a measure of quality control that is not typically found for other types of publications (but see Ferreira et al. 2016). Peer-reviewed scientific literature can include articles, books, and government reports, both in print and online. These can be found via Google

Scholar or one of many subscription databases such as ScienceDirect and the Clarivate Web of Science. Many of these will be available digitally via a library's subscription service. Hard copy books and older journal articles can be found through these indexing services and then requested from libraries. Finding one source can lead to others by accessing other literature citing or being cited by that source; Google Scholar is effective for doing this. Key words, provided in every peer-reviewed journal article, can be used as search terms to find related sources as well. Asking biologists and naturalists for ideas on references, or from lists of papers they have compiled and posted on websites, is another way to locate peer-reviewed materials.

Another important distinction is the one between primary and secondary literature. **Primary literature** presents the results from original research, which is usually published in the form of a peer-reviewed article but may also be found in the other types of information listed here. It is written by the person or persons who collected and analyzed the data; they are often experts in their subjects. In contrast, **secondary literature** reports the findings of others, often in a literature review or a meta-analysis. This textbook is an example of secondary literature. If you are using information that you found in secondary literature, it is important to seek out the original primary literature to avoid the risk of misinterpretation.

"GRAY" LITERATURE   Reports written by scientists from government fisheries and wildlife agencies, national and regional forest and park departments, and conservation organizations contain an enormous amount of information on conservation biology. It is referred to as **gray literature** because it is not printed or posted by a commercial or scientific publisher. That is, publishing is not the primary purpose of the organization that is distributing it. Gray literature can sometimes be of high quality and contain valuable information; it is often mentioned by leading authorities in conversations, lectures, and conference presentations.

OTHER ONLINE SOURCES   The internet provides ever-increasing access to databases, websites, electronic bulletin boards, news articles, specialized discussion groups, and other sources that can be valuable when searching for information about a particular species or ecosystem. Databases provided by governments, such as the US Department of Agriculture's PLANTS (plants.sc.egov.usda.gov), are easy to use and generally reliable, as are those from major international organizations such as the International Union for Conservation of Nature (IUCN; www.iucnredlist.org) and NatureServe (explorer.natureserve.org), both discussed below, as well as Birdlife International (www.birdlife.org). Wikis, such as Wikipedia, are sites that allow multiple users, or even the public at large, to post and edit information, including listing other sources. This crowdsourcing is a different type of quality control, with little if any centralized management of content. Searching online indexes of newspapers, magazines, and popular journals is also an effective strategy because results of important scientific research are often covered in the popular news media. When using material online, always be sure to validate the accuracy and source of the information.

New data through fieldwork

New information about a species usually must be learned through careful **fieldwork**—observations and data collection in the natural environment. Unfortunately, in the current culture of "publish or perish" and limited funding, fieldwork-based research is becoming less common in the peer-reviewed literature than modeling and literature review papers (Ríos-Saldaña et al. 2018). However, conservation science depends on new data being collected in the field. The natural history of a species can really only be learned through fieldwork. Only in the field can the conservation status of a species be determined, as well as its relationships to the biological and physical environment. Fieldwork for species such as polar bears, humpback whales, and tropical trees can be time-consuming, expensive, and physically arduous, but it is crucial for developing conservation plans for endangered species, and it can be exhilarating and deeply satisfying as well.

Methods for collecting data in the field can be very low-tech, such as those shown in Figure 7.1, or may employ specialized equipment (**FIGURE 7.2**). Increasingly, data collected in the field includes physical samples that are later analyzed in a laboratory using state-of-the-art tools (see "Laboratory Analysis of Biological Material" later in this chapter).

In cases where research methods are straightforward, scientific observations can often be aided by volunteers from the general public, as described in Chapter 2. For example, there is a long tradition, particularly in Britain, of dedicated amateurs conducting excellent studies of species in their immediate surroundings with minimal equipment or financial support. The geographical range and intensity of monitoring have been greatly extended through the use of citizen scientists. Training and educating volunteers not only expands the volume of data available to scientists but often transforms these individuals into advocates for conservation (Forrester et al. 2017). Four programs that rely

FIGURE 7.2 Sometimes specialized equipment is necessary to study organisms, particularly when they are very small. Here, Drs. Gina Wimp and Shannon Murphy sample invertebrates in a salt marsh at the Rutgers University Marine Station in New Jersey using D-Vacs. These backpack-mounted machines vacuum insects and other arthropods into a receptacle, which is then emptied, and the specimens are carefully sorted and identified.

Photo by Doug McCaskill, McCaskill Photography

FIGURE 7.3   Researchers and volunteers release a collared argali sheep ram (*Ovis ammon*) in Gun Galuut Nature Reserve, Mongolia. The argali are the largest mountain sheep in the world and are globally endangered. Data collected from collared animals are used to understand range requirements of and threats to this understudied species. This project is also an example of collaboration between a zoo and local groups.

© Richard P. Reading

heavily on volunteers for monitoring are Budburst, Zooniverse, NestWatch, and FrogWatch USA. Many of the methods for investigating populations are very specialized, such as the use of satellite tracking collars, and are best learned by studying under the supervision of an expert (**FIGURE 7.3**).

The need for fieldwork is highlighted by a study of the hooded vulture (*Necrosyrtes monachus*) in the countries of southern Africa (see the chapter opening photo). Vultures perform critical ecosystem services by removing carcasses that would otherwise spread disease and increase populations of pests such as rats. Of the 22 species of vultures in the world, 15 are considered **threatened** with extinction; the hooded vulture is classified as critically endangered by the IUCN. Globally, vultures face many threats, including poisoning from eating wild animals that have been killed with pesticides or lead shot or domesticated animals treated with anti-inflammatory drugs (Jimenez-Lopez et al. 2021). Other threats to vultures are loss of habitat, collisions with vehicles when feeding on roadkill, and electrocution from power lines, just to name a few. Models suggest that vultures are also threatened by global climate change, due to habitat becoming unsuitable (Phipps et al. 2017). In many parts of Africa, a significant threat is intentional poisoning by poachers, who kill the birds using drugged carrion because circling vultures give away their position. According to BirdLife South Africa (2017), more than 1500 vultures were killed this way between 2011 and 2014.

GLOBAL CHANGE
CONNECTION

Vultures are an important part of any ecosystem they inhabit; as "nature's cleanup crew," (sensu BirdLife.com) they help cycle nutrients and decrease the spread of disease. However, very little is known about the ecology of most of these species. To learn more about the behavior of the hooded vulture in northern Botswana, satellite transmitters on specially designed backpacks were placed on four vultures, and data about their movements were collected for up to four years (Reading et al. 2019). These data revealed that hooded vultures have much larger ranges than was previously believed—up to 30,000 km²—and that home range sizes vary significantly according to season. This type of information is important for management and conservation of the vulture, because it reveals the geographical range and international aspect of potential threats.

To truly understand a species, scientists must monitor populations repeatedly over time. Monitoring plays an important role in conservation biology for several reasons:

1. *Identifying when populations may be in trouble*   Monitoring allows us to determine whether effective population size is declining and helps determine reasons why, such as in response to levels of harvest or the arrival of invasive species (Homan et al. 2018). Observing a long-term decline in a species often provides motivation to take vigorous action to conserve it, such as in the case of the American bald eagle (**FIGURE 7.4**). Long-term monitoring records can help to distinguish long-term population increases or declines (possibly caused by human disturbance) from short-term fluctuations caused by variations in weather or unpredictable natural events (Swengel and Swengel 2018).

2. *Determining response to conservation efforts*   Monitoring records can also gauge whether an endangered species is showing a positive response to conservation management, even to the point when it no longer needs special protections.

3. *Measuring the health of a system*   Monitoring sensitive species, such as certain insects, can be used to indicate the condition of the entire ecosystem (Stork et al. 2017) (see "The species approach" later in this chapter).

4. *Providing a witness*   Fieldwork can sometimes bring researchers into confrontations with people who are engaged in illegal activities or otherwise harming species and ecosystems. Sometimes the simple presence of researchers engaged in the work of monitoring a population helps to protect it; if people are present, covert activities such as poaching will be more difficult to carry out. However, this also means that fieldwork can be dangerous, particularly for those who may already be a member of an at-risk group or otherwise perceived as a threat (Demery and Pipkin 2021).

For all these reasons (especially the first two), monitoring may be mandated by law as part of management efforts for endangered species and is often conducted in partnership with conservation organizations or university researchers. The most common types of monitoring are censuses, surveys, population demographic studies, and molecular monitoring.

CENSUS   A **census** is a count of the number of individuals present in a population. By repeating a census over successive time intervals, biologists can determine whether a population is stable, increasing, or decreasing in number. Censuses of a biological community can be conducted to determine what species are currently present in a locality. Censuses conducted over a wide area can help to determine the range of a species and its areas of local abundance. It is a comparatively inexpensive and straightforward

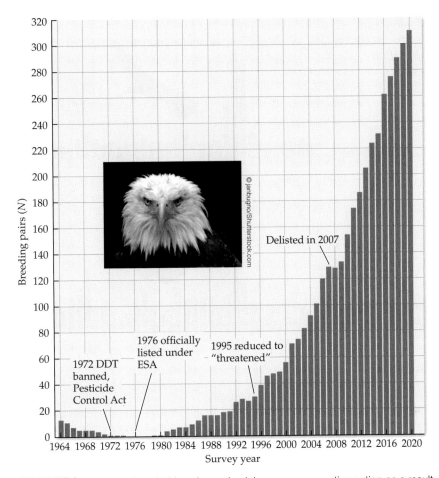

FIGURE 7.4 The American bald eagle received rigorous conservation action as a result of population monitoring. Breeding pairs observed along the James River in Virginia, are shown here over time, as they relate to conservation activity. Alarmingly low numbers of nesting pairs observed across the United States in the 1960s and 1970s spurred both research and policy action, including banning the pesticide DDT (which was harming the birds through biomagnification), increasing the penalties for hunting and trapping eagles, and listing for protection under the Endangered Species Act (ESA). Recovery after these actions meant that the species could be delisted, although it is still protected. The IUCN now lists them as "least concern." (After B. Watts. 2014. *James River Eagles Continue Historic Rise*. The Center for Conservation Biology [website]. https://ccbbirds.org/2014/09/04/james-river-eagles-continue-historic-rise. Accessed 4/16/19. Data from CCB.)

method for organisms that are easy to detect or are not mobile, such as for a small population of conspicuous wildflowers, but can be very difficult for especially large populations and small or highly mobile species. Censuses of orangutans, a highly elusive animal, can be done by counting the sleeping platforms that they build each night. In one study, the density of Bornean orangutans (*Pongo pygmaeus morio*) in a protected area was found to fluctuate in accordance with the fruiting patterns of trees in the area, providing valuable information about the behavior of these primates (**FIGURE 7.5**).

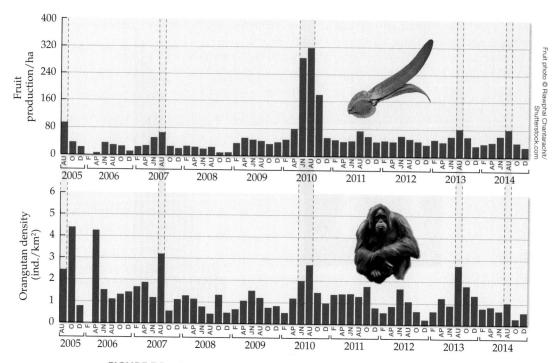

FIGURE 7.5 Census data can be used to compare environmental factors to factors regulating a population. Here we can see 10 years of census data for orangutans (*Pongo pygmaeus morio*) in the Danum Valley Conservation Area in the northern Borneo region of Sabah, Malaysia. It is compared here with fruit production by dipterocarp trees in the same area. Censusing the populations of orangutans is difficult because they live in trees, are often solitary, and have low densities. Their numbers are therefore estimated based on counting the sleeping platforms, or "nests," that they make daily. Shaded areas show periods of both high fruit production and large numbers of nests, which suggests that more orangutans moved into the protected area at these times. This is important for conservation because it shows that this primate has a strong feeding preference for dipterocarp fruits. (After T. Kanamori et al. 2017. *Primates* 58: 225–235.)

SURVEY   A **survey** of a population involves using a repeatable sampling method to estimate the abundance or density of a population or species in a part of a community. Survey methods are used when a population is very large or its range is extensive, making a direct measure of all individuals impossible. An area can be divided into sampling segments, and the number of individuals in certain segments can be counted. These counts can then be used to estimate the actual population size. For example, a rabbit population can be surveyed by walking **transects** (lines, often designated with measuring tape or string, along which biological data is collected) and recording the number of rabbits observed (Dieter and Schaible 2014). Boats and planes can be used to census whales and dolphins along transects in the ocean. New technologies are being employed to survey difficult-to-reach species, such as the use of drones to survey populations by taking photographs (Ore et al. 2015). Camera traps, which take a digital image when a motion sensor is tripped, are also used for difficult-to-survey animals.

As part of the North American Breeding Bird Survey (www.pwrc.usgs.gov/bbs), thousands of participants have been recording bird abundance at thousands of locations over the past 35 years along transects. This information is used to determine the stability of populations of over 400 bird species over time (**FIGURE 7.6**). Data from these surveys were the source for more than 380 publications in the decade before 2020 (https://www.pwrc.usgs.gov/BBS/Bibliography/). For example, a comparison of current occurrences with past censuses can be used to measure changes in species ranges or detect

(A)

FIGURE 7.6 (A) The North American Breeding Bird Survey (BBS) is one of the most comprehensive biological surveys in the world, as shown by this map of survey routes. Intensity of the shading represents the density of observer routes, each of which includes predetermined stops to record the occurrence of each bird species. (B) BBS data between 1966–2019 for the ruby-throated hummingbird shows a positive trajectory for populations (Smith and Edwards 2021). (A, from J. R. Sauer et al. 2013. *North Am Fauna* 79: 1–32; B, after USGS. BBS trends 1966–2019. Patuxent Wildlife Research Center—Bird Population Studies. Courtesy of the U.S. Geological Survey.)

(B)

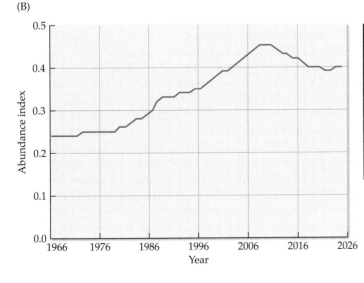

© Brian A Wolf/Shutterstock.com

the loss of populations (Smith and Edwards 2021). These data can also be used to determine which environmental variables (such as temperature) are influencing population numbers (Elsen et al. 2021).

DEMOGRAPHIC STUDY   **Demographic studies** follow known individuals of different ages and sizes in a population to determine their rates of growth, reproduction, and survival (Homan et al. 2018). Either the whole population or a subsample can be followed. In a complete population study, all individuals are counted, aged if possible, sized, sexed, and marked for future identification; their position on the site is mapped; and tissue samples may be collected for molecular analysis. Techniques vary depending on the characteristics of the species and the purpose of the study. Each discipline has its own techniques for following individuals over time: ornithologists band birds' legs, mammalogists often attach tags to animals' ears, and botanists nail aluminum tags to trees. Due to concern that the stress of being captured and tagged can have harmful or even fatal effects on animals (Jewell 2013), gentler techniques and guidelines are being developed.

> Demographic studies provide data on the numbers, ages, sexes, conditions, and locations of individuals within a population. These data indicate whether a population is stable, increasing, or declining and are the basis for statistical models used to predict the future of a species.

Information from demographic studies can be used in standard mathematical formulas (*life-history formulas*) to calculate the rate of population change and to identify critical stages in the life cycle (McCaffery et al. 2015). Demographic studies can also provide information on the age structure of a population. A stable population typically has an age distribution with a characteristic ratio of juveniles, young adults, and older adults. The absence or low representation of any age class, particularly juveniles, may indicate that the population is declining. Conversely, the presence of a large number of juveniles and young adults may indicate that the population is stable or even expanding.

LABORATORY ANALYSIS OF BIOLOGICAL MATERIAL   With the advent of molecular methods, new tools became available for monitoring populations (**FIGURE 7.7**). **Genetic monitoring** uses markers in DNA or proteins to detect the presence of, estimate the abundance of, and even detect rates of inbreeding or immigration among populations (Milligan et al. 2018). Genetic monitoring employs techniques from the field of *population genetics*. As genetic analysis becomes progressively less expensive, our capability expands to detect changes in population data over time or space. Collecting DNA samples from individuals can give us information about parentage, population structure, and genetic diversity (see Chapter 2). These samples can therefore also provide information on the potential for the population to undergo evolutionary adaptation to environmental change. An example of the application of genetic data for conservation can be found in the work of researchers who used a combination of genetic and demographic data from a threatened Mediterranean coral species to designate and prioritize conservation units within marine protected areas (Arizmendi-Mejía et al. 2015).

(A)

(B)

FIGURE 7.7   Increasingly, samples collected through fieldwork are brought back to a laboratory for analysis. (A) Dr. Simon Dures of Imperial College London, with other researchers, collects skin plugs from a sedated male lion in order to determine the structure of its metapopulation through genetic analysis. (B) Dr. Catarina C. Ferreira does an analysis on hair samples from Canada lynx to measure stress via cortisol concentration at Trent University, Peterborough, Ontario.

Samples for analysis may be taken directly from organisms (in the form of blood, skin, or other tissues), or even from scat and hair samples (Hedges et al. 2013). In some cases, specially trained dogs are used to locate scat samples of rare animal species (Matthew and Relton 2021). DNA studies using scats have revealed that population size is often larger than previous estimates made using traditional survey methods of direct observation, because some of the more elusive individuals have never been seen by observers. Alternatively, for organisms that are especially rare or small, environmental DNA (eDNA) can be used to estimate the presence, diversity, and density of a population in water, soil, or even air (see Chapter 2) (Taberlet et al. 2018). For example, eDNA has been used to monitor populations of imperiled West Indian manatees; surveys done this way found higher population estimates than those using aerial estimates (i.e., counting from helicopters or drones), which are also more costly to do (Hunter et al. 2018).

Analysis of samples collected in the field can also include other, nongenetic, methodologies such as isotope analysis (see Chapters 1 and 5), tests for the presence of disease, and measurements of hormone concentrations, just to name a few. For example, hair samples can be used to measure cortisol concentration, a measure of the stress an animal is experiencing (Terwissen et al. 2013) (see Figure 7.7B). Hormone analysis can also provide information about reproductive status, most importantly whether a female is pregnant or ovulating.

> Samples collected in the field may be brought back for laboratory analysis, providing insights into the genetic structure of populations or physiology of individuals.

Population viability analysis (PVA)

Monitoring data can allow us to calculate average mortality rates, average recruitment rates, the current age or size distribution of the population, and the area it occupies. This information can then be used to construct a mathematical model that estimates the ability of a population to persist in the future, a process known as **population viability analysis** (**PVA**). This is an extension of demographic analysis (e.g., Sæther and Engen 2015). There are many mathematical and statistical models used for PVA, which can be thought of as risk assessment—predicting the probability that a population or a species will go extinct at some point in the future (see Figure 8.5). It can also be used to identify vulnerable stages in the natural history of the species and to consider the effects of habitat loss, fragmentation, and deterioration (Meissen et al. 2017). In an analysis of 21 long-term ecological studies, researchers were able to validate the predictive power of this modeling tool (Brook et al. 2000). However, a more recent assessment of 160 published studies on birds that used PVAs found that many were poorly done, which could lead to misleading results (Chaudhary and Oli 2020).

> PVA uses mathematical and statistical methods to predict the probability that a population or species will go extinct within a certain time period. PVA is also useful in modeling the effects of habitat degradation and management efforts.

When done correctly, another important aspect of PVA is estimating how management efforts, such as changing hunting quotas or the area of protected habitat, will affect the probability of extinction (**FIGURE 7.8**). PVA can also model the effects of augmenting a population through the release of additional individuals caught in the wild elsewhere or raised in captivity.

One example of the use of PVA is a study of the western lowland gorilla (*Gorilla gorilla gorilla*), a slow-reproducing primate that is critically endangered (**FIGURE 7.9**). Reintroduction of captive-bred or rehabilitated wild, orphaned gorillas appeared to be successful in terms of survival of the releases, but it was not known whether these reintroduced populations were likely to persist over

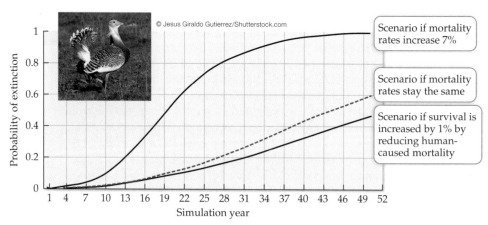

FIGURE 7.8 Population viability analysis (PVA) can be used to predict extinction rates under different scenarios. Here, the probability of extinction into the future is modeled for great bustard (*Otis tarda*) populations in Morocco. We can see that in 50 years the probability of extinction is approximately 55% if conditions stay the same, will be near 100% if mortality increases by just 7%, and can be reduced to near 40% if human-induced mortality is reduced so survival is increased by 1%. (After C. Palacin et al. 2016. *Endang Species Res* 30: 73–82. doi.org/10.3354/esr00726. CC BY 4.0, creativecommons.org/licenses/by/4.0.)

time. Demographic studies of reintroduced populations in the Batéké Plateau region of Congo and Gabon, supplemented with data from captive populations, were used to develop a population model for PVA (King et al. 2014). The analysis revealed that for this species, annual birthrates and adult female mortality rates were parameters that particularly affected the probability of a population going extinct within 200 years. For example, increasing the birthrate from 0.18 to 0.20 decreased the probability of extinction from 29% to 9%. Given the current population structure, the PVA results also suggested that these populations had a greater than 90% likelihood of surviving for 200 years. This is important support for the re-

FIGURE 7.9 Population viability analysis can be especially useful for long-lived and slowly reproducing species such as lowland gorillas.

introduction program, but it also emphasizes the necessity of monitoring over time because parameters such as birthrate and adult female survival may change in response to environmental stochasticity—random events in the environment, such as a drought or severe storm.

A key feature of PVA models, and a reason why they are especially useful for analyzing small populations, is that they can incorporate both environmental and demographic stochasticity (as explained in Chapter 6). For example, if the birthrate of a population has been determined to be 0.5, this means that each breeding individual has a 50% chance of producing young, which is analogous to the results of a coin flip. But we know that when we flip a coin 10 times, only occasionally will we see exactly 5 heads and 5 tails; this is simply due to random chance. There is even a chance (albeit very small) that we will flip 10 heads in a row. Similarly, even when we are confident of the birthrate, we cannot be sure how many young will actually be born in any given year, which can have especially dramatic implications for smaller populations. With a PVA, hundreds or thousands of simulations of individual populations can be run using this random variation, allowing us to determine the probability of population extinction within a certain period of time, the median time to extinction, changes in population size, and other metrics.

These probabilities will differ based on birth and death rates and other parameters that may be included in a PVA, such as rates of immigration and the frequencies of rare storms and disease outbreaks. For this reason, different populations within the same species may have different extinction probabilities. In one study of Pitcher's thistle (*Cirsium pitcheri*), a rare plant endemic to the Great Lakes area of the midwestern United States, it was found that, although most populations greater than 100 individuals were likely to survive over the next century, even some very large populations had a high probability of going extinct, based on poor population growth (Nantel et al. 2018).

Generally, about 10 years of monitoring data are needed to obtain a PVA with good predictive power. The results of some models can often change dramatically with different model assumptions and slight changes in parameters; thus, they must be interpreted with caution. Nevertheless, PVA is widely used and has value in demonstrating the possible effectiveness of alternative management strategies.

Minimum viable population (MVP)

PVA is also used to calculate the minimum number of individuals of a given species required for a population to persist over time. Some researchers argue that PVA should be used in the establishment of **recovery criteria**, predetermined thresholds that signal that an endangered species can be removed from protection under the Endangered Species Act (Beissinger 2015). In a groundbreaking paper, Shaffer (1981) defined the number of individuals necessary to ensure the long-term survival of a species as the **minimum viable population (MVP)**:

> *A minimum viable population for any given species in any given habitat is the smallest isolated population having a 99% chance of remaining extant for 1000 years despite the foreseeable effects of demographic, environmental, and genetic stochasticity, and natural catastrophes.*

In other words, the MVP is the smallest population size that can be predicted to have a very high chance of persisting for the foreseeable future. Shaffer emphasized the tentative nature of this definition, saying that the survival probabilities could be set at 95%, 99%, or any other percentage and that the time frame might similarly be adjusted—for example, to 100 or 500 years. The key point is that the MVP size allows a quantitative estimate to be made of how large a population must be to ensure long-term survival. In general, protecting a larger population increases the chance that the population will persist for a longer period of time (**FIGURE 7.10**).

FIGURE 7.10 If the goal is persistence for a greater number of years, then a larger minimum viable population (MVP) size is needed. A greater MVP is needed to ensure a higher probability of persistence, as illustrated in this case by a 50% probability of survival and a greater than 90% probability of survival. Both axes are on log scales. The values were derived from changes in population size and persistence of 1198 species. (After L. W. Traill et al. 2010. *Biol Conserv* 143: 28–34. © 2009. Reprinted with permission from Elsevier.)

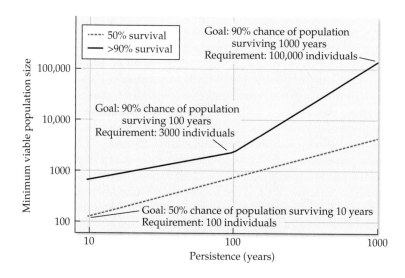

Shaffer (1981) compared MVP protection efforts to flood control. It is not sufficient to use average annual rainfall as a guideline when planning flood control systems; instead, we must plan for extreme situations of high rainfall and severe flooding, which may occur only once every 50 or 100 years. In the same way, when attempting to protect natural systems, we understand that certain catastrophic events, such as hurricanes, earthquakes, forest fires, epidemics, and die-offs of food items, may occur at even greater intervals. To plan for the long-term protection of endangered species, we must provide for their survival not only during average years but also during exceptionally harsh years. Consequently, an accurate estimate of the MVP size for a species requires an analysis of its environment. This can be expensive and require months, or even years, of research. Analyses of over 200 species for which adequate data were available (mainly vertebrates) indicated that most MVP values for long time periods fall in the range of 3000–5000 individuals, with a median of 4000 (Flather et al. 2011). For species with extremely variable population sizes, such as certain invertebrates and annual plants, protecting a population of about 10,000 individuals may be the ideal strategy.

Unfortunately, many species, particularly endangered species, have population sizes smaller than these recommended minimums. For instance, 77% of the reserves that host elephant populations in South Africa have fewer than 100 individuals, a number considered to be inadequate for long-term survival of the population (Pretorius et al. 2019).

Once an MVP size has been established for a species, the **minimum dynamic area (MDA)**—the area of suitable habitat necessary for maintaining the MVP—can be estimated by studying the size of the home range of individuals and colonies of endangered species (Pe'er et al. 2014; Thiollay 1989). It has been estimated that reserves in Africa of 100–1000 km² are needed to maintain many small mammal populations (**FIGURE 7.11**). To preserve populations of large carnivores, such as lions, reserves of 10,000 km² are needed.

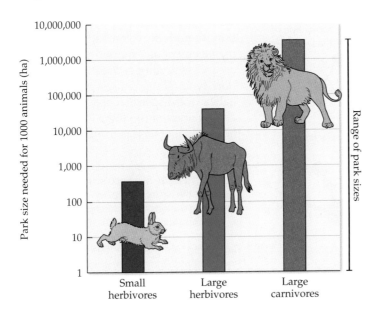

FIGURE 7.11   By considering the population size of various species at several parks of different sizes in Africa, it was determined how large a park must be to support 1000 individuals for each of three trophic groups. Note that the largest park measured was approximately 6.43 million ha—the same size as the minimum necessary to support 1000 large carnivores. Herbivores needed considerably less space for the same number of individuals. The minimum dynamic area (MDA) will differ between trophic guilds because carnivores need more space than herbivores, and large animals will need more space than small animals. (After C. M. Schonewald-Cox. 1983. In *Genetics and Conservation*, C. M. Schonewald-Cox et al. [eds.], p. 414. Benjamin/Cummings, Menlo Park, CA.)

Metapopulations

Over time, populations of a species may become extinct on a local scale while new populations may form nearby on other suitable sites. Often a species that lives in an ephemeral habitat, such as a streamside herb, is better characterized in terms of a **metapopulation** (a "population of populations") that is made up of a shifting mosaic of populations linked by some degree of migration. In some species, every population in the metapopulation is short-lived and the distribution of the species changes dramatically with each generation. In other species, the metapopulation may be characterized by one or more **source populations** (core populations) with fairly stable numbers and several **sink populations** (satellite populations) that fluctuate in size with arrivals of immigrants. Populations in the satellite areas may become extinct in unfavorable years, but the areas may be recolonized, or rescued, by migrants from the more permanent core population when conditions become more favorable (**FIGURE 7.12**). Metapopulations may also involve relatively permanent populations between which individuals occasionally move.

> Populations of a species are often connected by dispersal and can be considered a metapopulation. In such a system, the loss of one population can negatively affect other populations.

Bighorn sheep (*Ovis canadensis*) offer a well-studied example of metapopulation dynamics. These sheep have been observed dispersing between mountain ranges and occupying previously unpopulated sites, while mountains that previously had sheep populations are now unoccupied. Migration and gene flow occur primarily between populations less than 15 km apart and are greater when the intervening countryside is hilly rather than flat (Creech et al. 2014). Maintaining dispersal routes between existing population areas and potentially suitable sites, including across international borders, is important in managing this species (Buchalski et al. 2015).

The persistence of metapopulations often depends on habitat availability. For example, destruction of the habitat of one central, core population might result in the extinction of numerous smaller populations that depend on the core population for periodic colonization. Effective management of a species often requires an understanding of these metapopulation dynamics and a restoration of lost habitat and dispersal routes. For example, a population of the Indo-Pacific bottlenose dolphin (*Tursiops aduncus*) in coastal

(A) Three independent populations

(B) Simple metapopulation of three interacting populations

(C) Metapopulation with a large core population and three satellite populations

(D) Metapopulation with complex interactions

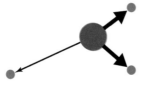

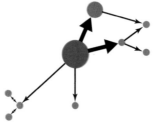

FIGURE 7.12 Possible metapopulation patterns, with the size of a population indicated by the size of the circle. The arrows indicate the direction and intensity of migration between populations. (After P. S. White. 1996. In *Restoring Diversity*, D. A. Falk, C. K. Millar, and M. Olwell [eds.], pp. 49–86. © 1996 by Island Press. Reproduced by permission of Island Press, Washington, DC. Fig 3-2 A,B,E,G.)

waters near Bunbury, Western Australia, was recently discovered through genetic analysis to be a source population for adjacent sink populations (Manlik et al. 2018). This is problematic because it had already been determined through PVA that the Bunbury population was not reproducing enough to sustain itself, much less the entire region (Manlik et al. 2016). Taken together, these findings suggest that the source-sink dynamics may reverse in the future, thus threatening the adjacent populations.

Long-term monitoring

To understand the reasons behind population changes, monitoring of populations needs to be combined with monitoring of other environmental parameters. The long-term monitoring of ecosystem processes (e.g., temperature, rainfall, humidity, soil acidity, water quality, discharge rates of streams, and soil erosion) and community characteristics (e.g., species present, percentage of vegetative cover, amount of biomass present at each trophic level) allows scientists to determine the health of the ecosystem and the status of species of special concern (Haase et al. 2018). The Long-Term Ecological Research (LTER) programs in the United States and in Europe (eLTER) focus on such changes on timescales ranging from months and years to decades and centuries (**FIGURE 7.13**). Another program to collect long-term data in the United States is the National Ecological Observatory Network (NEON); NEON is centrally operated, whereas the LTER programs are collections of individual investigator projects.

	Years	Research discipline	Physical events	Biological phenomena
10^5	100 Millennia			Evolution of species
10^4	10 Millennia	Paleoecology and limnology	Continental glaciation	Bog succession / Forest community migration
10^3	Millennium		Climate change	Species invasion / Forest succession
10^2	Century		Forest fires / CO_2-induced climate warming	Cultural eutrophication
10^1	Decade			Population cycles
10^0	Year		Sun spot cycle / El Niño events / Prairie fires / Lake turnover / Ocean upwelling	Prairie succession / Annual plants / Seasonal migration / Plankton succession
10^{-1}	Month			
10^{-2}	Day		Storms / Daily light cycle / Tides	Algal blooms / Daily movements
10^{-3}	Hour	Most ecology		

(LTER brackets the range from 10^2 Century to 10^{-1} Month)

FIGURE 7.13 The Long-Term Ecological Research (LTER) program focuses on timescales ranging from months to centuries in order to understand changes in the structure, function, and processes of ecosystems that are not apparent from short-term observations. (After J. J. Magnuson. 1990. *BioSci* 40: 495–501. By permission of Oxford University Press.)

As an example of the need for long-term monitoring, certain amphibian, insect, and annual plant populations are highly variable from year to year, so many years of data are required to determine whether a particular species is declining in abundance over time or merely experiencing a number of low population years that are in accord with its regular pattern of variation. In one instance, more than 50 years of observation of populations of two flamingo species (*Phoenicopterus ruber*, the greater flamingo, and *Phoeniconaias minor*, the lesser flamingo) in southern Africa revealed that large numbers of chicks fledged only in years with high rainfall, making it appear that reproduction was simply highly variable (**FIGURE 7.14**). In fact, there was a 31-year gap in which no major hatchings occurred, and the populations appeared to be in peril. Fortunately, the populations began to increase in 2008 and 2011, when numerous chicks hatched.

The fact that environmental effects may lag for many years behind their initial causes creates a challenge to understanding change in ecosystems. For example, acid rain, nitrogen deposition, and other components of air pollution

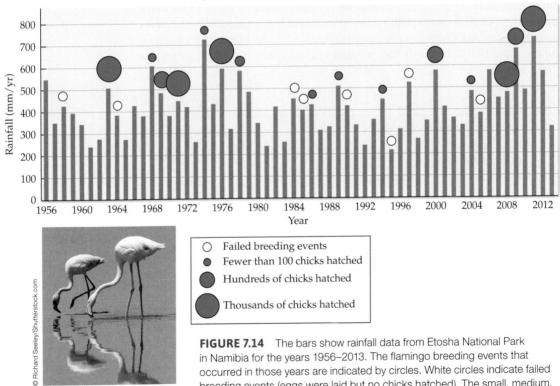

FIGURE 7.14 The bars show rainfall data from Etosha National Park in Namibia for the years 1956–2013. The flamingo breeding events that occurred in those years are indicated by circles. White circles indicate failed breeding events (eggs were laid but no chicks hatched). The small, medium, and large green circles indicate, respectively, fewer than 100 chicks hatched, hundreds of chicks hatched, and thousands of chicks hatched. There was a 31-year gap between 1976 and 2008 in which no large hatching event occurred. (After R. E. Simmons 1996. *Cons Bio* 10: 504–515 © 1996 Wiley, with updates from R. E. Simmons, pers. comm.)

© Richard Seeley/Shutterstock.com

may gradually change the water chemistry, algal community, and oxygen content of forest streams, ultimately making the aquatic environment unsuitable for the larvae of certain insect species. In this case, the cause (air pollution) may have occurred years or even decades before the effect (insect decline) becomes detectable. Even habitat fragmentation can have delayed effects on losses via gradual environmental degradation and metapopulation extinction. When a population has more individuals than predicted based on current levels of habitat loss or degradation, it is called **extinction debt** (Kuussaari et al. 2009) (also see Chapter 6). Such a determination is not possible without monitoring data.

GLOBAL CHANGE
CONNECTION

A major purpose of monitoring programs is to gather essential data on biological communities and ecosystem functions that can be used to document changes over time. Monitoring in these studies allows managers to determine whether the goals of their projects are being achieved or whether adjustments must be made in the management plans (called adaptive management), as discussed in Chapter 11. It is important to note that sometimes conservation decisions must be made before data collection is complete, or with only limited data. As we will see in the next section, complete evaluations are only available for some species and are more common in some taxa than in others. With time and effort, our data on species and populations will continue to expand and improve.

7.2 Conservation Categories

LEARNING OBJECTIVES

By the end of this section you should be able to:

7.2.1 Interpret the meaning of a report on any species in the IUCN Red List.

7.2.2 Contrast measures of threat used by the IUCN, Natural Heritage, and Living Planet Index.

Once the data collected on populations and species allow us to identify those species most vulnerable to extinction, it is useful to create a system to categorize extinction risk so that we can prioritize conservation efforts. To mark the status of rare and endangered species for conservation purposes, the IUCN has established conservation categories (see the IUCN Red List, www.iucn.org) (**FIGURE 7.15**). These categories have proved useful in establishing protection for threatened species at the national and international levels and directing attention toward species of special concern. The conservation categories are as follows:

> The IUCN uses quantitative information, including the area occupied by the species and the number of mature individuals currently alive, to assign species to conservation categories.

- *Extinct (EX)*   The species (or other taxon, such as subspecies or variety) is no longer known to exist. The IUCN currently lists 779 animal species and 123 plant species as extinct.

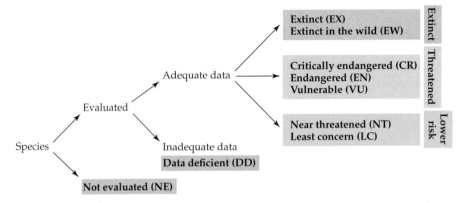

FIGURE 7.15 The IUCN categories of conservation status. This chart shows the distribution of the categories. Reading from left to right, they depend on (1) whether a species has been evaluated or not and (2) how much information is available for the species. If data are available, the species is then put into a category of lower risk, threatened, or extinct. (After IUCN. 2012. IUCN Red List Categories and Criteria. Version 3.1, 2nd ed. Gland, Switzerland and Cambridge, UK. https://portals.iucn.org/library/sites/library/files/documents/RL-2001-001-2nd.pdf.)

- *Extinct in the wild (EW)*   The species exists only in cultivation, in captivity, or as a naturalized population well outside its original range. The IUCN currently lists 38 animal species and 42 plant species as extinct in the wild.
- *Critically endangered (CR)*   The species has an extremely high risk of going extinct in the wild, according to any of the criteria A–E (**TABLE 7.1**).
- *Endangered (EN)*   The species has a very high risk of extinction in the wild, according to any of the criteria A–E.
- *Vulnerable (VU)*   The species has a high risk of extinction in the wild, according to any of the criteria A–E.
- *Near threatened (NT)*   The species is close to qualifying for a threatened category but is not currently considered threatened.
- *Least concern (LC)*   The species is not considered near threatened or threatened. (Widespread and abundant species are included in this category.)
- *Data deficient (DD)*   Inadequate information exists to determine the risk of extinction for the species.
- *Not evaluated (NE)*   The species has not yet been evaluated against the Red List criteria.

When used on a national or other regional level, there are two additional Red List categories:

- *Regionally extinct (RE)*   The species no longer exists within the country or region but is extant in other parts of the world.

TABLE 7.1	IUCN Red List Criteria for the Assignment of Conservation Categories
Red List criteria A–E	Quantification of criteria for Red List category "critically endangered"[a]
A. Observable reduction in numbers of individuals	The population has declined by ≥90% over the last 10 years or 3 generations (whichever is longer), either based on direct observation or inferred from factors such as levels of exploitation, threats from introduced species and disease, or habitat destruction or degradation.
B. Total geographical area occupied by the species	The species has a restricted range (<100 km^2 at a single location) *and* there is observed or predicted habitat loss, fragmentation, ecological imbalance, or heavy commercial exploitation.
C. Predicted decline in number of individuals	The total population size is less than 250 mature, breeding individuals and is expected to decline by 25% or more within 3 years or 1 generation.
D. Number of mature individuals currently alive	The population size is less than 50 mature individuals.
E. Probability the species will go extinct within a certain number of years or generations	Extinction probability is greater than 50% within 10 years or 3 generations.

Source: After IUCN. 2012. IUCN Red List Categories and Criteria. Version 3.1, 2nd ed. Gland, Switzerland and Cambridge, UK. https://portals.iucn.org/library/sites/library/files/documents/RL-2001-001-2nd.pdf

[a] Species that meets the described quantities for *any one* of criteria A–E may be classified as critically endangered. Similar quantification for the Red List categories "endangered" and "vulnerable" can be found at www.iucnredlist.org.

- *Not applicable* (NA)   The species is not eligible for the regional Red List because, for example, it is not within its natural range in the region (i.e., it has been introduced) or because it is only a rare migrant to the region.

Species in the critically endangered, endangered, and vulnerable categories are classified as "threatened with extinction," which is quantified based on the probability of extinction. These **Red List criteria**, described in Table 7.1, are based on PVA models, which in turn depend on data collected in the field. These criteria focus on population trends and habitat condition. The advantage of this system is that it provides a standard method of classification by which decisions can be reviewed and evaluated according to accepted quantitative criteria, using whatever information is available. The disadvantage is that they depend on data provided by countries, which may have a vested interest in reporting species numbers as increasing or decreasing. For example, although giant pandas (*Ailuropoda melanoleuca*) were down-listed from endangered to vulnerable, some have questioned whether China may have inflated numbers for political reasons, while others have questioned whether the downgrading was premature simply given the ongoing threat of habitat loss (Ma et al. 2018).

The IUCN system has been used to identify Red Lists of threatened species and to determine whether species are responding to conservation efforts.

Using habitat loss as a criterion in assigning categories is particularly useful for many species that are poorly known biologically, because species can be listed as threatened if their habitat is being destroyed even if scientists know little else about them. In practice, a species is most commonly assigned to an IUCN category based on the area it occupies, the number of mature individuals it has, or the rate of decline of the habitat or population; the probability of extinction is least commonly used.

Using the criteria in Table 7.1 and the categories in Figure 7.15, the IUCN has evaluated and described the threats to plant and animal species in its series of **Red Lists** (sometimes referred to as **Red Data Books**) of threatened species; these detailed lists of endangered species by group and by country can be seen at www.iucnredlist.org. As of 2021, species listed as threatened include 1323 of 5940 described mammal species, 1481 of 11,158 bird species, and 20,883 of 52,077 flowering plant species (see Table 6.1).

Most bird, amphibian, and mammal species have been evaluated using the IUCN system, but the levels of evaluation are lower for reptiles, fishes, fungi, and flowering plants, resulting in low apparent levels of threat. Even though numerous species of fishes, reptiles, molluscs, insects, crustaceans, and plants are designated as threatened with extinction, most species in these groups have still not yet been evaluated (see Table 6.1). The evaluations of other invertebrates, mosses, algae, fungi, and microorganisms are even less adequate. While in most cases the lack of data leads to underestimates of extinction risk, it has been argued that for some groups the risk may be overestimated when based on only presence-absence data, such as for amphibians (Cruickshank et al. 2016).

By tracking the conservation status of species over time, it is possible to determine whether species are responding to conservation efforts or are continuing to be threatened. One such measure is the **Red List Index**, which demonstrates that the conservation status of certain animal groups has continued to decline since 1988, with particularly sharp declines for corals and cycads (see Figure 6.2). Another measure based simply on numbers of individuals, the **Living Planet Index**, follows population sizes for more than 22,000 vertebrate species; this index has shown an average decline of 68% in species abundances from 1970 to 2016 (WWF 2020) (**FIGURE 7.16A**). The Neotropical area—South and Central America—has had the greatest decline (94%), with freshwater fishes leading losses among vertebrates (84%). Although the losses are deeply troubling, it is also important to note that the rate of decline has slowed since the 1990s and has been nearly flat over the last decade globally, with population sizes increasing on average in at least one region (**FIGURE 7.16B**).

A program similar to the efforts of the IUCN is the NatureServe network (www.natureserve.org). First organized by The Nature Conservancy, this network was made up of regional Natural Heritage Programs in the US and Canada (https://nhnm.unm.edu/about/network). Today, the NatureServe network spans both North and South America and includes over 100 partners representing over 1,000 conservation professionals. Together, the network gathers, organizes, and manages information on the occurrence of "elements

GLOBAL CHANGE
CONNECTION

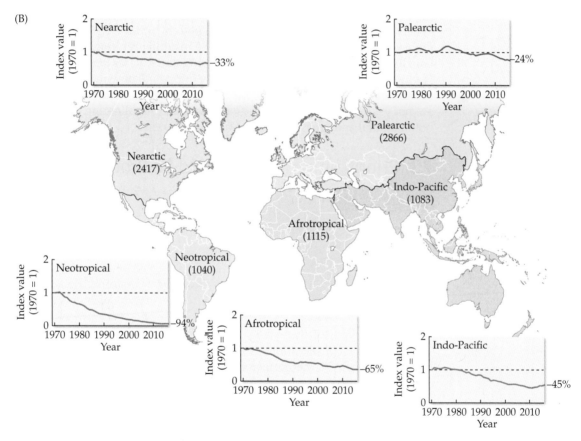

FIGURE 7.16 The Living Planet Index reflects the monitoring data of 27,569 populations representing 4801 vertebrate species across the globe (A) and shown by region (B). The number of species monitored is shown. The index value represents the average relative change in population abundance since 1970. The index was first published in 1998 and uses 1970 as a baseline (index = 1) because this is the period for which there is enough information available for a reliable index. Although the most recent index was published in 2020, it includes data only to 2016 because of the time it takes to collect, process, and publish monitoring data. Although an average drop across the globe of well over half in less than 50 years is rightfully alarming, the fact that the rate of decline (the slope of the line) has decreased in recent years of significant conservation activity (even showing a slight increase in the Nearctic) may be cause for optimism. (After WWF. 2020. *Living Planet Report 2020: Bending the Curve of Biodiversity Loss*. M. Grooten and R. E. A. Almond [eds.] WWF, Gland, Switzerland.)

of conservation interest"—more than 70,000 species, subspecies, and biological communities, in addition to more than a million precisely located populations. Elements are given status ranks based on a series of standard criteria: number of remaining populations or occurrences, number of individuals remaining (for species) or extent of area (for communities), number of protected sites, degree of threat, and innate vulnerability of the species or community. On the basis of these criteria, elements are assigned an imperilment rank from 1 to 5, ranging from critically imperiled (1) to demonstrably secure (5), on a global (G), national (N), and state/province (S) basis. Species are also classified as "X" (extinct), "H" (known historically, with searches ongoing), and "unknown" (uninvestigated). Data on these conservation elements are available on the NatureServe website (explorer.natureserve.org). When methods of the Natural Heritage system and the IUCN categories are used to evaluate the same species, the resulting rankings of threat are usually similar. For example, the whooping crane (*Grus americana*) has a NatureServe classification of G1 (globally imperiled) and an IUCN rank of endangered. However, there are many species on the NatureServe list that are not yet included in the IUCN global database, highlighting the need for both national and international evaluation programs.

7.3 Prioritization: What Should Be Protected?

LEARNING OBJECTIVES

By the end of this section you should be able to:

7.3.1 Evaluate the likelihood that a particular species will be protected.

7.3.2 Compare different approaches for making conservation priorities.

Although some conservationists would argue that no species should ever be lost, the reality is that numerous species are in danger of going extinct and there are too few resources available to save them all. The challenge to conservation efforts lies in finding ways to minimize the loss of biodiversity during a period of limited financial and human resources (Meyer et al. 2016).

The species approach

One approach to establishing conservation priorities involves protecting particular species—and in doing so, protecting an entire biological community and associated ecosystem processes. Protected areas are often established to protect individual species of special concern, such as rare species, endangered species, keystone species, and culturally significant species. Species such as these, which provide the impetus to protect an area and ecosystems, are known as **focal species**. One type of focal species is an **indicator species**, a species that is associated with an endangered biological community or set of

FIGURE 7.17 The Komodo dragon (*Varanus komodoensis*) of Indonesia is an example of a species that fits all three DUE categories: it is the world's largest lizard (distinctiveness), it has major potential as a tourist attraction in addition to being of great scientific interest (utility), and it occurs on only a few small islands of a rapidly developing nation (endangerment).

unique ecosystem processes. For instance, the endangered northern spotted owl is a forest indicator species in the Pacific Northwest of the United States. Many national parks have also been created to protect **flagship species**, such as tigers and pandas, which capture public attention, have symbolic value, and are crucial to ecotourism. Flagship and indicator species are also known as **umbrella species** because protecting them automatically protects other species and aspects of biodiversity.

Another way in which the prioritization of species for conservation can be determined is by evaluating the degree to which it is "DUE": distinctive, utilitarian, and endangered (**FIGURE 7.17**):

1. *Distinctiveness (or irreplaceability)*   A species is often given high conservation value if it is taxonomically distinctive—that is, it is the only species in its genus or family. Similarly, a population of a species having unusual genetic characteristics that distinguish it from other populations of the species might be a high priority for conservation. An ecosystem composed primarily of rare endemic species or with other unusual attributes (small area, scenic value, unique geological features) is given a high priority for conservation.

2. *Utility*   Species that have present or potential value to people, such as wild relatives of wheat, are given high conservation priority. Species with major cultural significance, such as tigers in India and the bald eagle in the United States, are also given high priority.

3. *Endangerment (or vulnerability)*   Species in danger of extinction are of greater concern than species that are not; thus, the whooping crane, with only about 249 individuals, requires more protection than the sandhill crane (*Grus canadensis*), with approximately 650,000 individuals (Dubovsky 2020).

The species approach follows from creating survival plans for individual species, which are developed by governments and private conservation organizations. In the Americas, the NatureServe network uses information on rare and endangered species to target new localities for conservation—areas where there are concentrations of endangered species or where the last populations of a declining species exist. In the IUCN Species Survival Commission, over 100 specialist groups provide action plans for endangered animals and plants.

The ecosystem approach

A number of conservationists have argued that rather than targeting species, ecosystems and the biological communities they contain should be targeted for conservation (Amel et al. 2017). They claim that spending $1 million or so on habitat protection and management of a self-maintaining ecosystem might preserve more species and provide more value to people in the long run than spending the same amount of money on an intensive effort to save just one conspicuous species. It often is easy to demonstrate an ecosystem's economic value to policymakers and the public in terms of flood control, clean water, and recreation, whereas arguing for a particular species may be more difficult. Thus, combining species and ecosystem approaches may be a good conservation strategy in many circumstances.

When using this ecosystem-based approach, conservation planners should try to ensure that representative sites of as many types of ecosystems as possible are protected. A **representative site** includes the species and environmental conditions characteristic of the ecosystem. Although no site is perfectly representative, biologists working in the field can often identify suitable sites for protection.

Where immediate decisions must be made to determine park boundaries and which species and ecosystems to protect, biologists are being trained to make **rapid biodiversity assessments**, also known as **rapid assessment programs**, or RAPs. RAPs involve mapping vegetation, making lists of species, checking for species of special concern, estimating the total number of species, and looking out for new species and features of special interest. A bioblitz (explained in Chapter 2) is one type of RAP.

The wilderness approach

Wilderness areas are a related priority, in part because they are more likely to contain species and populations that need protecting, such as old-growth trees (**FIGURE 7.18**). Large blocks of land that have been minimally affected by human activity, have a low human population density, and are not likely to be developed in the near future are also perhaps the only places in the world where large mammals can survive in the wild. These

FIGURE 7.18 Old-growth giant sequoias are protected in this designated wilderness area within Sequoia and Kings Canyon National Parks in California.

wilderness areas can also serve as reference areas for restoration (see Chapter 11). It is worth emphasizing that even these so-called wilderness areas have had a long history of human activity, and people have often affected the structure of the biological communities they contain.

The hotspot approach

Certain taxa can be used as **biodiversity indicators** to highlight new areas where concentrations of species can be protected. For example, a site with a high diversity of flowering plants often, but not always, will also have a high diversity of mosses, spiders, fungi, and other organisms. Further, areas with high diversity often have a high percentage of **endemism**—species occurring there and nowhere else (Ibanez et al. 2017). BirdLife International (www.birdlife.org) has identified over 13,000 Important Bird and Biodiversity Areas (IBAs) in over 200 different countries, including marine environments. Many of these localities are in urgent need of protection. Conditions that foster high bird diversity and endemism are likely to promote diversity and endemism in other groups.

Using a similar approach, Conservation International, WWF, and others have designated **hotspots** that have great biological diversity and high levels of endemism and that are under immediate threat of species extinctions and habitat destruction (see Biodiversity A–Z, www.biodiversitya-z.org/content/biodiversity-hotspots) (**FIGURE 7.19**). Using these criteria, 35 global hotspots have been targeted for new protected areas. These hotspots together encompass 50% of the world's endemic plant species and 42% of all terrestrial vertebrates—all on only 2.3% of Earth's total land surface.

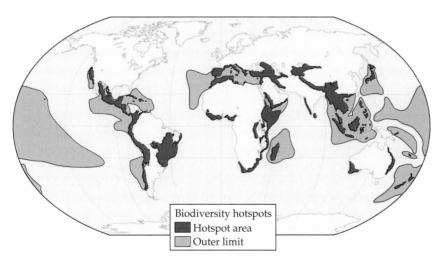

FIGURE 7.19   Hotspots are targets for protection because of their high biodiversity, endemism, and significant threat of imminent extinctions. (From WildArk.org. 2017. Biodiversity hotspots—Why are they so critical? *WildArk Journal* (online), adapted from R. A. Mittermeier et al. 2011. In *Biodiversity Hotspots*, F. Zachos and J. Habel [eds.]. p. 3. Springer, Berlin, Heidelberg. link.springer.com/chapter/10.1007/978-3-642-20992-5_1.)

7.4 Legal Protection of Species

LEARNING OBJECTIVES

By the end of this section you should be able to:

7.4.1 Use a case study of a species to explain why both national laws and international agreements are necessary for protection.

7.4.2 Defend the position that industrialized nations should help pay for biodiversity in other countries.

Once conservation biologists have identified a species as needing protection, laws can be passed and treaties can be signed to implement conservation efforts. National laws protect species within individual countries, while international agreements provide a broader framework for conservation.

National laws

People in many countries recognize that preserving a healthy environment and protecting species are linked to sustaining human health and a thriving economy. National governments and national conservation organizations in such countries acknowledge this and play an important role in the protection of all levels of biological diversity. Laws are passed to establish national parks and other protected areas; to regulate activities such as fishing, logging, and grazing; and to limit air and water pollution. International treaties that restrict trade in endangered animals are implemented at the national level and enforced at the borders. The true measure of a nation's commitment to protecting biodiversity is the effectiveness with which these laws are enforced.

In European countries, endangered species conservation is accomplished through domestic enforcement of international agreements such as the Convention on International Trade in Endangered Species (see "International Agreements") and the Ramsar Convention on Wetlands (see Chapter 12). Species that occur on the IUCN's international Red Lists of endangered species and in national lists are also protected. To indicate where these species can be found, the Fauna Europaea database (fauna-eu.org) provides information on the distribution of 235,708 terrestrial and freshwater species. Countries in Europe protect species and habitats through directives adopted by the European Union, with a special focus on migratory species. Some countries may have additional laws, such as the Wildlife and Countryside Act of 1981 in the United Kingdom, which protects habitat occupied by endangered species.

Even though many countries have enacted legislation to preserve biodiversity, national governments are sometimes unresponsive to requests from conservation groups to protect the environment. In some cases, national governments have acted to decentralize decision-making, relinquishing control of natural resources and protected areas to local governments, village councils, and conservation organizations.

National governments protect designated endangered species within their borders, establish national parks, and enforce legislation on environmental protection.

In the United States, the principal conservation law protecting species is the **Endangered Species Act** (**ESA**), passed in 1973 and subsequently amended in 1978 and 1982. The ESA was created by the US Congress to "provide a means whereby the ecosystems upon which endangered species and threatened species depend may be conserved [and] to provide a program for the conservation of such species." Species are protected under the ESA if they are on the official list of endangered and threatened species (**FIGURE 7.20**). In addition, a recovery plan is generally required for each listed species (Himes Boor 2014).

> The ESA mandates such strong protection for species that conservation and business groups often agree to compromises that allow some species protection along with limited development.

The language and categories used by the ESA are different from those in the IUCN or NatureServe systems. As defined by law, "endangered species" are those likely to become extinct as a result of human activities or natural causes in all or a significant portion of their range, whereas "threatened species" are those likely to become endangered in the near future. The secretary of the Interior Department, acting through the US Fish and Wildlife Service, and the secretary of the Commerce Department, acting through the National Marine Fisheries Service, add and remove species from the list based on information available to them. Since 1973, more than 2360 species have been added, including many well-known American species such as sockeye salmon (*Oncorhynchus nerka*) and the manatee (*Trichechus manatus*) and including 694 endangered species from elsewhere in the world that face special restrictions when they are imported into the United States. Many species are listed under the ESA only when

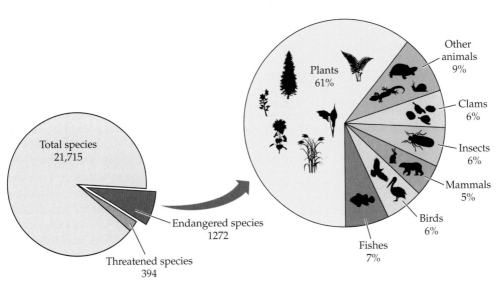

FIGURE 7.20 Of a total of approximately 21,700 species in the United States, 1666 have a legal designation as endangered or threatened, as defined by the Endangered Species Act. Of these, the largest number are plants. (After US Fish and Wildlife Service ECOS. 2021. *Listed Species Summary (Boxscore)*, as of March 29, 2021. https://ecos.fws.gov/ecp/report/boxscore; animals © glyph/shutterstock.com; plants © ntnt/shutterstock.com.)

they have fewer than 100 individuals remaining, making recovery difficult (see Chapter 6). An early listing of a declining species might allow it to recover and thus become a candidate for removal from the list sooner than if authorities wait for its status to worsen before adding it to the list. The great majority of species in the United States listed under the ESA are flowering plants and vertebrates, even though most of the world's species are insects and other invertebrates (ecos.fws.gov/ecp/report/boxscore) (see Figure 7.20). Only 92 insect species are currently listed; if the same proportion of insects as of vertebrates were protected, more than 30,000 species would be protected under the ESA, an awesome number to contemplate. Clearly, greater efforts must be made to study the lesser-known and underappreciated invertebrate groups and extend listing to those endangered species whenever necessary.

The protection afforded to species listed under the ESA is so strong that business interests and landowners often lobby strenuously against listing species in their area. At the extreme are landowners who destroy endangered species or species being considered for listing on their property to evade the provisions of the ESA, a practice informally known as "shoot, shovel, and shut up." Financial incentives exist to help private landowners improve habitats for endangered species (see Chapter 12), but they may be too small to be effective in many cases.

Another important obstacle to listing is the difficulty of species recovery—rehabilitating species or reducing the threats to species to the point where they can be removed from listing under the ESA, or "delisted." As of 2021, 63 species in the United States had been delisted because of recovery, and 11 had gone extinct (ecos.fws.gov/ecp/report/species-delisted). The most notable successes include the brown pelican (*Pelecanus occidentalis*), the American peregrine falcon (*Falco peregrinus*), and the American alligator (*Alligator mississippiensis*). In 2007, the bald eagle (*Haliaeetus leucocephalus*) was removed from the federal list of threatened and endangered species because its numbers in the lower 48 states had increased from 400 breeding pairs in the 1960s to over 316,700 individuals (USFWS 2020). The estimated increase in numbers of nesting pairs for one population in Virginia can be seen in Figure 7.4.

Overall, most listed species are still declining in range and abundance. Unfortunately, for around 20% of species the data are insufficient to determine whether their populations are changing over time. Due to their low numbers and consequent vulnerability, there is now recognition that even species that are candidates for delisting will still require some degree of conservation management to maintain their populations (Henson et al. 2018) (**FIGURE 7.21**).

The difficulty of implementing recovery plans for so many species is often not primarily biological but, rather, political, administrative, and ultimately financial. For example, an endangered river clam species might need to be protected from pollution and the effects of an existing dam. Installing sewage treatment facilities and removing a dam are theoretically straightforward actions, but they are expensive and difficult to carry out in practice. Total

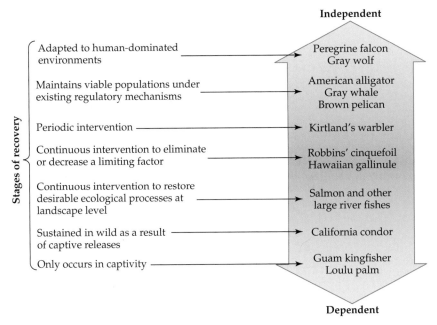

Independent

Adapted to human-dominated environments — Peregrine falcon
Gray wolf

Maintains viable populations under existing regulatory mechanisms — American alligator
Gray whale
Brown pelican

Periodic intervention — Kirtland's warbler

Continuous intervention to eliminate or decrease a limiting factor — Robbins' cinquefoil
Hawaiian gallinule

Continuous intervention to restore desirable ecological processes at landscape level — Salmon and other large river fishes

Sustained in wild as a result of captive releases — California condor

Only occurs in captivity — Guam kingfisher
Loulu palm

Dependent

Stages of recovery

FIGURE 7.21 Endangered species often require active management and intervention as part of the recovery process. There is a continuum, with some species independent of humans and others dependent on human intervention. (After J. M. Scott et al. 2005. *Front Ecol Environ* 3: 383–389. © The Ecological Society of America.)

expenditures reported for fiscal year 2016 were nearly $1.48 billion, of which more than $1.3 billion was reported by federal agencies and $99 million was reported by the states (USFWS 2018). The cost would be much higher if the US government granted private landowners financial compensation for ESA-imposed restrictions on the use of their property, an option that is periodically discussed in the US Congress.

Funding for the ESA has been growing steadily over the past 20 years, but the number of protected species has been growing even faster. As a result, there is less money available per species. The importance of adequate funding for species recovery is shown by a study demonstrating that species that receive a higher proportion of requested funding for their recovery plans have a higher probability of reaching a stable or improved status than species that receive a lower proportion of funding (Miller et al. 2002). The longer a species has been protected under the ESA, the higher is the probability that it is improving in status (Taylor et al. 2005) (**FIGURE 7.22**). Also, species have a higher probability of improving if critical habitat and a recovery plan have been designated for them (Gray et al. 2016).

Concerns about the implications of ESA protection force business organizations, conservation groups, and governments to develop compromises that reconcile both conservation and business interests (Evans et al. 2016). To provide a legal mechanism to achieve this goal, Congress amended the

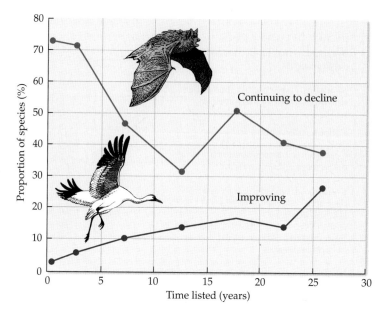

FIGURE 7.22 The longer species have been listed, protected, and managed under the Endangered Species Act, the greater is their probability of improving in status (as shown by the whooping crane) and the lower is their probability of continuing to decline in status (with the Indiana bat as an example). The numbers do not add up to 100% because some species are not changing in status and others are of unknown status. (After M. F. J. Taylor et al. 2005. *BioSci* 55: 360–366. By permission of Oxford University Press.)

ESA in 1982 to allow the design of **habitat conservation plans (HCPs)**. HCPs are regional plans that allow development in designated areas but also protect remnants of biological communities or ecosystems that contain groups of actually or potentially endangered species. These plans are drawn up by the concerned parties—developers, conservation groups, citizen groups, and local governments—and are given final approval by the US Fish and Wildlife Service. For example, the endangered Karner blue butterfly (*Lycaeides melissa samuelis*) is being protected under an HCP covering over 250,000 acres of habitat in Wisconsin by 38 partners, including timber companies, The Nature Conservancy, state agencies, local highway departments, power companies, and other landowners. While HCPs are not perfect, they are attempts to create the next generation of conservation planning. They seek to protect many species, entire ecosystems, or whole communities; extend over a wide geographical region; and include many projects, landowners, and jurisdictions.

International agreements

The protection of biodiversity must be addressed at multiple levels of government. International agreements have provided a framework for countries to cooperate in protecting species, ecosystems, and genetic variation. Treaties are negotiated at international conferences and come into force when they are ratified by a certain number of countries, often under the authority of international bodies, such as the United Nations Environment Programme (UNEP), the Food and Agriculture Organization of the United Nations (FAO), and the IUCN. It is important to note that while critically important, international agreements only have power through the implementation and enforcement of national laws.

Although the major control mechanisms that currently exist are based within individual countries, international agreements among countries are an important tool to protect species, ecosystems, and genetic variation. International cooperation is an absolute requirement for several reasons:

- *Species migrate across international borders.* Conservation efforts must protect species at all points in their ranges; efforts in one country will be ineffective if critical habitats are destroyed in a second country to which an animal migrates (see Chapter 9). For example, efforts to protect migratory bird species in northern Europe will not work if the birds' overwintering habitat in Africa is destroyed. Efforts to protect whales in US coastal waters will not be effective if these species are killed or harmed in international waters. Species are particularly vulnerable when they are migrating, as they may be more conspicuous, more tired, or more desperately in need of food and water. Globally, international parks, often called "peace parks," have been created to protect species living and moving through border areas, such as the Waterton-Glacier International Peace Park on the border of the United States and Canada, which protects grizzly bears (*Ursus arctos horribilis*) and lynx (*Lynx canadensis*).

- *International trade in biological products is commonplace.* A strong demand for a product in one country can result in the overexploitation of the species in another country to supply this demand (see Chapter 5). When people are willing to pay high prices for exotic pets, plants, or wildlife products such as rhino horn, poachers looking for easy profits, or poor and desperate people looking for any source of income, will take or kill even the very last animal to obtain this income. If overexploitation is to be prevented, the people who collect and trade wildlife products, as well as the consumers who buy them, need to be educated about the consequences of overuse of wild species. When poverty is the root of overexploitation, it is sometimes possible to provide people with economic alternatives while strictly controlling resource use (see Chapter 12). Where exploitation stems from greedy people seeking to make a profit, laws and enforcement efforts such as border checks should be strengthened.

- *Biodiversity provides internationally important benefits.* The community of nations benefits from the species and genetic variation used in agriculture, medicine, and industry; the ecosystems that help regulate climate; and the national parks and other protected areas of international scientific and tourist value. It is also widely recognized that biodiversity has intrinsic value, existence value, and option value (see Chapter 3). The developed countries of the world that use and rely on biodiversity and ecosystem services from poor tropical countries provide limited, inadequate funding to help these less-wealthy countries manage and protect globally significant resources. Funding levels need to be increased, and the funds must be used more effectively.

- *Many environmental pollution problems that threaten ecosystems are international in scope.* Such threats include atmospheric pollution and acid rain; the pollution of lakes, rivers, and oceans; greenhouse gas production and global climate change; and ozone depletion. Additionally, the environmental costs of many of these problems do not fall on countries in proportion to their role in causing them. For example, the United States and China are the world's leading producers of greenhouses gases (see Chapter 5), but many low-lying countries such as Bangladesh and the Maldives will be most affected by the rising sea levels associated with climate change. Because biodiversity conservation is important at both the national and global levels, it is fair for the developed countries of the world (including the United States, Canada, Japan, Australia, and many European nations) to help pay to protect biodiversity. Or consider the River Danube, which carries the pollution of a vast agricultural and industrial region that spans 10 countries before it empties into the Black Sea, another international body of water bordered by 4 additional countries. Problems such as these can only be solved by countries working together.

To address the protection of biodiversity, countries worldwide have signed a number of key international agreements. These agreements have provided a framework for countries to cooperate in protecting species, habitats, ecosystem processes, and genetic variation. Treaties are negotiated at international conferences and come into force when they are ratified by a certain number of countries, often under the authority of international bodies such as the UNEP, the FAO, and the IUCN.

As was mentioned in Chapter 5, one of the most important treaties protecting species at an international level is the **Convention on International Trade in Endangered Species** (**CITES**; www.cites.org), established in 1973 in association with UNEP. Currently there are 179 member countries. CITES, headquartered in Switzerland, establishes lists (known as Appendices) of species for which international trade is to be controlled or monitored (see Table 5.2). Member countries agree to restrict trade in these species and halt their destructive exploitation. Regulated plants include important horticultural species such as orchids, cycads, cacti, carnivorous plants, and tree ferns; timber species and wild-collected seeds are increasingly being considered for regulation as well. Closely regulated animal groups include parrots, large cats, whales, sea turtles, birds of prey, rhinos, bears, and primates. Species collected for the pet, zoo, and aquarium trades and species harvested for their fur, skin, or other commercial products are also closely monitored.

International treaties such as CITES are implemented when a country signing the treaty passes laws to enforce it. Nongovernmental organizations such as the IUCN, the Wildlife Crime Program, the TRAFFIC network (run by the WWF and the IUCN), and UNEP's World Conservation Monitoring Centre provide technical advice regarding legal and enforcement aspects of CITES to national governments. Countries may also protect species listed

FIGURE 7.23   A fur reference collection in northern China. For some products, such as the zebra skin, the type of animal involved can be easy to identify, but for other products, such as bags, coats, rugs, and shoes, the type of animal used to make them may be hard to determine; often microscopic analysis of hairs is required.

by national Red Data books. Once species protection laws are passed within a country, police, customs inspectors, wildlife officers, and other government agents can arrest and prosecute individuals possessing or trading in protected species and can seize the products or organisms involved (**FIGURE 7.23**). For example, Thai authorities at the Bangkok airport seized hundreds of endangered turtles being shipped illegally in passenger baggage in 2013. What made this story particularly unusual is that the seizure and news story took place just a day after a major CITES conference in Bangkok.

Member countries are required to establish their own management and scientific authorities to implement CITES obligations within their own borders (see Chapter 12). CITES is particularly active in encouraging cooperation among countries, in addition to fostering conservation efforts by development agencies. The CITES Secretariat periodically sends out bulletins aimed at publicizing specific illegal activities. For example, in recent years, the CITES Secretariat has recommended that its member countries halt wildlife trade with Vietnam because of that country's unwillingness to restrict the illegal export of wildlife from its territory.

CITES has been instrumental in restricting trade in certain endangered wildlife species. Its most notable success was a global ban on the ivory trade after poaching caused severe declines in African elephant populations

© Carl de Souza/AFP/Getty Images

FIGURE 7.24 Burning of ivory in Kenya. To keep ivory off the international market and hopefully reduce the killing of wild elephants, wildlife authorities in Kenya burned more than 15 tons of elephant tusks seized from poachers in 2015, and they made plans to burn an additional 120 tons of both elephant tusks and rhino horns in 2016.

(**FIGURE 7.24**). Recently, countries in southern Africa with increasing elephant populations have been allowed to resume limited ivory sales, resulting in an unfortunate increase in illegal harvesting.

One difficulty with enforcing CITES is that shipments of both living and preserved plants and animals are often mislabeled, due to either ignorance of species names or deliberate attempts to avoid the restrictions of the treaty (see Chapter 5). Also, sometimes countries fail to enforce the restrictions of the treaty because of corruption or a lack of trained staff. Finally, many restrictions are difficult to enforce because some international borders are remote, rugged, and difficult to monitor, such as that between Laos and Vietnam. As a result, the illegal wildlife trade continues to pose one of the most serious threats to biodiversity, particularly in Asia.

Another key treaty is the **Convention on the Conservation of Migratory Species of Wild Animals** (www.cms.int), often referred to as the **Bonn Convention**, which focuses primarily on bird species. This convention, which has been signed by 132 countries, complements CITES by encouraging international efforts to conserve bird species that migrate across international borders and by emphasizing regional approaches to research, management, and hunting regulations. The convention now includes protection of bats and their habitats and cetaceans in the Baltic and North Seas. Other important international agreements that protect species include the following:

- Convention for the Conservation of Antarctic Marine Living Resources (www.ccamlr.org)
- International Convention for the Regulation of Whaling, which established the International Whaling Commission (www.iwc.int)
- International Convention for the Protection of Birds, and the Benelux (Belgium/Netherlands/Luxembourg) Convention Concerning Hunting and the Protection of Birds
- Convention for the Conservation and Management of Highly Migratory Fish Stocks in the Western and Central Pacific Ocean (www.wcpfc.int)
- Additional agreements protecting specific groups of animals, such as prawns, lobsters, crabs, fur seals, Antarctic seals, salmon, and vicuña

A number of more broadly focused international agreements are also increasingly seeking direct protection of endangered species. For example, the Convention on Biological Diversity, described in Chapter 12, now includes recommendations for the protection of species on the IUCN Red List (www.iucnredlist.org).

A weakness of all these international treaties is that they operate through consensus, so strong measures often are not adopted if one or more countries oppose them. Also, any nation's participation is voluntary, meaning that countries can choose to ignore these conventions and pursue their own interests if they find the conditions of compliance too difficult. This flaw was highlighted when several countries decided not to comply with the International Whaling Commission's 1986 ban on whale hunting (see Chapter 5). Persuasion and public pressure are the principal means used to induce countries to enforce treaty provisions and prosecute violators, though funding through treaty organizations can also help. An additional problem is that many conventions are underfunded and are consequently ineffective in achieving their goals. Unfortunately, there are often no monitoring mechanisms in place to determine whether countries are even enforcing the treaties.

Summary

- Protecting and managing a rare or endangered species requires a firm grasp of its ecology and its distinctive characteristics (sometimes called its natural history). Long-term monitoring of a species in the field can determine whether it is stable, increasing, or declining in abundance over time.
- Population viability analysis (PVA) uses demographic, genetic, and environmental data to estimate how various management actions will affect the probability that a population will persist until some future date. It can be used to calculate the minimum viable population (MVP) size: the smallest population size that can be predicted to have a high chance of persisting for the foreseeable future. The MVP for many species is at least several thousand individuals.

■ A species may be best described as a metapopulation made up of a shifting mosaic of populations that are linked by some degree of migration.

■ The IUCN has developed quantitative criteria for populations and ecosystems to assign species to conservation categories: extinct, extinct in the wild, critically endangered, endangered, vulnerable, near threatened, least concern, data deficient, and not evaluated.

■ Priorities for protection can be determined in several ways, including the species approach, the ecosystem approach, the wilderness approach, and the hotspot approach.

■ National governments protect biodiversity by establishing national parks and refuges, controlling imports and exports at their borders, and creating regulations for air and water pollution. The most effective law in the United States for protecting species is the Endangered Species Act (ESA).

■ International cooperation is needed to protect biodiversity because species migrate across borders, there is an international trade in species, and biodiversity provides advantages to all countries.

■ At the international level, the Convention on International Trade in Endangered Species (CITES) allows governments to regulate, monitor, and sometimes prohibit trade in individuals and products from endangered species.

For Discussion

1. How might you monitor populations of a species of fish over time? Would your methods differ if you monitored a butterfly species? Why or why not?

2. Choose a threatened species in your region. Weigh the merits and limitations of a species-centered approach, an ecosystem approach, a wilderness approach, and a hotspot approach for protecting it. Which approach will be most effective for protecting this species? Which approach is most feasible? Which approach will protect the most species in addition to your target?

3. There are many laws to protect endangered species. Why don't species covered by such laws quickly recover?

Suggested Readings

Biodiversity A–Z. www.biodiversitya-z.org/content/biodiversity-hotspots. A tool for identifying hotspots of biodiversity; global summaries created by the United Nations Environment Programme World Conservation Monitoring Centre.

Cindy, B., et al. 2021. Passive eDNA collection enhances aquatic biodiversity analysis. *Communications Biology* 4: 1–12. An innovative technique for sampling biodiversity in marine systems that expands potential applications of environmental DNA (eDNA) and metabarcoding.

Demery, A. J. C., & Pipkin, M. A. 2021. Safe fieldwork strategies for at-risk individuals, their supervisors and institutions. *Nature Ecology & Evolution*, *5*(1): 5-9. An important discussion of the challenges of field work for people simply because of how others perceive their identity, and how we can be allies to keep them safe.

Milligan, B. G., et al. 2018. Disentangling genetic structure for genetic monitoring of complex populations. *Evolutionary Applications* 11(7): 1149–1161. A good introduction to the use of genetic tools in conservation, as well as the challenges it presents.

NatureServe. 2021. http://natureserve.org. This website organizes and presents data on biodiversity surveys from North America.

Reading, R. P., et al. 2019. Home-range size and movement patterns of Hooded Vultures *Necrosyrtes monachus* in southern Africa. *Ostrich* 90(1): 73–77. An example of monitoring and the importance of international cooperation for conservation of species.

Ríos-Saldaña, C. A., M. Delibes-Mateos, and C. C. Ferreira. 2018. Are fieldwork studies being relegated to second place in conservation science? *Global Ecology and Conservation* 14: e00389. A thoughtful look at the importance of fieldwork.

The titan arum (*Amorphophalus titanium*) is a globally endangered species endemic to Malaysia. It is a flagship species for ex situ conservation at botanical gardens across the globe, where blooms attract large crowds.

Establishing New Populations

and Ex Situ Conservation

In Chapters 4 through 7, we discussed the problems conservation biologists face in preserving naturally occurring populations of endangered species. This chapter discusses some exciting conservation methods used to establish new wild and semi-wild populations of rare and endangered species and increase the sizes of existing populations. These methods include breeding species in zoos, aquariums, and botanical gardens—organizations that also assist conservation through education and research programs. Captive breeding and other approaches to augment or establish new populations may allow species that have persisted only in captivity or in small, isolated populations to regain their ecological and evolutionary roles within their ecosystems. Furthermore, simply increasing the number and size of its populations generally lowers the probability a species will go extinct.

Population establishment programs are unlikely to be effective, however, unless the factors leading to the decline of the original wild populations are clearly understood and eliminated, or at least controlled. For example, endangered amphibians are increasingly being raised in captivity, but they cannot be released back into the wild if they lack resistance to the chytrid fungal pathogen *Batrachochytrium dendrobatidis*, a now widespread fungus that is killing amphibians worldwide (Kolby et al. 2015). One possibility is to breed captive amphibian populations for fungal resistance before attempting to establish new populations in the wild.

8.1 Establishing and Reinforcing Populations

LEARNING OBJECTIVES

By the end of this section you should be able to:

8.1.1 Distinguish between a reintroduction program, a reinforcement program, and a conservation introduction program.

8.1.2 Evaluate a conservation translocation program using guidelines from the IUCN.

8.1.3 Compare and contrast the challenges associated with translocating animal versus plant populations.

> Establishing new populations of endangered species can benefit the species itself, other species, and the ecosystem. However, such programs must identify and eliminate the factors that led to the original population's decline.

Three basic approaches, all involving the transport and release of existing captive-bred or wild-collected individuals, have been used for **conservation translocations**—the deliberate placement of organisms to achieve "measurable conservation benefit at the levels of a population, species, or ecosystems" (IUCN/SSC 2013). We will be discussing translocations that are intended to reduce extinction risk for species in this chapter, and then again in the context of whole ecosystems in Chapter 11. The IUCN's Conservation Translocation Specialist Group (www.iucnsscrsg.org), formerly the Reintroduction Specialist Group, coordinates many of these efforts. Consider three types of conservation translocations to establish and expand populations:

- A **reintroduction program** involves releasing captive-bred or wild-collected individuals at an ecologically suitable site within their historical range where the species no longer occurs (Carter et al. 2017).[1]

- A **reinforcement program** involves releasing individuals into an existing population to increase its size and/or gene pool to adjust age or sex structures; this approach is also referred to as restocking or augmentation. These released individuals may be raised in captivity or may be wild individuals collected elsewhere.

- A conservation **introduction program** involves moving captive-bred or wild-collected animals or plants to areas suitable for the species outside their historical range.

The main objectives of a reintroduction program are to create a new population in its original environment and to help restore a damaged ecosystem. For example, a program initiated in 1995 to reintroduce gray wolves (*Canis lupus*) into Yellowstone National Park aims to restore the equilibrium of predators, herbivores, and plants that existed prior to human intervention in the region (**FIGURE 8.1**). Wild-collected animals are also sometimes

---

[1]Some confusion exists about the terms denoting the establishment of populations. Reintroduction programs are sometimes called reestablishments or restorations.

caught and then released elsewhere within the range of their species when a new protected area has been established, when an existing population is under a new threat in its present location, or when natural or artificial barriers to the normal dispersal tendencies of the species exist. If possible, individuals are released near the site where they or their ancestors were collected to ensure genetic adaptation to their environment. In addition, genetic diversity of released individuals can ensure that they help reduce inbreeding and ensure the potential for future evolution of the population (Malone et al. 2018).

(A)

Courtesy of William Campbell/US Fish and Wildlife Service

(B)

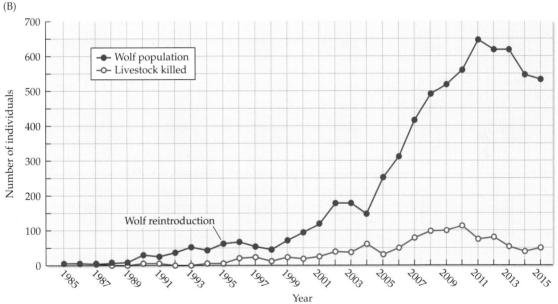

FIGURE 8.1 Gray wolf (*Canis lupus*) reintroduction in the Northern Rocky Mountains of North America is one example of a successful translocation program. Starting from zero in the 1980s, wolf numbers increased after reintroduction until they were delisted in 2017. (A) A gray wolf in Yellowstone National Park wears a radio transmitter collar that allows researchers to follow its movements. (B) Reintroduction has not been without cost to local ranchers; here we see both an increase in wolf numbers and increasing losses of livestock to wolf predation in Montana. Although these losses were very small relative to wolf numbers, they cost the program public support; targeted removal of the wolves that had learned to kill livestock was successful in decreasing predation events. (After N. J. DeCesare et al. 2018. *J Wildl Manag* 82(4): 711–722. © 2018 Wiley.)

In contrast to reintroduction, *introduction* to an entirely new location outside the existing range of a species might be an appropriate conservation tool when the environment within the known range of a species has deteriorated to the point at which the species can no longer survive there, or when a reintroduction is impossible because the factor causing the original decline is still present in the species' range. This approach is sometimes referred to as *assisted colonization*. For example, the New Zealand kakapo (*Strigops habroptilus*) is a large, flightless parrot that was introduced to offshore islands because nonnative predators had decimated its populations on the mainland (Moro et al. 2015). Conservation introductions may be necessary for many species due to global change if those species can no longer survive within their current ranges because of the changing climate (Hendricks et al. 2016).

GLOBAL CHANGE
CONNECTION

There are several important considerations for any conservation translocation program (Hunter-Ayad et al. 2020). The IUCN outlined these in its *Guidelines for Reintroductions and Other Conservation Translocations* (IUCN/SSC 2013), summarized as follows (adapted from Seddon et al. 2014b):

1. *Threats must be understood and addressed.* Both the cause of the original loss and potential future threats must be identified and sufficiently managed so as to ensure the success of the translocation.

2. *The ecology of the species must be sufficiently understood.* This is important for site selection (in the case of reintroductions and introductions) and also as a component of identifying threats. All biotic and abiotic needs for each life history stage must be met.

3. *Genetics of the species and released individuals must be sufficiently understood.* Captive populations may have lost much of their genetic variation or, when raised for several generations in captive conditions, may have become adapted to the benign captive environment. Individuals have to be carefully selected to guard against inbreeding depression and to produce a genetically diverse release population that will be adapted to the release site (Ottewell et al. 2014). Wild-caught individuals must be selected from a population living in an environment and climate that is as similar as possible to that of the release site.

4. *There must be enough viable habitat available to support a population.* As discussed in Chapter 7, the threshold of a minimum dynamic area (MDA) must be met, both for the released individuals and into the foreseeable future. This is important not just for habitat needs but also for evolutionary needs; in South Africa, cheetah translocated to fenced, private reserves must be managed as a metapopulation to maintain genetic diversity (Boast et al. 2018).

5. *Ecological risks must be assessed and accounted for.* Every translocation carries some measure of risk, not only to the species itself, but also to the ecosystem into which it is being released. These must be weighed against the potential benefits during planning stages. The reintroduction or introduction of a species must be carefully thought out so that the released species does not damage its new ecosystem or harm

populations of any local endangered species (Stringer and Gaywood 2016). These risks may include competition with or predation on existing species, impacts on food webs, modifications of the habitat, and even hybridization. Care must be taken that released individuals have not acquired any diseases that could spread to and decimate wild populations (see "Special considerations for animal programs").

6. *Human concerns must be addressed.* This includes any relevant policies or legislation. People who live in or near a translocation site may be affected and their needs must be considered, including aspects of culture, economics, and health. For example, farmers may be justifiably worried for their livestock or even their children if wolves are released nearby (see Figure 8.1B). As discussed in more detail in this chapter, public support is critical for successful translocations.

7. *In the event of a failure, it must be possible to remove or destroy all offspring of the released individuals.* If there are unacceptable levels of negative impacts for either humans or ecosystems, it must be feasible to reverse the translocation. This may be especially difficult with small and highly dispersible species such as rodents, insects, and some kinds of plants.

Special considerations for animal programs

Establishing new populations is often expensive and difficult because it requires a serious, long-term commitment. The programs that capture, raise, release, and monitor sea turtles, peregrine falcons, and whooping cranes, for example, have cost millions of dollars and have required years of work. When the animals involved are long-lived, the program may have to continue for many years before its outcome is known (Canessa et al. 2016). It is therefore unsurprising that the majority of translocation programs are of charismatic species; mammals represent more than 80%, followed by birds and reptiles, with less than half that many (Resende et al. 2020). One evaluation found that popularity was as important as threat status in predicting whether a species would be selected for a conservation translocation program (Díaz et al. 2018).

Population establishment programs for animals can become highly emotional public issues, as demonstrated by the programs for the California condor, the black-footed ferret, and the gray wolf in the United States and comparable programs in Europe. These programs are often criticized on many different fronts. They may be attacked as a waste of money ("Millions of dollars for a few ugly birds!"), unnecessary ("Why do we need wolves here when there are so many elsewhere?"), intrusive ("We just want to go about our lives without the government telling us what to do!"), poorly run ("Look at all the ferrets that died of disease in captivity!"), or unethical ("Why can't the last animals just be allowed to live out their lives in peace without being captured and put into zoos?"). The answer to all of these criticisms is straightforward. Although not appropriate for every endangered species, a well-run, well-designed captive breeding and population establishment program may be the best hope for a species' preservation.

Because of the conflicts and high emotions involved, it is crucial that population establishment programs include local people so that (ideally) the community has a stake in the program's success. (Indeed, this is true of any conservation project.) At a minimum, it is necessary to explain the need for the program and its goals and to convince local people to support it—or at least not to oppose it (Yochim and Lowry 2016). Programs are often more successful if they provide incentives or compensation to affected people rather than imposing rigid restrictions and laws.

Successful reintroduction programs often have considerable educational value. In Brazil, conservation and reintroduction efforts for the golden lion tamarin (*Leontopithecus rosalia*) have become a rallying point for the protection of the last remaining fragments of the Atlantic forest. In the Middle East and northern Africa, captive-bred Arabian oryx (*Oryx leucoryx*) have been successfully reintroduced into many desert areas that they formerly occupied, providing a source of national pride and opportunities for employment (Fisher 2016).

Disease is a special concern for released animals, both for the risk they pose to wild animals and vice versa. Animals raised in captivity may be especially vulnerable to disease once released. Care must also be taken to ensure that released individuals have not acquired any diseases while in captivity that could spread to and decimate wild populations. For example, captive black-footed ferrets (*Mustela nigripes*) must be carefully handled and quarantined so that they do not acquire diseases from people or dogs that they might transfer into wild populations upon their release in North American grasslands (**FIGURE 8.2A**).

Some animal species may require special care and assistance immediately after release to increase their survival prospects. This approach is known as **soft release**. Animals may have to be fed and sheltered at the release point until they are able to subsist on their own, or they may need to be caged temporarily at the release point and introduced gradually, once they become familiar with the sights, sounds, smells, and layout of the area (**FIGURE 8.2B**). Another benefit of a soft release is the ability to study the behavior of captive-bred animals to ensure that they will be able to survive in the wild. For example, a Spanish-Portuguese reintroduction program for Iberian lynxes (*Lynx pardinus*) released 10 individuals in an 85,000 ha enclosure in southeastern Portugal in 2015 (**FIGURE 8.2C**). Subsequent translocations have been successful in increasing populations (Bencatel et al. 2018). Studying their behavior helped predict that released individuals would be able to survive on their own in the wild (Sarmento et al. 2019; Rueda et al. 2021).

Human contact should be kept to a minimum during a soft release (**FIGURE 8.2D**). Animals that have acclimated to humans will be more likely to seek them out once released, increasing dependence and leading to conflicts. For example, initial soft releases of the brown bear (*Ursus arctos*) in Białowieża Forest of Poland and Belarus were unsuccessful for this reason (Samojlik et al. 2018). Captive-bred bears were held in cages in the forest, where they were fed and monitored before being released. Bears that had contact with people while in captivity in the early part of the project were less likely to

FIGURE 8.2 (A) A black-footed ferret (*Mustela nigripes*) raised at the captive colony in Colorado. (B) Fenced areas allow black-footed ferrets to experience the environment into which they will eventually be released. (C) Iberian lynx (*Lynx pardinus*) are considered by some to be the most endangered felid in the world. A reintroduction program by Spain and Portugal was initiated in 2013 funded by EU Life, a European Union program for funding environmental programs. (D) When transporting Iberian lynxes, it was important not only to protect them from disease but also to make sure that they had limited contact with humans.

feed themselves and more likely to get shot after release, whereas cubs that had had no contact with humans at all were more successful after release.

Animals can also be released without assistance such as food supplementation (**hard release**); a global review of the literature found that hard releases are more common than soft releases (Resende et al. 2020). Reintroductions of this type are more likely to succeed with wild-caught than captive-bred individuals (Bocci et al. 2016). Intervention may be necessary if animals appear unable to survive, particularly during episodes of drought or low food abundance. Even when animals appear to have enough food to survive, supplemental feeding may help by increasing reproduction and allowing the population to persist and grow. Outbreaks of diseases and pests may have to be monitored and dealt with. The effects of human activities in the area, such as farming and hunting, need to be observed and possibly controlled. In every case, a decision has to be made about whether it is better to give occasional temporary help to the species or to force the individuals to survive on their own (Harrington et al. 2013).

One special method used in establishing new populations or reinforcing existing populations is *head-starting*, an approach in which animals are raised in captivity during their vulnerable young stages and then released into the wild. The release of sea turtle hatchlings produced from eggs collected from the wild and raised in nearby hatcheries is an example of this approach (see Chapter 1). Another is the kakapo parrot, mentioned earlier; scientists harvest the birds' actual eggs and incubate them away from predation risk, replacing them with 3D-printed "smart eggs" that make noise like a newly hatched kakapo to prepare the mothers for the return of their chicks (Feldberg 2019). They also use drones to transport semen from "genetically important" males to receptive females on other parts of the island. In 2019, the project had a record-breaking year, with 218 eggs and 52 live chicks. This is incredible for an endangered species that had only 51 birds in 1995, but now has 201 individuals as the result of conservation programs (https://www.doc.govt.nz/our-work/kakapo-recovery).

Clearly, monitoring of establishment programs is crucial in determining whether the programs are achieving their stated goals (McCleery et al. 2014; Seddon et al. 2014a). Monitoring may need to be carried out over many years, even decades, because many reintroductions that initially appear successful eventually fail. The key elements of monitoring are determining whether released individuals survive and establish a breeding population, then following that population over time to see whether it increases in numbers of individuals and geographic range. In one example of long-term monitoring, the success of an introduction of the endangered white-faced darter dragonfly (*Leucorrhinia dubia*) in Eastern Moravia, Czech Republic, was documented 15 years after the release of 80 larvae (Dolný et al. 2018). Total population size was estimated using a mark-recapture method that identified individuals by numbering the wings of adult dragonflies (**FIGURE 8.3**); effective population size (N_e) and genetic diversity were determined from DNA microsatellites using discarded larval shells and portions of forelegs. With these methods, researchers found that the population was growing over time and actually had a greater genetic diversity than the source population.

Monitoring of important ecosystem elements is also needed to determine the broader impact of a reintroduction; for example, when a predator species is introduced, it will be crucial to determine its effect on prey species and competing species, as well as its indirect effect on vegetation (Baker et al. 2016).

The costs of reintroduction also need to be tracked and published so it can be determined whether reintroduction represents a cost-effective strategy. In the case of the orangutan (*Pongo* spp., see Figure 7.5), it was found that reintroduction costs 12 times as much per animal as the protection of forest habitat. Reintroduction was effective in the short term, but at time scales longer than 10–20 years, habitat conservation was much more cost-effective (Wilson et al. 2014).

FIGURE 8.3 *Leucorrhinia dubia* (Odonata: Libellulidae) introduced to artificially created bog pools in the Czech Republic. Although no dragonflies had been observed there previously, it was within historical ranges for the species. A permanent marker is used to carefully number individuals on their wings for scientific study, including mark-recapture methods that can be used to estimate population sizes over time.

Animal translocations appear to be a successful conservation strategy; according to a recent global review of nearly 600 published reports including both peer reviewed papers and those from the IUCN, 72% were categorized as successes and only 28% as failures (Morris et al. 2021). However, there is a bias in the literature in which the results of successful projects are more likely to be published, so the overall rate of success of translocations might be much less than these figures suggest. It is notable that success rates for reintroduction projects involving wildlife managed for hunting, including mammals, birds, reptiles, amphibians, and fish, are generally greater than that for endangered animals (Harding et al. 2016). In statistical analyses to identify factors most associated with success for terrestrial vertebrates, the most important were the number of individuals translocated and geographic location; also, translocations in Oceana were less successful than North America or Europe, likely because of the increased threat of invasive species on islands (Morris et al. 2021).

GLOBAL CHANGE
CONNECTION

Behavioral ecology of released animals

As mentioned, both introduction and reintroduction programs must often address the behaviors of animals that are being released (West et al. 2019). *Behavioral ecology* is the study of an animal's behavior in the context of its environment and considers the adaptive significance of those behaviors. When social animals, which include many mammals and some bird species, grow up in the wild, they learn from other members of their population, particularly their parents, how to interact with their environment and with other members of their species. They learn how to search for food and how to gather, capture, and consume it. When mammals and birds are raised in captivity, their environment is limited to a cage or pen, so exploration is unnecessary. Searching for food and learning about new food sources is not required because the same food items come to them

> Captive-bred mammals and birds must learn predator avoidance and species-appropriate social behavior if they are to survive and reproduce after being released into the wild. They may also require some support after release.

day after day, on schedule. For example, when the European bison (*Bison bonasus*), previously extinct in the wild, was reintroduced, it was necessary to determine whether the animals were successfully exploring and using their new, wild environment (Schmitz et al. 2015). In a review of conservation projects that use behavioral ecology, it was found that while foraging and dispersal behaviors were often considered, others, such as antipredator behavior and social behaviors, were not, even though these are important issues for reintroductions (Berger-Tal et al. 2016).

Social behavior may become highly distorted when animals are raised alone or in unnatural social groupings (i.e., in small groups or single-age groups). In such cases, the animals may lack the skills to survive in their natural environment and the social skills necessary to cooperatively find food, sense danger, find mating partners, and raise young (Parlato and Armstrong 2013). The greatest threats to the survival of these animals, and the primary reasons many such establishment projects fail, are predation ("Why are these guys trying to eat me?"), starvation ("Why aren't they feeding me anymore?"), and habitat quality ("My old home was way better!").

To overcome these behavioral problems, captive-raised mammals and birds may require extensive training before and after release. In some cases, human trainers use puppets or wear costumes to mimic the appearance and behavior of wild individuals so that young animals learn to identify with their own species rather than with humans (**FIGURE 8.4**). In other cases, wild individuals serve as "instructors" for captive individuals of the same species. For example, wild golden lion tamarins are caught and held with captive-bred tamarins so that the captive-bred tamarins will learn appropriate behavior from the wild ones. After they form social groups, they are released together.

Establishing plant populations

Methods used to establish new populations of rare and endangered plant species are fundamentally different from those used to establish animal species. Animals can disperse to new locations and actively seek out the most suitable microsites. The seeds of plants, however, are dispersed to new sites by agents such as wind, animals, water, or the actions of conservation biologists (e.g., Rood et al. 2015). Once a seed lands on the ground or an adult is planted at a site, it is unable to move, even if a suitable microsite exists just a few meters away. The immediate microsite is crucial for plant

FIGURE 8.4   California condor chicks (*Gymnogyps californianus*) raised in captivity are fed by researchers using puppets that look like adult birds. Conservation biologists hope that minimizing human contact with the birds will improve their chances of survival when they are returned to the wild.

survival: if the environmental conditions are in any way too sunny, too shady, too wet, or too dry, the seed will not germinate, or the resulting plant will not reproduce or will die.

Plant ecologists investigate the effectiveness of site treatments before and after planting—such as burning the leaf litter, removing competing vegetation, digging up the ground, and excluding grazing animals—as a means of enhancing population establishment (Reiter et al. 2016). They also research the most effective way to establish a species: whether seeds should be sown or adult individuals or seedlings should be transplanted into the enhanced site. For example, during the introduction of Mead's milkweed (*Asclepias meadii*), a threatened perennial prairie plant in the midwestern United States, it was found that survivorship was greater for older juvenile plants than for seedlings, and survival was more likely in burned habitat than in unburned habitat. Seedling survival rate was also higher in years with greater rainfall than average. This information can help to guide planting and management. Researchers also use population viability analysis (PVA) (see Chapter 7) and other modeling tools to determine best practices for plant reintroduction, such as how many new populations should be established to reduce probability of metapopulation extinction (**FIGURE 8.5**).

> New plant populations are established by sowing seeds or transplanting seedlings or adults. Site treatments such as burning off or physically removing competing plants are often necessary for success.

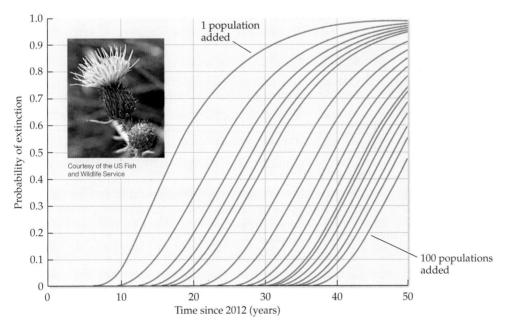

FIGURE 8.5 Metapopulation viability analysis (a modeling tool similar to PVA that considers number of populations rather than individuals) of Pitcher's thistle (*Cirsium pitcheri*; inset), a federally threatened species that grows in the Indiana Dunes. The graph shows that the probability of extinction over time was greatly reduced by increasing numbers of reestablished populations. After 40 years, the extinction probability is almost 100% with one population added, but it is reduced to less than 1% with more than 100 populations added. (From S. J. Halsey et al. 2015. *Restor Ecol* 24(4): 375–384. © 2015 Society for Ecological Restoration.)

Plant reintroductions frequently fail, although, as with animals, this fact is not captured in the literature because researchers do not usually publish failed experiments. Furthermore, many apparently successful reintroductions have either failed after additional years or have never established a second generation of plants, which is an important indicator of success. In general, success increases with the numbers of individuals or populations reintroduced (Liu et al. 2015) (see Figure 8.5), but this conclusion assumes that plantings are taking place in suitable habitat. Genetic analysis of source populations can assist in determining resistance to environmental stress and resilience during environmental change (He et al. 2016). Because of climate change, certain plant species may no longer be genetically suited to their present sites, and conservation biologists may have to look elsewhere in the range of a species for suitable genotypes to plant (Anderson and Song 2020).

GLOBAL CHANGE CONNECTION

The status of new populations

The establishment of new populations raises some novel issues at the intersection of scientific research, conservation efforts, government regulation, and ethics. These issues need to be addressed because reintroduction, introduction, and reinforcement programs will increase in the coming years as the biodiversity crisis eliminates more species and populations from the wild. In addition, assisted colonization may be needed for many species if their present ranges become too hot, too dry, or otherwise unsuitable because of global climate change or some other change in the environment (see Chapter 5) (Butt et al. 2020).

GLOBAL CHANGE CONNECTION

Many of the reintroduction programs for endangered species are mandated by official recovery plans set up by national governments. If such plans are to be formulated and implemented, conservation biologists must be able to explain the benefits and limitations of reintroduction and introduction programs in a way that government officials and the general public can understand, and they must address the legitimate concerns of those groups (**FIGURE 8.6**).

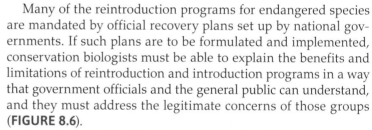

The establishment of new populations through reintroduction programs in no way reduces the need to protect the original populations of endangered species.

Sometimes stakeholders' concerns can be addressed by giving varying degrees of protection to new populations. In the United States, for example, populations can be designated *experimental essential* or *experimental nonessential* (for an example, see Figure 6.10). Experimental essential populations are regarded by the US Endangered Species Act as critical to the survival of the species, and they are as rigidly protected as naturally occurring populations. Experimental nonessential populations have less protection under the law; designating populations as nonessential often helps to overcome the fear of local landowners that having endangered species on their property will restrict how their land can be managed and developed.

Legislators and scientists alike must understand that the establishment of new populations through reintroduction programs in no way reduces the need to protect the original populations of the endangered species. Original populations are likely to have the most complete gene pool of the species and the most intact interactions with other members of the biological community.

FIGURE 8.6 Residents of southern Taiwan were surveyed to assess attitudes toward a restoration program for Formosan sika deer (*Cervus nippon taiouanus*), which had gone extinct in the wild, in Kenting National Park. After its reintroduction, the deer was associated with threats to crops (48% of respondents), and 18% of respondents had actually suffered losses to the deer. Even so, the majority of those surveyed believed the deer was a tourism resource (87%) and supported the reintroduction program (75%) (Yen et al. 2015).

© Imagemore Co., Ltd./Getty Images

In many cases, proposals are made by developers or government departments to compensate for habitat damage or eradication of endangered populations that happens as a result of development projects by requiring the creation of new habitat or the establishment of new populations. This activity is generally referred to as **mitigation**. Mitigation plans, which are often directed at legally protected species and habitats, often include (1) adjustments to the development plan to reduce the extent of damage, (2) establishment of new populations and habitat as compensation for what is being destroyed, and (3) enhancement of populations and habitat that remain after development. Given that the ability to create new populations of rare species is far from assured, mitigation plans for threatened or endangered species should be carefully and critically evaluated.

8.2 Ex Situ Conservation Strategies

LEARNING OBJECTIVES

By the end of this section you should be able to:

8.2.1 Create a conservation plan for a rare plant that includes both in situ and ex situ methods.

8.2.2 Explain why ex situ conservation alone is not sufficient for the preservation of species.

8.2.3 Evaluate the role of botanical gardens, zoos, and aquariums in conservation.

The best strategy for the long-term protection of biodiversity is the preservation of biological communities and populations (called **in situ conservation**, as discussed in Chapter 7). However, in the face of increasing threats to biodiversity, relying solely on in situ conservation is not currently a viable option for many rare and endangered species. For example, even under in situ conservation management and protection programs, species may still decline and go extinct in the wild for any of the reasons we have already discussed: habitat destruction, loss of genetic variation, demographic and environmental stochasticity, and so forth. Likewise, if a remnant population is too small to maintain the species, if it is still declining despite conservation efforts, or if the remaining individuals are found outside of protected areas, then in situ conservation may not be adequate. It is likely that the only way to prevent species in such circumstances from going extinct is to maintain individuals in artificial conditions under human supervision (Canessa et al. 2016).

Ex situ conservation—off-site conservation—used in place of or to complement in situ conservation can mean the difference between persistence and extinction for some species. Already a number of species that have gone extinct in the wild have survived because of propagation in captive colonies, a concept sometimes referred to as a "Zoo ark" (in reference to Noah's Ark; Odum 1995). The beautiful Franklin tree (*Franklinia alatamaha*), for example, grows only in cultivation and is no longer found in the wild. The same is true for the Spix's macaw (*Cyanopsitta spixii*), the species featured in the animated movies Rio and Rio II.

> When integrated with efforts to protect existing populations and to establish new ones, ex situ conservation is an important strategy for protecting endangered species and educating the public.

Ex situ facilities for animal preservation include zoos, game farms, and aquariums as well as the facilities of private breeders. Plants are maintained in botanical gardens, arboretums, and seed banks. However, ex situ conservation by these groups has evolved from simple "ark-ing" to addressing conservation challenges in at least five different ways (adapted from Traylor-Holzer et al. 2018):

1. *As an opportunity for research*   Research on individuals ex situ can provide insight into the basic biology, physiology, and genetics of a species that would not be possible with wild animals—information that can be useful to in situ conservation. Studying captive populations can also be critical for both identifying and addressing primary issues such as disease.

GLOBAL CHANGE
CONNECTION

2. *By addressing threats*   In addition to identifying threats, ex situ work can address threats through education, such as warning the public about the risk microplastics pose to marine life. Experts at zoos are an important resource for individuals and organizations seeking to pass legislation or work with industry to mitigate threats. Long-term, self-sustaining ex situ populations can also reduce the need to collect individuals from the wild for display or research.

3. *By offsetting impacts of stochastic demographic and/or environmental impacts*   Supporting specific life stages by increasing survival or increasing reproduction can be critical for small populations (see Chapter 6). Ex situ work can also protect genetic diversity and/or facilitate gene flow between populations to safeguard potential adaptation.

4. *As a way of "buying time" for populations facing significant decline*   This can include rescue programs in which wild populations in dire condition

are translocated, or it can create insurance populations that can be used for supplementation.

5. *By restoring populations after threats have been mitigated* In the case of Przewalski's horse (*Equus ferus przewalskii*), which had been declared extinct in the wild, there are now small herds descended from 14 captive-bred founder individuals released in a national park in Mongolia (**FIGURE 8.7A**).

Thus, we can see that ex situ and in situ conservation can be complementary strategies (**FIGURE 8.8**). Scientists and practitioners in both in situ and ex situ communities are increasingly working together to develop processes to coordinate and plan conservation efforts (Traylor-Holzer et al. 2019). The IUCN has developed a program they call the "One Plan Approach." This is the idea that all in situ and ex situ management and conservation activities for a given species or population should be integrated, involving all responsible parties and resulting in a clear, single plan. Importantly, these large conservation projects are more likely to consider meta population dynamics, genetics, and other aspects sometimes overlooked in smaller scale efforts; a variety of these projects are now occurring across the globe (https://www.cpsg.org/our-approach/one-plan-approach-conservation).

An example of a program that combines both ex situ and in situ approaches is the one for the rare Columbia Basin pygmy rabbit (*Brachylagus idahoensis*), which does not breed well in captivity (**FIGURE 8.7B**). As part of the conservation strategy for this species, groups of pygmy rabbits are released into large (2.5 ha) fenced enclosures that keep out predators, keep the rabbits in familiar surroundings, and encourage breeding. On occasion, tagged individuals from the breeding program are released back into the wild. Monitoring and management of populations of rare and endangered species in small protected areas are still somewhat wild, but occasional human intervention may be necessary to prevent population decline.

Despite these successes, ex situ conservation has several limitations. First, it is not cheap. The cost of maintaining zoos is enormous in comparison with

(A)

© iStock.com/pilipenkoD

(B)

Courtesy of the U.S. Fish and Wildlife Service

FIGURE 8.7 (A) The IUCN had declared Przewalski's horse to be extinct in the wild. Several zoos around the world have maintained populations and have been successful in breeding these animals. Reintroductions of the wild horses into their natural range in Mongolia have been successful and populations are increasing. (B) Rare Columbia Basin pygmy rabbits breed well in large seminatural enclosures. Individuals are then released into the wild.

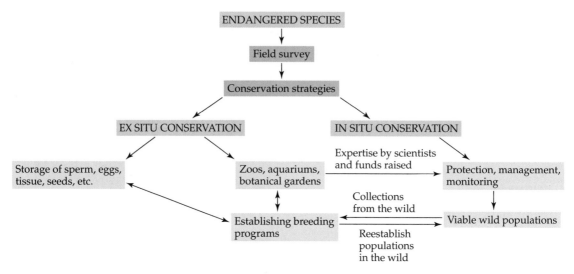

FIGURE 8.8 This model shows ways in which in situ (on-site) and ex situ (off-site) conservation efforts can benefit each other and provide alternative conservation strategies. While no species conforms exactly to this idealized model, the giant panda program (see Figure 8.11) has many of its elements. (After N. Maxted. 2001. In S. A. Levin [ed.], *Encyclopedia of Biodiversity, Vol. 2*, pp. 683–696. Academic Press, San Diego, CA.)

many other conservation activities, such as the budgets of national parks in developing countries. For example, Los Angeles Zoo has an annual operating cost of more than $26 million, and the National Zoo in Washington, DC, spends $2.6 million per year on the panda exhibit alone (Josephson 2018). Furthermore, ex situ programs protect only one species at a time. In contrast, when a species is preserved in the wild, an entire community—perhaps consisting of thousands of species—may be preserved, along with a range of ecosystem services (i.e., "umbrella species," as discussed in Chapter 7).

There are also several ecological and evolutionary problems with ex situ conservation. Only in natural biological communities are species able to continue their process of evolutionary adaptation (Olivieri et al. 2016). Ecosystem-level interactions among species, as discussed in Chapter 2, are often crucial to a rare species' continued survival; these interactions can be quite complex and probably cannot be replicated under captive conditions. Furthermore, captive animal populations are generally not large enough to prevent the loss of genetic variation through genetic drift; the same may be true of cultivated plant species when they have special requirements for pollination that might make it difficult to ensure adequate cross-fertilization among individuals. For such species, the best solution may be in situ conservation involving careful habitat protection and management. Despite these problems, ex situ conservation remains an important approach, making it possible for many species to have recovered from near extinction.

In addition to the significant contributions of ex situ conservation mentioned so far, captive individuals on display can serve as ambassadors for their species and help to educate the public about the need to preserve the species in the wild (see the chapter opening photo). Zoos, aquariums, botanical gardens, and the people who visit them regularly contribute money and expertise to in situ conservation programs (see Figure 7.3 for an example). In situ preservation of species, in turn, is vital to the survival of species that are difficult to maintain in captivity, as well as to the continued ability of zoos, aquariums, and botanical gardens to display species that do not have self-sustaining ex situ populations.

Zoos

A current goal of most major zoos is to establish viable, long-term captive breeding populations of rare and endangered animals. Zoos, along with affiliated universities, government wildlife departments, and conservation organizations, currently maintain over 10 million animals, representing more than 22,000 species and subspecies of mammals, birds, reptiles, amphibians, and even insects worldwide (Species360/ZIMS 2021). The major groups housed in zoos can be found in **TABLE 8.1**.

Zoos frequently partner with conservation organizations, lending their expertise and facilities for both ex situ and in situ conservation. For instance, seven North American zoos joined with universities and the Defenders of Wildlife to form the Panama Amphibian Rescue and Conservation Project (www.amphibianrescue.org). A major goal of this collaboration is to establish breeding populations of frogs and other amphibians that are being decimated

TABLE 8.1	**Number of Terrestrial Vertebrates Maintained in Zoos**[a]				
Location	Mammals	Birds	Reptiles	Amphibians	Total
Europe	111,607	137,619	44,616	40,376	334,218
N. America	53,513	65,688	32,737	40,252	192,190
S. America	3,612	8,447	2,701	3,859	18,619
Asia	94,325	49,091	16,913	1,355	161,684
Australia (Oceania)	10,496	15,420	5,163	2,096	33,175
Africa	5,179	9,818	3,025	322	18,344
WORLDWIDE TOTAL					
Individuals	278,731	286,106	105,155	88,260	758,252
Number of taxa[b]	2,142	3,946	2,184	627	8,899
Individuals of rare species[c]	121,642	67,114	38,752	43,880	271,388
Taxa of rare species	664	568	376	165	1,773
Percent of individuals born in wild	4%	7%	12%	2%	

Source: Data from Species360.org with assistance from Elizabeth Hunt and Nanette Reece.
[a]Excludes domestics, breeds, and varieties.
[b]This includes animals listed as high as class, and as low as subspecies.
[c]Rare species are those listed by IUCN as Extinct in the Wild, Critically Endangered, Endangered, Vulnerable, or Near Threatened.

in the wild (**FIGURE 8.9**). In other cases, zoos have partnered with local conservation groups and governments; the Kemp's ridley sea turtle conservation example described in Chapter 1 is one such case. As valuable as these ex situ breeding programs are, the in situ conservation work done by scientists at zoos is no less important; a recent review of recovery programs for species listed under the US Endangered Species Act found that zoos frequently are responsible for the monitoring and assessment of wild populations (Che-Castaldo et al. 2018). Zoos in one country often assist in situ conservation in other countries; for example, the hooded vulture monitoring program in Botswana discussed in Chapter 7 was a collaboration with scientists from the Denver Zoo in the United States (see the Chapter 7 opening photo).

Zoos also play an important advocacy role; animals in their collections make people aware of the threats to biodiversity and help the public feel connected to the fate of species across the globe (Sampaio et al. 2020). As part of the World Zoo Conservation Strategy, which seeks to link zoo programs with conservation efforts in the wild, the world's 2000 zoos and aquariums are increasingly incorporating ecological themes and information about the threats to endangered species into their

(A)

(B)
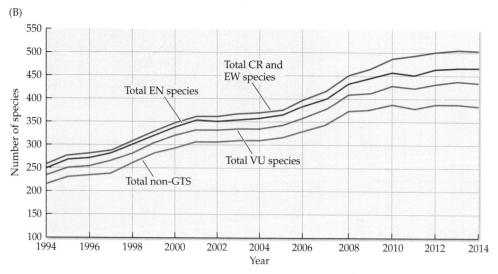

FIGURE 8.9 (A) A boy observes a Chinese giant salamander (*Andrias davidianus*) in a freshwater tank. Critically endangered in the wild, the Chinese giant salamander is the largest living species of amphibian. (B) The number of amphibian species at zoos has been steadily increasing since 1994, not only for non–globally threatened species (non-GTS), but also for vulnerable (VU), endangered (EN), and critically endangered (CR) species, and species extinct in the wild (EW). (B, after J. Dawson et al. 2016. *Conserv Biol* 30(1): 82–91. © 2015 Society for Conservation Biology.)

public displays and research programs. The potential educational and financial impacts of zoos are enormous, considering that they attract over 200 million visitors per year, according to the Association of Zoos and Aquariums (www.aza.org) (**FIGURE 8.10A**). Importantly, funds raised by zoos from visitor fees and other programs directly fund in situ conservation activities; for example, $28.9 million was spent by zoos between 2013 and 2015 on species listed under the Endangered Species Act in the United States (Che-Castaldo et al. 2018).

Zoos have the necessary knowledge and experience in animal care, veterinary medicine, animal behavior, reproductive biology, and genetics to establish captive animal populations of endangered species (**FIGURE 8.10B**). Zoos and affiliated conservation organizations have embarked on major efforts to build the facilities and develop the technology necessary to establish and house breeding colonies of endangered animals and to develop the new methods and programs needed to reintroduce species in the wild (**FIGURE 8.11**). Some of these facilities are highly specialized, such as that run by the International Crane Foundation in Wisconsin, which is attempting to establish captive breeding colonies of all crane species. These collaborative efforts have paid off.

The success of captive breeding programs has been enhanced by efforts to collect and disseminate knowledge about the maintenance of rare and endangered species. The Species Survival Commission's Conservation Breeding Specialist Group, a division of the IUCN, and affiliated organizations, such as the Association of Zoos and Aquariums, the European Association of Zoos and Aquaria, and the Australasian Regional Association of Zoological Parks and Aquaria, provide zoos with the necessary information on proper care and handling of these species as well as updates on the status and behavior of the animals in the wild (www.aza.org). This information includes nutritional requirements, anesthetic

(A)

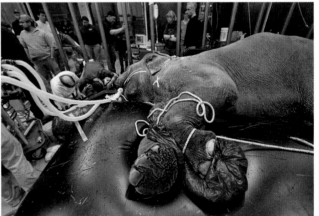

(B)

FIGURE 8.10 (A) Zoos can educate the public about the need to protect wildlife. Here, visitors to the Asahiyama Zoo on Japan's northern island of Hokkaido enjoy a parade of king penguins (*Aptenodytes patagonicus*). (B) Veterinarians carry out dental surgery on a captive Asian elephant. The knowledge gained by caring for captive animals can be applied to helping the species in the wild.

© LMR Group/Alamy

FIGURE 8.11 China's giant panda (*Ailuropoda melanoleuca*) is one of the world's most charismatic animals and has become emblematic of the fight to save endangered species. Wolong National Nature Reserve and other facilities in China have been successful at breeding giant pandas using artificial insemination and hand rearing. The reserve has established a reintroduction program, but the loss of habitat and the fact that captive-raised individuals may lack the behavioral skills needed to survive in the wild combine to make reintroduction of pandas particularly problematic.

techniques to immobilize animals and reduce stress during transport and medical procedures, optimal housing conditions, information about vaccinations and antibiotics to prevent the spread of disease, and breeding records. The collective effort is being aided by a central database called the Zoological Information Management System, maintained by Species360 (formerly the International Species Information System), which keeps track of over 220 million animal husbandry records from more than 1200 member institutions in 99 countries (Species360/ZIMS 2019).

The long-term success of breeding programs varies widely. Population viability analyses (PVA) conducted on 137 captive breeding programs in zoos found that without imports or exports of animals and under current management practices, 64% of populations are predicted to decline in size over the next 25 years (Che-Castaldo 2019). However, the models also suggest that these numbers can be significantly improved with management changes such as increasing the size of enclosures, supplementing genetic diversity through exchange programs and other measures. As we will see in the next section, cooperation among zoos is critical for successful breeding programs.

> Zoos often use the latest methods of veterinary medicine to establish healthy breeding colonies of endangered animals.

Many species provided with humane captive conditions reproduce with abandon—so much so that the use of contraceptives to control populations is required. However, some rare animal species, such as rhinoceroses, do not adapt or reproduce well in captivity. Still other animals will have too few individuals at a single zoo to support a breeding program on their own.

Zoos conduct extensive research to identify management methods that can overcome these problems and promote successful reproduction of genetically appropriate mates. In many cases, this can simply involve bringing in an individual, usually a male, from another zoo. These exchanges are facilitated by the publication of *studbooks* by zoos; these are records of births, deaths, parentage, source (e.g. whether acquired from the wild), and any current transfers of individuals for each species.

Some of the more technologically advanced methods for breeding come directly from human and veterinary medicine, while others are novel techniques developed at special research facilities such as the Smithsonian Conservation Biology Institute in Virginia and the San Diego Zoo Institute for Conservation Research. These techniques include **cross-fostering**, in which common species raise the offspring of rare species; **artificial incubation** of eggs under ideal hatching conditions; **artificial insemination** when adults do not show interest in mating or are living in different locations; and **embryo transfer**, which involves implanting fertilized eggs of rare species into surrogate mothers of common species (**FIGURE 8.12**). One of the most unusual techniques, known as **genome resource banking** (**GRB**), involves the freezing of purified DNA, eggs, sperm, embryos, and other tissues so that they can be used to contribute to future breeding programs, to maintain genetic diversity, and for scientific research. Many of these approaches have the added benefit that these materials are easier to transport than mature animals. However, these techniques are often expensive and species-specific. In any case, GRB and similar advanced methods are not substitutes for in situ and ex situ conservation programs that preserve behaviors and ecological relationships that are necessary for survival in the wild.

Even when successful, raising animals in captivity sometimes reduces the animals' conservation value. Animals living in captive breeding facilities may lose the behaviors they need to survive in the wild. Furthermore, populations raised in captivity may undergo genetic, physiological, and morphological changes that make them less able to tolerate the natural environment if they

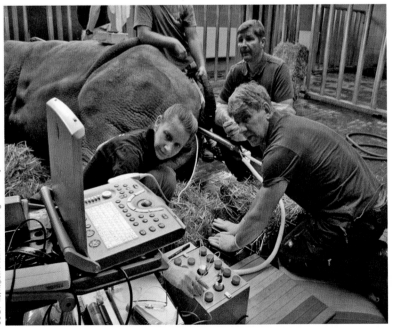

FIGURE 8.12 Oocyte (egg) collection from a southern white rhino (*Ceratotherium simum* ssp. *simum*) to test an in vitro fertilization method. If successful, the approach will be applied to the more highly endangered northern white rhino (*C. simum* ssp. *cottoni*).

are returned to the wild. Diseases acquired in captivity may render them unsuitable for release (Minuzzi-Souza et al. 2016). Consequently, when researchers establish an ex situ program to preserve a species, they must address a series of ethical questions:

- How will establishing an ex situ population benefit the wild population?
- Is it better to let the last few individuals of a species live out their days in the wild or to breed a captive population that may be unable to adapt to wild conditions?
- Does a population of a rare species consisting of individuals that have been raised in captivity and do not know how to survive in their natural environment really represent preservation of the species?
- Are rare individuals being held in captivity primarily for their own benefit, for the benefit of their entire species, for the economic benefit of zoos, or for the pleasure of zoo visitors?
- Are the animals in captivity receiving appropriate care based on their biological needs?
- Are sufficient efforts being made to educate the public about conservation issues?

Aquariums

Both wild-caught and aquarium-bred fish and other animals are maintained in public aquariums that are open to visitors (**FIGURE 8.13**). Major efforts are being made to develop breeding techniques so that rare species can be maintained in aquariums without further collection in the wild and in the hope that some can be released back into the wild.

Fish breeding programs use indoor aquarium facilities, seminatural water bodies, and fish hatcheries and farms. Many fish breeding techniques were originally developed by fisheries biologists for large-scale stocking operations involving trout, bass, salmon, and other commercial species. Other techniques were discovered in the aquarium pet trade when dealers attempted to propagate tropical fish for sale. These techniques are now being applied to endangered freshwater fauna. Currently, both public and private groups are making impressive efforts to unlock the secrets of propagating some of the more difficult species.

Aquariums have an increasingly important role to play in the conservation of endangered cetaceans, manatees, sea turtles, and other large marine animals. Aquarium personnel often respond to public requests for assistance in handling large animals stranded on beaches or disoriented in shallow waters. The aquarium community can use lessons learned from working with common species to develop programs to aid endangered species.

The ex situ preservation of aquatic biodiversity takes on additional significance due to the dramatic recent increase in aquaculture, which currently represents approximately 52% of total fish production for food (FAO 2020).

© Jeff Greenberg/Alamy Stock Photo

FIGURE 8.13   Public aquariums participate in both in situ and ex situ conservation programs and provide a valuable function by educating people about marine conservation issues.

This aquaculture includes the extensive salmon, carp, and catfish farms in the temperate zones, the shrimp and fish farms in the tropics, and the 12 million tons of aquatic products grown in China and Japan. As fish, frogs, molluscs, and crustaceans are increasingly domesticated and raised to meet human needs, it becomes necessary to preserve the genetic stocks needed to maintain and increase commercial production of these species and to protect them against disease and unforeseeable threats. A challenge for the future will be balancing the need to increase human food production through aquaculture with the need to protect aquatic biodiversity from increasing human threats. According to the Food and Agriculture Organization of the United Nations 2020 report, climate change is expected to decrease productivity of wild-caught fish, particularly in tropical regions, increasing the need to further develop technologies to support the demands of a growing human population (FAO 2020).

GLOBAL CHANGE
CONNECTION

Botanical gardens

The world's 1775 botanical gardens (also known as botanic gardens) contain major collections of living plants and represent a crucial resource for plant conservation through ex situ conservation, in situ conservation, research, and education (**FIGURE 8.14**). An **arboretum** is a specialized botanical garden focusing on trees and other woody plants. The world's botanical gardens currently contain about 4 million living plants, representing at least 105,634 species (approximately 27% of the world's flora), and help conserve over 41% of known threatened species (Mounce et al. 2017; www.bgci.org). One of the world's largest botanical gardens, the Royal Botanic Gardens, Kew, in England, has over 50,000 species of plants under cultivation; at least 2700 of these species are listed as threatened by the IUCN.

Botanical gardens have living collections and seed banks that provide ex situ protection and knowledge of endangered and economically important plants.

(A)

(B)

(C)

(D)

Photo by Richard B. Primack

Photo by Anna Sher

Courtesy of the Missouri Botanical Garden

Courtesy of the Royal Botanic Gardens, Kew

FIGURE 8.14 (A) The beautiful displays at Munich Botanical Gardens in Germany play an important role in conservation by connecting the public with plants and preserving those species that no longer occur in the wild. (B) Signage at the Kirstenbosch National Botanical Garden, South Africa, educates visitors about how to promote local biodiversity in their yards. (C) Field biologists with the Madagascar program of the Missouri Botanical Garden identify plants that may have medicinal benefits, work to safeguard the biodiverse areas where they grow, and raise plants for reintroductions. (D) A scientist at the Royal Botanic Gardens, Kew, United Kingdom, inspects preserved seeds as a part of the Millennium Seed Bank Partnership.

Botanical gardens increasingly focus their efforts on cultivating rare and endangered plant species, and many specialize in particular types of plants (see chapter opening photo). Many botanical gardens are involved in plant conservation, especially in the reintroduction of rare and endangered plant species and the restoration of degraded ecosystems. In Yunnan, China, only 52 individuals of a rare magnolia tree, *Magnolia sinica*, are left in the wild, but the species has been successfully cultivated at Kunming Botanical Garden, providing hope that reintroductions may be possible.

Staff members at botanical gardens are often recognized authorities on plant identification, distributions, and conservation status. Botanical gardens are able to educate an estimated 200 million visitors per year about conservation issues. At an international level, Botanic Gardens Conservation International (www.bgci.org) represents and coordinates the conservation efforts of over 800 botanical gardens. The priorities of this program include the creation of a worldwide database to support collecting activity and identification of important species that are underrepresented or absent from collections of living plants. One of its projects is the online PlantSearch database (www.bgci.org/plant_search.php), which currently lists over 555,500 plant species and cultivated varieties growing in 1116 botanical gardens and other contributing institutions, of which about 2400 are rare or threatened.

Seed banks

In addition to growing plants, botanical gardens and research institutes have developed collections of seeds, sometimes known as **seed banks**, obtained from the wild and from cultivated plants. These seed banks provide a crucial backup to their living collections (**FIGURE 8.15**). The seeds of most plant species can be kept dormant in cold, dry conditions for long periods and later germinated to produce new plants. This ability of seeds to remain dormant allows the seeds of large numbers of rare species to be frozen and stored in a small space, with minimal supervision and at a low cost. Seed banks are

(A)

(B)

Both photos: Courtesy of the U.S. Department of Agriculture

FIGURE 8.15 (A) At seed banks, seeds of many plant varieties are sorted, cataloged, and stored at freezing temperatures. (B) Seeds come in a wide variety of sizes and shapes. Each such seed represents a genetically unique, dormant individual.

especially important for rare and endangered species that may need to be re-introduced into the wild. Efforts are made to include the full range of genetic variation found in a species by collecting seeds from populations growing across the range of the species.

More than 1000 seed banks exist worldwide, with their activities coordinated by international groups. One of these is the Consultative Group on International Agricultural Research (www.cgiar.org), which works with over 3000 partners in over 70 countries. The Millennium Seed Bank Partnership, with partners in over 108 countries, claims to be the largest network for ex situ conservation in the world. Among these, the Millennium Seed Bank in Wakehurst, managed by the Royal Botanic Gardens, Kew, is the largest individual bank, with over 2.4 billion seeds representing over 40,000 species (Royal Botanic Gardens Kew 2021). The Center for Plant Conservation coordinates a network of 39 botanical gardens and arboretums in the United States and Canada to preserve seeds from more than 800 rare and endangered plants. One of these is a special facility at the Shaw Nature Preserve outside St. Louis that focuses on Missouri plants.

Typically, seeds are dried before storage in deep-freeze vaults that keep the seeds between –18°C and –20°C. If power supplies fail or equipment breaks down, however, an entire frozen collection could be damaged. To prevent such a loss, the Svalbard Global Seed Vault in Norway stores 400,000 frozen seed samples below permafrost. Because of the location of the samples, they are protected from the risk of loss of power that would threaten most frozen collections.

GLOBAL CHANGE
CONNECTION

Seed banks have been embraced by agricultural research institutes and the agricultural industry as an effective resource for preserving and using the genetic variation that exists in agricultural crops and their wild relatives. Preserving this genetic diversity is crucial to maintaining and increasing the high productivity of modern crops and their ability to respond to changing environmental conditions such as acid rain, global climate change, and soil erosion (Zhang et al. 2017). Researchers are in a race against time to preserve this genetic variation because traditional farmers throughout the world are abandoning their diverse local crop varieties in favor of standard, high-yielding varieties (Gliessman 2015). For example, it is estimated that less than half of the original genetic diversity in the cultivated soybean still exists, and it continues to decline (Zhou et al. 2015). This worldwide phenomenon is also illustrated by Sri Lankan farmers, who grew 2000 varieties of rice until the late 1950s, when they switched over to just 5 high-yielding varieties.

A major controversy in the development of agricultural seed banks is who owns and controls the genetic resources of crops. The genes of **landraces** of crop plants (local species that have been adapted by humans over time) and wild relatives of crop species represent the building blocks needed to develop elite, high-yielding varieties suitable for modern agriculture. Approximately 96% of the raw genetic variation necessary for modern agriculture comes from developing countries such as India, Ethiopia, Peru, Mexico, Indonesia, and China (**FIGURE 8.16**), yet most corporate breeding programs for elite strains are located in North America and Europe. In the past, genetic material

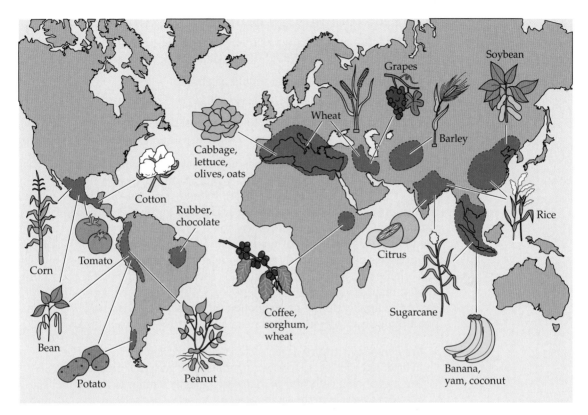

FIGURE 8.16 Crop species show high genetic diversity in certain areas of the world (shown in blue), often where the species were first domesticated or where they are still grown in traditional agricultural settings. This genetic diversity is of international importance in maintaining the productivity of agricultural crops. (Map provided by G. Wilkes, 1993; data based on Vavilov's centers of origin as discussed in J. R. Harlan. 1976. *Sci Am* 235(3): 88–97.)

was perceived as free for the taking. New strains were developed through sophisticated breeding programs and field trials. Seed companies sold the resulting seeds at high prices to maximize their profits, which often totaled hundreds of millions of dollars a year. However, neither the communities nor the countries from which the original seeds were collected received any profit from this activity.

One solution to this controversy involves negotiating agreements, using the framework of the Convention on Biological Diversity, in which countries agree to share their genetic resources in exchange for receiving new products and a share of the profits (see Chapter 12). A few such contracts have been negotiated, such as the one between Merck and the government of Costa Rica for the development of products based on species collected from the wild in that country (see Chapter 3 for similar examples). These agreements will be followed carefully to determine whether they are mutually satisfactory and can serve as models for future contracts.

8.3 Can Technology Bring Back Extinct Species?

LEARNING OBJECTIVES

By the end of this section you should be able to:

8.3.1 Explain the methods of de-extinction.

8.3.2 Debate arguments both for and against de-extinction.

In September of 2021, a bioscience and genetics company announced that they planned to ressurect the woolly mammoth, a species that has been extinct for 10,000 years (Loreno 2021). Indeed, the resurrection of versions of extinct species is now within our grasp due to the "coming together" of various techniques and approaches, some new and others well established (Seddon 2017). The idea of using technology to bring species back from extinction is not particularly new. Medical scientists have recreated viruses and other pathogens in the past to study their behavior in hope of preventing similar diseases. However, in recent years, there has been a growing interest in the application of **de-extinction** for conservation purposes. This can potentially be accomplished by one of at least three methods (Shapiro 2017):

1. *Back-breeding*   In this case, relatives of the extinct species may be hybridized or selected for specific traits. The goal is to create an organism morphologically and perhaps ecologically similar to the extinct species. An example of this is the attempt to bring back the extinct quagga, a relative of the zebra that has been extinct since the 1800s. By selective breeding of zebra, a team of researchers in South Africa are creating an animal that looks very similar to the extinct species, even if it is not genetically a quagga (Page and Hancock 2016).

2. *Cloning*   This method involves taking the nucleus of an intact cell from an extinct species and inserting it into the egg of a living, close relative. In 2003, scientists used this method with the Pyrenean ibex or bucardo (*Capra pyrenaica pyrenaica*), the last of which died in 2000. The process involved fusing frozen bucardo cells with goat egg cells from which the nuclei had been removed and implanting the resulting embryo into a domesticated goat, which served as a surrogate mother. Even though after many trials one embryo did develop to term, the resulting cloned ibex was badly deformed and died within minutes of its birth. Because of the necessity of an intact cell, this method is only possible for recently extinct species for which such cells have been preserved.

3. *Genomic reconstruction*   Using the molecular tool CRISPR, it is theoretically possible to take genes from an extinct species and insert them into the genome of a living relative. For example, genes taken from museum specimens of the extinct passenger pigeon could be inserted into the genome of the egg of an extant band-tailed pigeon. Thus, the resulting offspring would have at least some genes from the passenger pigeon. Rather than requiring a whole nucleus, this method only requires some

intact DNA. This requirement still limits the candidate species to those that have not been extinct so long that their genetic material has completely degraded. Such DNA may be obtained not only from carefully preserved museum specimens but also from parts or whole organisms preserved by being buried under a glacier or in an anoxic bog or oceanic mud, or even in the freezer of researchers.

But is it a good idea to create these versions of extinct species? The possibility brings up questions of ethics, practicality, and the fundamental goals of conservation. Some argue that we have an ethical responsibility to bring back species we have lost due to anthropogenic impact, while others may argue that de-extinction is unethical (Bennett et al. 2017). Would it be acceptable to bring back a species if its habitat no longer exists? Other questions include the following: If re-created passenger pigeons were brought up by another pigeon species and lacked the flocking, feeding, and other behaviors of extinct passenger pigeons, should they really be considered passenger pigeons? Would resurrected passenger pigeons function in an ecosystem in the same ways as past passenger pigeons? In part because of these concerns as well as the associated cost, careful selection of de-extinction candidates is paramount. Although it has been proposed that we should bring back such charismatic species as the wooly mammoth, other species may be more feasible and/or more ecologically valuable to resurrect (Seddon et al. 2014b).

> Benefits of de-extinction must be carefully weighed against its costs and risks on a case-by-case basis.

It is also unclear how de-extinction programs would affect and interact with existing conservation efforts. Would the ability to bring species back from extinction reduce the incentive to preserve species in the wild? Perhaps most importantly, will de-extinction research and efforts divert funding from research and conservation projects aimed at preserving existing species and ecosystems (Bennett et al. 2017)?

The thought-provoking questions raised by these exciting new techniques highlight the interdisciplinary nature of conservation biology—its considerable overlap with biotechnology, genetics, ecology, ethics, and economics. Just because we can do something does not necessarily mean that we should; whether or not we use de-extinction to bring back an extinct species or restore its ecosystem functions will depend on the development of new conservation technologies and the thoughtful discussions that they inspire.

Summary

- New populations of rare and endangered species can be established in the wild using either captive-raised or wild-caught individuals; new populations can also be established outside the historical range of a species.

- Identifying and reducing threats to a species is a key element in the success of reintroduction programs.

- Animals sometimes require behavioral training before release, as well as maintenance after release.

- Ex situ conservation is the preservation of individuals and populations in captivity. Zoos, aquariums, botanical gardens, and seed banks play an important role in conservation, highlighting the complementary roles of in situ and ex situ conservation.

- Technologies are being developed that could potentially revive versions of extinct organisms; however, this may be neither practical nor advisable.

For Discussion

1. How do you judge whether a reintroduction project is successful? Develop simple and then increasingly detailed methodologies to evaluate a specific project's success.

2. Would it be a good idea to create new wild populations of African rhinoceroses, elephants, and lions in Australia, South America, the southwestern United States, or other areas outside their current range? What would be some of the legal, economic, and ecological issues involved?

3. Would biodiversity be adequately protected if every species were raised in captivity? Is this possible? Is it practical? How would freezing a tissue sample of every species help to protect biodiversity? Again, is this possible or practical?

Suggested Readings

Butt, N., et al. 2020. Importance of species translocations under rapid climate change. *Conservation Biology* 35(3): 775-783. A literature review with some specific recommendations.

Coz, D. M., and J. C. Young. 2020. Conflicts over wildlife conservation: Learning from the reintroduction of beavers in Scotland. *People and Nature* 2(2): 406–419. A case study of the complexities of species reintroductions.

Fukano, Y., Y. Tanaka., and M. Soga. 2020. Zoos and animated animals increase public interest in and support for threatened animals. *Science of the Total Environment* 704: 135352. An elegant study that demonstrates the important role of zoos in conservation messaging.

Genovesi, P., and D. Simberloff. 2020. "De-extinction" in conservation: Assessing risks of releasing "resurrected" species. *Journal for Nature Conservation* 56: 125838. An argument against using this technology by comparing costs of protecting extant species versus bringing back those that are extinct.

Malone, E. W., et al. 2018. Which species, how many, and from where: Integrating habitat suitability, population genomics, and abundance estimates into species reintroduction planning. *Global Change Biology* 24(8): 3729–3748. An example of reintroduction planning for three species of fish, including genetic diversity considerations.

Westwood, M., et al. 2021. Botanic garden solutions to the plant extinction crisis. *Plants, People, Planet* 3(1): 22–32. A call to action on the important role that botanical gardens can play in conservation.

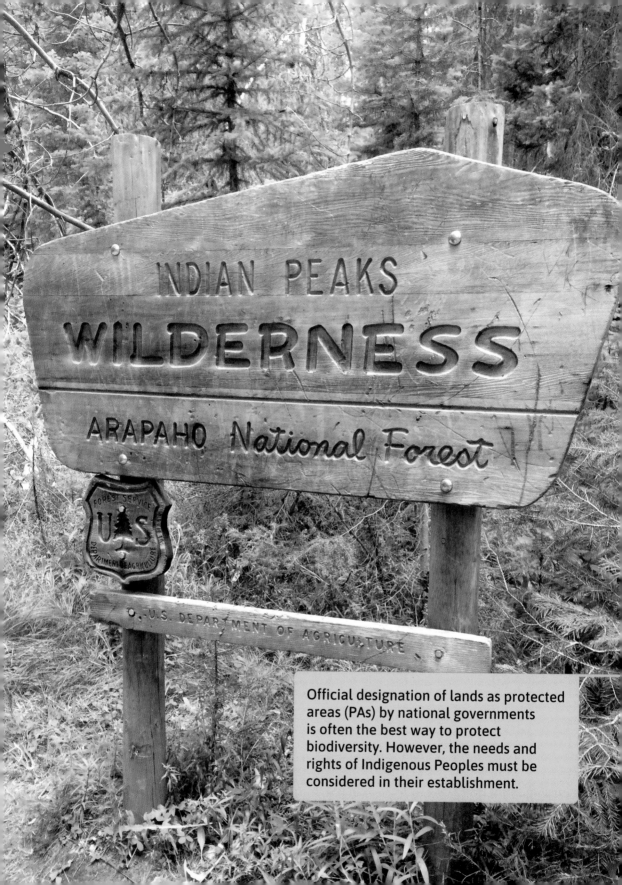

Official designation of lands as protected areas (PAs) by national governments is often the best way to protect biodiversity. However, the needs and rights of Indigenous Peoples must be considered in their establishment.

Protected Areas

9

One of the most used, successful, and surprisingly flexible of the tools for conservation is the creation of protected areas. A **protected area** (**PA**) is a clearly defined geographical space, recognized, dedicated, and managed, through legal or other effective means, to achieve the long-term conservation of nature with associated eco-system services and cultural values. Given that the biggest threat to most taxa is loss of habitat (Chapter 4), it should be no surprise that PAs are an international priority. Furthermore, protection of carbon-fixing forests and other ecosystems could potentially provide up to 37% of the reductions in greenhouse gas emissions needed to stabilize global warming to 2°C by 2030. According to the World Database on Protected Areas, there are 257,889 designated protected areas; most of these are on land (UNEP-WCMC and IUCN 2021). These data and other information about protected areas can be found on Protected Planet (www.protectedplanet.net), a website managed by the United Nations Environment Programme-World Conservation Monitoring Centre. Protected areas are covering more area and a wider range of ecosystems than ever before, with particular progress in marine areas, but there is more work to be done.

GLOBAL CHANGE
CONNECTION

The value of protecting areas from human development has long been recognized. Some of this history can be found in Chapter 1. For example, in the Arabic world, *hema*, literally "protection," refers to the practice of managing land by restricting human use in areas for certain periods; on the Arabian Peninsula, hema practice dates back at least 2000 years (IUCN WCPA 2021). Many countries and cultures have learned that in order to maintain use of resources, some level of protection is required, usually by a governing body.

Photo by Anna Sher

FIGURE 9.1
American bison (*Bison bison*) graze at Yellowstone National Park, the first national park established in the United States.

However, the practice of protecting lands for the sake of the wild plants and animals that live on them is a more recent development. The Yosemite Act of 1864, signed by President Abraham Lincoln, reserved the Yosemite Valley in California as a protected wilderness area, making it the first formal protected area in US history. This was followed by the establishment of Yellowstone National Park in 1872, which in turn is said to have sparked a worldwide national parks movement (**FIGURE 9.1**). Yosemite and Yellowstone are not unique, however; other early nationally protected lands include the Bogd Khan Uul Protected Area in Mongolia (1778), the Royal National Park in Australia (1879), and the Virunga National Park in the Democratic Republic of the Congo (established as Albert National Park in 1925) (Djossa 2018).

Some argue that establishing protected areas is ultimately the only way to preserve many species and ecosystems, particularly those species and habitats that do not adapt well to human impacts. Many species have become completely dependent on protected areas for their existence (Pacifici et al. 2020). According to the World Resources Institute (www.wri.org), preserving ecosystems involves the following:

1. Establishing individual protected areas
2. Creating networks of protected areas
3. Managing those areas effectively
4. Implementing conservation measures outside the protected areas
5. Restoring biological communities in degraded habitats

The first three of these topics are discussed in this chapter; the final two are covered in Chapter 10 and Chapter 11, respectively.

9.1 Establishment and Classification of Protected Areas

LEARNING OBJECTIVES

By the end of this section you should be able to:

9.1.1 Determine the level of protection for a given protected area, given its IUCN rank.

9.1.2 Contrast the benefits of strict protection with those of "multi-management" of protected areas.

9.1.3 Compare the growth of marine and terrestrial protected areas, and provide explanations for why they differ.

Protected areas can be established in a variety of ways, but the most common mechanisms are these (roughly in decreasing order of significance):

- Government action, usually at a national level, but often on regional or local levels

- Land purchases and easements by private individuals and conservation organizations

- Actions of Indigenous Peoples and traditional societies

- Development of biological field stations (which combine biodiversity protection and research with conservation education) by universities and other research organizations

Although legislation and land purchases alone do not ensure habitat protection, they can lay the groundwork for it. Partnerships among governments of developing countries, international conservation organizations, multinational banks, research and educational organizations, and governments of developed countries are another way to bring together funding, training, and scientific and management expertise to establish new protected areas.

Traditional societies also have established protected areas to maintain their ways of life or simply to preserve their land (Langton et al. 2014). Many of these protected areas have been in existence for a long time and have cultural or religious significance (see Figure 12.6). National governments in many countries, including the United States, Canada, Colombia, Brazil, and Australia, have recognized the rights of traditional societies to own and manage the land on which they live, hunt, and farm. However, in some cases, land is simply appropriated by governments (see chapter opening photo), or the recognition of land rights results only after significant conflict in the courts. Land owned or managed by Indigenous Peoples is discussed further in Chapter 10.

The International Union for Conservation of Nature (IUCN) has developed a six-category system for classifying protected areas (**TABLE 9.1**). The conservation of nature in protected areas is a primary management objective in all six categories, with lands in categories I–IV considered strictly protected. The total amount of area protected has increased over time, particularly in category VI (areas with sustainable use of natural resources). However, the importance of higher-protection categories cannot be overstated: a recent analysis using plants and invertebrates as a proxy for biodiversity estimated that the protection of pristine habitat halved extinction risk. Unfortunately, the categories that protect such land (Ia and Ib) are underrepresented among PAs (Sayre et al. 2020).

Protected areas in categories V and VI are considered multiple-use or multi-management protected areas, as they are administered not only to

TABLE 9.1	IUCN Protected Area Designations I–VI
Category	**Description**
Ia. Strict nature reserves	Category Ia are strictly protected areas set aside to protect biodiversity and also possibly geological/geomorphical features, where human visitation, use and impacts are strictly controlled and limited to ensure protection of the conservation values. Such protected areas can serve as indispensable reference areas for scientific research and monitoring.
Ib. Wilderness Area	Category Ib protected areas are usually large unmodified or slightly modified areas, retaining their natural character and influence without permanent or significant human habitation, which are protected and managed so as to preserve their natural condition.
II. National Park	Category II protected areas are large natural or near natural areas set aside to protect large-scale ecological processes, along with the complement of species and ecosystems characteristic of the area, which also provide a foundation for environmentally and culturally compatible, spiritual, scientific, educational, recreational, and visitor opportunities.
III. Natural Monument or Feature	Category III protected areas are set aside to protect a specific natural monument, which can be a landform, sea mount, submarine cavern, geological feature such as a cave or even a living feature such as an ancient grove. They are generally quite small protected areas and often have high visitor value.
IV. Habitat/Species Management Area	Category IV protected areas aim to protect particular species or habitats and management reflects this priority. Many Category IV protected areas will need regular, active interventions to address the requirements of particular species or to maintain habitats, but this is not a requirement of the category.
V. Protected Landscape/ Seascape	A protected area where the interaction of people and nature over time has produced an area of distinct character with significant, ecological, biological, cultural and scenic value: and where safeguarding the integrity of this interaction is vital to protecting and sustaining the area and its associated nature conservation and other values.
VI. Protected area with sustainable use of natural resources	Category VI protected areas conserve ecosystems and habitats together with associated cultural values and traditional natural resource management systems. They are generally large, with most of the area in a natural condition, where a proportion is under sustainable natural resource management and where low-level non-industrial use of natural resources compatible with nature conservation is seen as one of the main aims of the area.

Source: IUCN. 2021. *Protected Area Categories*. https://www.iucn.org/theme/protected-areas/about/protected-area-categories

conserve biodiversity but also to produce natural resources, such as timber and cattle, for human use. These multi-management protected areas can be particularly significant for several reasons:

- They are often much larger in area than other categories of protected areas.
- They may contain many or even most of their original species.
- They often adjoin or surround other protected areas.
- They are more likely to benefit local people than strictly protected areas, and therefore they are more likely to earn local support.

A review of 171 published reports on 165 protected areas found that all types (I–VI) were more likely to have a positive impact on the local people than a negative impact and that the multiple-use protected areas (V and VI) were the most likely to have positive conservation and socioeconomic outcomes (Oldekop et al. 2015). This is likely because local people benefit from direct use of the protected areas (see Chapter 3) and are more likely to be involved with their management.

More than 258,000 protected areas in IUCN categories I–VI have been designated worldwide, as documented by The World Database on Protected Areas (WDPA; Protected Planet 2021). Unfortunately, much of the protected land is concentrated in areas that are at high elevations, on steep slopes, and in remote areas far from roads and cities, where disturbance from humans is often already minimal (Sayre et al. 2020). For example, the world's largest terrestrial park is in Greenland on inhospitable terrain and covers 972,000 km^2. Far more advantageous for conservation is when the establishment of protected areas is guided by one of the approaches for species protection described in Chapter 7. The *Protected Planet Report* states that as of 2018, only 21% of key biodiversity areas were completely covered by protected areas, while 35% had no protection at all (UNEP-WCMC, IUCN, and NGS 2018).

The importance of increasing the number and extent of protected areas motivated commitments made by almost 130 countries and territories that signed the Convention on Biological Diversity Aichi Biodiversity Target 11 in 1993 (see Chapter 12):

> *By 2020, at least 17 percent of terrestrial and inland water areas and 10 percent of coastal and marine areas, especially areas of particular importance for biodiversity and ecosystem services, are conserved through effectively and equitably managed, ecologically representative and well-connected systems of protected areas and other effective area-based conservation measures and integrated into the wider landscape and seascape.*

Unfortunately, these ambitious goals were not reached by their target date, although most countries did increase their networks of PA since Aichi, resulting in global expansion in both terrestrial and marine protected area (**FIGURE 9.2A**). In fact, according to the Convention on Biological Diversity, only one reporting country has not had progress on the Aichi Target 11, with several exceeding their national goals, including Canada, India, Niger, and Peru (see www.cbd. int/aichi-targets/target/11). There are, in fact, now protected areas in every country, although the amount of area protected varies considerably among countries and regions (UNEP-WCMC, IUCN, and NGS 2018). South America, Africa, Australia, and northern Asia contain some very large reserves, while places like Europe have a comparable area protected by having larger numbers of smaller protected areas (**FIGURE 9.2B**). The coverage of protected areas has been increasing monthly, and as of May 2021, global coverage of protected areas

(A)

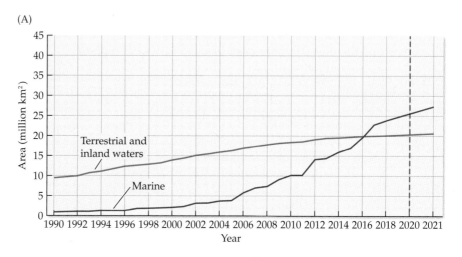

(B)

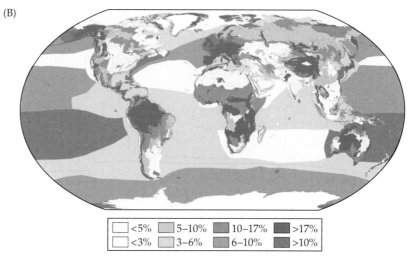

| <5% | 5–10% | 10–17% | >17% |
| <3% | 3–6% | 6–10% | >10% |

FIGURE 9.2 (A) The increase in amount of area in protected areas in terrestrial (in green) and marine protected areas (blue) since 1990. Graph also shows the targets that were not met for 2020 based on commitments from countries and territories to meet targets according to Convention on Biological Diversity Aichi Biodiversity Target 11. (B) The percentage of the world's terrestrial (in green) and marine (in blue) ecoregions that are protected. The darkest color for each map indicates those that have reached the targets stipulated by the Convention on Biological Diversity. (After UNEP-WCMC, IUCN, and NGS [2018]. *Protected Planet Report 2018*, based on *Protected Planet: The World Database on Protected Areas [WDPA]*, updated with data from February 2021. Available at https://www.protected-planet.net/en/resources/february-2021-update-of-the-wdpa-and-wd-oecm. UNEP-WCMC, IUCN, and NGS: Cambridge UK; Gland, Switzerland; Washington, D.C., USA, with unpublished data from the CBD Secretariat.)

was 16.64% of total land area and 7.74% of marine (UNEP-WCMC and IUCN 2021). As we can see from Figure 9.2, coverage of marine areas has increased faster than that of terrestrial.

In Europe, the area protected has increased exponentially: by 2021 there were over 120,000 sites, more than in any other region (IUCN WCPA 2021). These protected areas represent more than 26% of the area in Europe, encompassing terrestrial and inland waters. Within Europe, the European Union (EU) has reached the target of protecting at least 10% of coastal and marine areas by 2020, although these are over-represented by coastal waters and many protected areas are very small (<30 km^2).

A mechanism to promote and increase the number and size of protected areas is the IUCN Green List of Protected and Conserved Areas, the first global standard to recognize best practices in protected areas, established in 2014. To receive IUCN Green List certification, protected areas must meet a global standard based on four components: (1) good governance, (2) sound design and planning, and (3) effective management, which together lead to (4) successful conservation outcomes. Sites that do not meet standards initially can remain on a candidate list until they do; as of 2019 there were 40 Green List sites, with another 250 candidate sites. In 2018 alone, 15 new sites were added, including Ras Mohammed National Park, a marine site in Egypt near Sharm El-Sheikh (**FIGURE 9.3**). In this marine protected area 90% of corals are alive, a remarkable conservation feat given the impacts of climate change (see Figure 5.4); average coverage is only 30%–40% live corals for nonprotected reefs in the Red Sea.

> The IUCN has developed a classification system for protected areas, ranging from strict nature reserves to managed-resource protected areas, depending on the level of human impact and the needs of society for resources.

GLOBAL CHANGE
CONNECTION

© Hemis/Alamy Stock Photo

FIGURE 9.3 Ras Mohammed National Park is one of two protected areas in Egypt that received IUCN Green List certification in 2018. Certification requires implementation of global standards, including demonstrated conservation effectiveness. So far only 40 sites have achieved this certification worldwide, but an additional 250 sites are candidates.

Marine protected areas

Marine conservation has generally lagged behind terrestrial conservation efforts; however, the greatest growth in protected area is now being made in marine protected areas (MPAs; see Figure 9.2B). MPAs now cover nearly 28 million km^2 (Protected Planet 2021). The number of marine ecoregions with less than 1% protected area coverage has decreased from 22% to 17%, a positive development (see Figure 9.2B). Marine areas with national jurisdiction (0–200 nautical miles from the coast) have more protection (16.8%) than open ocean (1.2%), highlighting the need for international cooperation.

There are over 5000 marine and coastal protected areas, with a number of countries having recently created very large marine protected areas covering hundreds of thousands of square kilometers of marine ecosystems. The two largest of these were established in 2017: Marae Moana surrounding the Cook Islands (1.9 million km^2) and Ross Sea Region Marine Protected Area in Antarctica (1.6 million km^2).

The international goal of protecting 10% of the entire marine environment by the year 2020 was motivated in part by a desire to manage declining commercial fishing stocks (Rife et al. 2013); however, even stronger measures may be required to conserve the full range of coastal and marine biodiversity. Some of the current proposals for new MPAs are receiving pushback from groups that fear they will harm the fishing industry. However, MPAs can benefit the industry by providing opportunities for fish to reproduce and grow larger. Many MPAs preserve the nursery grounds of commercial species and maintain water quality and both the physical and biological features of ecosystems (Ban et al. 2017). In the process, high-quality protected areas can also provide recreational activities such as swimming and diving and the economic benefits associated with tourism. For example, the Fiji Locally Managed Marine Area Network protects the financial interests of that country by helping both the fishing and the ecotourism industry there (see Chapter 9 opening photo). Community-based ecotourism is one way that local people can benefit from MPAs; in Malaysia it was found that local people were highly motivated to be involved with community-based ecotourism, particularly when they had some environmental knowledge and/or understanding of the financial benefits (Masud et al. 2017).

Unfortunately, many MPAs exist only on maps and receive insufficient regulation, funding, and enforcement to prevent overharvesting and pollution. In the well-funded Phoenix Islands Protected Area, located in the central Pacific, fishing was banned in only 3% of the total area until international pressure led to an almost complete closure of fishing in 2015. Now this 408,250 km^2 MPA (nearly the size of California) bans all commercial fishing, with only a small sustainable use zone for consumptive use by the local population (UNESCO 2021b).

The effectiveness of protected areas

Studies show that protected areas generally are effective at protecting ecosystems, particularly when regulations are enforced and when areas are geographically isolated (Geldmann et al. 2013). For terrestrial protected areas, vertebrate abundance was found to be positively correlated with resources available, as measured by adequacy of staffing, budgets, and equipment (Geldman et al. 2018). A meta-analysis of the impact of terrestrial protected areas, using 156 studies including 13,669 species of vertebrates, invertebrates, and plants, found increases in both species richness (10.6% higher) and species abundance (14.5% higher) compared with samples taken outside the protected areas (Gray et al. 2016). Similarly, a global assessment of the efficacy of MPAs found that those that banned fishing had a higher abundance of several functional groups of fishes, which also indirectly benefited corals by decreasing algal cover (**FIGURE 9.4**).

Unfortunately, according to the Global Database on Protected Area Management Effectiveness, only a small percentage (20%) of the area covered by protected areas has been formally evaluated, making it difficult to fully assess

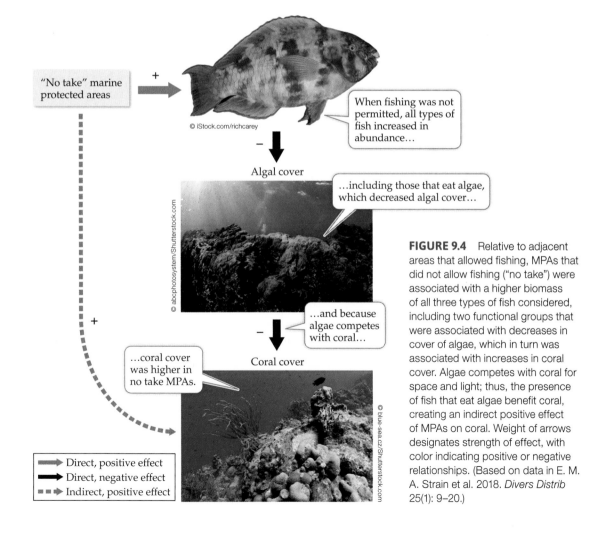

FIGURE 9.4 Relative to adjacent areas that allowed fishing, MPAs that did not allow fishing ("no take") were associated with a higher biomass of all three types of fish considered, including two functional groups that were associated with decreases in cover of algae, which in turn was associated with increases in coral cover. Algae competes with coral for space and light; thus, the presence of fish that eat algae benefit coral, creating an indirect positive effect of MPAs on coral. Weight of arrows designates strength of effect, with color indicating positive or negative relationships. (Based on data in E. M. A. Strain et al. 2018. *Divers Distrib* 25(1): 9–20.)

the global impact of protected areas on biodiversity. There are also cases where the protected status is violated by the very governments charged with enforcing it, such as the extensive logging in the Sochi National Park in the Western Caucasus to enable construction for the Olympic Games (Bragina et al. 2015).

According to a recent evaluation, the average ecological footprint within protected areas is 50% less than the global mean, and IUCN areas I and II have significantly less than this (Jones et al. 2018). However, it was estimated in this same study that a full third of protected areas across the globe are being impacted by "intense human pressure," including farming, roadbuilding, and human settlement within protected area boundaries. They found that mean human pressure has increased in recent years but that newer protected areas had less human pressure than those created during or before the 1990s.

Nevertheless, areas officially protected today will not necessarily be protected tomorrow. Protected areas can be reduced in size by the government, be opened up for exploitation, or even have their protected status removed (known as **degazettement**), particularly if they are found to contain valuable natural resources (Mascia et al. 2014). Furthermore, it is important to recognize that the long-term survival of some species in protected areas, and even of the ecosystems themselves, remains in doubt because populations of many species and the area of the ecosystems may be so reduced in size that their eventual fate is extinction.

GLOBAL CHANGE
CONNECTION

Although the number of species living within a protected area is an important indicator of the area's potential to protect biodiversity, protected areas need to maintain healthy ecosystems and viable populations of important species.

Measuring effectiveness: Gap analysis

Gap analysis compares biodiversity priorities with existing and proposed protected areas (see, for example, Simaika et al. 2013), but it can consider several factors. In the past, conservationists used informal means to ensure that the high-priority areas were protected, for example, by establishing national parks in different regions with distinctive ecosystems and ecological features. Now, however, conservationists are using more systematic planning processes involving gap analysis (Carrizo et al. 2017). Gap analysis generally consists of the following steps:

1. Data are compiled describing the presence and distribution of species, ecosystems, and physical features of the region, which are sometimes referred to as conservation units. Information on human densities and economic factors can also be included.

2. Conservation and social goals are identified, such as the amount of area to be protected for each ecosystem, the number of individuals of rare species to be protected, or the desired balance between wilderness and mixed resource management.

3. Existing conservation areas are reviewed to determine what is protected already and what is not (known as "identifying gaps in coverage").

4. Additional areas are identified to help meet the conservation goals ("filling the gaps").

5. These additional areas are reviewed in more detail and, if appropriate and practical, protected in some way (often by being directly purchased or designated as national parks). Management plans are then developed and implemented.

6. The new protected areas are monitored to determine whether they are meeting their stated goals. If not, the management plan can be changed or, possibly, additional areas can be acquired to meet the goals.

Gap analysis can also be used to identify holes in conservation at international scales. For example, a group of researchers from 27 institutions and organizations did an analysis of all existing protected areas and 25,380 species across the globe (Butchart et al. 2015). They calculated that protected areas currently include only 77%–78% of important sites for biodiversity, and there was insufficient coverage for 57% of the species they evaluated (**FIGURE 9.5**). They determined that in order to reach the Aichi Biodiversity Targets, an additional 3.3 million km^2 outside existing protected areas were needed, plus 14.8 million km^2 to cover all threatened species. To cover these plus all documented important sites for biodiversity and individual country and biome targets, the researchers estimated that the amount of area protected needed to be roughly doubled.

In another example, a global analysis considered PA coverage since the Aichi Target 11 in the context of climate change (Carrasco et al. 2021). Among

GLOBAL CHANGE
CONNECTION

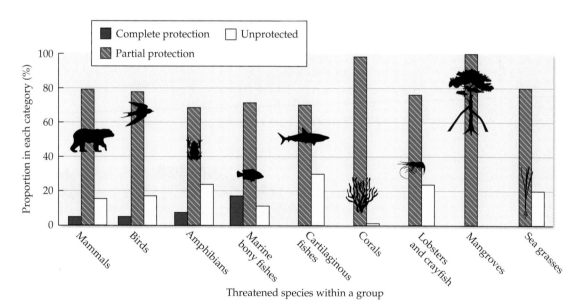

FIGURE 9.5 A visualization of steps 1–3 of a gap analysis to identify where protection is needed for various groups of animals and plants. This graph shows the proportion of threatened species that receive any sort of protection. Depending on the group, 10%–30% of threatened species lack protection anywhere in their range; most threatened species have at least partial protection. Relatively few threatened species are completely protected. (After S. H. M. Butchart et al. 2015. *Conserv Lett* 8(5): 329–337.)

other findings, it identified which specific countries needed to expand PAs to include more climate refugia for species that would likely have to move in response to global change (Chapter 5).

To evaluate where new protected areas should be established, as well as where gaps exist, **geographic information system (GIS)** mapping is a vital tool (see Chapter 4). In gap analysis, GIS can help researchers integrate the wealth of data on the natural environment with information on species distributions (Iannella et al. 2018). The basic GIS approach involves storing, displaying, and manipulating many types of spatial data involving factors such as vegetation types, climate, soils, topography, geology, hydrology, species distributions, human settlements, and resource use (**FIGURE 9.6**). This approach can point out correlations among the abiotic, biotic, and

> GIS is an effective tool for gap analysis. It uses a wide variety of information to pinpoint critical areas and species that are priorities for protection.

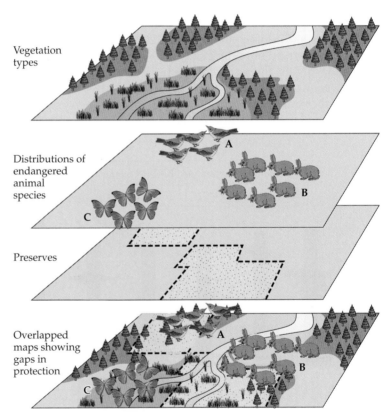

Vegetation types

Distributions of endangered animal species

Preserves

Overlapped maps showing gaps in protection

FIGURE 9.6 Geographic information systems (GIS) provide a method for integrating a wide variety of data for analysis and display on maps. In this example, vegetation types, distributions of endangered animal species, and preserved areas are overlapped to highlight areas that need additional protection. The overlapped maps show that the distribution of Species A is predominantly in a preserve, Species B is protected only to a limited extent, and Species C is found entirely outside the preserves. Establishing a new protected area to include the range of Species C would be the highest priority. (After J. M. Scott et al. 1991. In D. J. Decker et al. [Eds.]. *Challenges in the Conservation of Biological Resources: A Practitioner's Guide*, pp. 167–179. Westview Press, Boulder, CO.)

human elements of the landscape; help plan parks that include a diversity of biological communities; and even suggest sites that are likely to support rare and protected species. GIS is also useful for tracking global climate, to identify regions at risk (see Figure 5.3 and Figure 5.4B). Aerial photographs and satellite imagery are additional sources of data for GIS analysis, and they can highlight patterns of vegetation structure and distribution over local and regional scales (see Chapter 4 opening photo). These images can dramatically illustrate when current government policies are not working and need to be changed.

GLOBAL CHANGE
CONNECTION

9.2 Designing Protected Areas

LEARNING OBJECTIVES

By the end of this section you should be able to:

9.2.1 Evaluate the effectiveness of a protected area on a map, given its shape, dimensions, proximity to other protected areas, and other features.

9.2.2 Consider what species would benefit from a larger versus smaller protected area.

The size and placement of protected areas throughout the world are often determined by the distribution of people, potential land values, the political efforts of conservation-minded citizens, and historical factors (Mills et al. 2014). Ideally, the selection of new protected areas will follow conservation priorities, such as to protect a particular species, ecosystem, or biodiversity hotspot (see Chapter 7); however, there are many other possible scenarios. In urban and suburban areas, the fund-raising skills of private conservation groups, wealthy individuals, and government departments often represent the most important factors in determining what land is acquired. In other cases, certain parcels of land may be purchased to protect a critical water supply or a charismatic species, and others may be acquired simply because they adjoin the property of an influential citizen. Moreover, sometimes lands are set aside for conservation protection because they have no immediate commercial value, as in the case of the protected area in Greenland mentioned in Section 9.1; they are effectively the lands nobody wants (Venter et al. 2018).

The design of reserves is of great interest to governments, corporations, and private landowners, who are being urged, or even mandated, to manage their properties for both the commercial production of natural resources and the protection of biodiversity. However, conservation biologists must be cautious about using simplistic, overly general guidelines for designing protected areas, because every conservation situation requires individual consideration. Everyone benefits from increased communication between the academic scientists, who are developing theories of nature reserve design, and the managers, planners, and policymakers, who are actually creating new reserves (Reed and Abernathy 2018).

Conservation biologists often start by considering the **four Rs**:

1. *Representation*   Protected areas should contain as many features of biodiversity (species, populations, habitats, etc.) as possible.

2. *Resiliency*   Protected areas must be large enough to maintain all aspects of biodiversity in a healthy condition for the foreseeable future, including as climate conditions change.

3. *Redundancy*   Protected areas must include enough examples of each aspect of biodiversity to ensure its long-term existence in the face of future uncertainties.

4. *Reality*   There must be sufficient funds and political will not only to acquire and protect lands but also to regulate and manage the protected areas.

GLOBAL CHANGE
CONNECTION

The following, more specific, questions about reserve establishment and design are also useful for discussing how best to construct and link protected areas:

- Given a particular amount of funding to spend on a protected area or network of areas, what is the most effective way to spend it?

- How large must a nature reserve be to effectively protect biodiversity?

- Is it better to have a single large protected area or multiple smaller reserves?

- When a network of protected areas is created, should the areas be far apart or close together, and should they be isolated from one another or connected by corridors?

- How many individuals of an endangered species must be included in a protected area to prevent its local extinction?

- What is the most cost-effective way to design a protected area to achieve its conservation goals?

- What is the best shape for a nature reserve?

Conservationists explore many of these issues using the island biogeography model of MacArthur and Wilson (1967), which describes the relationship between the size of an area and the number of species it supports (see Chapter 6). Researchers that study the island biogeography model and data from protected areas have proposed various principles of reserve design, which are still being debated (**FIGURE 9.7**).

FIGURE 9.7   Principles of reserve design that are based in part on theories of island ▶ biogeography. Imagine that the reserves are "islands" of the original biological community surrounded by land that has been made uninhabitable for the original species by human activities such as farming, ranching, or industrial development. The practical application of these principles is still being studied and debated, but in general, the designs shown on the right are considered preferable to those shown on the left. (After C. L. Shafer 1997. In M. W. Schwartz [Ed.], *Conservation in Highly Fragmented Landscapes*, pp. 345–378. Chapman and Hall: New York, link.springer.com/chapter/10.1007/978-1-4757-0656-7_15, based on Shafer 1994. In E. A. Cook and H. N. van Lier [Eds.], *Landscape Planning and Ecological Networks*, pp. 201–223. Elsevier: Amsterdam.)

	Worse	Better
(A)	Ecosystem partially protected	Ecosystem completely protected
(B)	Smaller reserve	Larger reserve
(C)	Fragmented reserve	Unfragmented reserve
(D)	Fewer reserves	More reserves
(E)	Isolated reserves	Corridors maintained
(F)	Isolated reserves	"Stepping-stones" facilitate movement
(G)	Uniform habitat protected	Diverse habitats (e.g., mountains, lakes, forests) protected
(H)	Irregular shape	Reserve shape closer to round (fewer edge effects)
(I)	Only large reserves	Mix of large and small reserves
(J)	Reserves managed individually	Reserves managed regionally
(K)	Humans excluded	Human integration; buffer zones

When applied to protected areas, the model frequently assumes that parks are **habitat islands**—intact habitat surrounded by an unprotected matrix of inhospitable terrain. In fact, many species are capable of living in and dispersing through the surrounding habitat matrix (see Chapter 10). This may explain the findings that species diversity on oceanic islands increases more rapidly with size than it does on habitat islands and that among habitat islands, this relationship is stronger for urban islands than for forest islands (Matthews et al. 2016).

> Principles of design have been developed to guide land managers in establishing and maintaining networks of protected areas.

Also, aspects of protected area models differ depending on whether the targets are terrestrial or marine species; vertebrates, higher plants, or invertebrates; or whole ecosystems. Many aspects of marine species' life cycles and dispersal mechanisms are largely unknown (Burgess et al. 2014). MPAs and complementary conservation efforts outside protected areas must be designed to accommodate the high mobility of some species and the limited dispersal of other target species, thus ensuring that corridors and genetic populations are preserved.

Protected area size and characteristics

An early debate in conservation biology centered on one question: Is species richness maximized in one large nature reserve or in several smaller ones of an equal total area (McCarthy et al. 2006; Soulé and Simberloff 1986)? This debate is known in the literature as the **SLOSS debate** (single *l*arge *or* *s*everal *s*mall). Is it better, for example, to set aside one reserve of 10,000 ha or four reserves of 2500 ha each? Proponents of the "single large" option argue that only large reserves can maintain sufficient numbers of large, wide-ranging, low-density species (such as large carnivores) over the long term. Large reserves also minimize the ratio of edge habitat to total habitat, encompass more species, and sometimes have greater habitat diversity than small reserves.

> Large reserves are generally better able to maintain many species because they support larger population sizes and a greater variety of habitats. However, small reserves are important in protecting particular species and ecosystems.

The advantage of large parks was effectively demonstrated by a classic analysis of 299 mammal populations in 14 national parks in western North America (**FIGURE 9.8**) (Newmark 1995). Extinction rates were very low or zero in parks larger than 1000 km^2 but much higher in parks smaller than 1000 km^2. The impact of the size of a protected area is notable across taxa (see Figure 7.11). Across multiple studies, a reduction of habitat area was consistently found to be associated with significantly decreased species persistence, species richness, succession rate, and trophic dynamics (see Figure 4.12) (Haddad et al. 2015).

On the other hand, once a park reaches a certain size, the number of new species added with each increase in area starts to decline. At that point, creating a second large park or another park some distance away may be a more effective strategy for preserving additional species than simply adding area to the existing park.

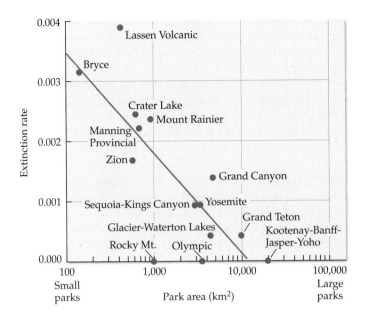

FIGURE 9.8 Mammals have higher extinction rates in smaller parks than in larger parks. If these same trends continue for the next 100 years, small parks such as Bryce are predicted to lose more than 30% of their mammal species, whereas large parks such as Yosemite will lose only 5%, and Kootenay-Banff-Jasper-Yoho is not expected to lose any species. Each dot represents the actual extinction rate of mammal populations (expressed as the proportion of species that have gone extinct per year) for a particular US national park, Canadian national park, or two or more adjacent parks. (After W. D. Newmark 1995. *Conserv Biol* 9: 512–527. © 1995 Wiley.)

The research on extinction rates of populations in large parks has three practical implications:

1. When a new park is being established, it should generally be made as large as possible (within the context of an overarching strategy for optimizing protected areas) in order to preserve as many species as possible, contain large populations of each species, and provide a diversity of habitats and natural resources. Keystone resources should be included, in addition to habitat features that promote biodiversity, such as elevational gradients.

2. When possible, land adjacent to protected areas should be acquired to reduce external threats to existing parks and maintain buffer zones. For example, terrestrial habitats adjacent to wetlands are often needed by semiaquatic species such as snakes and turtles. Moreover, protecting natural ecological units, such as entire watersheds or mountains, is often the best means to reduce external threats.

3. The effects of climate change, invasive species, and other threats are altering ecosystems within existing protected areas. These changes can reduce the area of habitat available for a species and lead to declines in population size and increased probability of extinction. These changes emphasize the need for preserving corridors or otherwise connecting protected areas to facilitate the dispersal of species among them.

GLOBAL CHANGE
CONNECTION

Although research suggests that large parks may be best able to achieve conservation goals in many situations, there are some instances where several small parks or a mixture of large and small ones is better for conservation.

For example, a study of wetlands in New Zealand determined that certain rare plant species were not necessarily present in the large protected areas and that several small reserves were more effective for their protection (Richardson et al. 2015).

Small reserves, even those less than a hectare in size, may effectively protect isolated populations of rare species, particularly if they contain a unique habitat type (Laguna et al. 2016). For example, the Erzi Nature Reserve, an IUCN category I protected area in the Republic of Ingushetia, North Caucasus, Russia, is only 5970 ha but supports 1100 species of plants, 18% of the Caucasian flora (Urbanavichus and Urbanavichene 2017). Also, regional biodiversity may depend on small natural features such as temporary pools, caves, single trees, or rock outcrops (Hunter et al. 2015). These features may contain keystone resources (mentioned in Chapter 7) or ecological processes and so may have disproportionately large conservation value, a phenomenon that has been named the *Frodo effect*.[1] For example, ephemeral desert springs should be prioritized for inclusion in protected areas because so many organisms depend on their cycles of wet and dry periods (Acuña and Ruhí 2017). Thus, several small protected areas that include critical small natural features will conserve more species than a single large one that does not contain these features.

Even if a large reserve would protect the most species in a given system, often officials have no choice but to accept the challenge of managing species and biological communities in small reserves. The 4335 nationally designated protected areas in England, for instance, have an average area of only 0.17 km^2 (**FIGURE 9.9**) (Shwartz et al. 2017). Numerous countries have many more small protected areas (less than 100 ha) than medium and large ones. This is particularly true in places such as Europe, China, and Southeast Asia that have been intensively cultivated for centuries.

Singapore provides an excellent example of how small reserves provide long-term protection for numerous species. A group of reserves that have been isolated since 1860 currently protect 5% of the original habitat on Singapore, yet they still contain around half of the country's original flora and fauna, including 350 of the original bird species and 26 of the mammal species (Corlett 2013). In addition, small reserves located near populated areas make excellent conservation education and nature study centers that further the long-range goals of conservation biology by developing public awareness of important issues. By 2050, over 68% of the world's population will live in urban areas (UN 2018); thus, there is a need to develop such reserves for public use and education.

It is generally agreed that protected areas should be designed to minimize edge effects. Conservation areas that are rounded in shape minimize the edge-to-area ratio, and the center is farther from the edge than in other park shapes. Long, linear parks have the most edge—all points in the park are close to the edge. Consequently, for parks with four straight sides, a square park is a better design than an elongated rectangle of the same area. However,

---

[1] The Frodo effect borrows its name from the character in J. R. R. Tolkien's *The Lord of the Rings*, who is small in stature yet responsible for the future of everyone.

FIGURE 9.9 Hardington Moor National Nature Reserve preserves some of the only remaining native grasslands in England. Although only 8.7 ha, it is one of England's larger protected areas. Total protected area in the United Kingdom has been increasing steadily since the 1950s. Although most reserves are relatively small, taken together, the more than 100,000 protected areas in the United Kingdom provide valuable protection for many species.

most parks have irregular shapes because land acquisition is typically a matter of opportunity rather than design. Many parks are linear because they follow streams and rivers.

As discussed in Chapter 4, the internal fragmentation of protected areas by roads, fences, farming, logging, and other human activities should be avoided as much as possible because fragmentation creates barriers to dispersal and often divides a large population into two or more smaller populations, each of which is more vulnerable to extinction than the large population. Fragmentation also provides entry points for invasive species, which may harm native species, and creates more undesirable edge effects.

9.3 Networks of Protected Areas

LEARNING OBJECTIVES

By the end of this section you should be able to:

9.3.1 Design a network of protected areas to benefit a particular guild of species.

9.3.2 Discuss the value of linking existing protected areas with corridors such as those shown in Figure 9.12.

To overcome some of the effects of fragmentation and limits on the size of protected areas, conservationists have developed strategies to aggregate small and large protected areas into larger conservation networks. Networks are also important to accommodate metapopulation dynamics (see Chapter 7). Currently the largest coordinated protected area network in the world is Natura 2000 (ec.europa.eu/environment/nature/natura2000), covering 18% of the European Union's land area and almost 6% of marine areas (**FIGURE 9.10**). The network includes federally protected parks and reserves, but most of the land is actually privately owned and in mixed use. Areas targeted for inclusion are those that are identified as ensuring the long-term survival of Europe's most valuable and threatened species and habitats. Member countries of the European Union are obliged to ensure that management of these lands supports the goals of the network.

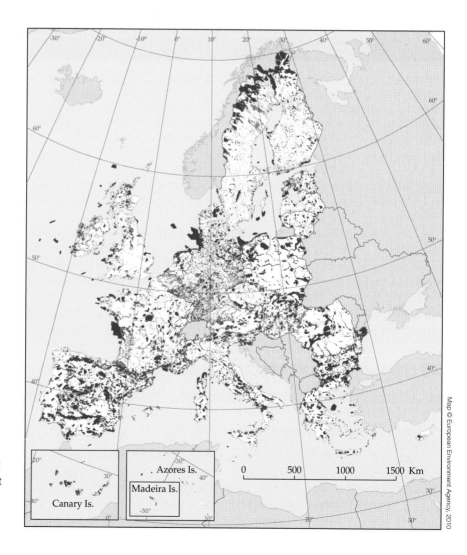

FIGURE 9.10
Natura 2000 sites as of 2019. This is the largest network of protected areas in the world, established in 1992.

Map © European Environment Agency, 2010

Cooperation among public and private landowners is important for creating these networks, particularly in developed metropolitan areas and other areas where there are many small, isolated parks controlled by different government agencies and private organizations. An excellent example of cooperation intended to network urban protected areas is the Chicago Wilderness project (www.chicagowilderness.org), which consists of more than 500 organizations and 2205 km^2 of protected area across three US states in the metropolitan Chicago region. This collaboration protects shoreline, tallgrass prairies, woodlands, rivers, streams, and other wetlands and coordinates conservation efforts and education in a large industrial city.

Habitat corridors

Often a protected area is embedded in a larger matrix of habitat managed for human uses, such as timber forest, grazing land, and farmland. If the protection of biodiversity is valued in these areas and is included in their management plans, then these mixed-use areas can serve as buffers, corridors, or other key components of protected area networks. Habitat areas managed for resource extraction can sometimes also be designated as important secondary sites for wildlife and as dispersal corridors between isolated protected areas. Using a network approach, groups of rare species can be managed as large metapopulations to facilitate gene flow and migration among populations (Andrello et al. 2015).

Growing numbers of conservationists argue that connectivity is important, and they are taking steps to link isolated protected areas into large systems through the use of **habitat corridors**—strips of protected land running between the reserves (Magrach et al. 2012). Such habitat corridors, also known as **conservation corridors** or **movement corridors**, can allow plants and animals to disperse from one reserve to another, facilitating gene flow and the colonization of suitable sites.

> Establishing habitat corridors can potentially transform a set of isolated protected areas into a linked network with populations interacting as a metapopulation.

Corridors are clearly needed to preserve animals that must migrate seasonally among different habitats to obtain food and water—for example, the large grazing mammals of the African savanna such as wildebeests, elephants, and zebras. If these animals were confined to a single reserve by fences, farms, and other anthropogenic factors, they might starve. The width required for effective corridors varies depending on the species, length of the corridor, and other factors. Increasingly, these corridors are threatened by human settlements and agriculture (Ojwang et al. 2017). Corridors that facilitate natural patterns of migration will probably be the most successful at protecting species. For example, large grazing animals often migrate in regular patterns across a rangeland in search of water and the best vegetation (**FIGURE 9.11**). In seasonally dry savanna habitats, animals often migrate along the riparian forests that grow along streams and rivers. In mountainous areas, many bird and mammal species regularly migrate to higher elevations during the warmer months of the year. For example, a corridor was established in Costa Rica to link two wildlife reserves, the Braulio Carrillo National Park and La Selva Biological Station, to protect migrating birds. A 7700 ha corridor of

FIGURE 9.11 Migration corridors in southern Kenya are important for many East African species, including wildebeest (inset), elephant, and zebra. Here, the status of major wildlife corridors is shown between protected areas. Human settlements and agriculture significantly threaten corridors to the east. (After G. O. Ojwang et al. 2017. *Wildlife Migratory Corridors and Dispersal Areas: Kenya Rangelands and Coastal Terrestrial Ecosystems.* Kenya Vision 2030, Kenya Wildlife Service.)

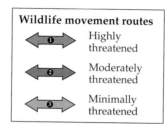

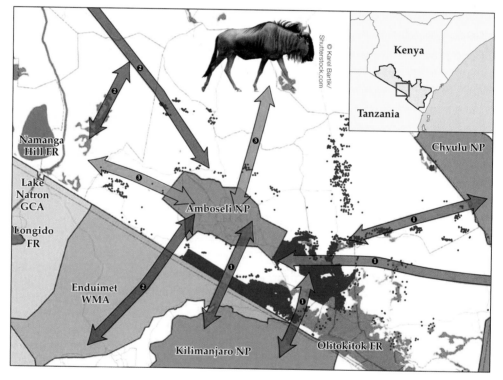

forest several kilometers wide and 18 km long, known as the Zona Protectora Las Tablas, was set aside to provide an elevational link that allows at least 75 species of birds to migrate between the two large conservation areas.

In many areas, roads are a primary obstruction to the creation of habitat corridors (**FIGURE 9.12A**). In these cases culverts, tunnels, and overpasses can create passages under and over roads and railways that allow reptiles, amphibians, and mammals to travel between habitat fragments or protected areas (Soanes et al. 2013). An added benefit of these passageways is that they reduce collisions between animals and vehicles, which saves lives and money. A meta-analysis found that mitigation efforts overall reduced roadkill by 40% versus the controls, with the combination of fencing and road-crossing structures reducing roadkill by 83% (Rytwinski et al. 2016).

(A)

© iStock.com/Steve_Gadomski

FIGURE 9.12 (A) An overpass above a
fenced-off divided highway allows animals to
migrate safely between two forested areas.
(B) Individuals of a species naturally disperse
between two large protected areas (areas
1 and 2, left-hand panel) by using smaller
protected areas as stepping-stones. The
right-hand panel shows that habitat destruction and a large edge effect zone caused by
a new road have blocked a migration route.
To offset the effects of the road, compensation sites (orange) have been added to the
system of protected areas, and an overpass
has been built over the highway to allow
dispersal. (B, after R. Cuperus et al. 1999.
Biol Cons 90: 41–51. © 1999. Reprinted with
permission from Elsevier.)

(B)   Before fragmentation   After fragmentation

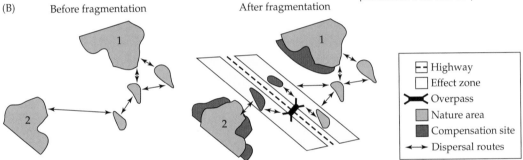

Some conservation biologists have started to plan habitat corridors on a truly huge scale. Wildlands Network has a detailed plan, called the Western Wildway Network (previously the "Spine of the Continent Initiative"), that would link all large protected areas in the western United States and Canada by habitat corridors, creating a system that would allow large and currently declining mammals, such as bears and mountain lions, to coexist with human society.

As the global climate changes, many species are moving to higher elevations and higher latitudes (see Chapter 5). Creating corridors to protect expected routes—such as north–south river valleys, ridges, and coastlines— would be a useful precaution. Extending existing protected areas in the direction of anticipated species movements could help to maintain long-term populations (Nuñez et al. 2013). Corridors that cross gradients of elevation, rainfall, and soils could also allow the local migration of species to more favorable sites.

GLOBAL CHANGE CONNECTION

Although the idea of corridors is intuitively appealing, there are some possible drawbacks (Ogden 2015). In particular, corridors may facilitate the movement of pest species and disease; a single infestation could quickly spread to all of the connected nature reserves and cause the extinction of all populations of a rare species. Also, animals dispersing along corridors may be exposed to greater risks of predation because human hunters as well as animal predators tend to concentrate on routes used by wildlife. Finally, there is a risk that the corridors, which in some cases can be expensive and difficult to construct, will not be used by the intended species.

Some studies published to date support the conservation value of corridors, while others show no effect (Gilbert-Norton et al. 2010; Pardini et al. 2005). There would also be value in leaving small clumps of original habitat between large conservation areas to facilitate movement in a stepping-stone pattern (**FIGURE 9.12B**). An example is the network of protected areas that birds use as stopping points along the flyways of their annual migration routes.

9.4 Landscape Ecology and Park Design

LEARNING OBJECTIVES

By the end of this section you should be able to:

9.4.1 Consider how a landscape diversity perspective will inform protected area design differently than a species diversity focus.

The interaction of human land use patterns, conservation theory, and park design is evident in the discipline of **landscape ecology**, which investigates patterns within the mosaic of the physical environment, ecological communities, ecosystem processes, and human-ecosystem interactions on local and regional scales (**TABLE 9.2**). It has focused on terrestrial systems, but the same approach can apply to marine and other aquatic systems (Jelinski 2015).

Landscape ecology has been most intensively studied in Europe and Asia, where long-term practices of traditional agriculture and forest management

TABLE 9.2	A Proposed Framework for Landscape Design That Promotes Biodiversity Conservation

1. Distinguish and delineate different patches of land covers (e.g., rice fields, hay fields, forest, villages) in the selected landscape.

2. Categorize patches as unaltered (lower human use) versus altered (higher human use) land covers.

3. Identify the constraints on land-use planning (e.g., economic, social, political).

4. Given these constraints, create a landscape plan that maximizes the total amount and diversity of unaltered land cover, especially near water (e.g., by restoring some abandoned hay fields to forest);

5. ... minimize human disturbance within altered land cover, especially near water (e.g., by removing roads through forest);

6. ... and aggregate altered land covers associated with high-intensity land uses, especially away from water (e.g., by planning future development to be near existing villages, rather than in currently unaltered areas).

Source: After S. A. Gagné et al. 2015. *Landsc Urban Plan* 136: 13–27.

have shaped landscape patterns. In the European countryside, cultivated fields, pastures, woodlots, and hedges alternate to create a mosaic that affects the distribution of wild species. Likewise, in the traditional Japanese landscape known as *satoyama*, flooded rice fields, hay fields, villages, and forests provide a rich diversity of habitat for wetland species, such as dragonflies, amphibians, and waterfowl (Jiao et al. 2019) (**FIGURE 9.13**). These heterogeneous landscapes, which include a mix of human-created and natural features, are critical for the survival of some species. However, in some places, rural people have switched to farming practices that are more intensive, involving more machinery and the application of fertilizer. To protect species and ecosystems in such situations, the design and management of protected areas frequently include strategies to maintain the traditional landscapes, in some cases by subsidizing traditional practices or having volunteers manage the land.

To increase the number and diversity of animals, wildlife managers sometimes create the greatest amount of landscape variation possible within the confines of some protected areas, particularly refuges or other areas managed primarily for hunting and fishing. Fields and meadows are created and maintained, small thickets are encouraged, groups of fruit trees and crops are planted, patches of forests are periodically cut, small ponds and dams are developed,

> In some cases, long-term traditional human use has created landscape patterns that preserve and even increase biodiversity.

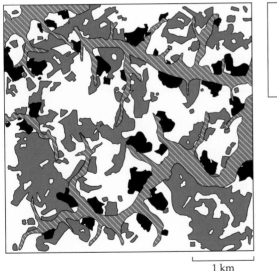

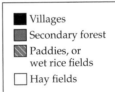

1 km

FIGURE 9.13 Schematic of a traditional rural landscape near Tokyo, Japan, with an alternating pattern of villages (black); secondary forest (green); paddies, or wet rice fields (brown); and hay fields (beige). Such landscapes were common in the past but are now becoming rare because of the increasing mechanization of Japanese agriculture, the movement of the population away from farms, and the urbanization of the Tokyo area. The area covered is approximately 4 km × 4 km. (After H. Kobori and R. B. Primack. 2003. *Ambio* 32: 307–311.)

and numerous trails and dirt roads meander across and along all the patches. Such landscaping is often appealing to the public, who are the main visitors and financial contributors to the park. However, the species in these landscapes are likely to be principally common species that depend on human disturbance—and, in some cases, invasive species. To remedy this localized approach, large animals, such as bears, mountain lions, and tigers, generally are best managed on the level of a regional landscape, in which the sizes of the landscape units more closely correlate with the natural population sizes and migration patterns of the species (Wikramanayake et al. 2011).

9.5 Managing Protected Areas

LEARNING OBJECTIVES

9.5.1 Argue both for and against a "nature knows best" perspective to managing protected areas.

9.5.2 Design a monitoring plan for a protected area that has a goal of maximizing plant species diversity.

9.5.3 Do a cost-benefit analysis of whether some part of a given protected area should be a converted to a transition zone with mixed use.

Some people believe that "nature knows best" and that humans do not need to actively manage biodiversity—that once protected areas are legally established, the work of conservation is largely complete. The reality, however, is often very different. Management is required because humans have already modified many local environments so much that the remaining species and ecosystems need human monitoring and intervention in order to survive. The world is littered with "paper parks" created by government decree but left to flounder without any management. These protected areas have gradually—or sometimes rapidly—lost species as their habitat quality has degraded. In some countries, people readily farm, log, mine, hunt, and fish in protected areas because they feel that government land is owned by "everyone," so anybody can take whatever they want and nobody is willing to intervene (see "tragedy of the commons" in Chapter 3).

Alternatively, many parks are actively managed according to carefully prepared **management plans** designed to prevent deterioration. Important considerations in management plans include protecting biodiversity, maintaining ecosystem services and health, preserving historical landscapes, and providing resources and experiences of value to local inhabitants and visitors (Hobbs et al. 2010). Part of the management plan involves making the public aware of which activities are encouraged (for example, wildlife photography) and which are prohibited (for example, hunting) and then enforcing the rules. The public can even be encouraged to engage in behaviors that help mitigate the negative impacts of ecotourism (**FIGURE 9.14**).

Both photos by Anna Sher

FIGURE 9.14 Management of protected areas may involve advising visitors how to avoid harming wildlife, as does this sign in the parking lot of Table Mountain National Park, on the Cape of Good Hope in South Africa.

In many European, Asian, and African countries with well-established traditions of cultivation, ranching, and grazing, hundreds (and even thousands) of years of human activity have shaped habitats such as woodlands, meadows, and hedges. These habitats support high species diversity as a result of traditional land management practices, which must be maintained if the species are to persist (Baiamonte et al. 2015). Similarly, grasslands that have been grazed in the past by large wild animals or domesticated animals, such as cattle, still need to be grazed; they have the highest richness and productivity with this type of disturbance (**FIGURE 9.15**). If protected areas that include these types of habitats are not managed, they will undergo **ecological succession**—a predictable, gradual, and progressive change in species over time—and many of their characteristic species will disappear as a few species of shrubs and trees become dominant.

It is important, though, to be cautious in taking management actions. In some cases, often because of a lack of complete understanding of an ecosystem or conflicting management objectives, management practices may be ineffective or even detrimental. Some protected areas are managed to promote the abundance of a game species, such as deer, for hunting. Similarly, management has frequently involved eliminating top predators, such as wolves and cougars. However, without predators to control them, game populations, such as deer (and, incidentally, rodents that feed on seeds and can spread disease), sometimes increase far beyond expectations, resulting in overgrazing, habitat degradation, and a collapse of animal and plant communities.

Overenthusiastic park managers who remove hollow or dead trees, rotting logs, and underbrush to "improve" a park's appearance may unwittingly remove a critical keystone resource needed by certain animal species for nesting and overwintering, by rare plants for seed germination, and by all species as an integral part of nutrient cycling. In these instances, a

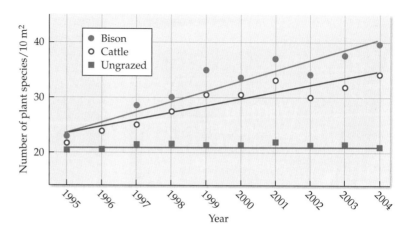

FIGURE 9.15 Large herbivores originally grazed the tallgrass prairies of the midwestern United States. The loss of these herbivores has altered the ecology of this ecosystem, with a resulting loss of plant species. Management involving grazing by cattle and bison resulted in a gradual increase in plant species in prairie research plots over a 10-year period, compared with ungrazed control plots. (After E. G. Towne et al. 2005. *Ecol Appl* 15: 1550–1559. © 2005 by the Ecological Society of America.)

Management plans are needed that articulate conservation goals and practical methods for achieving them. Management activities can include controlled burns, enforcement of restrictions on human use, and the maintenance of keystone resources, especially water.

"clean" park can become a biologically sterile park. Likewise, in many parks, fire is part of the natural ecology of the area (Nimmo et al. 2013). Attempts to suppress fire completely are expensive and waste scarce management resources. Suppressing the normal fire cycle may eventually lead to the loss of fire-dependent species and to massive, uncontrollable fires.

The most effectively managed parks are usually those whose managers have the benefit of research and monitoring programs and have funds available to implement their management plans.

Managing sites

Park managers often must actively manage sites to ensure that all or particular successional stages are present so that species characteristic of each stage have a place to persist and thrive. In some wildlife sanctuaries, grasslands and fields are maintained by livestock grazing, burning, mowing, tree cutting, or shallow plowing in order to retain open habitat in the landscape. One common way to do this is to set localized, controlled fires periodically in grasslands, shrublands, and forests to reinitiate the successional process (Middleton 2013). Obviously, such burning must be done in a legal and carefully controlled manner to prevent damage to nearby property (**FIGURE 9.16A**). Also, prior to burning, land managers must develop a program of public education to explain to local residents the role of fire in maintaining the balance of nature. In other situations, parts of protected areas must be carefully managed to *minimize* human disturbance and fire (**FIGURE 9.16B**).

KEYSTONE RESOURCES   In many protected areas, it may be necessary to preserve, maintain, and supplement keystone resources on which many species depend (see Chapter 2). These resources include trees that supply fruit when little or no other food is available, pools of water during a dry season, exposed mineral licks, and so forth. Keystone resources and keystone species can be enhanced in managed conservation areas to increase the populations of species whose numbers have declined.

By planting areas with food plants and building an artificial pond, it may be possible to maintain species in a smaller conservation area and at higher densities than would be predicted based on studies of species distribution in undisturbed habitat. Artificial ponds, for example, can provide habitat for insects that people enjoy watching, such as dragonflies, and also serve as centers of public education in urban areas (Oertli 2018). Likewise, nest boxes or drilled nesting holes in living and dead trees can provide substitute shelter resources for birds and mammals where there are few dead trees with nesting cavities (Rueegger 2017).

(A)

Photo by Elise Smith, US Fish and Wildlife Service

(B)

Courtesy of U.S. National Park Service

FIGURE 9.16   Conservation management: intervention versus leave-it-alone management. (A) Heathland in protected areas of Cape Cod, Massachusetts, is burned on a regular basis in order to maintain the open vegetation habitat and to protect wildflowers and other rare species. (B) Sometimes management involves keeping human disturbance to an absolute minimum. Muir Woods National Monument is a forest of old-growth coast redwoods, protected in the midst of the heavily urbanized San Francisco Bay area.

In each case a balance must be struck between, at one extreme, establishing nature reserves free from human influence and, at the other extreme, creating seminatural gardens in which the plants and animals are dependent on people. Further, in some cases human replacements or augmentations of resources may be ineffective, such as placing nest boxes on small trees that lack other features hollow-nesting birds need (Le Roux et al. 2016).

Rivers, lakes, swamps, estuaries, ponds, lakes, and all the other types of wetlands must receive a sufficient supply of clean water to maintain ecosystem processes. Yet protected areas may end up directly competing for water resources with agricultural irrigation projects, demands for residential and industrial water supplies, flood control schemes, and hydroelectric dams. Wetlands are often interconnected, so a decision affecting water levels and quality in one place will have ramifications for other areas. In particular, the construction of dams on major rivers often completely alters the environmental conditions, eliminating or reducing the abundance of many of the native fish and other aquatic species (Soukhaphon et al. 2021). Consequently, one strategy for maintaining wetlands is to include entire watersheds within given protected areas.

Monitoring sites

An important aspect of managing protected areas involves monitoring components that are crucial for biodiversity, such as the quality and quantity of water in ponds and streams; the number of individuals of rare and endangered species; and the density of herbs, shrubs, and trees (Pocock et al. 2015). Methods for monitoring these components include recording standard observations, carrying out surveys, and taking photographs from fixed points. Monitoring an area's biodiversity is sometimes combined with monitoring social and economic aspects of surrounding communities because of the linkages between people and conservation. In particular, it is often important to monitor the amount and value of plant and animal materials that people obtain from nearby ecosystems.

Protected areas must be monitored to determine whether their goals are being met, and management plans may need to be adjusted based on new information from monitoring.

Managers must continually assess the information they gain from monitoring and adjust park management practices in an adaptive manner to achieve their conservation objectives, a process sometimes referred to as **adaptive management** (**FIGURE 9.17**). Unfortunately, monitoring programs are often discontinued after a few years due to a lack of funding or interest. The lack of long-term monitoring and baseline inventories can make it difficult or impossible to design and evaluate management plans and detect the effects of relatively subtle, chronic problems, such as acid rain and climate change, which can dramatically alter ecosystems over time (see Chapter 7). To remedy this situation, many scientific research organizations and government agencies have begun to implement programs to monitor and study ecological change over the course of decades and centuries. Programs include the system of Long-Term Ecological Research (LTER) sites established by the US National Science Foundation (see Figure 7.13), the National Ecological Observatory Network, and the World Network of Biosphere Reserves of the United Nations Educational, Scientific and Cultural Organization (UNESCO).

GLOBAL CHANGE CONNECTION

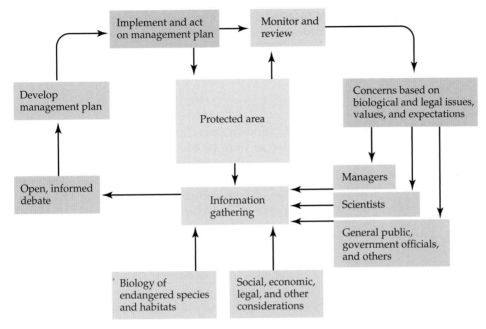

FIGURE 9.17 Model of an adaptive management process for protected areas, emphasizing the decision-making stages. Input is solicited from many sources, and then the plan is developed, implemented, and monitored. (After S. J. Cork et al. 2000. *Conserv Biol* 14: 606–609. © 2000 Wiley.)

The scale and methods of monitoring have to be appropriate for management needs. For large parks in remote areas, remote sensing using satellites, drones, and airplanes may be an effective method for monitoring logging, shifting cultivation, mining, and other activities (Fawzi et al. 2018; Grecchi et al. 2017). Monitoring that does not require specialized equipment can even utilize citizen scientists (see Chapter 7).

Management and people

A central part of any park's management plan must be policies on the use of the park resources by different groups of people. In developing countries especially, restricting access to protected areas can cost people—who may have traditionally used the area and its resources—access to the basic resources that they need to stay alive. Such displaced people may oppose conservation in these areas (Brockington and Wilkie 2015). Many parks flourish or are destroyed depending on the degree of support, neglect, or hostility they receive from the people who live in or near them. In the best-case scenario, local people become involved in park management and planning, are trained and employed by the park authority, and benefit from the protection of biodiversity and regulation of activity within the park (**FIGURE 9.18**).

> The involvement of local people is often the crucial element missing from conservation strategies. Local people need to be involved in conservation programs, as participants, employees, and leaders.

FIGURE 9.18 When local people are involved in park management and benefit from it, the protected area has a greater chance of success. Here we see a ranger at Bontebok National Park, a small protected area in the South African park system about 200 km east of Cape Town.

Unfortunately, it is sometimes necessary to exclude local people from protected areas, especially in cases when resources are being overharvested, either legally or illegally, to the point that the health of the ecosystem and the existence of endangered species are threatened (Packer et al. 2013). Such degradation and loss of biodiversity can result from overgrazing by cattle, excessive collection of fuelwood, or hunting with guns. In such cases the only solution may be a strategy of "fences and fines."

In the worst-case scenario, if there is a history of bad relations and mistrust between local people and the government or if the purpose of the park is not explained adequately, the people may reject the park's concept and ignore its regulations. In this case, the local people may come into conflict with the park personnel, to the detriment of the conservation goals. Park personnel and even armed soldiers may have to patrol constantly to prevent illegal activity. Escalating cycles can lead to outright conflict in which park personnel and the local inhabitants may be threatened, injured, or even killed. In one alarming example, a reporter uncovered a WWF report in which Indigenous people feared "repression" and other mistreatment from park rangers at Messok Dja in the Republic of the Congo (Baker and Warren 2019).

Zoning as a solution to conflicting demands

A possible way to deal with conflicting demands on a protected area is **zoning**, which considers the overall management objectives for a park and sets aside designated areas that permit or give priority to certain activities. For example, some areas of a forest may be designated for timber production, hunting, wildlife protection, nature trails, or watershed maintenance. Other zones may be established for the recovery of endangered species, restoration of degraded communities, and scientific research. The challenge in zoning is to find a compromise that people are willing to accept that provides for the long-term, sustainable use of natural resources. Managers often need to spend considerable effort informing the public about what activities are acceptable in particular areas of a park and then enforcing park regulations (Pristupa et al. 2018).

For example, an MPA might allow fishing in certain areas and strictly prohibit it in others; certain areas might be designated for surfing, waterskiing, and recreational diving, but these sports may be prohibited elsewhere. The enforcement of zoning is often a major

> Zoning allows the separation of mutually incompatible activities. MPAs are often zoned with no-fishing areas where fish and other marine organisms can recover from harvesting.

© Peter Titmuss/Alamy Stock Photo

challenge in MPAs because fishermen tend to fish on the edges of no-fishing zones, as those are the areas where the fishing is best, leading to overfishing at the margins of MPAs. A combination of local involvement, publicity, education, clear posting of warning signs, and visible enforcement significantly increases the success of any zoning plan, especially in the marine environment (Fox et al. 2012).

Biosphere reserves

UNESCO has pioneered another zoning approach, termed **biosphere reserves**, under its World Network of Biosphere Reserves program, which integrates traditional land use patterns (such as farming, grazing, and managing forests), research, protection of the natural environment, and sometimes tourism at a single location. There are currently 714 biosphere reserves in 129 countries. These locations often have well-established human settlements and scenic landscapes.

A desirable feature of the biosphere reserve program is a system in which there are zones delineating varying levels of use (**FIGURE 9.19A**). At the center is a core area in which ecosystems are strictly protected, with all human activity either prohibited or tightly regulated. This core is surrounded by a buffer zone in which traditional human activities, such as the collection of edible plants and small fuelwood, are monitored and nondestructive research is conducted. Surrounding the buffer zone is a transition zone in which some forms of sustainable development, such as small-scale farming, are allowed. In addition, some extraction of natural resources, such as selective logging, and experimental research are also permitted. This general strategy of surrounding core conservation areas with buffer and transitional zones can encourage local people to support the goals of the protected area. However, although these zones are easy to draw on paper, in practice it has been difficult to inform and gain agreement from residents who live in or near biosphere reserves about where the zones are and what uses are allowed in them.

The value of the strategy of surrounding core conservation areas with buffer and transition zones is still being debated. The approach has benefits: local people may be more willing to support park activities if they are allowed zoned access to the park, and certain desirable features of the landscape created by human use may be maintained (such as farms, gardens, and early stages of succession). Also, buffer zones may prevent parks from becoming isolated islands of nature and may create corridors that facilitate animal dispersal between highly protected core conservation areas. Yet zoning for multiple uses and resource extraction may only work if the core area is large enough to protect viable populations of all key species and if people are willing to respect the zones and their designated uses. Respect for zones varies greatly in different parts of the world and among different social situations. In places where park management, political will, and land tenure are weak, buffer zones often are seen as a commons or as unowned and unmanaged lands that are up for grabs, which greatly reduces their effectiveness.

One example of a biosphere reserve is Mont Saint-Hilaire, near Montreal, Canada, established in 1978 (**FIGURE 9.19B**). The reserve has a core area of 500 ha that protects some of the last remaining old-growth deciduous forest in Quebec (UNESCO 2021a). The core area is surrounded by a transition zone of 600 ha, which in turn is surrounded by a landscape of suburban development and agriculture. The biosphere reserve contains many important organisms, including 33 threatened plant species and individual trees over 400 years old (**FIGURE 9.19C**). A survey found that the number of plant species present in the reserve had increased over 50 years, from 485 species before the reserve was established to 683 species (Elliott and Davies 2019). However, most of the

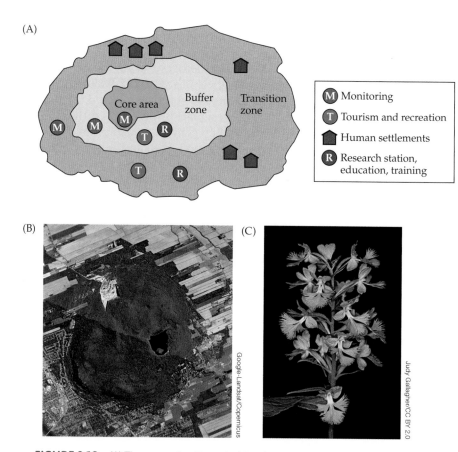

FIGURE 9.19 (A) The general pattern of a biosphere reserve: a core protected area is surrounded by a buffer zone, where human activities are monitored and managed and where research is carried out; this, in turn, is surrounded by a transition zone, where sustainable development and experimental research take place. (B) The UNESCO Biosphere Reserve of Mont Saint-Hilaire, Quebec, Canada, includes the Gault Nature Reserve, which is owned and managed for research and recreation by McGill University and includes more than 1000 ha of primary forest. The dark patch is a lake. The reserve is surrounded by development and agriculture. (C) The large purple fringed orchid (*Platanthera grandiflora*) is one of the rare species that was found in the reserve during a 2015 survey.

species gains were nonnative plants with low evolutionary distinctiveness, while the local extinctions had much higher distinctiveness. These findings point to both the benefits and limitations of protected areas; so long as there is pressure from outside, there can still be species losses.

9.6 Challenges to PA Management

Managers of protected areas face many challenges. Perhaps the most basic and unavoidable of these challenges are increased pressures from growing human populations and demand for the use of natural resources. Although all threats to species discussed in Chapter 4 are relevant to protected areas, the following are some areas of particular concern.

Poaching

Human populations will continue to increase dramatically in the coming decades, while resources such as fuelwood, medicinal plants, and wild meat will become harder to find. Similarly, people who are poor and hungry will enter the nearby protected areas to take what they need to live, regardless of whether they have permission (see "Overexploitation" in Chapter 5). Within protected areas, if park rangers are underpaid, even they may be motivated to begin illegally harvesting and selling the very resources they are charged with protecting. Addressing poverty and enforcing regulations are the most important factors to address poaching (see Chapter 13).

Trophy hunting

There is considerable debate about the role of hunting in conservation. On one side, many countries finance conservation primarily from the sale of hunting licenses. For developing countries, the focus is on large game, such as the hunting of rhinos or buffalo by wealthy foreigners (see Chapter 3). Trophy hunting, when strictly regulated, has minimal effects on the overall number of individuals within a park and can even double as a management tool. One example of this is selling expensive licenses to hunt specific, individual old male rhinos that are killing young rhinos and preventing younger males from mating. However, critics argue that allowing sport hunting of any type not only sends the wrong message—that killing these animals is acceptable—but also supports a market that, in turn, promotes illegal trophy hunting. This is illustrated by the killing of Cecil the Lion (see the introduction to Chapter 3); because South Africa permits lion hunting, the American hunter may not have known that the lion he was shooting had been illegally taken from one of the protected areas. The financial stakes are high enough to motivate local hunting guides to steal from the parks. Making all trophy hunting illegal and removing these animal products from circulation, as was done when the Kenyan government burned tons of ivory, has helped in some cases (see Figure 7.24).

Human-animal conflict

Problems are inevitable as more people live and farm closer to high concentrations of wildlife. When animals are hungry, they will leave the park and go into nearby agricultural fields and villages in search of food. Elephants, primates, and flocks of birds can all be significant crop raiders, while carnivores such as tigers pose a different set of challenges to nearby residents. Some nonprofit organizations and governments address these problems by creating opportunities for local people to also benefit from the animals, compensating them for their losses, and helping to build fences or other deterrents (**FIGURE 9.20**).

Degradation

Multiple-use areas can suffer from the negative effects of mining, cattle grazing, and oil exploration due to the lack of management or poor enforcement of policies. Even strictly protected areas are at risk of degradation from recreation, including wildfires, littering, fragmentation and erosion from off-road driving, and the habituation of wild animals, to name just a few (see Chapter 3). Furthermore, strictly protected areas may suffer from the same degradation as multiple-use areas, including instances when lack of oversight leads to illegal use. Generally speaking, degradation from human use is minimized in multiple-use areas where there is oversight and policies that regulate it.

GLOBAL CHANGE CONNECTION

Climate change

The extent to which existing protected areas will allow species and ecosystems to persist in the face of climate change is an important question and is being investigated (Regos et al. 2016). Species may not be able to persist in a protected area if the climate or the associated vegetation changes significantly. For example, the Doñana National Park in southern Spain is a **World Heritage site** that is already rated as under "very high threat" by UNESCO (Scheffer et al. 2015) due to mining, agriculture, and the diversion of water. The site contains some of the most important overwintering habitat in Europe for migrating waterbirds, yet these wetlands may dry out in coming decades due to an increasingly warmer and drier climate. A global analysis of climate conditions projected to the year 2070 found that the majority of countries will fail to protect >90% of their climate conditions under present coverage by PAs (Elsen

© Steve Taylor ARPS/Alamy Stock Photo

FIGURE 9.20 Elephants from protected areas trample local farms and lower support for the animals' conservation. In one response, beehive boxes established along the perimeters of farms create anti-elephant "fences."

et al. 2020). In this rapidly changing environment, it is important to preserve elevational and environmental gradients, corridors, and climate refugia so that species and ecosystems can gradually spread in response to a changing climate.

Funding and personnel

For park management to be effective, there must be adequate funding for a sufficient number of well-equipped, properly trained, reasonably paid, and highly motivated park personnel who are willing to carry out park policy (Uwayo et al. 2020). Buildings, communications equipment, and other appropriate elements of infrastructure are necessary to manage a park. In many areas of the world, particularly in developing, but also in developed, countries, protected areas are understaffed, and the park staff lack the equipment to patrol remote areas. Without enough radios and vehicles, they may be restricted to the vicinity of headquarters and unaware of what is happening in their own park. International conservation organizations and government agencies regularly assist in providing funds for managing protected areas in developing countries, but often this funding remains insufficient. Increasing funding for the management of protected areas needs to be a priority for government agencies and conservation organizations (Watson et al. 2014).

Finally, at the end of the day, conservation biologists need to account for whether their management of protected areas achieved the stated goals and whether the funds were spent effectively. As human impact increases in most areas of the world, the importance of protected areas for the protection of biodiversity is expected only to increase (Regos et al. 2016).

Summary

- Protecting habitat is the most effective method of preserving biodiversity. Almost 15% of the Earth's land surface and freshwater and more than 7% of the global ocean is now protected; most of the increases in protected area are currently marine.

- Conservation biologists are developing guidelines for designing protected areas: the areas should be large whenever possible, they should not be fragmented, and managers should create networks of conservation areas for maximum protection.

- Habitat corridors connecting protected areas may allow species dispersal to take place and may be particularly important in maintaining known migration routes.

- Protected areas often must be actively managed in order to maintain their biodiversity. Monitoring provides information that is needed to evaluate whether management activities are achieving their intended objectives or need to be adjusted.

- Management might involve zoning to establish areas where certain activities are allowed or prohibited. Managing interactions with local people is critical to the success of protected areas and should be part of a management plan.

- Adequate staffing and funding are necessary for park management.

For Discussion

1. Obtain a map of a town, state, or nation that shows protected areas (such as nature reserves and parks) and multiple-use managed areas. Who is responsible for each parcel of land, and what is the goal in managing it? Consider the same issues for aquatic habitats (ponds, lakes, rivers, coastal zones, etc.).

2. If you could protect additional areas on the map, where would they be and why? Show their exact locations, sizes, and shapes, and justify your choices.

3. Think about a national park or nature reserve you have visited. In what ways was it well run or poorly run? What were the goals of this protected area, and how could they be achieved through better management?

4. Can you think of special challenges in the management of aquatic preserves such as coastal estuaries, islands, or freshwater lakes that would not be faced by managers of terrestrial protected areas?

Suggested Readings

Carrasco, L., et al. 2021. Global progress in incorporating climate adaptation into land protection for biodiversity since Aichi targets. *Global Change Biology* 27(9): 1788–1801. A gap analysis of the success of countries for establishing protected areas that will accommodate the needs posed by climate change.

Di Marco, M., et al. 2019. Wilderness areas halve the extinction risk of terrestrial biodiversity. *Nature* 573(7775): 582–585. Models using plant and insect data demonstrate the importance of having pristine areas with minimal to no human interference.

Elsen, P. R., et al. 2020. Keeping pace with climate change in global terrestrial protected areas. *Science Advances* 6(25): eaay0814. We must consider not just the areas of protected land but also the climates that they represent, in order to protect the organisms that have evolved to live there.

Jones, K. R., et al. 2018. One-third of global protected land is under intense human pressure. *Science* 360(6390): 788–791. Just because an area is designated as protected doesn't mean that it is safe from anthropogenic impacts.

Pacifici, M., M. Di Marco, and J. E. Watson. 2020. Protected areas are now the last strongholds for many imperiled mammal species. *Conservation Letters* 13(6): e12748.

Strain, E. M., et al. 2019. A global assessment of the direct and indirect benefits of marine protected areas for coral reef conservation. *Biodiversity Research* 25(1): 9–20. MPAs had clear impacts on guilds of fish, which indirectly positively impacted coral cover.

UNEP-WCMC and IUCN 2021. *Protected Planet Report 2020*. UNEP-WCMC and IUCN: Cambridge UK; Gland, Switzerland. https://livereport.protectedplanet.net A comprehensive overview of the state of protected areas across the world.

Venter, O., et al. 2018. Bias in protected-area location and its effects on long-term aspirations of biodiversity conventions. *Conservation Biology* 32(1): 127–134. We must be willing to protect those areas that humans use.

During the quarantine in response to COVID-19 in 2020, across the world there were sightings of wildlife in urban areas, such as this sika deer seen at the UNESCO World Heritage Todaiji Temple in Nara, Japan. In some cases, the animals have long existed there and were just rarely seen.

Conservation Outside Protected Areas

10

Establishing protected areas with intact ecosystems is essential for species conservation. It is, however, shortsighted to rely solely on government-mandated protected areas (PAs) to preserve biodiversity. That kind of reliance can create a paradoxical situation in which species and ecosystems inside the PAs are preserved while the same species and ecosystems outside are allowed to be damaged, which in turn results in the decline of biodiversity within the protected areas (Kinnaird and O'Brien 2013). This decline has at least three causes:

1. Most PAs are too small to protect viable, long-term populations of many species, especially large animals and migratory species.

2. Many species are attracted to resources available outside PAs. For example, it is not uncommon for primate species to feed on nutrient-rich crops in villages that are adjacent to PAs in which they live.

3. Many species migrate between PAs seasonally to avoid freezing temperatures or other climate extremes or to access mates, water, food, and other resources.

In general, the smaller a PA, the more dependent it is on neighboring unprotected lands for the long-term maintenance of biodiversity. A crucial component of conservation strategies must be the protection of biodiversity inside and outside—both immediately adjacent to and away from—PAs (Troupin and Carmel 2014). According to even the most optimistic predictions, in the future at least 70% of the world's land will remain outside PAs. Given this reality, the UN Convention on Biological Diversity (see Chapter 12) established global conservation targets that include both PAs and what they call "other effective area-based conservation measures" (OECMs; Aichi Target 11). Conservation biologists must help define "effective conservation" and determine to what degree OECMs contribute to global targets (Watson et al. 2016).

The COVID-19 pandemic created a glimpse into the diversity and abundance of wildlife both inside and out of protected areas (see chapter opening photo). During the quarantine that significantly decreased the human presence in urban areas for months at a time, a flurry of astonishing wildlife photographs were taken in what had typically been busy streets, leading some to surmise that this represented changes in species movements and ranges. Indeed, there was some evidence that this was so (Manenti et al. 2020). The "global human confinement experiment" posed by the COVID-19 lockdown did create opportunities to study the impacts of humans on wildlife (Bates et al. 2020). However, wildlife has always existed in urban areas, bringing into question whether COVID-19 brought any change other than our increased awareness of how many species we share our cities with (Zellmer et al. 2020). Camera traps have often documented wildlife in urban areas (Magle et al. 2019).

Some conservation biologists argue that, given how much land is now occupied by people, we must find ways to promote biodiversity in human-dominated landscapes (**FIGURE 10.1**). This is an idea ecologist Michael Rosenzweig (2003) named **reconciliation ecology**. This generally involves increasing environmental complexity, such as creating habitat in urban settings by enhancing parks or planting *green roofs*—gardens or other plantings on roofs or buildings (Loke et al. 2015). Another example of reconciliation ecology is the creation of pollinator habitat under solar panel arrays (Semeraro et al. 2018). Other examples of how nature is promoted in highly human-altered environments, and also some of the problems of this approach, are discussed in more detail later in the chapter.

Jeff McNeely (1989), an International Union for Conservation of Nature (IUCN) expert in PAs, suggested that the park boundary "is too

© iStock.com/Roschetzky

FIGURE 10.1 California sea lions (*Zalophus californianus*) are an example of a species that has done very well outside of protected areas, adapting well to human-altered environments. Once they were protected from overhunting, populations tripled.

often also a psychological boundary, suggesting that since nature is taken care of by the national park, we can abuse the surrounding lands, isolating the national park as an 'island' of habitat which is subject to the usual increased threats that go with insularity." In the worst case, a devastated landscape polluting the air and water will strangle the PA it surrounds and block the movement of dispersing animals and plants. In the best case, however, unprotected areas surrounding PAs will provide additional space for ecosystem processes and new populations. In many ways, conservation outside PAs should strive to blur the distinctions between protected and unprotected ecosystems as much as possible by maintaining unprotected areas in a state of reasonable ecological health.

> Many endangered species and unique ecosystems are found partly or entirely on unprotected lands. Consequently, the conservation of biodiversity in these places must be considered.

In this chapter, we explore strategies to include biodiversity protection as a management objective outside formal PAs as a complementary strategy for conservation.

10.1 The Value of Unprotected Habitat

LEARNING OBJECTIVES

By the end of this section you should be able to:

10.1.1 Identify potential habitat outside of PAs for a given species.

10.1.2 Use the map in Figure 10.3 to determine what proportion of potential habitat falls outside of PAs for the Florida panther, and explain what panther movement patterns tell us about these areas.

10.1.3 Identify what types of species will be best adapted to use of military land.

The human use of ecosystems varies greatly in unprotected lands, but in almost every country, numerous rare species and ecosystems exist primarily or exclusively on unprotected public lands or on lands that are privately owned (**FIGURE 10.2**). Even land managed primarily for timber harvesting, grazing, mining, or other economic uses can support endangered species. For example, 45 bird species were found to use unprotected shoreline along the Yellow Sea in China (Chan et al. 2019), and 28 mammal species use forest fragments in Sumatra (Weiskopf et al. 2019). Clearly, strategies for reconciling human needs and conservation interests in unprotected areas are critical to the success of conservation plans.

GLOBAL CHANGE
CONNECTION

The situation of the Florida panther (*Felis* [= *Puma*] *concolor coryi*) provides an excellent example of the importance of unprotected habitat. This endangered subspecies of mountain lion lives in South Florida and has a population of only 120–230 adult individuals (USFWS 2017). Although small, these numbers represent a potentially significant increase since 2014 (100–180 adults) and progress toward the recovery goal of 240 individuals (USFWS

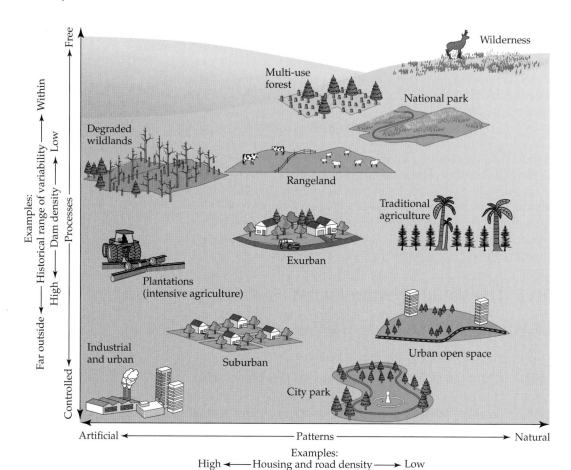

FIGURE 10.2 Landscapes vary in the extent to which humans have altered the patterns of species composition and natural vegetation through activities such as agriculture, road construction, and housing; ecosystem processes (water flow, nutrient cycling, etc.) also vary because of fire control activities, dam construction, and other activities that alter plant cover. Wilderness areas retain most of their original patterns and processes, urban areas retain the least, and other landscapes retain various intermediate amounts. (After D. M. Theobald, 2004. *Front Ecol Environ* 2: 139–144.)

GLOBAL CHANGE
CONNECTION

2008). Given that there were only 20 individuals when they were first listed as endangered with the Endangered Species Act in 1967, this represents an illustration of the importance of formal protection and the success of conservation efforts by both state and federal agencies. Nonetheless, Florida panthers still face significant threats, including from loss of habitat to development, vehicle collision, and feline distemper, which is spread by domestic cats (Chiu et al. 2019). It has also been shown that Florida panthers will lose habitat as a result of sea level rise associated with climate change (Catano et al. 2015). All of these threats highlight the risks outside of PAs. Roads, in particular, have been shown to significantly impact the movement of Florida panthers (Criffield et al. 2018). Southern Florida is a patchwork of public and private land, and animals tracked with radio collars show that many spend some or most of their time on private lands (FFWCC 2018) (**FIGURE 10.3**). Approximately 30% of the land area of the state of Florida is protected in some capacity, with 2.6% of that protected by private organizations or individuals (Florida Natural Areas Inventory 2019). However, as Figure 10.3B shows, much of the natural range of the Florida panther is on unprotected land. Viable possibilities for protecting the Florida panther in these areas include educating private landowners on the value of conservation, and paying willing landowners to practice management options that allow for the

(A)

Courtesy of Larry Richardson, USFWS

FIGURE 10.3 (A) The Florida panther (*Puma concolor coryi*) is found on both public and private lands in South Florida. (B) The blue shaded portions show suitable breeding habitat, and the orange dots represent radio-telemetry records of 17 collared panthers in 2017–2018. Protected areas are outlined in black. (B, after Florida Fish and Wildlife Conservation Commission. 2018. *Annual report on the research and management of Florida panthers: 2017–2018*. Fish and Wildlife Research Institute & Division of Habitat and Species Conservation, Naples, FL, USA; R. A. Frakes et al. 2015. *PLOS ONE* 10[7]: e0133044/CC0. doi.org/10.1371/journal.pone.0133044; Florida Natural Areas Inventory. Florida Conservation Lands, April 2019. https://www.fnai.org/webmaps/ConLandsMap/.)

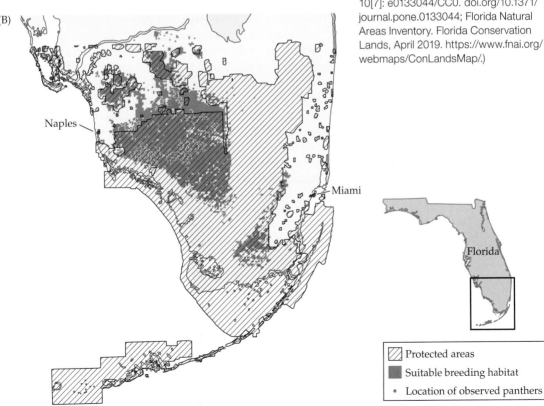

panthers' continued existence—specifically, minimizing habitat fragmentation and maintaining their preferred habitats of hardwood forest and cypress swamp. In addition, special road underpasses have been built in the hopes of reducing panther deaths from collisions with motor vehicles.

Other species of big cats also frequently live outside PAs. For example, lions in southern Africa must traverse unprotected land surrounding the Kavango-Zambezi Transfrontier Conservation Area, and minimizing conflict with humans in these areas is a high priority for conservation planning (Cushman et al. 2018).

Next we will discuss the importance of several types of land that are not contained within traditional PAs but nonetheless are important for biodiversity.

Military land

Native species can often continue to live in unprotected areas, especially when those areas are set aside or managed for some other purpose that is not harmful to the ecosystem, such as security zones surrounding government installations and military reservations. For example, the US Department of Defense (DOD) manages more than 8.8 million ha, much of it undeveloped; 80% of its large military installations (>2 km^2) have active management plans due to significant natural resources (CRS 2020). In fact, the US DOD has the highest density of federally listed species of any federal land management agency (DOD NR Program 2016). Rare and endangered desert tortoises (*Gopherus agassizii*), manatees (*Trichechus manatus*), red-cockaded woodpeckers (*Picoides borealis*), bald eagles (*Haliaeetus leucocephalus*), Atlantic white cedars (*Chamaecyparis thyoides*), and the least Bell's vireo (*Vireo bellii pusillus*) all have found a safe haven on military lands. The White Sands Missile Range in New Mexico alone is almost 1 million ha in area, about the same size as Yellowstone National Park. While certain sections of military reservations may be damaged by military activities, much of the habitat remains as an undeveloped buffer zone with restricted access. After these areas are no longer needed for military purposes, they also make excellent candidates for protected areas; for example, nearly 31,000 ha of former military bases were converted to nature reserves by the German Federal Agency for Nature Conservation (Ellwanger and Reiter 2019). It is being reported that five years after these were established, wolves are returning to Germany in record numbers (Corbley 2020).

Conservation goals are now regularly incorporated into planning and management of military training activities of many countries; the United Kingdom's Ministry of Defense even has its own environmental magazine, *Sanctuary*. The US Fish and Wildlife Service, which is responsible for enforcement of the Endangered Species Act, works actively with the US DOD; from 1991 to 2015, the DOD spent $1.22 billion on 571 listed species (DOD NR Program 2016).

In addition to formal military operations focused on habitat protection, there are affiliated groups that do conservation, such as the Army Ornithological Society's Ascension Island seabird conservation project in the United Kingdom. This volunteer operation was formed by "soldier naturalists" to collect high-quality population data on the Falkland Islands (Sanctuary 2018). It is run by service members who give up leave time and contribute personal funds in the interest of conservation.

Military lands, particularly training grounds but also those associated with armed conflict, can be advantageous for conservation for several reasons (adapted from Hanson 2018):

- Large tracts of land that are reserved for military use are protected from development and resource extraction.
- Some species benefit from the disturbance regimes created by training exercises.
- There is reduced human activity, including in buffer zones or "peace parks" along disputed borders.

On the other hand, many military bases contain toxic waste dumps and high levels of chemical pollutants. In addition, severe disturbance in the form of bomb explosions, artillery practice, and the use of heavy vehicles can have significant negative effects on the resident wildlife.

Unprotected ecosystems

There are several ecosystem types that typically are unprotected but can offer valuable habitat for plants, animals, and other organisms. Type and quality of management of these areas can significantly impact their value for conservation.

FORESTS Forests that either are selectively logged on a long cutting cycle or are cut down for farming using traditional shifting cultivation methods may still contain a considerable percentage of their original biota and maintain most of their ecosystem services (MacKay et al. 2014). This is particularly true if fires and erosion have not irreversibly damaged the soil and if native species can migrate from nearby undisturbed lands, such as steep hillsides, swamps, and river forests, and colonize the sites. The value of the habitat also depends on management; a study done in Spain found that forested lands managed by neighborhood councils (*juntas vecinales*) had the same or higher levels of plant diversity as those inside the adjacent national park, Picos de Europa (Guadilla-Sáez et al. 2019) (**FIGURE 10.4**). In another example, in African tropical forests, gorillas, chimpanzees, and elephants can tolerate selective logging and other land uses that involve low levels of disturbance, though only when hunting levels are controlled by active antipoaching patrols (Stokes et al. 2010).

GRASSLANDS The mown edges of roadsides often provide an open grassland community that is a critical resource for many species, such as butterflies (Zielin et al. 2016). A similar habitat is provided by the surprisingly large amount of mown fields occupied by power lines. In the United States, corridors for power line rights-of-way occupy over 2 million ha. Power line corridors managed with infrequent mowing and without herbicides maintain high densities of birds, insects, and other animals (King et al. 2009). If such

FIGURE 10.4 Plant biodiversity can be just as high in locally managed areas as in federally managed forest. This area is one of those controlled by neighborhood councils (*juntas vecinales*) in Spain.

management practices could be extended over a greater proportion of power line rights-of-way, these areas could become additional habitat for insects and a wide range of other species.

UNPROTECTED WATERS   Many heavily altered aquatic ecosystems can also have value for conservation. For example, in estuaries and seas managed for commercial fisheries, many of the native species remain because commercial and noncommercial species alike require an undamaged chemical and physical environment. It has been determined that the greatest diversity of marine plants (such as sea grass and mangroves) occurs outside existing marine protected areas (MPAs; see Chapter 9) (Daru and le Roux 2016). Also, many marine animals such as salmon, whales, and sea turtles migrate great distances, including across areas that are not protected. Even for smaller organisms, movement between MPAs can be critical (**FIGURE 10.5**); while movement patterns are often affected by habitat fragmentation,

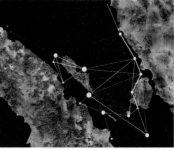

Connectivity today   Connectivity with ocean warming

FIGURE 10.5 Leopard grouper (*Mycteropcera rosacea*) was a species used to study movement between MPAs in the Gulf of California, Mexico. Global warming will significantly hinder connectivity between MPAs (indicated by the dots and lines) because higher temperatures will reduce the length of time that larvae can survive and disperse (Álvarez-Romero et al. 2018).

warming sea temperatures can also create barriers to dispersal between MPAs for larval stages of fish and other marine animals (Álvarez-Romero et al. 2018). Fortunately, given the charismatic nature of marine megafauna, there is often broad support and even direct local involvement in protecting these species. Examples include the work of volunteers to rescue sea turtles (see Chapter 1) and a dramatic effort by the navy and local residents to push 100 short-finned pilot whales back into the sea after Sri Lanka's largest beaching event (The Guardian 2020).

Even though dams, reservoirs, canals, dredging operations, port facilities, and coastal development harm native aquatic communities, some bird, fish, and other aquatic species are capable of adapting to the altered conditions, particularly if the water is not polluted. For example, California sea lions flourish in heavily trafficked coastal waters (see Figure 10.1). However, there is abundant research that suggests that even when there is no obvious pollution, many species are likely to have higher abundance within MPAs than outside of them (e.g., Pikesley et al. 2016).

Land that is undesirable to humans

Other areas that are not protected by law may retain species because the human population density and degree of use are typically very low. As mentioned, border areas, such as the demilitarized zone between North Korea and South Korea, often have an abundance of wildlife because they remain undeveloped and generally unoccupied by people. Governments frequently manage mountain areas, which are often too steep and inaccessible for development, as valuable watersheds that produce a steady supply of water and prevent flash flooding and erosion. They also harbor important natural communities. Likewise, desert and tundra species and ecosystems may be at less risk than other unprotected communities because such regions are marginal for human habitation and use. However, it is also true that in areas like the Arctic region, the current warming of the climate will result in further development of transportation infrastructure and a greater interest in mining deposits of oil, gas, and minerals in these so-called undesirable lands.

GLOBAL CHANGE
CONNECTION

Private land

In many parts of the world, wealthy individuals have acquired large tracts of land for their personal estates and for private hunting. These estates are frequently used at very low intensity, often in a deliberate attempt by the landowners to maintain large wildlife populations. In particular, some estates in Europe preserve unique old-growth forests that have been owned and protected for hundreds of years. An increasing number of individuals are purchasing property specifically for conservation, in a cross between philanthropy and investment (Gooden and Grenyer 2019). Such privately owned lands, whether owned by individuals, families, corporations, or tribal groups, often contain important aspects of biodiversity.

There is also increasing popularity of private game reserves in many parts of Africa and elsewhere, where wildlife is managed for ecotourism or game hunting. Although there are many such reserves that are beneficial for biodiversity, as we discussed in Chapter 3, these activities do not come without costs. This will be discussed in more detail in the section on extractive reserves.

Management for biodiversity can vary a great deal between landowners, of course; a study of private landowners enrolled in the Indiana Classified Forest and Wildlands Program found that landowners with large holdings, those with environmental motives, and those who had seen improvements occur on their land were more likely to be good stewards (Farmer et al. 2017). Strategies that encourage private landowners and government land managers to protect rare species and ecosystems are obviously essential to the long-term conservation of biodiversity. This chapter and Chapter 12 describe these strategies.

Even small yards and home gardens can be useful for supporting biodiversity, particularly of insects (Ribeiro et al. 2016). The National Wildlife Federation has a backyard "wildlife certification program" in which homeowners can receive a certificate and a sign once they ensure that their property contains all the elements of wildlife habitat, including a food source, water, and sheltering plant cover for protection and reproduction. Homeowners' associations may require the use of native plants or a minimum number of trees in landscaping to support biodiversity and may even supervise natural areas that increase the value of their properties (**FIGURE 10.6**). Often higher levels of insect diversity can be achieved by practices as simple as mowing grasses less frequently.

(A)

(B)

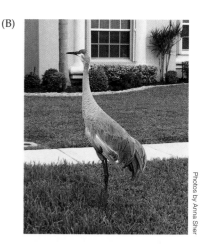

Photos by Anna Sher

FIGURE 10.6 (A) This housing development in Lake Worth, Florida, has wetland conservation areas that are managed largely by the local homeowners' association. The creation of areas such as this are one way that developers can legally mitigate their negative impact on the habitat that was displaced, while also increasing the value of the houses they build. (B) The importance of these conservation areas is apparent from the diversity of plants and animals that live there and, like this sandhill crane (*Grus canadensis*), often appear in people's yards. Many sandhill cranes migrate from protected areas in the north each winter and are dependent on such remnants of habitat, while others are year-round residents.

10.2 Conservation in Urban and Other Human-Dominated Areas

LEARNING OBJECTIVES

By the end of this section you should be able to:

10.2.1 Evaluate the conservation value of a given urban landscape.

10.2.2 Design a certification program for "environmentally friendly" tea. What might the requirements be?

Human-dominated areas can pose unique challenges to biodiversity. Although many species have evolved with us and thus do well in our company, these tend to be (some of) the same species all over the world, leading to a "strip mall" phenomenon; that is, the unique characteristics of a local ecosystem are lost as it becomes dominated by ubiquitous pigeons, rats, weeds, and other human-adapted species. Indeed, the urban environment has the capacity to exert considerable selective pressure (Schilthuizen 2019). For many species, however, human-dominated areas mean competition with exotic species, loss of habitat, pollution, and disturbance regimes that have been altered to suit human preferences. For example, both fire and flooding, to which native species may be adapted and which may thus promote diversity, are typically suppressed. However, even amidst these challenges, areas with biodiversity can often be found—habitat that can be enhanced with proper management.

GLOBAL CHANGE
CONNECTION

Urban areas

Many native species can persist in the less altered habitats found within urban areas, including public parks, streams, and ponds (Meffert and Dziock 2012). In other cases cities simply provide suitable habitat for a variety of species. Preserving and managing these remnants of biodiversity within a human-dominated matrix not only presents special challenges but also provides unique opportunities to educate the public about biodiversity conservation. For example, in Europe, storks (*Ciconia ciconia*) often nest in chimneys and towers, and endangered raptors such as the peregrine falcon (*Falco peregrinus*) and bald eagle (*Haliaeetus leucocephalus*) make nests and raise their young in the skyscrapers of downtown New York, where numerous small animals (including the ubiquitous pigeons and rats common to urban centers) provide abundant food sources (**FIGURE 10.7**). Even ponds at golf courses in urban areas and gravel pits dug for construction materials may be suitable habitats for certain newts, dragonflies, and other wetland species, provided the water is not polluted. In one study of 27 artificial water bodies in Australia, researchers found that greater than 70% of the regional diversity of fish species could be found in these human constructions (Davis and Moore 2016). Whether intentional or not, these are examples of reconciliation ecology because they demonstrate ways in which humans and other species can coexist.

FIGURE 10.7 "Pale Male" (on the left) is a famous red-tailed hawk that lived on a Fifth Avenue residence in New York City. The unusually light-colored hawk had at least eight mates and fathered several broods. Some believe he is still alive, over 30 years old, although the use of poisons to control rats in the surrounding area is a significant risk to all rodent-eating raptors in the city. Pale Male inspired a website (www.palemale.com), at least three children's books, and even a movie. He is one of the first hawks known to have built a nest on a building in this city, and when the homeowners attempted to displace the hawks by removing the support structures for the nest, local bird lovers successfully protested to keep it. Pale Male is seen here with his then-current mate, Lola.

© D. Bruce Yolton

As exciting as such examples of urban adaptations might be, we cannot assume that all species have the potential to live within human-dominated landscapes. For example, the value of urban parks for biodiversity found in some developed countries may not apply to rapidly growing megacities; in South America, urban parks were found to be dominated by European weeds (Fischer et al. 2016). We have a lot to learn about just what habitat and disturbance features are important for various species and how to integrate those features into our urban and suburban landscapes. In general, increasing the intensity of land use will decrease the number of native species found in a location, and adaptable, generalist species (often nonnative invasives) will tend to do best. The size and configuration of landscape features will determine which species and ecosystem processes are maintained. More work is needed to evaluate how general conservation principles apply in specific locations.

GLOBAL CHANGE
CONNECTION

Increasing the presence of wild animals in the urban landscape comes with fairly serious consequences for both animals and humans. For example, as woodland areas and mountain canyons become urbanized or suburbanized in California and the American Southwest, people tend to create yards and gardens that attract deer. Deer bring with them a host of problems: they can carry ticks that transmit illnesses to humans, such as Lyme disease and Rocky Mountain spotted fever; they are a significant potential road hazard; and the bucks can become aggressive toward humans during mating season. In some areas, deer that live within developments also attract predators, including cougars, thus increasing the potential for human-wildlife conflicts with a scarce and ecologically important top carnivore.

Understanding the ecology, the ecosystem processes, and the characteristics of the human use of a location is critical for implementing policies to promote conservation in unprotected urban areas. Deciding on the proper tools, though, requires good information on ecology and complex urban human-natural systems and knowledge of how best to motivate people to behave in conservation-friendly ways. These areas of research are growing and beginning to provide insights that are improving urban conservation.

Other human-dominated landscapes

Most of the world's landscapes have been affected in some way by human activity, but fortunately, considerable biodiversity can be maintained in well-managed and low-intensity traditional agricultural systems, grazing lands, hunting preserves, forest plantations, and recreational lands (Carrière et al. 2013). Birds, insects, and other animal and plant species are often abundant in traditional agricultural landscapes, with their mixture of small fields, hedges, and woodlands. Some species are found almost exclusively in these traditional human-dominated habitats. In comparison with more-intensive, so-called modern, agricultural practices (which emphasize high yields of crops for sale in the market, mechanization, and external inputs), these traditional landscapes experience less exposure to herbicides, fertilizers, and pesticides and have more heterogeneity of habitat. Similarly, farmlands worked using organic methods support more birds than farmlands worked using non-organic methods, in part because organic farms have more insects for the birds to eat. In many areas of the world, however, the best agricultural lands are being more intensively used while less optimal lands are abandoned as people leave for urban areas (Phelps et al. 2013).

Conservation biologists are increasingly discussing the value of the strategy of **land sharing**, in which low-intensity human activities, such as traditional or organic agriculture, can coexist with some elements of biodiversity. The alternative is **land sparing**, in which intensive human activities, such as modern agriculture, are practiced on some of the lands while allowing the rest to remain in their natural state. The best strategy for any given location will depend on the local circumstances, the price of land and crops, and potential financial incentives (Baudron and Giller 2014).

One notable example of preserving biodiversity in an agricultural setting comes from tropical countries and their traditional shade coffee plantations, in which coffee is grown under a wide variety of shade trees, with often as many as 40 tree species per farm (Philpott et al. 2007) (**FIGURE 10.8**). Worldwide, coffee plantations cover approximately 99 million ha (FAO 2019). Shade coffee plantations have structural complexity created by multiple vegetation layers and a diversity of birds and insects comparable to the adjacent natural forest, and they represent a rich repository of biodiversity (Rodrigues et al. 2018). The presence of such coffee plantations can also potentially slow the pace of deforestation (Hylander et al. 2013). Therefore, programs are being developed to encourage and subsidize farmers to maintain their shade-grown coffee plantations and to market the product at a premium price as "environmentally friendly" shade-grown coffee. But let the buyer beware: there are currently no uniform standards for shade coffee. Thus, some coffee marketed as "environmentally friendly shade coffee" may actually be grown as sun coffee with only a few small, interspersed trees.

GLOBAL CHANGE
CONNECTION

Even ecosystems that are managed primarily for the production of natural resources can retain considerable biodiversity, and they are important to the success of conservation efforts.

(A)

(B)

FIGURE 10.8 Two types of coffee management systems. (A) Shade coffee is grown under a diverse canopy of trees, providing a forest structure in which birds, insects, and other animals can live. (B) Sun coffee is grown as a monoculture, without shade trees. In a monoculture system, animal life is greatly reduced.

Shade-grown chocolate and other tropical tree crops are similarly unregulated. However, there are third-party auditors that will certify growers who adhere to certain standards; one of the most robust certifications for coffee has been developed at the Smithsonian Migratory Bird Center. Criteria include that the dominant tree species are native and that there is at least 40% shade cover that is at least 12 m high, and the coffee must be certified organic (www.coffeehabitat.com/certification-guide/).

In developing countries, conservation biologists have started innovative programs in which local people living in rural areas are paid directly for protecting individuals and populations of flagship species, including rhinos, tigers, gorillas, and other species of conservation interest (Dinerstein et al. 2013). When the animals do well, the people are paid directly or receive money for village improvements (see the case studies in Section 10.5).

In many countries, large parcels of government-owned land are designated as **multiple-use habitat**; that is, they are managed to provide a variety of goods and services. An emerging and important research area involves the development of innovative ways to reconcile competing claims on land use, such as logging, mining, ranching, species conservation, and tourism. This will require careful analyses and consideration of the trade-offs of pursuing alternative development options in regard to both environmental and socioeconomic priorities (Koh and Ghazoul 2010). A different approach is to use regulations, the legal system, and political pressure to prevent government-approved activities on public lands if these activities threaten the survival of endangered species.

In the United States, the Bureau of Land Management oversees more than 245 million ha, including 67% of the state of Nevada and large amounts of Utah, Wyoming, Oregon, and Idaho (https://www.blm.gov/about/

what-we-manage/national). National forests cover over 83 million ha, including much of the Rocky Mountains, the Cascade Range, the Sierra Nevada, the Appalachian Mountains, and the southern coast of Alaska. In the past, these lands have been managed for logging, mining, grazing, wildlife, and recreation. The challenge is that often each one of these activities is managed by itself but their cumulative effects threaten biodiversity. Increasingly, multiple-use lands also are being valued and managed for their ability to protect species, biological communities, and ecosystem services (Kemp et al. 2013). The US Endangered Species Act of 1973 and other similar laws, such as the 1976 National Forest Management Act, require landowners, including government agencies, to avoid activities that threaten listed species. One such activity is overgrazing by cattle; when cattle grazing is reduced or eliminated on overgrazed rangelands, some of these ecosystems can recover in a few years or decades (Earnst et al. 2012).

Another approach to protecting biodiversity in human-dominated landscapes has been to define standards of best practices so that the use of resources does not harm biodiversity. The Forest Stewardship Council has been one such organization by working to promote the certification of timber produced from sustainably managed forests. For the Forest Stewardship Council and similar organizations to grant certification, the forests need to be managed and monitored in the interests of their long-term environmental health, and the rights and well-being of local people and workers need to be protected. The certification of forests is increasing rapidly in many areas of the world, especially in response to buyers in Europe, who often request certified wood products. At the same time, major industrial organizations representing such industries as logging, mining, and agriculture are lobbying for their own alternative certification programs, which generally have lower requirements for monitoring and weaker standards for judging whether practices are sustainable.

10.3 Ecosystem Management

LEARNING OBJECTIVE

By the end of this section you should be able to:

10.3.1 Evaluate the land management of your area to determine whether it qualifies as "ecosystem management." Why or why not?

Resource managers around the world are increasingly being urged by their governments and conservation organizations to think on larger geographical scales, particularly given climate-change-driven shifts in the distributions of species and makeup of ecosystems. Traditionally, these managers may have focused on the production of goods and services that could be managed on the local scale, such as volume of timber or number of park visitors. But today, these managers are being asked to expand their emphasis to a broader perspective

GLOBAL CHANGE CONNECTION

Ecosystem management links private and public landowners, businesses, and conservation organizations in a planning framework that facilitates acting together on a large scale.

that includes the conservation of biodiversity and the protection of ecosystem processes (Altman et al. 2011). That is, they are shifting to **ecosystem management**, a system of large-scale management involving multiple stakeholders, the primary goal of which is preserving ecosystem components and processes for the long term while still satisfying the current needs of society (**FIGURE 10.9**). Rather than having each government agency, private conservation organization, business, or landowner act in isolation and in its own interests, ecosystem management envisions them cooperating to achieve common objectives (Redpath et al. 2013). For example, in a large forested watershed along a coast, ecosystem management would link all owners and users located from the tops of the hills to the seashore, including foresters, farmers, business groups, townspeople, and the fishing industry.

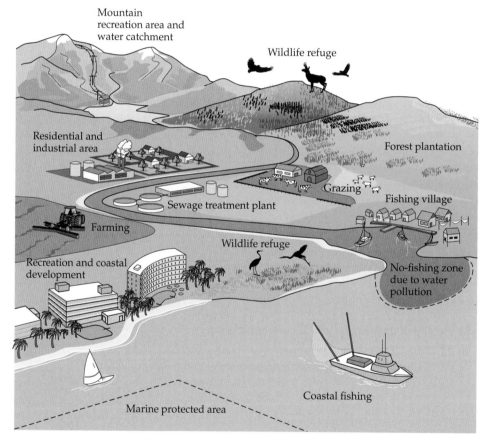

FIGURE 10.9 Ecosystem management involves bringing together all the stakeholders that affect a large ecosystem and receive benefits from it. In this case, a watershed needs to be managed for a wide variety of purposes, many of which influence one another. (After wtf/IUCN/UNEP. 1992. *Global Biodiversity Strategy: Guidelines for Action to Save, Study, and Use Earth's Biotic Wealth Sustainably and Equitably.* World Resources Institute, Washington, DC.)

Important themes in ecosystem management include the following:

- Using the best science available to develop a coordinated plan for the area that is sustainable; includes biological, economic, and social components; and is shared by all levels of government as well as business interests, conservation organizations, and private citizens

- Ensuring viable populations of all species, representative examples of all biological communities and successional stages, and healthy ecosystem functions

- Seeking and understanding connections between all levels and scales in the ecosystem hierarchy—from the individual organism to the species, community, ecosystem, and even regional and global scales

- Monitoring significant components of the ecosystem (numbers of individuals of significant species, vegetation cover, water quality, etc.), gathering the needed data, and then using the results to adjust management in an adaptive manner—a process sometimes referred to as adaptive management (see Figure 9.17)

One successful example of ecosystem management is the work of the Malpai Borderlands Group (www.malpaiborderlandsgroup.org), a nonprofit cooperative enterprise formed by ranchers and other local landowners who promote collaboration among conservation organizations such as The Nature Conservancy, private landowners, scientists, and government agencies (Malpai Borderlands Group 2010). The Malpai Borderlands region contains many rare and federally listed species and is home to the endemic Gould's turkey and white-sided jackrabbit. This is also the home of fewer than 100 human families, primarily ranchers. The group is developing a network of cooperation across the Malpai planning area, which comprises nearly 400,000 ha of unique, rugged mountain and desert habitat along the Arizona and New Mexico border. The Malpai Borderlands Group uses controlled burning as a range management tool, reintroduces native grasses, applies innovative approaches to cattle grazing, incorporates scientific research into management plans, and takes action to avoid habitat fragmentation by using **conservation easements** (agreements not to develop land) to prevent residential development.

A logical extension of ecosystem management is **bioregional management**, which integrates protection with human use and often focuses on a single large ecosystem, such as the Caribbean Sea or the Great Barrier Reef of Australia, or on a series of linked ecosystems, such as the protected areas of Central America. A bioregional approach is particularly appropriate where there is a single, continuous, large ecosystem that crosses international boundaries or when activity in one country or region will directly affect an ecosystem in another country. For the European Union and the 21 individual countries that participate in the United Nations Environment Programme (UNEP) Mediterranean Action Plan (MAP Phase II), for example, bioregional cooperation is necessary because the enclosed Mediterranean Sea has large

FIGURE 10.10 Protection of both biodiversity and human needs is integrated into an international, large-scale management plan in the UNEP Mediterranean Action Plan. Twenty-one partner countries collaborate for sustainable use of the region (inset).

GLOBAL CHANGE
CONNECTION

human populations along the coasts, heavy oil tanker traffic, and weak tides that cannot quickly remove pollution resulting from cities, agriculture, and industry (UNEP/MAP 2016) (**FIGURE 10.10**). This combination of problems threatens the health of the entire Mediterranean ecosystem, including the sea, its surrounding lands, and its associated tourist and fishing industries. Cross-boundary management is also necessary because pollution from one country can significantly damage the natural resources of neighboring countries.

10.4 Working with Local People

LEARNING OBJECTIVES

By the end of this section you should be able to:

10.4.1 List several means by which ecocolonialism can be avoided.

10.4.2 Contrast the outcomes and effectiveness of two different programs that involve Indigenous Peoples.

10.4.3 Explain how payments for ecosystem services (PES) operates, with examples.

Even remote regions that are considered "wilderness" by governments and the general public often have small, sparse human populations. Societies that practice a traditional way of life in rural areas with relatively little outside influence in terms of modern technology are variously re-ferred to as "tribal people," "Indigenous people," "native people," or, more generally, "traditional people" (IWGIA 2021). Lands conserved by these groups are referred to as Indigenous and community conserved territories and areas (ICCAs). A database dedicated to capturing information about these lands is the ICCA Registry (www.iccaregistry.org). Worldwide, there are 370–500 million traditional people living in more than 90 countries (World Bank 2021a). It is necessary to distinguish these established tradi-tional peoples from more recent settlers, who may not be as concerned with the health of surrounding biological communities or as knowledge-able about the species living there and the land's ecological limits. Rather than being a threat to a pristine environment, in some cases traditional peoples have been an integral part of these environments for thousands of years (Middleton 2013).

Traditional ecological knowledge (TEK) refers to the understanding by Indigenous Peoples of ecosystems and management, gained over millen-nia of experience in direct contact with the land (Berkes 1993). It is also referred to as Indigenous and local knowledge (ILK) or simply traditional knowledge (TK) (IUCN 2016). Several IUCN specialist groups consult TEK to help inform Red List assessments, and TEK is increasingly being recognized for its value for improving restoration practices (Robinson et al. 2021). The present mixture and relative densities of plants and animals in many biological communities may reflect the historical activities—such as fishing, selective hunt-ing of game animals, and planting or encouraging of useful plant species in fallow agricultural plots—of people in the area. In some cases, the conservation of species depends on TEK. For example, a study of the conservation of the skywalker hoolock gibbon (*Hoolock tianxing*) outside nature reserves found that this highly endangered species has maintained population stability for more than a decade in large part due to the TEK practice of limiting poaching (Zhang et al. 2020).

> In many parts of the world, areas with high biodiversity are inhabited by Indigenous Peoples with long-standing systems for resource protection and use. People with traditional knowledge are critical to conservation efforts in those areas.

Many traditional societies have strong conservation ethics. These ethics are often subtler and less clearly stated than Western conservation beliefs, but they tend to influence people's actions in their day-to-day lives, perhaps more than Western beliefs (Ban et al. 2013). It is important that conservation programs be developed in partnership with these groups, who are often walking a fine line between expectations regarding biodiversity preservation and economic pressures (Kohler and Brondizio 2017).

Local people who support conservation as an integral part of their liveli-hood and traditional values often take the lead in protecting biodiversity. This is further strengthened when they are able to obtain **legal title**—a right to ownership that is recognized by the government—to their traditionally owned lands; empowering these communities to obtain these rights is often an important component of efforts to establish locally managed protected areas in developing countries (Rai and Bawa 2013). Today, Indigenous com-munities own or manage at least 38 million km², spanning 87 countries or

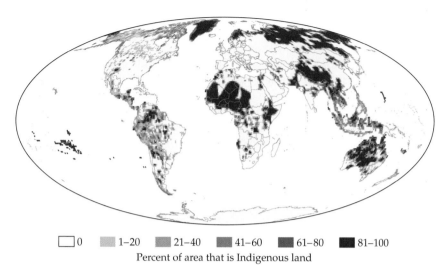

0   1–20   21–40   41–60   61–80   81–100

Percent of area that is Indigenous land

FIGURE 10.11 Indigenous Peoples manage or have tenure rights over at least 38 million km², spanning every continent. Here we see the percentage of area that is Indigenous land. (From S. T. Garnett et al. 2018. *Nat Sustain* 1: 369–374. doi. org/10.1038/s41893-018-0100-6.)

regions (Garnett et al. 2018) (**FIGURE 10.11**). This area includes 97% of the land in Papua New Guinea and 100 million ha (22%) of the Amazon basin of Brazil. In Australia, tribal people control 90 million ha, including many of the most important areas for conservation. It is estimated that globally these lands encompass a substantial percentage of the world's biodiversity.

By its nature, TEK can even help inform our responses to climate change. Indigenous communities in Sarawak, Malaysian Borneo, have adjusted the way that they manage their land and resources to ensure food and resource security in direct response to rising temperatures and increasing drought (Hosen et al. 2020). The structure of the community and mechanisms to distribute and pass down information facilitate adaptive management and sustainable land use.

The challenge, then, is to develop strategies for including these local peoples in conservation programs and policy development both outside and inside protected areas (Gavin et al. 2015). The partnership of traditional people, government agencies, and conservation organizations working together has been termed **co-management** (Borrini-Feyerabend et al. 2004). Co-management involves sharing management decisions and their consequences. The new strategies have been developed in an effort to avoid **ecocolonialism**, the practice by some governments and conservation organizations of disregarding the traditional rights and practices of local people in order to establish new conservation areas (see Chapter 9). The practice is called ecocolonialism because of its similarity to the historical abuses of native rights by colonial powers of past eras. The involvement of local people in the conservation of their lands is an issue of social justice (see Chapter 12).

In many new conservation projects, the economic needs of local people are included in conservation management plans, to the benefit of both the

GLOBAL CHANGE
CONNECTION

people and the reserves (Roe et al. 2013). Such projects, known as **integrated conservation and development projects** (**ICDPs**), are now regarded as worthy of serious consideration, though in practice they are often problematic to implement, as described later in the chapter. There are many possible strategies that could be classified as ICDPs, ranging from wildlife management projects to ecotourism, and these strategies may or may not include formal PAs. These projects attempt to combine the protection of biodiversity and the customs of traditional societies with aspects of economic development, including reducing poverty, creating jobs, improving health, and ensuring food security. A large number of such programs have been initiated over the last 25 years, and they have provided opportunities for evaluation and improvement. Involving local people in ongoing monitoring efforts may increase information and also help to determine how the people themselves perceive the benefits and problems of the project (Braschler 2009). The hope of such projects is that the local people will decide that sustainable use of their local resources is more valuable than destructive use of those resources and that these people will become involved in biodiversity conservation. The following are some examples of the types of ICDPs currently in practice.

> ICDPs involve local people in sustainable activities that combine biodiversity conservation and economic development.

Biosphere reserves

In UNESCO's World Network of Biosphere Reserves, traditional people are allowed to use resources from designated buffer zones around strictly protected core areas (see Chapter 9 and Figure 9.20). The Biosphere Reserves program recognizes the role of people in shaping the natural landscape as well as the need to find ways in which people can sustainably use natural resources without degrading the environment. The program includes the La Gomera Biosphere Reserve in the Canary Islands off the coast of northwestern Africa (**FIGURE 10.12**). This Spanish archipelago

FIGURE 10.12 La Gomera is a Biosphere Reserve located in the Canary Islands. The total area is 84,522 ha with a core area of 13,134 ha. It is the least populated island of the archipelago, with a population of approximately 23,000. The island contains more than 4000 species.

has a small Indigenous community known for their whistled language and crafts with palm leaves. The core area is home to 1021 endemic species (https://biosfera.lagomera.es/biodiversidad-en-la-reserva/). Another example is the Kopet Dag Biosphere Reserve in the Islamic Republic of Iran. The region includes two mountain ranges, and woodland covers 80% of the area. Several nomadic groups use the area for rangelands for summering and wintering grounds; the main Indigenous people here are the Kormanj (Kurdish) and Tukmen tribes. The reserve is a biodiversity hotspot that safeguards many endangered species.

In situ agricultural conservation

The long-term health of modern agriculture depends on the preservation of the genetic variability maintained in local varieties of crops cultivated by traditional farmers. One innovative suggestion has been for an international agricultural body, such as the Consultative Group on International Agricultural Research (CGIAR; www.cgiar.org), to subsidize villages as in situ (in-place) custodians of traditional varieties of crop species.

A different approach to linking traditional agriculture and genetic conservation is being used in arid regions of the American Southwest, with a focus on dryland crops with drought tolerance. A private organization, Native Seeds/SEARCH, collects the seeds of 2000 traditional crop cultivars for long-term preservation, representing both the cultural heritage and farming expertise of more than 50 Indigenous communities (Native Seeds/SEARCH 2009). The organization also encourages a network of thousands of farmers and other members to grow traditional crops, provides them with the seeds of traditional cultivars, and buys their unsold production. They also provide seed free or at reduced cost to members of Native American tribes. The value of such genetic conservation programs is being increasingly recognized (Jarvis et al. 2016).

Countries have also established special reserves to conserve areas containing wild relatives and ancient landraces of commercial crops. Species reserves protect the wild relatives of wheat, oats, and barley in Israel and of citrus in India.

Extractive reserves

In many areas of the world, traditional people have extracted products from natural communities for decades and even centuries. The use, sale, and barter of these natural products are a major part of people's livelihood (see Figures 3.5 and 3.14). Understandably, local people are very concerned about retaining their rights to continue collecting natural products from the surrounding countryside. In areas where such collection represents an integral part of traditional society, the establishment of a national park that excludes the traditional collection of products will meet with as much resistance from the local community as will a landgrab that involves the exploitation of the natural resources and their conversion to other uses. A type of protected area known as an **extractive reserve** may present a sustainable solution to this problem.

One such example is found in the Brazilian Amazon, where the government is addressing the legitimate demands of local citizens by establishing extractive reserves from which settled people collect natural materials, such as medicinal plants, edible seeds, rubber, resins, and Brazil nuts, in ways that minimize damage to the forest ecosystem (Duchelle et al. 2012) (**FIGURE 10.13**). These extractive reserves, which comprise about 3 million ha, guarantee the ability of local people to continue their way of life and guard against the possible conversion of the land to cattle ranching and farming. However, populations of large animals in extractive reserves are often substantially reduced by subsistence hunting by local people, and the density of Brazil nut seedlings is reduced by the intense collection of mature nuts.

Many countries in eastern and southern Africa have started aggressively applying community development and sustainable harvesting strategies in their efforts to preserve wildlife populations. Governments are attempting to develop programs to generate income from trophy hunting and wildlife tourism that can be operated at the village level and provide clear benefits to local people (Naidoo et al. 2016). One example is a community-based natural resource management program in which local communities working with the government sell opportunities to hunt high-value trophy species, such as lions and elephants, to safari companies (see Chapter 3). Revenue is also generated through operating tourist facilities. To maintain the needed densities of wildlife, the village community must work together with government officials to prevent illegal hunting.

There has been vocal support from some conservation biologists for selling hunting licenses as a means of conserving species, especially when local people are involved (e.g., Di Minin et al. 2016). However, trophy hunting is considered ethically questionable by those who believe that killing purely for sport (rather than for food) is morally wrong and point to the faulty reasoning behind consequentialism (that is, that the ends justify the means; Nelson et al. 2016). Furthermore, they argue that revenues from trophy hunting for conservation are insufficient, usually do not reach the local community, and are decreased via corruption (Lindsey et al. 2016). An evaluation of African game reserves found that wealth

FIGURE 10.13 Extractive reserves established in Brazil provide a reason to maintain forests. The trunks of wild rubber trees are cut for their latex, which flows down the grooves into the cup. Later the latex will be processed and used to make natural rubber products.

© Wolfgang Kaehler/Alamy

generation was often at the cost of wildlife conservation, primarily due to weaknesses in government management (Pitman et al. 2017). Finally, the market for hunting licenses creates pressure to "produce an animal" that inevitably leads to poaching, as was seen in the Cecil the Lion story (see Chapter 3). Thus, some question whether sport hunting belongs in the same category as extractive reserves that support locals with food, firewood, or other resources.

Community-based initiatives

In many cases, local people already protect natural areas and resources such as forests, wildlife, rivers, and coastal waters in the vicinity of their homes. Protection of such **community conserved areas**, sometimes called **community-based conservation** (**CBC**), is often enforced by village elders because of the clear benefits to the local people (see Chapter 12). These benefits include maintaining natural resources (e.g., food supplies and drinking water) and the use of the land for religious and traditional practices. The protection of biodiversity may even be an intrinsic aspect of local beliefs (see Figure 1.4). In this way, the goal of CBC is to align ecological, economic, and social goals. A review of 136 CBC projects across the globe found that degree of local participation, environmental education, and skills-training programs all significantly contributed to win-win outcomes for the people and to biodiversity (Brooks 2017). The most important feature, however, was institutional capacity building: efforts to improve infrastructure and communication and decision-making processes. Governments and conservation organizations can assist local conservation initiatives by providing access to scientific expertise, training programs, and financial assistance to develop needed infrastructure, in addition to simply offering legal title to traditional lands.

One example of a local initiative is the Community Baboon Sanctuary in eastern Belize, which was created by a collective agreement among a group of villages to maintain the forest habitat required by the local population of black howler monkeys (known locally as baboons). Ecotourists visiting the sanctuary pay a fee to the village organization, and additional payments are made if they stay overnight and eat meals with a local family. Conservation biologists working at the site have provided training for local nature guides, a body of scientific information on the local wildlife, funds for a local natural history museum, and business training for the village leaders.

In the Pacific islands of Samoa, much of the rain forest land and marine area is under "customary ownership": it is owned by communities of Indigenous people (Boydell and Holzknecht 2003). Villagers are under increasing pressure to sell logs from their forests to pay for schools and other necessities. Despite this situation, the local people have a strong desire to preserve the land because of the forest's religious and cultural significance, as well as its value for medicinal plants and other products. A variety of solutions are being developed to meet these conflicting needs. In 1988, in American (or Eastern) Samoa, where about 90% of the land is under customary ownership, the US government leased forest and coastal land from the villages to establish a new national park (Office of National Marine Sanctuaries 2021). Under this agreement, the villages gained needed income yet retained ownership of the land and their traditional hunting and collecting rights (NPS 2009).

Payments for ecosystem services

A creative strategy involves making direct payments to individual landowners and local communities that protect critical ecosystems and the services they provide, in effect paying the community to be a good steward of the land (Wunder 2013). This type of program is sometimes referred to as **payments for ecosystem services (PES)**, and PES projects are becoming increasingly popular (**FIGURE 10.14**). Governments, nongovernmental conservation organizations, and businesses develop markets in which local villagers and landowners can participate through protecting and restoring ecosystems. For example, owners of a forest may receive direct payments from a city government for the ecosystem services provided by the forest, such as controlling floods and providing drinking water. Local landowners

New markets are being developed in which local people and landowners are paid for providing ecosystem services by, for example, protecting forests to maintain water supplies and planting trees to absorb carbon dioxide. Programs that address climate change issues are expected to become more common in the coming years.

GLOBAL CHANGE
CONNECTION

(A)

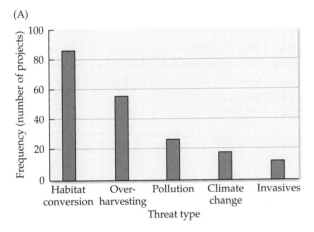

(B)

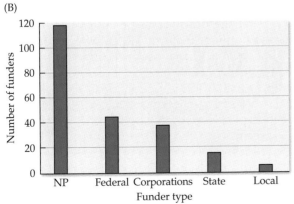

FIGURE 10.14 Patterns of payments for 103 ecosystem services in PES projects in 37 countries. (A) Number of projects addressing different types of threat. Most projects address issues of habitat conversion (from forest to agricultural land) and overharvesting of trees. (B) Funding sources for the projects are primarily nonprofit (NP) conservation organizations but also include government agencies at national, state, and local levels, as well as corporations. (After H. R. Tallis et al. 2009. *Front Ecol Environ* 7: 12–20. © The Ecological Society of America.)

and farmers can be paid for allowing large predators such as wolves, bears, tigers, and mountain lions to be on their land, with additional payments given as compensation if their livestock is attacked (Dickman et al. 2011).

Rural people can be drawn into newly developing international markets for ecosystem services (Ecosystem Marketplace 2010), especially programs that trade in carbon credits for reducing atmospheric levels of greenhouse gases. In the Kasigau Corridor Reducing Emissions from Deforestation and Forest Degradation (REDD) project in southeastern Kenya, villagers protect a natural migration corridor for elephants between Tsavo East National Park and Tsavo West National Park (**FIGURE 10.15**). The villagers earn funding from international programs and carbon credits, as determined by an independent outside evaluation, for protecting the forest and maintaining wildlife populations. Funds from the REDD program have been used to pay for wildlife patrols, to build schools and other infrastructure, and to start local businesses, while in the process creating 350 new jobs and generating a total of $1.2 million (Dinerstein et al. 2013). Programs addressing carbon sequestration and climate change are likely to expand greatly in coming years, and they may provide substantial funds for land protection. However, at present such programs are sometimes unable to pay enough money to prevent landowners from converting their land to other uses (Banerjee et al. 2013).

GLOBAL CHANGE
CONNECTION

PES programs have been effective for improving habitat for the giant panda (*Ailuropoda melanoleuca*) in China as a part of the Natural Forest Conservation Program (Tuanmu et al. 2016). In fact, an evaluation of the changes in vegetation revealed that often more improvement occurred in areas managed by locals than in areas managed by the government. However, this success only took place when the financial incentives were very high.

FIGURE 10.15 Elephants (*Loxodonta africana*) in Kenya depend on this wildlife corridor for their annual migration between two national parks. The corridor is maintained by local people, who benefit financially.

Evaluating conservation initiatives that involve traditional societies

Unfortunately, when external funding ends, and if the projected income stream fails to develop, many of these conservation projects involving local communities end abruptly. Even for projects that appear successful, there is often no monitoring of ecological and social parameters to determine whether the project goals are being achieved. Any conservation program design must include mechanisms for evaluating the progress and success of measures being taken.

A key element in the success of many of the projects discussed in the preceding sections is the opportunity for conservation biologists to complement and work with stable, flexible, local communities with effective leaders and competent government agencies (Baker et al. 2012). When these factors are present, certain community conservation projects do appear to be successful at combining the protection of biodiversity with sustainable development and poverty reduction. The Equator Initiative of the United Nations, cosponsored by many leading conservation organizations, businesses, and governments, is helping to fund and publicize such efforts (www.equatorinitiative.org).

However, in many cases a local community may have internal conflicts and poor leadership, making it incapable of administering a successful conservation program. Moreover, conservation initiatives involving recent immigrants or impoverished, disorganized local people may be difficult to carry out, and government agencies working on the project may be ineffective or even corrupt. Consequently, while working with local people may be a desirable goal, in some cases it simply is not possible.

10.5 Case Studies: Namibia and Kenya

LEARNING OBJECTIVES

By the end of this section you should be able to:

10.5.1 Evaluate CBNRM programs in Namibia.

10.5.2 Identify the challenges to conservation in sub-Saharan Africa and how current management approaches address them.

Throughout the world, the protection of biodiversity is being included as an important objective of land management. We conclude the chapter by examining two case studies of successful community-based natural resource management (CBNRM) programs, in Namibia and Kenya, that illustrate some of the challenges and successes of managing biodiversity outside protected areas.

CBNRM programs in Africa represent an approach in which local landowners and communal groups are given the authority to manage and profit from the wildlife on their own property. In many African countries, wildlife both inside and outside national parks is managed by government officials, often with no input from the local people, who gain little or no economic benefit from the wildlife on their own land and have no incentive to protect the wildlife. By changing the management system to CBNRM, government officials and conservation organizations hope to counterbalance pressures threatening local wildlife while simultaneously contributing to rural economic development. There is a long history of CBNRM programs in Africa, but it has been difficult to develop stable programs that are effectively managed and economically viable.

One of the most ambitious programs for local communities managing wildlife is found in Namibia in southern Africa. Namibia has nearly 2.5 million people, with 17.4% living in poverty (World Bank 2021b). There are 68,400 km^2 of forest, and 17% of total area is in protected areas (Republic of Namibia MET 2021). Namibia includes four different biomes, from desert to subtropical savanna (**FIGURE 10.16**).

Beginning in 1996, the Namibian government granted traditional communal groups the right to use and manage the wildlife on their own lands. To obtain these rights, a group needs to form a management committee, develop a management plan, and determine the boundaries of its land. The government then designates the group as a "community conservancy." The benefits of forming a conservancy and participating in wildlife management are fourfold:

1. The conservancy can form joint ventures with tour operators, with about 5%–10% of the gross earnings paid to the conservancy. A certain number of the employees in the tourist operation are hired from among the communal group. Revenues from the joint ventures are used to train and pay game guards, again hired from the communal group, who monitor the wildlife populations and prevent poaching.

2. Using funds from the joint ventures, the conservancy members can build and operate campsites for tourist groups, providing direct revenue, employment, and experience for the communal group.

3. The conservancy can apply to the government for a trophy-hunting quota, which will be granted if the wildlife populations are large enough, as indicated by monitoring. Hunting licenses can then be sold or auctioned off to professional hunters, who bring in wealthy foreign tourists willing to pay a high price for an African hunting experience. Payments to the conservancy for high-value animals such as lions and elephants can be as large as $11,000 per animal. Meat from the hunted animals is distributed to the group members as an added benefit. This approach to funding is not without controversy, however (see Chapter 3 and "Extractive reserves" in this chapter). Some economic analysis suggests that tourism alone (i.e., without hunting) is not sufficient to cover operating costs (Naidoo et al. 2016).

4. Once the conservancy has formed a wildlife management plan, four species of wildlife—gemsbok, springbok, kudu, and warthog—can be hunted for subsistence. In practice, the hunting is often done by game guards and professional hunters, and the meat is distributed to everyone in the community.

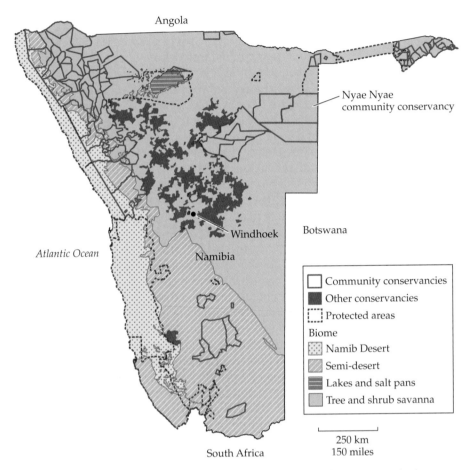

FIGURE 10.16 The distribution of current and emerging community conservancies in Namibia, in which communal groups agree to protect biodiversity. It can be seen here that some biomes, such as tree and shrub savanna, are more represented in these community conservancies than in formal protected areas. State-protected lands are also shown. The category "other conservancies" denotes areas where communal groups have not yet committed to forming conservancies to protect wildlife. (After NACSO 2008. *Namibia's communal conservancies: A review of progress in 2008.* NACSO, Windhoek, Namibia. Updated data from B. Riehl et al. 2015. *PLOS ONE* 10[5]: e0125531. doi.org/10.1371/journal.pone.0125531. CC BY 4.0, creativecommons.org/licenses/by/4.0.)

In the past two decades, 79 conservancies have been established in Namibia, covering 19% of the country's land surface (Riehl et al. 2015) (see Figure 10.16). Help in the initial establishment of the conservancies has come from external funding agencies, such as the US Agency for International Development. Conservancy members have received further training in tourism, finance, and marketing, along with effective advocacy to gain support from the government and the private sector. Although the financial gain by the CBNRM is well documented and social benefits such as improved health in these areas relative to adjacent areas have been documented, whether other expected social benefits, such as education, have improved is less clear (Riehl et al. 2015). This is due in part to the large degree of variability between conservancies with a great deal of tourism and those that have none. Some analyses also suggest that although hunting and tourism in conservancies could earn more per hectare than livestock rearing (Lindsey et al. 2013), conservancies may not be economically viable in the long term (Humavindu and Stage 2015). PES and other revenue-sharing systems have been proposed to address the problem of viability (Lapeyre 2015).

So far, the communal management system seems to be having positive effects on conservation. Namibia currently claims to host the world's largest populations of free-ranging cheetah and black rhino, both international species of concern. Many large mammal species, especially ungulates, have been observed as having higher numbers within the conservancies than in adjacent, unprotected land (Lindsey et al. 2013).

Other African countries have programs that are similar to Namibia's. In Kenya, for example, about two-thirds of the country's 650,000 large animals—including giraffes, elephants, zebras, and ostriches—live outside park boundaries in rangelands used by commercial ranches and as traditional grazing lands by local people (Western et al. 2009). The rangelands outside the parks are increasingly unavailable to wildlife, though, because of fences, poaching, and agricultural development, which have led to a gradual decline in wildlife numbers.

Kenya, like Namibia, has a combination of regulations, community involvement, and economic incentives contributing to the persistence of substantial populations of wildlife in spite of the challenges (Kinnaird and O'Brien 2013). In some places, private ranching in which wildlife and livestock are managed together for both meat and ecotourism is more profitable than managing livestock alone, because the livestock and the wildlife use different food resources. As in Namibia, many ranches have also developed facilities for foreign tourists who want to view wildlife, which creates an additional source of revenue and an incentive for protecting these species.

In general, community-based management programs in sub-Saharan Africa have had positive ecological outcomes, although the social and economic outcomes are not always as positive as those seen in Kenya and Namibia (Galvin et al. 2018). In many cases, a dependence on tourism and subsidies from outside donor governments and conservation and development organizations can make them vulnerable. When these outside subsidies cease, the wildlife programs often end as well, suggesting that the programs are often not really profitable on their own. The ineffectiveness and corruption of some local government agencies are additional factors that can cause such programs to fail. These community wildlife programs will be judged successful when they can demonstrate that they can both protect wildlife and provide a stable income source for the local people.

Summary

- Considerable biodiversity exists outside PAs, particularly in habitats managed for multiple-use resource extraction. Such unprotected habitats are vital for conservation because in almost all countries, PAs account for only a small percentage of the total area. Animals and plants living in PAs often disperse to unprotected land, where they are vulnerable to hunting and harvesting, habitat loss, and other threats from humans.

- Governments are increasingly encouraging the protection of biodiversity as a priority on multiple-use land, including forests, grazing lands, agricultural areas, military reservations, and urban areas. All of these can be managed for conservation, to increase populations of conservation interest.

- Government agencies, private conservation organizations, businesses, and private landowners can cooperate in large-scale ecosystem management projects to achieve conservation objectives and use natural resources sustainably. Bioregional management involves cooperation across large regions to manage large ecosystems, which frequently cross international borders.

- Community conservation projects involve local people in conservation activities in a way that protects biodiversity and provides benefits to the community. These projects might involve working with nearby PAs, maintaining traditional agriculture, and making payments for ecosystem services.

- In Africa, many of the characteristic large animals are found predominantly in rangeland outside the parks. Local people and landowners often maintain wildlife on their land for a variety of purposes. Local communities are now generating income by combining wildlife management and ecotourism, sometimes including trophy hunting.

For Discussion

1. Consider a national forest that has been used for decades for logging, hunting, and mining. If endangered plant species are discovered in this forest, should these activities be stopped? Can logging, hunting, and mining coexist with endangered species, and if so, how? If logging has to be stopped or scaled back, do the logging companies or their employees deserve any compensation? Explain your answer.

2. Do you think that trophy hunting on private reserves is a good means by which to preserve species? Why or why not? On what basis should this decision be made: economic, ethical, past success, or future potential for conservation?

3. Choose a large aquatic ecosystem that includes more than one country, such as the Black Sea, the Rhine River, the Caribbean, the St. Lawrence River, or the South China Sea. What agencies or organizations have responsibility for ensuring the long-term health of the ecosystem? In what ways do they, or could they, cooperate in managing the area?

Suggested Readings

Chan, Y. C., et al. 2019. Conserving unprotected important coastal habitats in the Yellow Sea: Shorebird occurrence, distribution and food resources at Lianyungang. *Global Ecology and Conservation* 20: e00724. It is important to document diversity in unprotected areas, in part to be able to justify the establishment of new PAs.

Gooden, J., and R. Grenyer. 2019. The psychological appeal of owning private land for conservation. *Conservation Biology* 33(2): 339–350. Increasingly, people are purchasing land specifically for conservation, motivated in part by a need for meaning.

Guadilla-Sáez, S., et al. 2019. Biodiversity conservation effectiveness provided by a protection status in temperate forest commons of north Spain. *Forest Ecology and Management* 433: 656–666. A rare example of diversity actually being higher outside of a PA than inside.

Magle, S. B., et al. 2019. Advancing urban wildlife research through a multi-city collaboration. *Frontiers in Ecology and the Environment* 17(4): 232–239. The establishment of the Urban Wildlife Information Network (UWIN) has the potential to expand our understanding of these areas as habitat.

Rodrigues, P., et al. 2018. Coffee management and the conservation of forest bird diversity in southwestern Ethiopia. *Biological Conservation* 217: 131–139. Bird diversity was high across a range of densities of trees in shade-grown coffee.

Schilthuizen, M. 2019. *Darwin Comes to Town: How the Urban Jungle Drives Evolution*. Picador/Macmillan, London. A book about urban biodiversity and evolutionary change.

Weiskopf, S. R., et al. 2019. The conservation value of forest fragments in the increasingly agrarian landscape of Sumatra. *Environmental Conservation* 46: 340–346.

Zellmer, A. J., et al. 2020. What can we learn from wildlife sightings during the COVID-19 global shutdown? *Ecosphere* 11(8): e03215.

Zhang, L., et al. 2020. Influence of traditional ecological knowledge on conservation of the skywalker hoolock gibbon (*Hoolock tianxing*) outside nature reserves. *Biological Conservation* 241: 108267. An example of how TEK can play an important role in species conservation.

Volunteers and students plant mangrove trees in Kondang Merak, located on the eastern coast of Java, Indonesia, in an effort to restore the local marine ecosystem.

Restoration Ecology

On World Environment Day 2021, the United Nations declared the UN Decade on Ecosystem Restoration (UN News 2021). **Ecosystem restoration** was defined as "the process of halting and reversing degradation, resulting in improved ecosystem services and recovered biodiversity. Ecosystem restoration encompasses a wide continuum of practices, depending on local conditions and societal choice" (UNEP 2021).

Among these practices, **ecological restoration** is the process of assisting or accelerating the recovery of degraded, damaged or destroyed ecosystems (SER 2021) (**FIGURE 11.1**). Recovery, in this context, is the return to a former state, but as we will see, this is not always possible. Restoration of some type is often necessary because simply establishing protected areas is often insufficient for conservation of biodiversity; damage from intensive human activities such as mining, ranching, and logging often results in ecosystems losing much of their **ecological resilience**, or natural ability to recover. In some cases, recovery would require centuries or even millennia, without human assistance. This assistance may be as basic as removing an invasive species or reintroducing a single species, or as complex as changing the course of a river or creating new land forms (SER 2021). At other times, ecological processes rather than ecosystems need to be restored. For example, annual floods disrupted by the construction of dams, or natural fires stopped by fire suppression may need to be reintroduced if the absence of these processes proves harmful to species and ecosystems.

GLOBAL CHANGE
CONNECTION

(A)

(B)

FIGURE 11.1 (A) Trout stream habitat that has been degraded by human activities. (B) Trout stream habitat that has been restored by installing fencing to exclude cattle, planting native species, and reinforcing stream banks with rocks.

The restoration of ecosystems can be motivated by the desire not only for the protection of species and the strengthening of ecosystem function, but also for the reestablishment of ecosystem services such as recreation, and for its role in creating jobs. Many restoration efforts are supported and even initiated because of the direct connection between a healthy environment and people's personal and economic well-being (Derak et al. 2018) (see Chapter 3). Restoration can even increase the absorption of greenhouse gases (Pugh et al. 2019). Rebuilding damaged ecosystems also can be used to enlarge, enhance, and connect protected areas, as well as to create buffers around them (see Chapter 9). Thus, ecological restoration activities can have far-reaching impacts, as reflected by the stated aim of the UN Decade on Ecosystem Restoration to "prevent, halt and reverse the degradation of ecosystems on every continent and in every ocean. It can help to end poverty, combat climate change and prevent a mass extinction" (Decade on Restoration 2021).

Some ecosystems have been so degraded by human activity that their resilience, or ability to recover on their own, is severely limited. Ecosystem restoration reestablishes functioning ecosystems, with some or all of the original species or, sometimes, a different group of species.

Because many degraded areas are unproductive and of little economic value, governments are often willing to restore them to increase their economic productivity and conservation value. For example, degraded areas may be subject to soil erosion and increased risk of flooding; restoration in these cases may be motivated by a desire to mitigate threats to human life or property. Restoration also can be part of **compensatory mitigation** or **biodiversity offsets**, in which a new site is created or rehabilitated in compensation for a site that has been destroyed elsewhere by development (Rohr et al. 2018).

This is particularly true for wetlands, for which a "no net loss policy" has been adopted by many jurisdictions (see Section 11.5).

These efforts to repair ecosystems are guided by both practical experience and scientific research (Clark et al. 2019). **Restoration ecology** is the science of restoration—the research and scientific study of restored populations, communities, and ecosystems. The process of ecological restoration provides useful scientific data, while restoration ecology interprets and evaluates restoration projects in a way that can lead to improved methods. Increased knowledge from this science allows us to repair at least some of the damage we have inflicted upon ecosystems.

Ecosystem restoration has its origins in older, applied technologies that attempted to restore ecosystem functions or species of known economic value, such as wetland creation (to prevent flooding), mine site reclamation involving adding soil and replanting vegetation (to prevent soil erosion and contamination of water sources), range management of overgrazed lands (to increase the production of grasses), and technologies to facilitate tree planting on cleared land (to reduce erosion, for timber, and to increase other ecosystem values). However, these approaches often produce biological communities that are overly simplified or cannot maintain themselves. As concern for biodiversity has grown, so too has an emphasis on reestablishment of original or historical species assemblages and processes, that is, ecological restoration. The input of conservation biologists is needed to achieve these goals.

11.1 Where to Start?

LEARNING OBJECTIVES

By the end of this section you should be able to:

11.1.1 Outline the principles of ecological restoration.

11.1.2 Contrast the four main approaches to restoration ecology, providing an example of each.

11.1.3 Use the flow diagram in Figure 11.2 to create a plan for a restoration project.

In the context of the Decade on Ecosystem Restoration, the UN announced that 115 countries have committed to restoring up to 1 billion ha of land—an area the size of China (UNEP 2021). To support these efforts, several international organizations have come together to publish a set of principles for successful restoration. These organizations include the Food and Agriculture Organization of the UN (FAO), the Society for Ecological Restoration (SER), and the IUCN's Commission on Ecosystem Management (CEM). The guidelines can be summarized as follows (adapted from FAO, IUCN CEM, and SER 2021):

GLOBAL CHANGE
CONNECTION

1. *Global contribution* Ecosystem restoration can have far-reaching impacts beyond those of biological conservation and should support the goals of other environmental international agreements, such as those that address desertification and climate change (see Chapter 12).

2. *Broad engagement and information sharing* All those who are affected by a restoration project (i.e., *stakeholders*) should have an equal opportunity to be engaged in a meaningful way and benefit from it fairly. This principle explicitly names groups who have been historically disenfranchised, including local communities, Indigenous Peoples, ethnic minorities, women, youth, and LGBTIQ+ people. In other words, restoration projects should adhere to the principles of environmental justice (see Chapter 3). Likewise, diverse information sources that include Indigenous, local, and scientific knowledge will help ensure long-term success.

3. *Benefits to nature and people* Although ecological restoration focuses on the recovery of the natural system, ecosystem restoration as defined in this context has the explicit goal of producing a "net gain" not only in biodiversity and ecosystem health but also in ecosystem services (see Chapter 3) and human well-being in general. The goal of an ecosystem restoration project should be the greatest possible improvement in all these areas, including mitigating climate change. It is important for these benefits to be communicated to the public (DeAngelis et al. 2020).

GLOBAL CHANGE
CONNECTION

4. *Measurable and achievable goals* A restoration project should clearly define its goals, based on a shared vision of the stakeholders (see #2). The desired ecosystem, cultural, and socioeconomic outcomes should be quantifiable such that progress can be measured. Plans for implementation should be made that keep in mind relevant policies, stakeholders, and resources. Desired outcomes may include an increase in numbers of species or the number of people employed by the project.

5. *Addressing causes of degradation* Restoration cannot be successful in the long term if the underlying reasons the area has lost biodiversity and ecosystem services are not addressed. These may include direct causes, such as pollution, or indirect causes such as poverty or other socioeconomic problems.

GLOBAL CHANGE
CONNECTION

6. *Monitoring and management* Managing and measuring the progress and impacts of a restoration project are essential and should be considered at multiple scales. For example, any restoration occurring on a river must consider what is occurring not only in the restored site itself, but also up- and downstream, as well as any effects on local people. Careful monitoring will also allow there to be modifications in management as needed throughout the lifetime of the project, called **adaptive management** (see Chapter 9). For example, native species may have to be reintroduced if they have not survived, and invasive species may have to be removed if they are still abundant or there has been a **secondary invasion** by a different invader (González et al. 2017).

Each of these elements works with the others to produce a successful restoration project, several of which will be discussed in this chapter. **FIGURE 11.2** shows how aspects of goal setting (per #4 of the guidelines), identifying sources of degradation (#5), and monitoring and management (#6) work together in the context of engagement (#2) and benefits (#3). Note that all elements of the project must follow from establishing an overarching purpose (#1). Often this ultimate goal of a restoration effort is to create ecosystems that are comparable in function or species composition to existing **reference sites** (McDonald et al. 2016). Reference sites are central to the very concept of restoration; they act as comparison sites, providing explicit

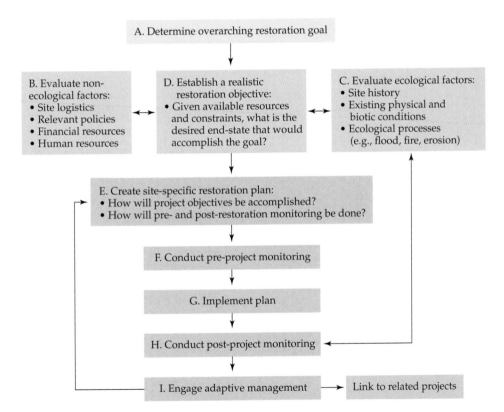

FIGURE 11.2 A flow diagram of a scientific approach to restoration. The first step is to (A) clarify the overarching goal of the restoration project, such as to increase biodiversity at a site. Then, to determine how this will be achieved, it is necessary to evaluate both (B) nonecological factors, such as how much money is available for the project, and (C) ecological factors, such as the condition of the soil and what plant species currently are found there. These factors will inform and be influenced by (D) the specific objectives for restoration, such as establishing plant species that will promote diversity at higher levels. Only after these steps have been taken can (E) a specific, realistic plan for how to implement restoration be created, such as removing weeds, improving the soil, or planting seeds. The implementation of the plan should accompany both pre- and postmonitoring (F–H). Monitoring progress of the restoration objectives will then facilitate (I) improvement over time at both that site and future projects. (After P. B. Shafroth et al. 2008. *Restor Ecol* 16(1): 97–112. © 2008 Society for Ecological Restoration International.)

restoration goals and allowing for quantitative measures of the project's success. Unrestored areas can act as "negative" reference sites or controls to further determine the impact of the restoration actions (González et al. 2015).

In a biodiversity conservation context, it is most common for the reference site to be dominated by native species, contain representatives of all key functional groups of species, have a physical environment suitable for native species and ecosystem processes, and be secure from detrimental outside disturbances. In some cases, this is not realistic. For example, at arid and cold sites, achieving such recovery may take decades or even centuries. Furthermore, site conditions or limitations of resources may make the establishment of native species undesirable or impossible. For this reason, the Society for Ecological Restoration states that ecological restoration means putting an ecosystem back on a trajectory towards historic conditions, not necessarily a return to an actual historic state. Climate change, in particular, may change the trajectory of not only a restoration site but its reference sites as well (SER 2021).

GLOBAL CHANGE
CONNECTION

There are four main approaches that define outcomes when considering the restoration of biological communities and ecosystems (**FIGURE 11.3**):

1. *No action*   Restoration is deemed too expensive, previous attempts have failed, or experience has shown that the ecosystem will recover on its own. Letting the ecosystem recover on its own, also known

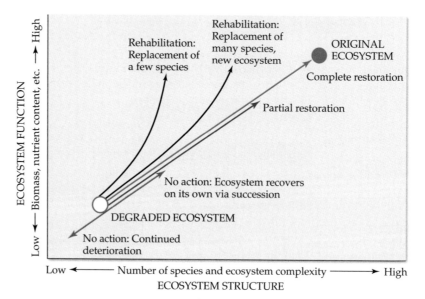

FIGURE 11.3 Decisions must be made about whether the best course of action is to completely restore a degraded site (green arrow), partially restore it (blue arrow), rehabilitate it by introducing different species (black arrows), or take no action (red arrow). (After A. D. Bradshaw. 1984. *Landsc Plan* 11: 35–48. Copyright 1984. Reprinted with permission from Elsevier.)

as **passive restoration**, is typical for abandoned agricultural fields or tree plantations, which may return to forest within a few decades (González et al. 2016). However, it should be noted that even in this case, the species composition may be quite different (Holmes and Matlack 2018).

2. *Rehabilitation*   A degraded ecosystem is replaced with a different but productive ecosystem type, one that might even include some nonnative species. For example, a degraded forest might be replaced with a productive pasture or a tree plantation. The term **rehabilitation** applies if some of the original ecosystem function is obtained without recovering most of the original species (SER 2021b). Just a few species may be replaced, some of them not even native species, or a larger-scale replacement of many species may be attempted. As the ultimate goal is not to restore the original ecosystem, some authors consider the term *restoration* inappropriate to refer to rehabilitated ecosystems (Aronson et al. 2018).

3. *Partial restoration*   At least some of the ecosystem functions and some of the original, dominant species are restored. An example is replanting a degraded grassland with a few species that can survive. Partial restoration typically focuses on dominant species or particularly resilient species that are critical to ecosystem function, delaying action on the rare and less common species that would be part of a complete restoration program.

4. *Complete restoration*   The area is completely restored to its original species composition and structure by an active program of site modification and reintroduction of the original species. Natural ecological processes must be reestablished because they help the system recover and contribute to long-term resilience. Although complete restoration is an ideal, it is rarely if ever actually achieved; some argue that it is impossible to ever fully return to a predisturbance state (González et al. 2018; Hobbs 2018).

Which of the above trajectories are selected for a given project will depend on a multitude of practical and scientific considerations, many of which we have already discussed in this chapter. Scientific factors include how natural systems work and functional roles of species (Palmer et al. 2016). However there are also much more basic considerations, such as of the cost and availability of seeds, when to water plants, how much fertilizer to add, how to remove invasive species, and how to prepare the surface soil. Fortunately, ecological restoration often involves professionals from different fields who can lend their expertise for these types of questions, ultimately enriching the restoration process. However, these practitioners sometimes have different goals than conservation biologists. For instance, civil engineers involved in major projects seek economical ways to permanently stabilize land surfaces, prevent soil erosion, make the site look better to the general public, and, if possible, restore the productive value of the land. This differing focus has led to such actions as planting nonnative species that do a good job of stabilizing the soil but later became invasive (such as tamarisk trees; see Section 11.3).

Thus, it may not be surprising that in a recent study of hundreds of restoration sites along rivers in the southwestern United States, it was found that that the greater the number of partners on a project, including both those with practical and scientific expertise, the better the ecological restoration outcomes (Sher et al. 2020). The same study was the first to find that a combination of characteristics of the management teams explained the actual composition of the plant communities they had restored better than environmental factors such as soil type or climate (Primack et al. 2021). In short, the importance of the human element of restoration cannot be overstated.

The decisions management teams make about the types of intervention required will heavily depend on the degree of alteration; the most degraded systems will likely require both biological and physical alterations to the habitat (**FIGURE 11.4**). If the damage has been caused by abiotic factors such as soil erosion or lowered water tables, then the source of the problem should be addressed or at least considered before any attempt is made to reestablish species (González et al. 2018).

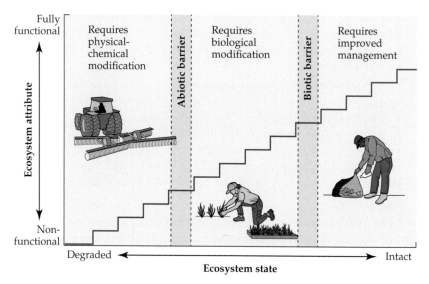

FIGURE 11.4 A conceptual model that considers thresholds for the restoration of ecosystem function. Generally, the most degraded, and therefore nonfunctional, sites will require overcoming abiotic constraints that contribute to the problem, such as by plowing a field that is dominated by nonnative invasive species, removing levees from a river, or building structures for coral to grow on. Biotic barriers can be overcome by planting native species, reintroducing missing trophic levels, or providing a food source. Of course, biotic changes can lead to abiotic ones, as in the case of an invasive plant species increasing the frequency or intensity of fires. If an ecosystem is mostly intact and functional, improved management rather than restoration is needed. (After Parks Canada, and Canadian Parks Council. 2008. *Principles and Guidelines for Ecological Restoration in Canada's Protected Natural Areas.* National Parks Directorate/Parks Canada Agency, Gatineau, Québec, based on S. Whisenant 1999. *Repairing Damaged Wildlands: A Process-Oriented, Landscape-Scale Approach.* Cambridge University Press, Cambridge.)

One quandary that occurs in the restoration of highly degraded ecosystems is ecological assembly order; that is, in what order and when should the species components of the system be put back together? When trophic order is apparent (e.g., predators require an established prey species), the decision seems obvious, but the overlap in functional relationships among many species complicates such decisions. For example, degraded parks in Africa may have contained a dozen or more ungulate species with similar ecological roles. In what order should these be reintroduced? Knowledge about the specific ecology of individual species can help guide such complicated decisions.

Another important issue is the genotypes of plant species that are being reintroduced to a restoration site. In general, it is advisable to follow the local-is-best (LIB) approach, which prioritizes locally adapted genotypes. This is important for at least two reasons. First, there may be local adaptations within a species that will make it either more or less likely to thrive at the restoration site. Second, there is a risk of outbreeding depression or genetic swamping of local strains by introducing nonlocal genotypes (see Chapter 6). There are also other problems with this approach, including that such genetic matching may be unfeasible or impossible (Smith et al. 2007). In other cases, harvesting wild native plants risks hurting the source populations (Meissen et al. 2015), or the local genotypes may have such low genetic variability that they risk inbreeding depression at the restoration site. For this reason, restoration practitioners may need to resort to less-related plant stock or even nonnative species. In at least some cases, the use of cheaper, commercial seed rather than seed that was hand collected or specially reared has no adverse ecological effects (Reiker et al. 2015). Restoration should also consider introducing genotypes from warmer regions in anticipation of climate change effects (see the section "Assisted migration" in Chapter 8) (Dumroese et al. 2015).

Restoration ecology is valuable to the broader science of ecology because it provides an acid test of how well we understand a biological community; the extent to which we can successfully reassemble a functioning ecosystem from its component parts demonstrates the depth of our knowledge and points out deficiencies (Bradshaw 1990; Palmer et al. 2016). For example, in their review of the literature on restoration of tropical forests, Catarina Jakovac and colleagues determined that land use prior to restoration strongly determined the species composition of early dominant species, which has implications for ecosystem function (Jakovac et al. 2021). However, they also found that there were very few studies of the microbial dynamics during restoration of tropical forests—a significant research gap, given our growing understanding of the importance of microbes in ecosystems.

Although efforts to restore degraded terrestrial communities have typically emphasized the plant community, some researchers argue that restoration ecology should devote more attention to the other major components of the community, including trophic interactions and food webs (Vander Zanden et al. 2016). Fungi and bacteria play vital roles in decomposition

> Restoration projects require monitoring to determine whether goals such as costs and speed of recovery are being met. Such projects may also provide new insights into ecological processes.

and nutrient cycling in soil (see Chapter 2); soil invertebrates are important in creating soil structure; herbivorous animals are important in reducing plant competition and maintaining species diversity; birds and insects are essential pollinators; and many birds and mammals have vital functions as insect predators, soil diggers, and seed dispersers (e.g., Thompson et al. 2018). Many birds, insects, and other animals may be able to recolonize the site on their own, but other large animals and aboveground invertebrates may have to be reintroduced from existing populations or captive breeding populations (see Chapter 8) if they are unable to disperse to the site on their own. Restoration efforts may also succeed by focusing on reestablishing ecological processes that support native communities, such as flooding events or fire cycles, rather than just planting specific plant taxa (Moreno-Mateos et al. 2015b).

11.2 Restoration in Urban Areas

LEARNING OBJECTIVES

By the end of this section you should be able to:

11.2.1 Identify a candidate site for restoration in a given urban area, with clearly defined goals.

11.2.2 Describe the special benefits of urban restoration projects.

Highly visible restoration efforts are taking place in many urban areas. These efforts seek to reduce the intense human impact on ecosystems and enhance the quality of life for city dwellers (Honold et al. 2016). Local citizen groups often welcome the opportunity to work with government agencies and conservation groups to restore degraded urban areas. Unattractive drainage canals in concrete culverts can be replaced with winding streams bordered with large rocks and planted with native wetland species (Neale and Moffett 2016). Vacant lots and neglected lands can be replanted with native shrubs, trees, and wildflowers. Gravel pits can be packed with soil and restored as ponds. Establishing native plant species in these urban areas often leads to increases in populations of native birds and insects (Archibald et al. 2017). These efforts have the additional benefits of fostering neighborhood pride, creating a sense of community, and enhancing property values (**FIGURE 11.5**). An example can be found in southern China, where it was determined that river restoration in Guangzhou would increase property values by 5% (Chen 2017). However, such restorations are often only partially successful for restoring habitat because of their small size and the fact that they are embedded in the highly modified urban environment. Developing urban places where people and biodiversity can coexist has been termed *reconciliation ecology* (see Chapter 10).

> Highly visible restoration efforts are taking place in many urban areas to reduce the intense human impact on ecosystems and enhance the quality of life for city dwellers.

FIGURE 11.5 A highly successful urban restoration project in the center of Seoul, Korea, rebuilt the Cheonggyecheon Stream, which had been buried under a road. This 10.9 km² public park now provides many ecosystem services, enhances the quality of life for residents, and is a major tourist attraction.

Restoring native communities on huge landfills presents one of the most unusual opportunities for urban restoration. In the United States, 150 million tons of trash are buried in more than 2600 active landfills each year (EPA 2021). When the landfills reach their maximum capacity, they are usually capped with sheets of plastic and layers of clay to prevent toxic chemicals and pollutants from seeping out. If these sites are left alone, they are often colonized by weedy, exotic species. However, these eyesores can instead be the focus of conservation efforts; planting native shrubs and trees attracts birds and mammals that will bring in and disperse the seeds of a wide range of native species. The Fresh Kills restored landfill site on Staten Island in New York City is a good example of such restoration practices (NYC Parks 2021). It occupies almost 1000 ha and has garbage mounds as tall as the Statue of Liberty, with a volume 25 times that of the Great Pyramid of Giza. The landfill was closed in 2001 and is now undergoing restoration to create a huge public park with many elements of a native ecosystem, a project that is being implemented in six phases to be completed by the year 2036 (**FIGURE 11.6**). The eventual goal is to create a large public parkland area (almost three times the size of New York City's Central Park) with abundant wildlife and many recreational, cultural, and educational amenities. Basketball courts, soccer fields, and a playground have already been completed, and the creation of a wildlife refuge is in progress.

FIGURE 11.6 Freshkills is an 8.9 km² public park being built over the world's largest landfill (Fresh Kills, in Staten Island, New York). It stopped accepting household garbage in 2001, and the restoration activities are ongoing. It is intended to provide both recreation opportunities for people and habitat for native species.

11.3 Restoration Using Organisms

LEARNING OBJECTIVES

By the end of this section you should be able to:

11.3.1 Explain the ecological basis of the concept of rewilding.

11.3.2 Consider costs and benefits of biological control, and identify situations when it is advantageous.

Restoration is often limited in geographical scope due to the associated cost, but there is a growing movement to consider ways to repair ecosystems at grand scales, made possible in some cases by reestablishing certain ecosystem dynamics. Introducing animals (and other types of organisms such as bacteria or fungi) can accomplish what would not be logistically or financially possible otherwise. **Rewilding** is a term introduced by Michael Soulé in the mid-1990s to describe the reintroduction of top carnivores in order to regulate the system from the top down (Svenning et al. 2018). Since that time, the term *rewilding* has been used in a variety of contexts, especially attempts to restore aspects of ecosystems that last existed in the Pleistocene Epoch, more than 11,000 years ago, termed *Pleistocene rewilding*. The most famous of these is the reintroduction of wolves in Yellowstone National Park (see Chapter 8 and

> Animals and other organisms can facilitate restoration at scales that might otherwise not be possible.

Figure 8.1). The idea that the restoration of an entire ecosystem can be facilitated by the reintroduction of one or more missing functional groups of animals is an approach that has also been referred to as *trophic rewilding* (Svenning et al. 2018).

A famous and controversial attempt at rewilding was in the Oostvaardersplassen ("eastward-sailing wetland"), a nature reserve outside Amsterdam, the capital of the Netherlands. The European landscape has arguably been more altered by human activities than any other in the world, but Frans Vera, a Dutch government scientist, believed that the reintroduction of large herbivores that had been absent from the landscape for hundreds of years could return the Oostvaardersplassen to a former, more functional state. Because many of these species are now extinct, he introduced modern mammals as ecological surrogates; beginning in the 1980s, he brought in Heck cattle in place of extinct aurochs (wild cattle) and Konik ponies in place of tarpans, the last of Europe's wild horses (**FIGURE 11.7**).

The rewilding effort had remarkable effects on the landscape. Populations of horses and deer exploded, grasslands and marshes began to thrive as woody vegetation retreated, and many endangered birds took up residence in the newly opened habitat. However, wolves had not yet reached the Oostvaardersplassen. The lack of significant predation at this site led to booming herbivore populations that had no opportunity to expand beyond the isolated reserve to other areas in search of food. Photos and videos of starving animals were shown on television and in other media, and people objected to such cruel "treatment" of animals, even though it was a natural process. This led to new management policies of shooting suffering animals (Barkham 2018). There was also public debate over why managers are waiting for top predators to arrive on their own rather than reintroducing them, although reintroducing carnivores has been found to be far from a guaranteed method to reduce herbivores (Alston et al. 2019).

Photo by Richard Primack

FIGURE 11.7 Reintroduced horses and other large herbivores in the Oostvaardersplassen in the Netherlands helped decrease the dominance of trees through grazing.

Another type of restoration by animals is the release of biological control and bioremediation organisms. Unlike rewilding, these biological introductions are not used to help restore ecological balance by mimicking historical conditions, but rather to remove unwanted elements that were introduced by humans. **Bioremediation** is the use of an organism to clean up pollutants, such as the use of prokaryotes to break down the oil in an oil spill or wetland plants to take up agricultural runoff to clean the water (see the section "Ecosystem services" in Chapter 3). **Biological control** (also known as **biocontrol**) is the use of one type of organism, such as an insect, to manage another, undesirable, species, such as an invasive plant. Historically, a focus on human needs has led to problems in some cases, when the released organism itself has become a pest (see Chapter 5). And in some cases, the control of an invasive species was followed by the dominance of a second invasive species. However, these experiences have informed a broader view that considers the whole ecosystem, facilitating the use of both bioremediation and biocontrol in restoration contexts with conservation-oriented goals (Seastedt 2015).

One example of the use of biocontrol for ecological restoration is the release of the tamarisk leaf beetle (*Diorhabda* spp.) along rivers in the western United States. Rivers and their riparian plant communities are frequently the objects of restoration efforts, but the geographical scale of the problems often limits what can be done (González et al. 2015). The focus of restoration efforts in Texas, New Mexico, Arizona, and other western states has frequently been the removal of exotic tamarisk (*Tamarix* spp.) trees, which now dominate many riparian zones. When it behaves invasively, this species is associated with a host of problems that negatively affect both plants and animals (Sher 2013). Efforts at removing the tree with bulldozers or herbicides risk harm to native species, are difficult to use in remote regions, and are too expensive to implement on a large scale.

In response to these problems, more than a decade of research on biological control of the tamarisk eventually led government scientists to release the tamarisk leaf beetle in the wild in 2003 (Bean and Dudley 2018). By 2018, the beetle had spread to cover hundreds of miles of rivers, feeding on tamarisk leaves and turning acres of the invasive tree brown. The goal was to facilitate the recovery of native trees and other species (**FIGURE 11.8**). In some cases, a clear relationship has been found between a decrease in the invasive tree and recovery by native species (Sher et al. 2018) (**FIGURE 11.9**). However, even 20 years after the introduction of the biocontrol may be too early to determine the ultimate response of the ecosystem, especially in the context of climate change (Henry et al. 2018; Nagler et al. 2018).

GLOBAL CHANGE
CONNECTION

Just as the rewilding projects are not without problems, in this case there have been criticisms about unintended effects of the biological control on wildlife, particularly some species of herpetofauna (reptiles and amphibians) (Bateman et al. 2014) and birds (Darrah and van Riper 2018). Concern for a federally listed bird that nests in the tamarisk even resulted in litigation by an environmental group against the agency that released the beetle. Several scientists have argued that the overall ecological benefits of reducing

(A) 2006

(B) 2013

(C) 2015

Photo by Anna Sher

Photo by Wayne Ranney; inset courtesy of Eric Coombs, Oregon Department of Agriculture, Bugwood.org

Photo by W. Wright Robinson

FIGURE 11.8 (A) The introduced tamarisk (salt cedar, *Tamarix*) forms monocultures, as shown here in a 2007 photograph of the Colorado River in Utah. (B) Such inaccessible areas are being restored by releases of a biological control insect, the tamarisk leaf beetle (*Diorhabda* spp., inset), which feeds on tamarisk leaves, turning them brown, as shown in a 2013 photo. (C) Native species, such as the green willows in this photo, can then recover.

the tamarisk are worth such problems (Bloodworth et al. 2016). This case illustrates the point that the benefits of any restoration effort must always be weighed against perceived, potential, and actual costs.

A frequent scientific critique of both rewilding and biocontrol restoration projects is that the practitioners spend too much of their resources in active conservation and not enough on monitoring, researching, or publishing findings. This problem is caused in large part by limited funding; these projects are chronically underfunded and often rely on volunteers and nonprofit support. The lack of scientific publications generated by some of the world's largest rewilding projects in particular may be a reason why they have not gained wider publicity and acceptance. Even though the pace is slow, long-term efforts like these will provide important lessons for restoration efforts elsewhere.

> In some cases, restoration may be inadvisable due to economic costs or possible negative impacts on the ecosystem.

FIGURE 11.9 Changes to plant communities where a biological control beetle was released to reduce cover in invasive *Tamarix* trees. Over a period of five years, the increase in native plant cover significantly corresponded to the reduction in *Tamarix*, suggesting that removing the invader was reducing competitive pressure. This effect was even greater when a "cut-stump" method was employed in addition to the biocontrol ("herbicide + biocontrol"), that is, one in which an herbicide was applied to the stumps of chainsawed *Tamarix* trees. (After A. A. Sher et al. 2018. *Ecol Eng* 111: 167–175. Copyright 2017. Reprinted by permission of Elsevier.)

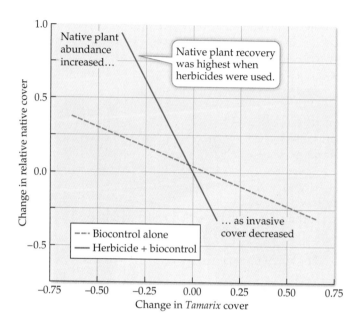

11.4 Moving Targets of Restoration

LEARNING OBJECTIVE

By the end of this section you should be able to:

11.4.1 Explain why both a forest invaded with an exotic species and a restored ecosystem might be considered "novel ecosystems."

Since ecosystems change over time in response to climate change, plant succession, the varying abundance of common species, and other factors, the goals of restoration may have to be changed over time as well or modified to include temporal dynamics to remain realistic. In many situations in which human activities have dramatically altered the environment, some biologists say, we will have to accept **novel ecosystems** in which there is a mixture of native and nonnative species coexisting in a community unlike the original or reference site (Higgs 2017; Hobbs et al. 2013). These novel ecosystems may differ from historical conditions in species composition and function, reflecting the shifting nature of species, the alteration of the environment, and even human values (Harris et al. 2006; Hertog and Turnhout 2018).

Restoration ecology is increasingly addressing the issue of the moving target, especially as it becomes evident that so many ecosystems simply cannot be restored in the traditional sense (Arthington et al. 2014). The soil chemistry or water availability may be too different, or the elimination of an introduced species may be impractical or even undesirable if that species can perform an ecological role similar to that of a missing native species. For example, research on native versus novel forests in Hawaii found that species richness was greater in the novel forest that included both native and exotic species and that total plant biomass and nutrient cycling were either the same or greater there than in the native forest (Mascaro et al. 2012). Even though the native plant species in this system are declining, proponents of novel ecosystems may accept such novel forests as acceptable targets for restoration efforts. In response to such a perspective, other researchers argue that advocating for novel ecosystems is problematic because advocates for development may use novelty as an excuse to not attempt complete restoration when it would have otherwise been possible (Miller and Bestelmeyer 2016).

11.5 Restoration of Some Major Communities

LEARNING OBJECTIVES

By the end of this section you should be able to:

11.5.1 Compare the challenges of restoring wetlands with those of restoring prairies.

11.5.2 Explain why restoration of prairies is particularly difficult.

11.5.3 Discuss the importance and challenges of a biocultural restoration program.

In addition to rivers, many efforts to restore ecosystems have focused on wetlands, lakes, prairies, and forests. These environments have been severely altered by human activities and are good candidates for restoration work.

Wetlands

Some of the most extensive restoration work has been done on wetlands, including swamps and marshes (Moreno-Mateos et al. 2015b). The United States has wetland protections under the Clean Water Act and the US government policy of "no net loss," meaning that the number of hectares of wetlands must either stay the same or increase. Thus, large development projects that damage wetlands must repair them or create new wetlands to compensate for those damaged beyond repair (Goldberg and Reiss 2016). The focus of these efforts has been on recreating the natural hydrology of the area and then planting native species. Many successful restoration projects have resulted from this legislation. However, it has fallen short of expectations due to a lack of monitoring and long-term management; short-term success is often followed a few years later by a failure to meet project goals (Clare and Creed 2014). Strategies to restore the biodiversity of rivers include the complete removal of dams and other structures (Foley et al. 2017) and controlled releases of water from dams (Glenn et al. 2017). For peatlands degraded by harvesting for horticultural peat, a well-recognized restoration approach is the moss layer transfer technique, which consists of spreading native plant material, collected from the top 10 cm of natural peatlands, over the restored area to facilitate reestablishment of species and ecosystem processes (Chimner et al. 2017).

Wetland restoration is motivated by more than just a concern for biodiversity, however. The 2005 destruction of New Orleans and other Gulf Coast cities by Hurricane Katrina, and to a lesser extent by Hurricane Rita soon after Katrina, was in part a result of development activities destroying the region's wetlands, which had protected the coast from the force of hurricanes. The ensuing natural disaster has become a classic example of the importance of ecosystem services to biological and human communities alike (see Chapter 3). Ironically, the damage that followed these hurricanes had been predicted seven years earlier by the Louisiana Coastal Wetlands Conservation and Restoration Task Force (1998), which had stressed the urgent need for immediate action to restore lost wetlands. Since the hurricanes, restoration projects have begun again, but if they are not adequately funded and large enough in scope, New Orleans will remain vulnerable to another destructive flood. Currently there is an $800 million wetland restoration project on the Mississippi River delta near New Orleans that involves moving vast amounts of sediment; the project is estimated to take at least a decade to complete (AP 2019).

Experience has shown that efforts to restore wetlands often fail to closely match the species composition or hydrologic characteristics of reference sites. The subtleties of species composition, water movement, and soils, as well as the site history, can be too difficult to match. Often the restored wetlands are dominated by exotic, invasive species (Cooper et al. 2017). However, the restored wetlands usually do have some of the wetland plant species, or at least similar ones, and can provide some of the functions of the reference sites (Baldwin et al. 2019). The restored wetlands also have some of the beneficial ecosystem characteristics, such as flood control and pollution reduction, and they are often valuable for wildlife habitat. Additional research into restoration methods may result in further improvement.

Aquatic systems

GLOBAL CHANGE
CONNECTION

Both freshwater and marine systems of all types are subject to degradation by pollution, overexploitation of resources, global warming, and other factors (see Chapters 4 and 5), making them candidates for restoration. Aquatic restoration may also need to deal with problems regarding water chemistry, trophic relationships with exotic species, and physical conditions of the shore or bank. Although aquatic systems are often considered more resilient than terrestrial systems, once damage has become severe, restoration can be more complex.

One of the most common types of damage to lakes and ponds is **cultural eutrophication**, the accumulation of excess nutrients in the water caused by human activity. Signs of eutrophication include an increased prevalence of algal species (particularly surface scums of blue-green algae), decreased water clarity and oxygen content, fish kills, and an eventual increase in the growth of floating plants and other water weeds. In many lakes, the eutrophication process can be reversed by reducing the amounts of mineral nutrients entering the water through better sewage treatment or by diverting polluted water. One of the most dramatic and expensive examples of lake restoration has been the effort to restore Lake Erie (Bullerjahn et al. 2016). Lake Erie was the most polluted of the Great Lakes in the 1950s and 1960s, suffering from deteriorating water quality, extensive algal blooms, oxygen depletion in deeper waters, declining indigenous fish populations, and collapsed commercial fisheries. To address this problem, the governments of the United States and Canada have invested billions of dollars since 1972 in wastewater treatment facilities, reducing the annual discharge of phosphorus into the lake from 15,000 tons in the early 1970s to less than 2000 tons today (Scavia et al. 2019).

> Lake restorations help to improve water quality and restore the original species composition and community structure. Ecological restoration is an important and growing tool for conservation, but the protection of existing biodiversity remains the first priority.

Marine restoration is a younger field than either terrestrial or freshwater restoration but is developing rapidly (Saunders et al. 2020). In a historical review of marine restoration programs across the globe, Megan Saunders and colleagues (2020) found many encouraging examples of success. A number of large-scale projects are ongoing in their restoration of estuaries and bays damaged by human activities, including the Chesapeake Bay in the eastern United States (**FIGURE 11.10**). Chesapeake Bay is one of the most important fishing grounds and recreational areas in the United States. However, pollution from residential, agricultural, and industrial lands bordering the bay has caused a dramatic decline in the water quality, which affects all aspects of biodiversity. The economic consequences of this pollution have also been apparent: harvests of fish and shellfish have declined, and the water has become unsafe for swimming in places. This type of general pollution from an entire landscape is referred to as **nonpoint source pollution**, and it requires a comprehensive restoration approach, as no single source of the pollution can be readily identified and contained. In 1987 the federal, state, and local government bodies responsible for the bay signed an agreement to reduce nutrient and sediment loads coming into the bay by 40%, to be achieved mainly through improving the health of streams and watersheds feeding water.

In 2014, an agreement between federal partners and the governments of six bordering US states committed to restoring native oyster habitat in 10 Chesapeake Bay tributaries by 2025. Representing 2000 acres, it is likely the largest oyster sanctuary restoration ever planned (Bruce et al. 2021). This work is being done through "planting" hatchery-produced oyster larvae on both existing oyster reefs and human-constructed reefs. It was found that unhealthy levels of nitrogen were reduced not only by reducing inputs, but also through nutrient cycling by restored populations of oysters and microorganisms they harbor in the sediment. A recent evaluation found that successfully restored reefs were estimated to remove about seven times as much nitrogen each day than unrestored sand/mud bottom areas (Bruce et al. 2021). Even low densities of oysters could remove nearly 64 kg/ha/y, which translates to an estimated annual removal of more than 9000 kg (~20,000 pounds) of nitrogen in a single tributary, and these rates could be almost tripled in places with high densities of oysters.

(A)

(B)

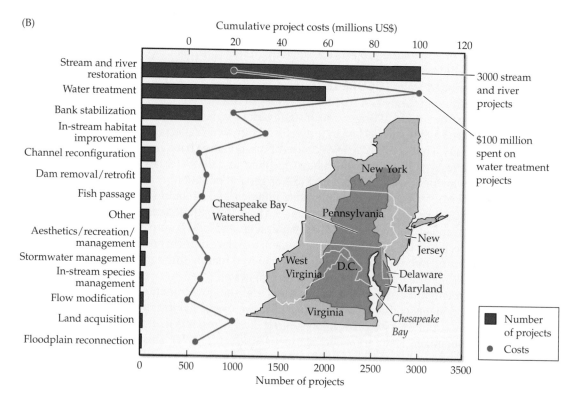

FIGURE 11.10 (A) A variety of measures have been taken to restore the health of the Chesapeake Bay ecosystem. (B) The map shows the watersheds that drain into the bay. The graph shows the cumulative costs for each type of project and the number of projects. Stream and river restoration is the most common type of project, and the most money has been spent on water treatment projects. (B, after B. M. Hassett et al. 2005. *Front Ecol Environ* 3: 259–267. © The Ecological Society of America.)

The largest number of projects involve stream and river restoration, which includes regrading slopes and planting native vegetation. However, the most money has been spent on water treatment projects. Given the capacity of some aquatic vegetation, microorganisms, and animals to filter water, the role of supporting these natural elements of the ecosystem in order to facilitate self-repair should not be overlooked.

Prairies and farmlands

Because they are species-rich, have many beautiful wildflowers, and can be established within a few years, prairies represent ideal subjects for restoration work (Denning and Foster 2018). Innovative thinkers at the University of Wisconsin, Madison, under the leadership of research director Aldo Leopold, were pioneers of prairie restoration in the 1930s (**FIGURE 11.11**). Many techniques have been used in attempts at prairie restoration, but the basic

(A)

(B)

FIGURE 11.11 (A) In the late 1930s, members of the Civilian Conservation Corps (one of the organizations created by President Franklin Roosevelt in order to boost employment during the Great Depression) participated in a University of Wisconsin project to restore the wild species of a midwestern prairie. (B) The prairie as it looks today.

method involves site preparation by shallow plowing, burning, and raking if prairie species are present, or by eliminating all vegetation by plowing or applying herbicides if only exotics are present. Native plant species are then established by transplanting them in prairie sods obtained elsewhere, planting individuals grown from seed, or scattering prairie seeds collected from the wild or from cultivated plants. The simplest method is to gather hay from a native prairie and spread it on the prepared site. Native species are more likely to become established in the absence of fertilizer, which tends to favor nonnative species. (Of course, reestablishing the full range of plant species, soil structure, and invertebrates could take centuries or might never occur.)

A conundrum for restoration ecology, illustrated by work in the prairies, is how to determine the target state of the ecosystem. Humans have had an impact on many ecosystems for centuries or even millennia; for example, early humans hunted many North American mammals to extinction more than 12,000 years ago. Should North American grasslands be restored to a state resembling those that existed either before European colonization (a few hundred years ago) or before human colonization (more than 12,000 years ago)?

One of the most ambitious proposed restorations involves recreating a short-grass prairie ecosystem, or "buffalo commons," on about 380,000 km^2 of the Great Plains states, from the Dakotas to Texas and from Wyoming to Nebraska (Davenport 2018). Some of this land is currently used for environmentally damaging and often unprofitable agriculture and grazing supported by government subsidies. The human population of this region is declining as farmers and townspeople go out of business and young people move away. From the ecological, sociological, and economic perspectives, the best long-term use of much of this region might be as a restored prairie ecosystem. The human population of the region could stabilize around nondamaging core industries such as tourism, wildlife management, and low-level grazing by cattle and bison, leaving only the best lands in agriculture. As mentioned in Section 11.3, some have argued for a process of North American rewilding whereby large game animals from Africa and Asia, such as elephants, cheetahs, camels, and even lions, would be released in an attempt to re-create the types of ecological interactions that occurred in North America before humans arrived on the continent (Hayward 2009). Both these proposed projects are controversial because many of the farmers and ranchers in the region want to continue their present way of life without alteration, and they tend to be highly resentful of unwanted advice and interference from scientists and the government. The projects are also controversial because of the proposed release of nonnative mammals in North American ecosystems.

Tropical dry forest in Costa Rica

An exciting restoration process has been ongoing for over 35 years in northwestern Costa Rica. The tropical dry forests of Central America have long suffered from large-scale conversion to cattle ranches and farms. This destruction has gone largely unnoticed, as international scientific and public attention has focused on more glamorous rain forests elsewhere in South and

Central America (e.g., see Chapters 2 and 4). In response to this problem, ecologists Dan Janzen and Winnie Hallwachs proposed the protection of 10,400 ha of dry forest in 1986, which was thereafter incorporated as a protected area by the Costa Rican government. An international effort followed to restore the biology and cultural connectivity of degraded forest in what is now known as the Área de Conservación Guanacaste (ACG) (ACG 2021; Janzen and Hallwachs 2021) **(FIGURE 11.12)**. Today, the ACG is a 163,000 ha wilderness protected area that contains four of the five main ecosystems of the tropics—marine/coastal, dry forest, cloud forest, and rain forest—making it unique among protected areas in the Americas. Over the years, it has received support from nine countries, 50 international organizations, and more than 10,000 private donors; some of this funding has been specifically for restoration ecology projects and other ecological research. The importance of the site for biodiversity cannot be overstated; it has been estimated that there are more terrestrial species in the ACG than all those that exist in northern Mexico, the United States, and Canada combined (ACG 2021).

(A)

(B)

Photographs by Brad Zlotnick

(C)

FIGURE 11.12 The Área de Conservación Guanacaste (ACG) is an experiment in restoration ecology—an attempt to restore the devastated and fragmented tropical dry forest of Costa Rica. (A) A barren grassland with scattered forest fragments was heavily grazed by cattle and frequently burned. (B) Native trees and other species became established once again in this young forest after 17 years without cattle and fire. Note the person in the lower left for scale. (C) Public education, including both adults and schoolchildren, is an important part of the Programa de Educación Biológica (PEB) at the ACG.

The effort to restore marginal ranchlands, low-quality farms, and forest fragments at the ACG has prioritized eliminating fires started by people and stopping logging and hunting. Although planting trees was incorporated at the beginning, it was discovered that simply reducing disturbances by people allowed the forest to naturally regenerate (Hulshof and Powers 2020). It was also found that restoration of the dry forest required protection of the surrounding rain forest, motivating the addition of more than 5000 ha (SER 2021). These activities have successfully converted tens of thousands of hectares of pastures and old fields to a species-rich, dense young forest with abundant and growing populations of native animals. It is anticipated that within 200 years the entire area will be fully recovered (ACG 2021).

A key element in this restoration plan is what has been termed **biocultural restoration**, meaning that ACG staff members teach basic biology and ecology on-site to thousands of students in grades 4 through 6 from dozens of neighboring schools and also give presentations to citizen groups, all as part of the ACG core mission in what is called the Programa de Educación Biológica (PEB) (**FIGURE 11.12C**). Due to its importance, this educational work even continued during much of the COVID-19 pandemic, albeit with masks and social distancing. These programs, combined with the fact that all members of ACG's staff and administration are resident Costa Ricans, has resulted in residents viewing ACG as if it were a large ranch producing "wildland resources" for the community rather than an exclusionary "national park."

11.6 The Future of Restoration Ecology

LEARNING OBJECTIVE

By the end of this section you should be able to:

11.6.1 Consider why complete restoration may not be a focus in coming years.

E. O. Wilson predicted in 1992, "Here is the means to end the great extinction spasm. The next century will, I believe, be the era of restoration in ecology." He may be right. Restoration ecology is an evolving and rapidly growing discipline, with its own scientific society (the Society for Ecological Restoration) and increasing numbers of journals: *Restoration Ecology, Ecological Restoration, Ecological Engineering: The Journal of Ecosystem Restoration, Ecological Management and Restoration,* and others. Ecosystems are being restored using methods developed by the discipline, books are being written about the subject, and new courses are being taught at more universities. Scientists are increasingly able to make use of the growing range of published studies and suggest improvements in restoration techniques. At its best, restored land can provide new opportunities for protecting biodiversity and generating an appreciation of nature.

Conservation biologists in this field must take care to ensure that restoration efforts are not simply public relations endeavors taken by environmentally damaging corporations that are only interested in continuing to conduct "business as usual." A 5 ha "demonstration" or "best practices" project in a highly visible location does not compensate for thousands or tens of thousands of hectares damaged elsewhere and should not be accepted as adequate by conservation biologists. Attempts to mitigate or offset the destruction of an intact biological community by the building of a similar species assemblage at a new location are almost certainly not going to provide homes for the same species or provide similar ecosystem functions (Moreno-Mateos et al. 2015a). However, new generations of restoration practitioners will need to find a compromise between the almost impossible complete recovery of pristine ecosystems and a more pragmatic pursuit of functional systems that provide some key ecosystem services to society. This will require better information sharing between managers and scientists, humility, objectivity, and innovation (Clark et al. 2019) and more effective restoration based on a systematic evaluation of monitoring and results of previous projects (Cooke et al. 2018). Also, the Society for Ecological Restoration and other related societies must continue working to reach multinational corporations, large organizations, and government departments that can implement change and restoration at larger scales (Perring et al. 2018).

The best long-term strategy remains protecting and managing biological communities where they are found naturally; many researchers argue that the requirements for the long-term survival of all species are most likely to be found there, and the protection of intact systems is generally cheaper and easier than repairing systems that have been degraded. In addition, we need to consider restoring ecosystems in anticipation of the impacts of climate change (see Chapter 5) (Cook-Patton et al. 2020). There are many technical, scientific, logistic, and economic challenges for future restoration ecologists and managers to address in our endeavor to repair the damage we have done as a species.

Summary

- Ecological restoration is the practice of reestablishing populations, ecosystems, and landscapes that include degraded, damaged, or even destroyed habitat. Restoration ecology provides methods for reestablishing species, whole biological communities, and ecosystem functions in degraded habitat.

- The establishment of new communities on degraded or abandoned sites provides an opportunity to enhance biodiversity and can improve the quality of life for the people living in the area. Restoration ecology can also provide opportunities for scientists to learn more about ecological processes and for the public to be involved in conservation efforts.

- Restoration projects begin by eliminating or neutralizing factors that prevent the system from recovering. Then some combination of site preparation, habitat management, and reintroduction of original species gradually allows the community to regain the species and ecosystem characteristics of designated reference sites. Attempts to restore habitat need to be monitored to determine whether they are reestablishing the composition of historical species and the functions of the ecosystem.

■ Biological control is a tool whereby organisms such as insects can be used to remove invasive species. Bioremediation uses organisms such as bacteria to clean up pollutants.

■ In some cases, restoration to a former state is impractical or impossible due to the nature or extent of the degradation, the presence of invasive species, or climate change. In such cases, a novel ecosystem that may have some of the same functionality of the original one may be considered, but it should not be valued over the native ecosystem.

■ Creating new habitat to replace lost habitat elsewhere, which is known as compensatory mitigation or biological offsetting, has value but should be regarded as only part of an overall conservation strategy that includes the protection of species and ecosystems where they naturally occur.

For Discussion

1. Restoration ecologists are improving their ability to restore biological communities. Does this mean that biological communities can be moved around the landscape and positioned in convenient places that do not inhibit the further expansion of human activities?

2. What methods and techniques could you use to monitor and evaluate the success of a restoration project? What timescale would you suggest using?

3. What do you think are some of the easiest ecosystems to restore? The most difficult? Why?

4. Aldo Leopold encouraged humans to "keep every cog and wheel" in order to maintain healthy ecosystems. Is it necessary, or even possible, to return every missing species back to a restored ecosystem?

Suggested Readings

Alston, J. M., et al. 2019. Reciprocity in restoration ecology: When might large carnivore reintroduction restore ecosystems? *Biological Conservation* 234: 82–89. Rewilding with predators does not guarantee impacts on herbivores.

Hobbs, R. J. 2018. Restoration ecology's silver jubilee: Innovation, debate, and creating a future for restoration ecology. *Restoration Ecology* 26(5): 801–805. An argument in favor of a more practical and less idealistic approach.

Hulshof, C. M., and J. S. Powers. 2020. Tropical forest composition and function across space and time: Insights from diverse gradients in Área de Conservación Guanacaste. *Biotropica* 52(6): 1065–1075. The ACG restoration has provided a great deal of information for the field of ecology.

Pugh, T. A., et al. 2019. Role of forest regrowth in global carbon sink dynamics. *Proceedings of the National Academy of Sciences USA* 116(10): 4382–4387. Restored forests absorb even more carbon from the atmosphere than old-growth forests.

Santos, L. N., et al. 2019. Reconciliation ecology in Neotropical reservoirs: Can fishing help to mitigate the impacts of invasive fishes on native populations? *Hydrobiologia* 826(1): 183–193. Sometimes actions that help humans also support biodiversity.

Saunders, M. I., et al. 2020. Bright spots in coastal marine ecosystem restoration. *Current Biology* 30(24): R1500–R1510. An optimistic review of decades of research, failures, and successes.

Sher, A. A., et al. 2020. The human element of restoration success: Manager characteristics affect vegetation recovery following invasive *Tamarix* control. *Wetlands* 40(6): 1877–1895. Characteristics of the people and organizations involved in restoration can have surprisingly large effects on outcomes.

Fishermen in Oman benefit from a collaborative effort of the Ministry of Agriculture and Fisheries Wealth, the World Bank, and fishery stakeholders to make fishing ecologically sustainable and profitable again.

The Challenges of Sustainable Development

12

Increasingly, many conservation biologists, policymakers, and land managers are recognizing the need for sustainable development—economic development that satisfies both present and future needs for resources and employment while minimizing the impact on biodiversity and functioning ecosystems (Selomane et al. 2015). **Sustainable development**, a term sometimes used interchangeably with *sustainability*, can be contrasted with more-typical development that is unsustainable. Unsustainable development cannot continue indefinitely, because it destroys or uses up the resources on which it depends. As defined by some environmental economists, **economic development** implies improvements in the efficiency, organization, and distribution of resource use or other economic activity but not necessarily increases in resource consumption. Economic development is clearly distinguished from **economic growth**, which is defined as material increases in the amount of resources used. Sustainable development is a useful and important concept in conservation biology because it emphasizes improving current economic development and limiting unsustainable economic growth.

But will sustainable development actually help stop species extinctions? To find out, Leclère and colleagues (2020) used several different models to create future projections of conservation metrics such as species abundance and land conversion based on different scenarios of sustainable development. They found that while increased conservation measures alone could make a positive difference, an approach that combined conservation with increased sustainability in both supply and demand of agricultural products could effectively reverse our trend of terrestrial wildlife losses (**FIGURE 12.1**). While the different models they tested did not agree about how much improvement there would be, all of them predicted that we could return to above 2010 levels of wildlife abundance by implementing these practices. Perhaps the even more important lesson was the trajectory of loss if we do not make sustainable development a priority.

> The goal of sustainable economic development is to provide for the current and future needs of human society while at the same time protecting species, ecosystems, and other aspects of biodiversity.

Sustainability can be implemented in many different spheres of an economy, and it can overlap directly with conservation efforts. For example, investment in national park infrastructure that improves the protection of biodiversity and provides revenue opportunities for local communities is sustainable development, as is the implementation of less destructive logging and fishing practices. Especially in developing countries, sustainability is tightly intertwined with issues of social justice; one cannot have sustainable development without ensuring that all parties benefit. This is true in part because environmental degradation disproportionately affects the poor. It has been argued that the combination of environmental, economic, and social considerations explicit in sustainable development will inevitably lead to improved quality of life, corporate profits, and human health in addition to environmental benefits (Dernbach and Cheever 2015).

Unfortunately, the term *sustainable development* can also be misappropriated. Few politicians or businesses are willing to proclaim themselves to be against sustainable development. Thus, many large corporations, and the policy organizations that they fund, misuse the notion of sustainable development to "greenwash" their industrial activities, with only limited change in actual practice. In these cases, both ecosystems and people will likely suffer.

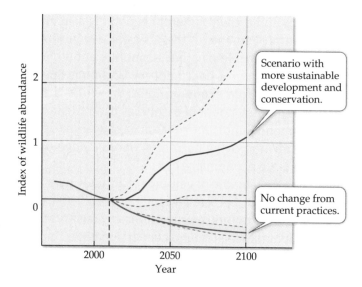

FIGURE 12.1 Models of past versus future scenarios show dramatically different results depending on what humans choose to do. This graph shows wildlife population densities as measured by an index that uses the year 2010 as the baseline, comparing future projections if we change nothing (in red) as compared with what is possible if we have more sustainable development (in blue). In this model, sustainable development includes the following: sustainable increased crop yields, increased trade of agricultural goods, reduced waste of agricultural goods, a diet shift to less meat, increased management of protected areas, and increased restoration and large-scale conservation planning. Shaded area represents the confidence interval of the model. (After D. Leclère et al. 2020. *Nature* 585: 551–556. doi.org/10.1038/s41586-020-2705-y.)

For instance, a plan to establish a huge mining complex in the middle of a forest wilderness cannot justifiably be called sustainable development simply because a small percentage of the land area is set aside as a park. Not only will precious habitat be lost, but local inhabitants who depend on resources from that forest will be impacted as well. Waste from the mine can poison fish and people alike. Similarly, building huge houses filled with "energy-efficient" appliances, and driving cars that boast the latest energy-saving technology but are routinely driven long distances, cannot really be called sustainable development or "green technology" when the net result is increased energy use. Alternatively, some people champion the opposite extreme, claiming that sustainable development means that vast areas must be kept off-limits to all development and should remain as, or be allowed to return to, wilderness. This may not be the best option for either conservation goals or people: some places require active restoration (see Chapter 11) or active management (see Chapters 9 and 10) to best protect biodiversity; barring all people will inevitably harm local interests.

The primary conflict of sustainable development is often not between people and nature so much as it is between the powerful and the vulnerable. As with all such disputes, informed scientists and citizens must study the issues carefully, identify which groups are advocating which positions and why, and then make careful decisions that best meet both the needs of human society and the protection of biodiversity and ecosystems. In many cases this involves compromise, and in most cases compromises form the basis of government policy and laws, with conflicts resolved by government agencies and the courts.

12.1 Sustainable Development at the Local Level

LEARNING OBJECTIVES

By the end of this section you should be able to:

12.1.1 Evaluate a conservation policy from your area, identifying ways that it could be improved.

12.1.2 Contrast the strategies of conservation leasing, conservation banking, conservation development, payments for ecosystem services, and conservation concessions.

Most efforts to find approaches that promote both the preservation of species and habitats and the needs of society rely on initiatives from concerned citizens, conservation organizations, and government officials. These efforts may take many forms, but they begin with individual and group decisions to prevent the destruction of habitats and species in order to preserve things of perceived economic, cultural, biological, scientific, or recreational value. The results of these initiatives often end up codified into environmental regulations or laws.

Local and regional conservation regulations

In modern societies, local (city and town) and regional (county, state, and provincial) governments pass laws to provide protection for species and habitats while at the same time allowing development for the continued needs of society. Often, but not always, these local and regional laws are comparable to, or stricter than, national laws, particularly for protections of clean water and air and, less often, endangered species. Such laws are passed because citizens and political leaders feel that they represent the will of the majority and provide long-term benefits to society. Important local laws typically govern when and where hunting and fishing can occur (**FIGURE 12.2**); the size, number, and species of animals that can be taken; and the types of weapons, traps, and other equipment that can be used. Restrictions are enforced through licensing requirements and patrols by game wardens and police. In some settled and protected areas, hunting and fishing are banned entirely. Similar laws affect the harvesting of plants, seaweed, and shellfish. Certification of the origin of biological products may be required to ensure that wild populations are not depleted by illegal collection or harvest. These restrictions have long applied to certain animals such as trout and deer and to plants of horticultural interest such as orchids, azaleas, and cacti. More recently, there are certification of origin programs for ornamental fish, timber, and other products.

Laws that control the ways in which land is used are another means of protecting biodiversity (Reed et al. 2014). For example, on a more local scale, vehicles and even people on foot may be restricted from habitats and resources that are sensitive to damage, such as birds' nesting areas, bogs, sand dunes, wildflower patches, and sources of drinking water. Uncontrolled fires may severely damage habitats, so practices, such as campfires, that contribute to accidental fires are often rigidly controlled. Zoning laws, among the strongest and most widely used restrictions, sometimes prevent construction in sensitive areas such as barrier beaches and floodplains. Even where development is

FIGURE 12.2
The 350 Colorado State Wildlife Areas cover more than 276,000 ha and are managed for hunting, fishing, hiking, and wildlife preservation.

permitted, building permits are often reviewed carefully to ensure that damage is not done to endangered species or ecosystems, particularly wetlands. For major regional and national projects, such as dams, canals, mining and smelting operations, oil extraction, and highway construction, environmental impact statements must be prepared that describe the damage that such projects will or could possibly cause so that these projects can be conducted in a more environmentally sensitive manner.

One of the most powerful strategies in protecting biodiversity at the local and regional levels is the designation of intact biological communities as nature reserves, conservation land, and state and provincial parks and forests (see Chapter 9 and Figure 12.2). Government bodies buy land and establish protected areas for various uses—local parks for recreation, conservation areas to maintain biodiversity, forests for timber production and other uses, and watersheds to protect water supplies.

The passage and enforcement of conservation-related laws on a local level can become an emotional experience that divides a community and can even lead to violence. To avoid such counterproductive outcomes, conservationists must be able to convince the public that using resources in a thoughtful and sustainable manner creates the greatest long-term benefit for the community. The general public must be made to look beyond the immediate benefits that can come with the rapid and destructive exploitation of resources. For example, towns often need to restrict development in watershed areas to protect water supplies; this may mean that houses and businesses are not built in these sensitive areas and landowners may have to be compensated for these lost opportunities. It is essential that conservation biologists clearly communicate the reasons for restrictions that protect biodiversity and ecosystems. Those affected by the restrictions can become allies in the protection of resources if they understand the importance and long-term benefits of reduced access. These people must be kept informed and consulted throughout the decision-making process. Conservation biologists must develop the necessary skills, combined with the best science, to negotiate and compromise; to encourage conservation actions in others and understand their perspectives; and to explain positions, regulations, and restrictions. These skills are part of the growing field of conservation psychology (Clayton and Myers 2015). Having a fervent belief in one's cause is no longer enough.

Land trusts and related strategies

In many countries, nonprofit, private conservation organizations are among the leaders in acquiring land for conservation (Hardy et al. 2018b). In the United States, over 22 million ha of land are protected at the local level by 1363 **land trusts**, which are private, nonprofit corporations established to protect land and natural resources (Land Trust Alliance 2021). Land trusts are common in Europe: In the Netherlands, about half the protected areas are privately owned. In Britain, the National Trust has nearly 6 million members and owns about 250,000 ha of land, much of it farmland, including 780 miles of coastline and 500 "heritage" properties that include sites of historical

Land trusts are private conservation organizations that purchase and protect land. Conservation easements and limited development agreements are also used by land trusts to increase the amount of land under protection.

or social importance. The Royal Society for the Protection of Birds (RSPB) has more than 1 million members and manages 200 reserves with an area of almost 130,000 ha. A major emphasis of many of these reserves is nature conservation and education, often linked to school programs.

It is also common for land trusts such as the RSPB to collaborate with scientists or even have researchers on their staff, to monitor and evaluate the state of the land and its biodiversity and also to capitalize on the opportunity to do basic science there. The United Kingdom publishes a *State of Nature* report that summarizes information from species surveys and other research on human impacts, much of which is collected on land trust properties (State of Nature Partnership 2019).

In the United States, the largest land trust is The Nature Conservancy (TNC; TNC.org), which protects land not only by buying it but also by assisting in a number of different roles, including transferring land to local governments and organizations that will protect it (**FIGURE 12.3**). TNC now has over a million members and 400 scientists on staff, impacting conservation in 72 countries. They help protect over 50.5 million ha of land.

While the simple purchase of land may seem the most straightforward approach to conservation, in practice, property law can be quite complex and

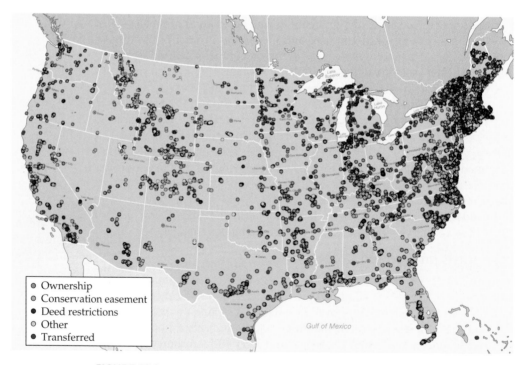

FIGURE 12.3 Land trusts may own and manage land or may give it to local or national governments in special agreements. The Nature Conservancy is currently the largest land trust organization in North America. This map shows the many ways and locations that TNC helps protect habitat for biodiversity. (After The Nature Conservancy, http://www.tnclands.tnc.org/.)

may not always be the best option for conservation of species. Fortunately, there are several approaches used by land trusts and governments to protect land, described later in this chapter.

In addition to purchasing land outright, both governments and conservation organizations such as TNC protect land through **conservation easements**, also called **conservation covenants**, in which landowners give up the right to develop, build on, or subdivide their property, typically in exchange for a sum of money, lower real estate taxes, or some other tax benefit (Graves et al. 2019). Conservation easements can have a variety of goals that may or may not explicitly include protection of species and/or habitats and that have important implications for how they are managed and the resulting effectiveness for conservation (**FIGURE 12.4**). For many landowners, accepting a conservation easement is an attractive option: they receive a financial gain while still owning their land and are able to feel that they are assisting conservation objectives. In general, landowners are most willing to consider conservation easements when they are well paid; the agreement is short-term, lasting just a few years, rather than permanent; and there is no legal contract involved (Sorice et al. 2013). Of course, the offer of lower taxes or money is not always necessary; many landowners will voluntarily accept conservation restrictions without compensation.

> Conservation lands require ongoing management and continuous vigilance, but they often provide extensive benefits to the local society and economy.

A related approach is the use of "revolving funds": conservation organizations purchase the land and then sell it to a conservation-minded owner with a conservation easement or covenant in place to protect that land in perpetuity. Several countries have policies in place to make this possible, including Australia, Canada, Chile, and the United States. These countries were found to have used US$384 million in revolving funds to protect over 684,000 ha (Hardy et al. 2018).

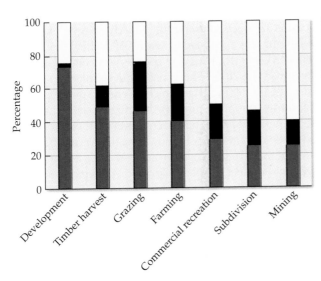

FIGURE 12.4   Conservation easements are intended to protect land that might otherwise be developed; however, a large proportion of them do allow some forms of development and other uses. These figures reflect a review of 269 conservation easement documents that spanned six US states. When activities were not specified, they were likely to be allowed by default. (After A. R. Rissman et al. 2015. *Conserv Lett* 8(1): 68–76. © 2014 The Authors.)

Another strategy that land trusts and local and regional governments use is **limited development**, also known as **conservation development** (Mockrin et al. 2017). In these situations, a landowner, a property developer, and a government agency and/or conservation organization reach a compromise that allows part of the land to be commercially developed while the remainder is protected by a conservation easement. Limited development allows the construction of necessary buildings and other infrastructure for an expanding human society; the projects are often successful precisely because being located adjacent to conservation land enhances the value of the developed land.

Governments and conservation organizations can further encourage conservation on private lands through other mechanisms, including compensating private landowners for desisting from some activity that damages the environment and for implementing conservation activity (Knoot et al. 2015). **Conservation leasing** involves providing payments to private landowners who actively manage their land for biodiversity protection. Tax deductions and payments can also be obtained for any costs of restoration or management, including weeding, controlled burning, establishing nest holes in trees for wildlife, and planting native species. In some cases, private landowners may still be allowed to develop their land later, even if endangered species come to live there.

A related idea is **conservation banking**, in which a landowner deliberately preserves an endangered species or a protected habitat type, such as wetlands, or even restores degraded habitat and creates new habitat. A conservation bank is like a financial bank in that it is intended to be a stable protector, but instead of money, it protects natural resource values. The US Fish and Wildlife Service (USFWS) evaluates proposals for new conservation banks for situations where negative impacts on listed species are anticipated. Once approved, the USFWS awards a landowner habitat or species credits in exchange for permanently protecting the land and managing it for these species. These credits can then be purchased by developers in compensation or as a biodiversity offset for a similar habitat that is being destroyed elsewhere by a construction project (Apostolopoulou and Adams 2017). For example, the Muddy Boggy Conservation Bank in eastern Oklahoma was expanded to nearly 230 ha in exchange for 522 credits for the endangered American burying beetle (*Nicrophorus americanus*; see Figure 6.10) (Business Wire 2015). This conservation bank was specifically established to assist Oklahoma industries such as energy development, pipeline construction, and transportation projects that might have otherwise been shut down if the endangered species had been found on their land. The industry can be relieved of the responsibility for how its actions may affect the beetle by purchasing credits that are then used to support the Muddy Boggy Conservation Bank's ability to protect the beetle on protected land.

A further conservation strategy is a **payments for ecosystem services** (PES) program (see Chapter 10), in which a landowner is paid for providing specific conservation services (**FIGURE 12.5**) (Chapman et al. 2020). Utility companies may also gain carbon credits by paying for habitat protection (e.g., paying a landowner for not cutting down a forest) and restoration (e.g., paying a landowner for planting trees and establishing a new forest); these carbon credits are then used to offset the carbon emissions produced through the burning

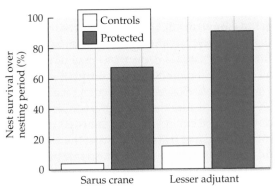

FIGURE 12.5 Fledgling success for nests of the sarus crane (*Grus antigone*) and the lesser adjutant (*Leptoptilos javanicus*) that were protected by villagers in the northern plains of Cambodia versus those that were not. Villagers participating in the program were paid by the Wildlife Conservation Society, an international conservation organization based in the United States. The local inhabitants were able to significantly supplement their incomes with payments that provided extra incentives for achieving successful nests. (After T. H. Clements et al. 2013. *Biol Conserv* 157: 50–59. Copyright 2012. Reprinted with permission from Elsevier.)

of fossil fuels. On a larger scale, carbon offset payments by governments and international corporations can be used to compensate for emissions of greenhouse gases (Gonçalves and Anselmi 2018). PES programs are also used to pay landowners for protecting the vegetation on their property, which is important for downstream flood control and drinking water supply.

Conservation concessions are an innovative approach in which conservation organizations outbid extractive industries such as logging companies, not for ownership of the land, but for the rights to use and protect it. The government or large landowner receives the same annual income from a conservation organization that would have been paid by a logging company, and the animals and plants of the area are protected rather than destroyed.

Enforcement and public benefits

The conservation measures described in this section and elsewhere in this book must be continuously monitored to make sure that regulations and laws are enforced and that agreements are being carried out, particularly in cases where destruction cannot be easily reversed. In one scenario, a developer may agree to limit the amount of development and conserve an area of forest but then obtain construction permits, ignore the agreement, and clear all the trees. By the time action can be taken to stop the developer, the trees and the habitat they provided are gone and cannot be easily replaced. Even if sanctions such as fines or forfeiting of bonds are imposed, the developer may feel that the potential profits outweigh such considerations, and managers and officials usually take a "what's done is done" approach and allow the cleared land to be developed. Conservation workers need to raise awareness so that the public and the judicial system view "breach of promise" against the environment with the same seriousness as similar crimes against personal property.

The most powerful tool for enforcement has been fining violators, especially when these funds are directly applied to conservation, such as the restoration of damaged land. When the terms of a conservation easement are broken or other environmental harm is done, there may be a restoration remedy; that is, the violator may be required to pay to return the degraded property to as close to its original state as possible. For example, if a land trust is obligated to preserve a stand of trees under the terms of a conservation easement but then allows those trees to be cut down, the land trust may be obligated to replant trees and nurture them to maturity at a cost that well exceeds that of the land itself.

Public perception can also be a source of problems. Local efforts by land trusts to protect land are sometimes criticized as being elitist because they provide tax breaks only to individuals wealthy enough to take advantage of them while decreasing the revenue collected from land and property taxes. Other analysts argue that land used in other ways, such as for agriculture or commercial activity, is more productive. Although land in trust may initially yield lower tax revenues, the loss is often offset by the increased value and consequently increased property taxes of houses and land adjacent to the conservation area. In addition, the employment, recreational activities, tourist spending, and research projects associated with nature reserves and other protected areas generate revenue throughout the local economy, which benefits local residents (Di Minin et al. 2013). Finally, by preserving important features of the landscape and natural communities, local nature reserves also protect and enhance the cultural heritage of the local society, a consideration that must be valued if sustainable development is to be achieved. For example, if a nature reserve contains a feature or space sacred to the local people, visitors must be excluded or instructed in how to act respectfully (**FIGURE 12.6**).

(A)

Photo by Pseudopanax

(B)

Photo by Anna Sher

FIGURE 12.6 (A) Te Waikoropupū Springs are located in Golden Bay, in New Zealand's South Island; they have cultural significance to the Māori people, who believe that both the spiritual and physical survival of all living things is dependent on the protection of the *mauri*, or life force, of the springs. (B) The sign says in Māori, "We welcome our visitors to the sacred Waikoropūpū Springs, welcome to the lands of the local iwi Ngāti Tama, Te Ātiawa, Ngāti Rārua." The New Zealand Department of Conservation has placed extensive interpretation at the entrance that explains the importance of the site and instructs visitors not to drink or even touch the water. One of these panels states, "Te Waikoropūpū provides our iwi with a spiritual and physical link to our tūpuna (ancestors). As a collective, we operate under a whakaruruhau (umbrella) entity called Manawhenua ki Mōhua to ensure our taonga tuku iho (treasures from our ancestors) is protected. Our role as Kaitiaki is to uphold the mana and maintain the wairua and the mauri of our taonga (treasure) and to ensure its integrity is protected for future generations."

12.2 Conservation at the National Level

LEARNING OBJECTIVE

By the end of this section you should be able to:

12.2.1 Describe the variety of approaches used by national governments to protect biodiversity.

Throughout the modern world, national governments play a leading role in conservation activities. The level of a government's conservation actions can substantially affect the conservation outcomes within its borders. Conflicts between government officials and local people during conservation implementation, and lack of government funding are two significant problems faced by parties actively working to reestablish endangered species (Crees et al. 2016). Conservation biologists contribute to solving these problems by providing government officials with key information on threats to biodiversity and the local situation on the ground, with the hope and expectation that resulting laws, regulations, and management actions will be used to protect biodiversity.

Similar to local and regional governments, national governments can use their revenues and authority to buy land for conservation. In the United States, special funding mechanisms exist at the national level, such as the Lands Legacy Initiative and the Land and Water Conservation Fund, to purchase land for conservation purposes. The establishment of national parks is a particularly important conservation strategy (see Chapter 9). National parks are the single largest source of protected lands in many countries. For example, in Costa Rica, 26% of the nation's land area is protected, with nearly half in national parks (Costa Rica National Parks 2021). Outside the protected areas, deforestation is occurring rapidly, and soon national parks may represent the only undisturbed habitat and source of natural products, such as timber, in the whole country. The US National Park Service protects about 34 million ha with 423 sites. The US government also protects biodiversity in more than 567 National Wildlife Refuges covering 324 million ha (95% of which is aquatic), including 63 with wilderness areas, and the US Bureau of Land Management's National Landscape Conservation System covers 33 million ha, including National Conservation Areas, National Monuments, and wilderness areas.

National legislatures and governing agencies are the principal bodies that develop policies that regulate environmental pollution. Laws are passed by legislatures and then implemented in the form of regulations imposed by government agencies. Laws and regulations affecting air emissions, sewage treatment, waste dumping, and the development of wetlands are often enacted to protect human health and property and resources such as drinking water, forests, and commercial and sport fisheries. The level of enforcement of these laws demonstrates a nation's

> National governments protect designated endangered species within their borders, establish national parks, and enforce legislation on environmental protection.

determination to protect the health of its citizens and the integrity of its natural resources. At the same time, these laws protect biological communities that would otherwise be destroyed by pollution and other human activities. The air pollution that exacerbates human respiratory disease, for instance, also damages commercial forests and biological communities, and pollution that ruins drinking water also kills terrestrial and aquatic species such as turtles, fish, and aquatic plants.

National governments can also have a substantial effect on the protection of biodiversity through the control of their borders, ports, industry, and commerce. To protect forests and regulate their use, governments can ban logging, as was done in Thailand following disastrous flooding; they can restrict the export of logs, as was done in Indonesia; and they can penalize timber companies that damage the environment. Certain kinds of environmentally destructive mining can be banned. Methods of shipping oil and toxic chemicals can be regulated. Conservation biologists can provide government officials with key information for developing the needed policy framework, and resource managers and others can then use the resulting laws and regulations to protect biodiversity.

Although many countries have enacted legislation to protect endangered species, forests, wetlands, and other aspects of biodiversity, national governments are sometimes unresponsive to requests from conservation groups to protect the environment. Governments have even acted to remove the legal protected status of national parks, sacred forests, and other conservation areas (degazettement; see Chapter 9) in order to facilitate the extraction of natural resources and economic development (Hardy et al. 2017). Governments sometimes do this because they feel that the needs of the broader regional and national society for natural resources and economic development are more important than the needs of the local people, the environment, or the ecosystem services they may lose. There are also cases where the downgrading, downsizing, and degazettement of protected areas are linked to certain government officials who profit personally from their actions. In some cases, national governments recognize that local people are best able to protect ecosystems close to where they live and have relinquished control of these resources to local governments, village councils, and conservation organizations.

12.3 International Approaches to Sustainable Development

LEARNING OBJECTIVES

By the end of this section you should be able to:

12.3.1 Identify the major accomplishments of international agreements to protect biodiversity.

12.3.2 Explain why countries need to act together to protect biodiversity.

In June of 2021, the leaders of the largest and wealthiest democratic countries met to discuss economic issues and international policy, in advance of the international climate conference to be held later that year. In the G7 (Group of Seven) Summit, the heads of Britain, Canada, France, Germany, Italy, Japan, and the United States joined leaders of other European Union countries, Australia, India, South Africa, and South Korea to make highly influential decisions for the global economy. The discussions in this and other international conferences necessarily involve natural resources and threats to them, including climate change. As we learned in Chapter 3, most major economic decisions both affect and are affected by biodiversity.

GLOBAL CHANGE
CONNECTION

Although about half of the world's wealth is controlled by the Western countries that make up the G7, the biological diversity needed for humanity's future well-being is actually concentrated in the tropical countries of the developing world, most of which are relatively poor and experiencing rapid rates of population growth, development, and habitat destruction (see Figure 2.18). Developing countries may be willing to preserve biodiversity, but they are often unable to pay for the habitat preservation, research, and management required for the task. Thus, it is imperative that there be communication and collaboration among all countries, for the benefit of all.

Furthermore, the newly industrializing countries must ensure that their development efforts do not have severely negative impacts on biodiversity; for example, China's Belt and Road Initiative will expand infrastructure, including railways and energy pipelines (Ascensão et al. 2018). Such development is likely to dramatically reduce and fragment habitat; according to a WWF report, the proposed trade corridors intersect 1,739 Important Bird Areas or Key Biodiversity Areas and 46 biodiversity hotspots or Global 200 ecoregions (WWF 2017). The developed countries of the world (including the United States, Canada, Japan, Australia, and many European nations) and newly industrializing countries, such as China, must watch the impacts of their own development while working with developing countries to preserve the biodiversity needed by the world as a whole.

> International cooperation and agreements to protect biodiversity are needed for migratory species and for occasions when threats occur across countries.

While the major legal and policing mechanisms that currently exist in the world are based within individual countries, international cooperation to protect biodiversity is a requirement for several reasons:

- *Species migrate across international borders.* Conservation efforts must protect species at all points in their ranges; efforts in one country will be ineffective if critical habitats are destroyed in a second country to which an animal migrates (Ripple et al. 2014). For example, efforts to protect migratory bird species in northern Europe will not work if the birds' overwintering habitat in Africa is destroyed.

- *International trade in biological products is commonplace.* A strong demand for a product in one country can result in the overexploitation of the species in another country to supply this demand (see Chapter 5).

- *Biodiversity provides internationally important benefits.* The community of nations benefits from the species and genetic variation used in agriculture, medicine, and industry; the ecosystems that help regulate climate; and the national parks and other protected areas of international scientific and tourist value. These benefits have been estimated to be in the trillions of dollars (McCarthy et al. 2012).

- *Many environmental pollution problems that threaten ecosystems are international in scope.* Such threats include atmospheric pollution and acid rain; the pollution of lakes, rivers, and oceans (see Chapter 4); greenhouse gas production and global climate change; and ozone depletion (Lin et al. 2014) (see Chapter 5).

International Earth summits

Given these realities, there is motivation to make progress on conservation issues at international meetings such as the G7 Summit. Historically, the most important of these meetings that have considered environmental issues in the context of sustainable development are the following:

- United Nations Conference on the Human Environment (1972)
- World Commission on Environment and Development (1987)
- United Nations Conference on Environment and Development (1992)
- General Assembly Special Session on the Environment (1997)
- World Summit on Sustainable Development (2002)
- United Nations Conference on Sustainable Development ("Rio + 20"; 2012)
- Paris Climate Conference (2015)

GLOBAL CHANGE
CONNECTION

Several of these conferences have resulted in significant international agreements, often termed "conventions," that provide frameworks for countries to cooperate in protecting species, habitats, ecosystem processes, and genetic variation. Treaties that are negotiated at international conferences come into force when they are ratified by a certain number of countries and then implemented and enforced at the national level (**FIGURE 12.7**). Those that specifically protect species, such as the Convention on International Trade in Endangered Species (CITES), were discussed in Chapter 7. Here, we elaborate on the most important conferences and other international agreements that impact habitat use more generally, especially in the context of sustainable development.

The United Nations Conference on Environment and Development (UNCED), held for 12 days in June 1992 in Rio de Janeiro, Brazil, was one of the most significant steps toward adopting a global approach to sound environmental management. Known unofficially as the **Earth Summit** or the **Rio Summit**, the conference brought together representatives from 178 countries, including heads of state, leaders of the United Nations, and individuals from major conservation organizations and other groups representing religions and Indigenous Peoples. Their purpose was to discuss ways of combining increased protection of the environment with sustainable economic development in less-wealthy countries (United Nations 1993).

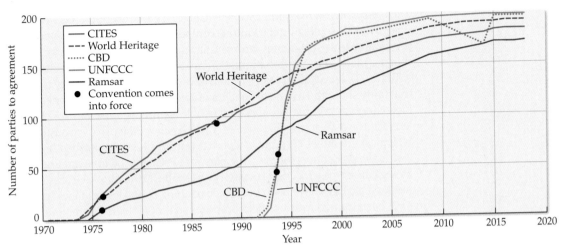

FIGURE 12.7 Major multinational environmental agreements (MEAs) are negotiated and then ratified by the governments of individual countries, which become "parties," or participants, in the provisions of the agreements or treaties. A treaty comes into force (i.e., countries begin to follow its provisions) when it has been signed by a certain number of countries (indicated by a dot). The plot lines show the numbers of countries that have ratified various treaties that provide for biodiversity protection by protecting habitat (the Ramsar Convention on Wetlands of International Importance, the World Heritage Convention Concerning the Protection of the World Cultural and Natural Heritage), species (the Convention on International Trade in Endangered Species [CITES], the Convention on Biological Diversity [CBD]), and the environment (the United Nations Framework Convention on Climate Change [UNFCCC]). (After WRI. 2003. *World Resources 2002–2004: Decisions for the Earth: Balance, Voice, and Power.* World Resources Institute, Washington, DC. www.wri.org/research/world-resources-2002-2004. CC BY 4.0, adapted from UNEP. 1999. GEO-2000. Earthscan Publication Ltd., London, UK, with updates from MEA websites.)

In addition to initiating many new projects, conference participants discussed, and most countries eventually signed, four major groundbreaking documents that still shape policy today:

1. *The Rio Declaration*   This nonbinding declaration provides general principles to guide the actions of both wealthy and poor nations on issues of the environment and development. The right of nations to use their own resources for economic and social development is recognized, as long as the environments of other nations are not harmed in the process. The declaration affirms the "polluter pays" principle, in which companies and governments take financial responsibility for the environmental damage that they cause.

2. *The United Nations Framework Convention on Climate Change (UNFCCC)*   Almost universally ratified (194 signatories), this agreement requires industrialized countries to reduce their emissions of carbon dioxide (CO_2) and other greenhouse gases and to make regular reports to the United Nations on their progress. While specific emission limits were not decided on, the convention states that greenhouse gases should be stabilized at levels that will not interfere with the Earth's climate.

GLOBAL CHANGE
CONNECTION

3. *The Convention on Biological Diversity (CBD)* This convention has three objectives: (1) protecting the various components of biodiversity, (2) using the components sustainably, and (3) sharing the benefits of new products that are made with the genetic resources of wild and domestic species (CBD 2021). Developing international laws for intellectual property rights that fairly share the financial benefits of biodiversity among countries, biotechnology companies, and local people is proving to be a major challenge to the convention. Because of concerns about the use or misuse of biological materials, certain developing countries have established highly restrictive procedures for granting permits to scientists who want to collect biological samples for their research (Watanabe 2015). In other cases, new research facilities have been built in developing countries, and local people have been trained in scientific procedures so that biological samples do not have to be exported. In 2010, the participants in the CBD developed a list of goals to achieve sustainability, called the **Aichi Biodiversity Targets** (www.cbd.int/sp/targets). The main goal was to slow or stop the loss of biodiversity by reducing the impact of human activities. This was to be achieved in part by changing governmental policies and increasing the percentage of land and ocean under protection (see Chapter 9). Targets have not yet been met, but significant progress has been made in recent years (see Figure 9.2).

4. *Agenda 21* This 800-page document is an innovative attempt to comprehensively describe the policies governments need to implement for environmentally sound development. Agenda 21 links the environment with other development issues, which are most often considered separately, such as child welfare, poverty, gender issues, technology transfer, and the unequal division of wealth. Plans of action address the problems of atmospheric, terrestrial, and aquatic pollution; land degradation and desertification; mountain development; unsustainable agriculture and rural development; and deforestation. Financial, institutional, technological, legal, and educational mechanisms that governments can use to implement the action plans are also described.

After the UNFCCC, international agreements to reduce global greenhouse gas emissions to below-1990 levels resumed with the Kyoto Protocol in December 1997. The agreement was ratified in 2004 under the UNFCCC (see #2), and many countries have established policies that have reduced their emissions of greenhouse gases, primarily CO_2. There were additional talks in Bali in December 2007, in Copenhagen in December 2009, in Warsaw in November 2013, and in Paris in December 2015. The **Paris Agreement** is the strongest international agreement on climate change yet, requiring action from all 191 signatory nations. Previous agreements did not require reductions from China, which has become the greatest contributor of greenhouse gases. Goals of the Paris Agreement (or *Accord de Paris*) included taking actions to prevent global temperature from increasing more than 2°C above preindustrial levels. The national climate action plans already shared by 186 nations alone are expected to cut global emissions in half (Davenport 2015). The parties to the UNFCCC now meets annually to negotiate issues, including funding. Each meeting is called a Conference of the Parties; the Twenty-fourth Conference of the Parties (COP24) was held in Katowice, Poland, in 2018 (**FIGURE 12.8**). The ability of countries to cooperate in dealing with climate change can be seriously compromised by partisan politics, such as when the US government temporarily withdrew from the Paris Agreement in 2020. It was re-signed hours after the next administration took office; however, the easing of regulations to slow emission of greenhouse gases during the previous four years will make it even more difficult for the United States to reach its goal of a 25% reduction by 2025 (Davenport 2020; Mai 2021).

FIGURE 12.8 A word cloud advertising the Conference of Parties (COP24) international meeting in Katowice, Poland, in 2018.

The **World Summit on Sustainable Development**, held in Johannesburg, South Africa, in 2002, emphasized achieving the social and economic goals of sustainability. This shift in focus from the Rio Summit highlights a significant, ongoing debate over whether the emphasis in conservation should be to promote sustainable use of natural resources for the benefit of poor people or to protect natural areas and biodiversity (Roe et al. 2013).

The **UN Conference on Sustainable Development** (unofficially, Rio + 20), held in 2012, linked biodiversity conservation to sustainable development and climate change and emphasized the need for market-based solutions (Carrière et al. 2013). It resulted in what is referred to as the 2030 Agenda, which includes 169 targets. A working group from that conference then developed a set of 17 sustainable development agenda items, presented at the UN Sustainable Development Summit in 2015. On January 1, 2016, the 17 goals from that meeting became officially in force, with the ambitious aims to eliminate poverty, ensure equitability (social justice), and stop climate change.

International agreements that protect habitat

Seven international conventions protect biodiversity either directly or through the protection of habitat (**TABLE 12.1**), and each seeks to balance conservation concerns with development needs. Those that include provisions for land use and sustainable development include (1) the **Ramsar Convention on Wetlands**, (2) the **World Heritage Convention** (WHC) (or the **Convention Concerning the Protection of the World Cultural and Natural Heritage**), and (3) the **Convention on Biological Diversity** (CBD).

Countries can gain international recognition for protected areas through the Ramsar Convention, the World Heritage Convention, and the World Network of Biosphere Reserves. Transfrontier parks in border areas provide opportunities for both conservation and international cooperation.

TABLE 12.1	The Seven Biodiversity-Related International Agreements, with Their Primary Objectives	
	Convention	**Primary objective**
	Convention on Biological Diversity (CBD)	Promotes the conservation of biological diversity, sustainable use, and equitable sharing of benefits
	Convention on International Trade in Endangered Species (CITES)	Ensures that trade in animals and plants does not threaten their survival
	Convention on the Conservation of Migratory Species of Wild Animals	Provides guidelines for the conservation and sustainable use of migrating animals throughout their ranges
	Ramsar Convention on Wetlands (RCoW)	Promotes the conservation and sustainable use of wetlands and their resources that contribute to both biodiversity and human well-being
	International Treaty on Plant Genetic Resources for Food and Agriculture	Promotes the conservation of plant genetic resources and the equitable sharing of the benefits that arise from them
	World Heritage Convention (WHC)	Mandates the identification and conservation of the world's cultural and natural heritage by protecting a specific list of sites

Source: Convention on Biological Diversity, www.cbd.int/brc/.

GLOBAL CHANGE CONNECTION

The Ramsar Convention on Wetlands (www.ramsar.org) was established in 1971 to halt the continued destruction of wetlands, particularly those that support migratory waterfowl, and to recognize the ecological, scientific, economic, cultural, and recreational values of wetlands (**FIGURE 12.9**). The Ramsar Convention covers freshwater, estuarine, and coastal marine habitats and includes 2422 sites with a total area of more than 254 million ha. The 171 countries that have signed the Ramsar Convention agreed to conserve and protect their wetland resources and designate at least one wetland site of international significance for conservation purposes. The fourth Ramsar Strategic Plan was launched August 21, 2015, to cover the period from 2016 to 2024.

The WHC's goal is to protect cultural areas and natural areas of international significance through its World Heritage Site program (whc.unesco.org). The convention is unique because it emphasizes the cultural as well as the biological significance of natural areas and recognizes that the world community has an obligation to financially support the sites. Limited funding for World Heritage Sites comes from the United Nations Foundation, which also supplies technical assistance. As with the Ramsar Convention, the WHC seeks to give international recognition and support to protected areas that are established initially by national legislation. The 250 World Heritage Sites protecting natural areas cover about 3.5 million ha and include some of the

FIGURE 12.9 The Djoudj National Bird Sanctuary is a Ramsar-listed site in Senegal noted for millions of waterbirds, such as these great white pelicans (*Pelecanus onocrotalus*), which stop here during their annual migration south from Europe in autumn.

world's premier conservation areas: Serengeti National Park in Tanzania, Sinharaja Forest Reserve in Sri Lanka, Iguaçu Falls in Brazil (**FIGURE 12.10**), Manu National Park in Peru, five sites in Queensland, Australia, Komodo National Park in Indonesia, and Great Smoky Mountains National Park in the United States, to name a few (see Figure 5.4B for a map of World Heritage coral reefs).

UNESCO's World Network of Biosphere Reserves was founded in 1971. Biosphere reserves are designed to be models that demonstrate the

FIGURE 12.10 World Heritage sites include some of the most revered and well-known conservation areas in the world, such as Iguaçu Falls in Brazil.

compatibility of conservation efforts and sustainable development for the benefit of local people, as described in Chapters 9 and 10 (see Figures 9.19 and 10.12). A total of 714 biosphere reserves have been created in 129 countries, including 302 reserves in Europe and North America, 157 in Asia and the Pacific, 130 in Latin America and the Caribbean, 85 in Africa, and 33 in the Arab states (UNESCO 2021). Countries also have the authority to withdraw sites from the agreement; for example, in 2017 the United States withdrew 17 sites from its previous total of 47. This is another example of the vulnerability of international agreements to changes in political leadership.

Other international agreements have been ratified to prevent or limit pollution that poses regional and international threats to the environment. For example, the Convention on Long-Range Transboundary Air Pollution in the European region recognizes the role that the long-range movement of air pollution plays in acid rain, lake acidification, and forest dieback; the Convention for the Protection of the Ozone Layer was signed in 1985 to regulate and phase out the use of chlorofluorocarbons; and the Convention on the Law of the Sea promotes the peaceful use and conservation of the world's oceans.

Conservation measures can also contribute to promoting cooperation between governments. Such is often the case when countries need to manage areas collectively. In many areas of the world, largely uninhabited mountain ranges mark the boundaries between countries. These areas often are designated as national parks, with each country managing its own wildlife and ecosystems. As an alternative, countries can establish transfrontier parks on both sides of boundaries to cooperatively manage whole ecosystems and promote conservation on a large scale (Thondhlana et al. 2015). An early example of this collaboration was the decision to manage Glacier National Park in the United States and Waterton Lakes National Park in Canada as the Waterton-Glacier International Peace Park. Today, intensive efforts are being made to link national parks and protected areas in Zimbabwe, Mozambique, and South Africa into the Great Limpopo Transfrontier Park and other, larger management units (**FIGURE 12.11**). An added advantage of this joint management is that the seasonal migratory routes of large animals will be protected. As another example, the establishment of the Red Sea Marine Peace Park between Israel and Jordan is important not only for conservation, but also for its potential for building trust in a war-ravaged region.

Marine pollution is another issue of vital concern because of the extensive areas of international waters not under national control and the ease with which pollutants released in one area can spread to another area. Agreements covering marine pollution include the Convention on the Law of the Sea, the Regional Seas Programme of the United Nations Environmental Programme (UNEP), and the Convention on the Prevention of Marine Pollution by Dumping of Wastes and Other Matter (see Figure 10.10 for the UNEP Mediterranean Action Plan). Regional agreements cover the northeastern Atlantic Ocean, the Baltic Sea, and other specific locations, particularly in the northern Atlantic region. The pelagic zone of the open ocean (the area of the ocean far from the shore) is still largely unexplored and unregulated at this point and is in urgent need of protection.

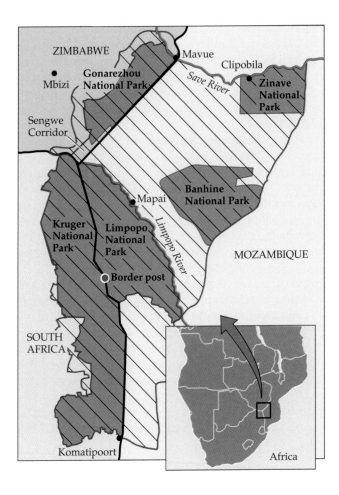

FIGURE 12.11   The Great Limpopo Transfrontier Park, which includes Gonar-ezhou National Park, Kruger National Park, Limpopo National Park, and several smaller conservation areas, has the potential to unite wildlife management activities in South Africa, Mozambique, and Zimbabwe. A larger transfrontier conservation area (hatched area) will include Zinave National Park, Banhine National Park, private game reserves, and private farms and ranches. (After K. Tinley and W. van Riet. 1991. Conceptual proposals for Kruger Banhine; A Transfrontier Natural Resource Area. Prepared for South African Nature Foundation, now WWF South Africa. Map available here: https://www.peaceparks.org/first-map-ever-of-great-limpopo-transfrontier-park/.)

12.4 Funding for Conservation

LEARNING OBJECTIVES

By the end of this section you should be able to:

12.4.1   List several sources of funding for conservation at local to international scales.

12.4.2   Evaluate the relative contributions of different countries to conservation, both in terms of total amounts and relative to GNP.

It is estimated that adequately protecting the world's biodiversity would cost $100–$440 billion per year (Barbier et al. 2018; UNDP 2018). As one small example, it was estimated that orchard farmers' profits would be 2.4 times less if they had to use hand pollination in the absence of bumblebees (Pérez-Méndez et al. 2020). While $100 billion is an enormous amount of money, it is dwarfed by the more than $686 billion spent on the military defense of the United States in 2019 alone. Similarly, while the conservation funds provided by the US government and the World Bank seem large, they are small compared with the other activities that these and other major international organizations support. Most importantly, even the higher estimate of $440 billion is dwarfed by the estimated value, $24 trillion, of these natural resources (**FIGURE 12.12**).

Unfortunately, actual spending to protect these resources, estimated at $4–$10 billion, is considerably less than is needed (Barbier et al. 2018). The sources of this funding are diverse, including national budgets and international development aid ("official development assistance"). In general, however, there has been movement away from funding conservation efforts solely with local taxpayer dollars, in favor of a variety of other approaches. For example, nongovernmental organizations (NGOs) and policies that require industry and development to pay for mitigation have become important sources of funding within developed countries.

Meanwhile, funding of conservation in developing countries has been less straightforward. One of the most contentious issues for international conferences and treaties has been deciding how to fund the proposals, particularly the CBD and other programs related to sustainable development, conservation, and climate change. Over the past 20 years, international funding for conservation by developed countries, foundations, and private donors has increased, though not as much as some say is necessary (Barbier et al. 2018). In anticipation of the UN Sustainable Development Summit in September 2015, a group of officials from various nations convened in Ethiopia's capital for the Third International Conference on Financing for Development. They

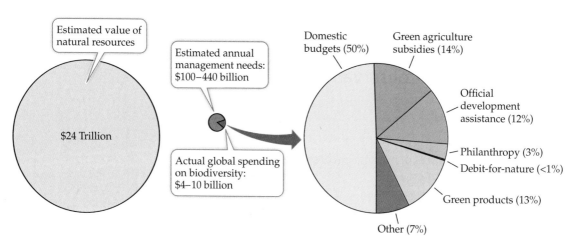

FIGURE 12.12 A comparison of the value of global biodiversity versus spending needs and realities (left), and a breakdown of the source of those funds (right). Note that the greatest proportion—nearly half—comes from national governments ("domestic budgets"). (After UNDP. 2018. *BIOFIN The Biodiversity Finance Initiative Workbook 2018: Finance for Nature*. New York: United Nations Development Programme and C. Parker et al. 2012. *The Little Biodiversity Finance Book*. Oxford: Global Canopy Programme.)

produced the Addis Ababa Action Agenda, intended as the global plan of how to implement and support the post-2015 agenda (Bhattacharya et al. 2015). However ambitious and well-intentioned, like many agreements of this type it was immediately criticized for lacking any concrete commitments or "teeth" to require increased funding (Barcia 2015).

International assistance and the World Bank

At the time of the Earth Summit, the cost of conservation programs was estimated to be about $600 billion per year, of which $125 billion was to come from developed countries as part of their **official development assistance** (**ODA**). Because the level of ODA from all countries in the early 1990s totaled approximately $60 billion per year, implementing these conventions would have required a tenfold increase in the aid commitment at that time. The developed countries did not agree to this increase in funding, but they offered an alternative: each country would increase its level of total foreign assistance to 0.7% of its gross national product (GNP) by the year 2000, which would have roughly doubled the ODA from developed nations. In the past 20 years, the ODA has been increasing steadily, now 66% more than in 2000. However, as of 2018, only a few wealthy northern European countries had met the GNP target: Luxembourg (0.99%), Sweden (1.02%), Norway (0.99%), Denmark (0.74%), and the United Kingdom (0.70%), along with United Arab Emirates (1.02%) and Turkey (0.95%) (OECD 2021a). The United States is the largest total contributor at $35.5 billion in 2020, but as this represents only 0.17% GNP, it is still well below the 0.70% target (OECD 2021b). In fact, the United States ranks near the bottom of the list of developed nations in terms of percentage of GNP contributed to international conservation programs. In the US government, funding for international conservation programs is spread across many departments, including the Agency for International Development, the National Science Foundation, the Smithsonian Institution, and the USFWS.

Through the 1980s, ODA funding for conservation projects was approximately $200 million per year, but starting in the early 1990s, it increased to $1 billion per year and is now estimated to be $4–$12 billion per year (United Nations 2021). In 2017, the United Nations agreed on a global indicator framework for the sustainability goals and targets of the 2030 Agenda for Sustainable Development (United Nations 2017), with 232 indicators on which agreement had been reached. Among these is "Target 15.a: Mobilize and significantly increase financial resources from all sources to conserve and sustainably use biodiversity and ecosystems." According to the Intergovernmental Science-Policy Platform on Biodiversity and Ecosystem Services (IPBES), an international and independent body dedicated to science-based approaches to conservation, conservation ODA is one of the handful of measures of the Agenda for Sustainable Development that has significantly increased over time (IPBES 2019; **FIGURE 12.13A**). Dozens of countries receive this biodiversity ODA, on almost every continent (**FIGURE 12.13B**). However, the proportion of total ODA that is committed to conservation has varied by country and by year, often as a function of political leadership; for example, the funding for conservation ODA from the United States decreased by 42% between 2016 and 2018 (**FIGURE 12.13C**).

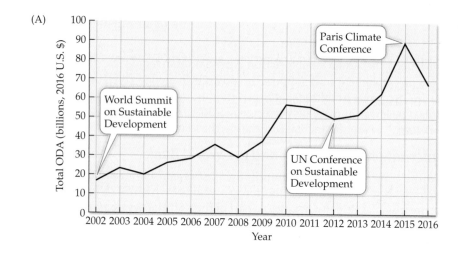

(A)

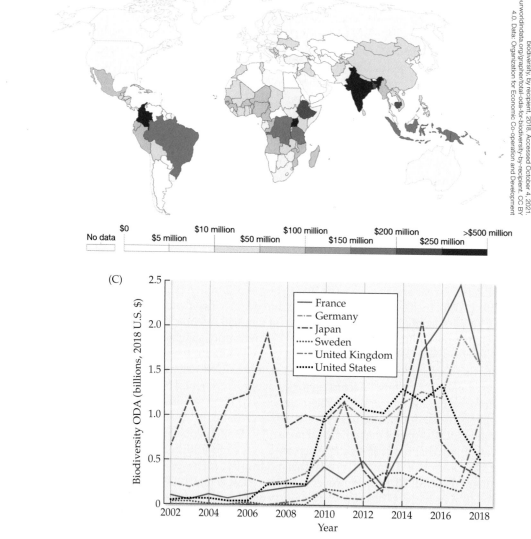

(B)

(C)

◀ **FIGURE 12.13** (A) Total official development assistance (ODA) received by developing countries specifically for conservation and sustainable use of biodiversity and ecosystems, in billions of US dollars, over time. Key international agreements are indicated. (B) Map of recipient countries, in millions of US dollars in 2017. (C) Biodiversity ODA by donor country. Big fluctuations in these numbers often represent changes in political leadership. (After United Nations. 2019. https://unstats.un.org/sdgs/indicators/database.)

Much of the conservation funding by developed countries in recent years has been channeled through the **World Bank** (www.worldbank.org) and the associated **Global Environment Facility** (GEF) (www.thegef.org/gef). The World Bank is a multilateral development bank established to promote international trade and economic activity. It is governed mainly by developed countries, and only a small portion of its activities are related to conservation. The related GEF was established specifically to channel money from developed countries to conservation and environmental projects in developing countries, with much of its funding distributed by the World Bank, other multilateral development banks with a regional focus, and the United Nations Development Programme.

The World Bank and other international funding mechanisms are often cited not for contributions to conservation, but rather for the opposite: many development programs, such as those involving road construction, mining, and forestry, are expressly detrimental to the environment. International aid can cause significant losses in habitat through the indirect effects of subsidies ("perverse subsidies") that encourage the excessive use of fossil fuels, water use, and environmentally harmful agricultural practices and accelerate the exploitation of fisheries and timber resources (see Chapter 3). These perverse subsidies greatly exceed investments in biodiversity (**FIGURE 12.14**). This does not mean that international aid for development always harms biodiversity; in fact, a review of 3534 projects implemented across 41,307 locations found that most projects sponsored by the World Bank had little impact on tree cover or conservation outcomes (Buchanan et al. 2018).

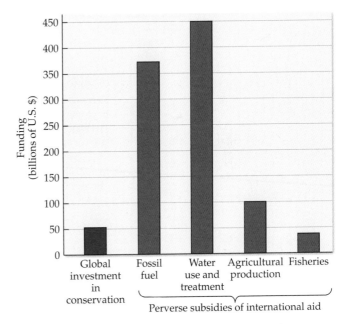

FIGURE 12.14 International aid can support projects that are detrimental to biodiversity conservation and can thus be considered "perverse subsidies." Here we can see relative amounts spent in various sectors on these often damaging activities, relative to the amount of international aid spent on conservation. Note that these data are compiled from several sources; annual spending is for specific years, from 2010 to 2016, and thus should be considered primarily in terms of relative, rather than absolute, values. (After UNDP. 2018. *The BIOFIN Workbook 2018: Finance for Nature.* New York: United Nations Development Programme, based on sources within.)

The potential of international funding to contribute to conservation goals is illustrated by the joining of the World Bank with the World Wildlife Fund (WWF; worldwildlife.org), an international conservation organization, to initiate a new program, the Forest Alliance, to protect and manage over 100 million ha of forest in countries around the world. The World Bank is also one of the leaders in efforts to reduce CO_2 emissions caused by deforestation in tropical countries, including Indonesia. Through its Forest Carbon Partnership Facility, companies and developed countries can offset their present production of greenhouse gases by purchasing carbon credits for maintaining tropical forests. The World Bank has partnered with the WWF and other large NGOs in implementing such programs.

International NGOs (also called INGOs) involved with conservation (e.g., the WWF, Conservation International, BirdLife International, TNC, and the Wildlife Conservation Society) implement conservation activities directly, often through a carefully articulated set of priorities and programs (Robinson 2012). These NGOs have also emerged as leading sources of conservation funding, raising funds from membership dues, donations from wealthy individuals, sponsorship from corporations, and grants from foundations and international development banks (**FIGURE 12.15**). Although foundations and conservation organizations provide

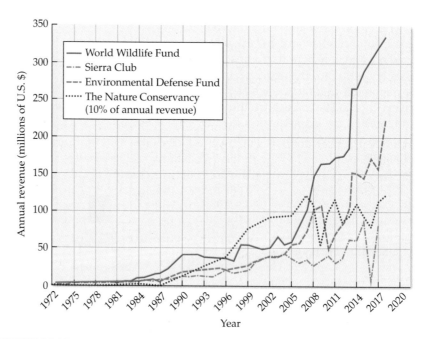

FIGURE 12.15 Over the past five decades there has been a dramatic increase in the annual contributions to many conservation organizations, as illustrated by the revenues of four large nongovernmental organizations (NGOs) in the United States: The Nature Conservancy (TNC), World Wildlife Fund (WWF), Environmental Defense Fund, and Sierra Club. Note that the revenue for TNC shown on the graph should be multiplied by 10; for 2018, contributions to TNC were approximately $1.3 billion. (After P. Zaradic et al. 2009. *PLOS ONE* 4: 37367/CC BY 4.0, with updates from P. Zaradic and M. Sher.)

only a small fraction of the funding for biodiversity projects, they can sometimes be more flexible and can fund innovative small projects and provide more intensive management than is typical from projects with government funding.

International NGOs are often active in establishing, strengthening, and funding both local NGOs and government agencies in the developing world that run conservation programs (Cohen-Shacham et al. 2015). From the perspective of an INGO, working with local organizations in developing countries is an effective strategy because it relies on local knowledge and it trains and supports groups of citizens within the country, who can then be conservation advocates for years to come. NGOs are often perceived to be more effective at carrying out conservation projects than government departments, but programs initiated by NGOs may end after a few years when funding runs out, and often they fail to have a lasting effect. Moreover, the income of NGOs can be quite variable, depending on the state of the economy.

Environmental trust funds

In addition to direct grants and loans for projects provided by the World Bank and other institutions, another important mechanism used to provide secure, long-term support for conservation activities in developing countries is a **national environmental fund** (**NEF**). An NEF is typically set up as a conservation trust fund or foundation in which a board of trustees—composed of representatives of the host government, conservation organizations, and donor agencies—allocates the annual income from an endowment to support inadequately funded government departments and nongovernment conservation organizations and activities. NEFs have been established in over 50 developing countries with funds contributed by developed countries and by major organizations such as the World Bank, the GEF, and the WWF.

One important early example of an NEF, the Bhutan Trust Fund for Environmental Conservation (BTFEC), was established in 1991 by the government of Bhutan in cooperation with the World Bank and the WWF. The BTFEC has already received $53.3 million (exceeding its goal of $20 million). In 2021, 208 projects were funded by BTFEC, with 48% specifically for biodiversity conservation or ecosystem management, and another 32% considered to be "cross cutting across conservation and social needs" (BTFEC 2021). The fund provides more than $1 million per year for surveying the rich biological resources of this eastern Himalayan country.

Debt-for-nature swaps

Many countries in the developing world have accumulated huge international debts that they cannot repay. As a result, some developing countries have rescheduled their loan payments, unilaterally reduced them, or stopped making them altogether. Because of the low expectation of repayment, the commercial banks that hold these debts have sometimes sold them at a steep discount on the international secondary debt market. For example, at times Costa Rican debt has traded for only 14%–18% of its face value.

In a creative approach, debt from the developing world is used as a vehicle for financing projects to protect biodiversity in so-called **debt-for-nature swaps**

(Cassimon et al. 2014). In one common type of debt-for-nature swap, an NGO in the developed world (such as Conservation International) buys up the debts of a developing country; the NGO agrees to forgive the debt in exchange for the country's carrying out a conservation activity. This activity could involve land acquisition for conservation purposes, park management, development of park facilities, conservation education, or sustainable development projects. In another type of swap, a government of a developed country that is owed money directly by a developing country may decide to cancel a certain percentage of the debt if the developing country will agree to contribute to an NEF or some other conservation project.

Debt-for-nature swaps have converted debt valued at more than $2.6 billion into conservation and sustainable development activities in over 30 countries, with the largest proportion in Latin America. However, the greatest overall debt-for-nature swap occurred in Poland in 1992 when $3 billion in debt was exchanged for environmental concessions. Currently, debt-for-nature swap is being considered to mitigate the negative effects of China's Belt and Road Initiative (Mengi and Wang 2021).

Other sources of funding

Although most funding for conservation is at least managed or collected by either NGOs or government agencies, there are, increasingly, cases of conservation actions funded directly by private contributions. The purchase of land for the purpose of conservation by individuals is one such example, discussed in Chapter 10. Another example is the use of social media; crowdfunding, in which individuals pool resources for a specific goal, has been applied to various aspects of conservation, including research, lobbying, and on-the-ground conservation work. A review of 72 fundraising platforms found 577 conservation-focused projects that raised $4,790,634 between 2009 and 2017 (Gallo-Cajiao et al. 2018). For example, more than $284,000 was raised through crowdsourcing to expand the Arcadia Marsh Nature Preserve on Lake Michigan. However, such fundraising will disproportionately help charismatic species; the majority of the 208 species addressed in these campaigns were birds and mammals. As a case in point, the highlighted value of Arcadia Marsh is as habitat for 17 species of endangered or threatened birds.

There is also a role to be played by conservation organizations and businesses working together to market "green products." These are playing an increasingly important role in global conservation funding (see Figure 12.12). For example, currently the Forest Stewardship Council and similar organizations are certifying wood products from sustainably managed forests and certifying sustainably grown cocoa (see Figure 13.3), and coffee companies are marketing shade-grown coffee (see Figure 10.8).

Still another funding mechanism to protect tropical forests is being adopted as part of international efforts to address climate change. Because about 20% of global greenhouse gas emissions result from destruction of the tropical forest, a funding mechanism called **Reducing Emissions from Deforestation and Forest Degradation (REDD+)** was developed to protect tropical forests (Munawar et al.

GLOBAL CHANGE
CONNECTION

2015). REDD+ rewards poorer nations for preserving forests by paying them for the carbon that is stored in their forests.

The United Nations adopted REDD+ (www.un-redd.org) as a flagship program in 2016 in partnership with the FAO, UN Development Programme (UNDP), and UNEP. Its purpose is to assist 65 partner countries across the world to achieve both climate and forest goals by supporting government efforts to stop deforestation. For example, in 2019 Paraguay's REDD+ program was approved to receive $72.5 million for their 26 million tons of CO_2 of forest emission reductions between 2015 and 2017 (UNEP 2019). This was the fourth payment under this program to developing countries that year. Organizations of many types and at all levels are involved in designing, implementing, and monitoring the outcomes of REDD+ projects. Over the last decade, the UN-REDD program implemented over 100 community-based REDD+ projects (UN-REDD Programme 2021). In these programs, the community receives funds not only to not cut down trees, but also to support the preservation of local ethnobotanical knowledge and monitoring and maintenance of the forest.

GLOBAL CHANGE
CONNECTION

Such efforts to reduce deforestation have other, cascading positive impacts; for example, modern slavery has been tied to deforestation in developing countries, and REDD+ has been proposed as a mechanism to address this significant human rights violation (Jackson and Sparks 2020).

How effective is conservation funding?

Conservation organizations have developed a number of tools to evaluate the effectiveness of funded projects. For example, the Independent Evaluation Office of the GEF publishes a detailed annual report of its findings; the one published in 2020 assessed 1706 completed GEF projects that accounted for $7.5 billion in GEF grants and $36.2 billion in promised cofinancing (GEF 2020). These programs are considered in terms of sustainability and successful outcomes not only for the environment but for the local people as well.

Many environmental projects supported by international aid do not provide lasting solutions to the environmental, economic, and social problems they purport to address, because of failure to deal with the "4 Cs"—concern, contracts, capacity, and causes. Environmental aid will be effective only if these conditions are met:

- Both donors and recipients genuinely want to solve the problems. Do key people really want the project to be successful, or do they just want the money?
- Mutually satisfactory and enforceable contracts for the project can be agreed on. Will the work actually be done once the money is given out? Will money be siphoned off into private hands?
- There is the capacity to undertake the project in terms of institutions, personnel, and infrastructure. Do people have the skills to do the work? Do they have the necessary resources—such as vehicles, research equipment, buildings, and access to information—to carry out the work?

- The causes of the problem are addressed. Will the project treat the underlying causes of the problem or just provide temporary relief of the symptoms?

The need for increased funding for biodiversity remains great at the local, national, and international levels, with significant implications for the global economy and international human rights. According to the WWF *Global Futures Report*, continuing at the current levels of conservation funding costs us $479 billion in terms of lost economic growth (Roxburgh et al. 2020).

As we have seen throughout this textbook, the economic issues inherent in conservation of natural resources mean that dynamics of power and privilege among human groups will come into play. Thus, the protection of biodiversity, environmental justice, and sustainable development go hand in hand; each requires the other.

Summary

- Sustainable development is defined as economic development that satisfies the present and future needs of human society while minimizing its impact on biodiversity. Achieving sustainable development is a challenge for conservation biology and society.

- Legal efforts are made to protect biodiversity at the local, regional, and national levels and to regulate activities affecting both private and public lands. Governments and private land trusts buy land for conservation purposes or acquire conservation easements and development rights for future protection. Laws can limit pollution, regulate or ban certain types of development, and set rules for hunting and other activities—all with the aim of preserving biodiversity and protecting human health.

- International agreements and conventions that protect biological diversity are needed because species migrate across borders, there is an international trade in biological products, the benefits of biological diversity are of international importance, and the threats to diversity are often international in scope and require international cooperation.

- One of the most important mechanisms for measuring progress on international conservation is the Aichi Biodiversity Targets, established at the Convention on Biological Diversity.

- Conservation groups, governments in developed countries, and the World Bank provide funding to protect biodiversity, especially in developing countries. National environmental funds and debt-for-nature swaps are also used to fund conservation activities. However, additional funding is still needed to address both biodiversity and sustainable development needs.

For Discussion

1. What are the roles of government agencies, private conservation organizations, businesses, community groups, and individuals in the conservation of biodiversity? Can they work together, or are their interests necessarily opposed to each other?

2. Why are the concerns for human rights, sustainable development, and biological conservation intertwined? Can one be addressed without impacting the other two? Are these impacts positive or negative?

3. When global targets established by the Earth Summit (for biodiversity) or the Paris Agreement (for global warming) are not met, does that mean there was no value to setting them? Why or why not?

Suggested Readings

Chapman, M., et al. 2020. A payment by any other name: Is Costa Rica's PES a payment for services or a support for stewards? *World Development* 129: 104900. Perceptions of payment for ecosystem services (PES) can differ greatly between leaders and participants.

Dinerstein, E., et al. 2019. A global deal for nature: Guiding principles, milestones, and targets. *Science Advances* 5(4): eaaw2869. An estimation of what it will take to protect 30% of nature—an ambitious target to halt climate change.

Gallo Cajiao, E., et al. 2018. Crowdfunding biodiversity conservation. *Conservation Biology* 32(6): 1426–1435. Creative approaches to funding conservation take advantage of social media and increased public awareness.

Leclère, D., et al. 2020. Bending the curve of terrestrial biodiversity needs an integrated strategy. *Nature* 585(7826): 551–556. A comparison of several models to predict conservation outcomes based on sustainable development scenarios.

Roxburgh, T., et al. 2020. *Global Futures: Assessing the global economic impacts of environmental change to support policy-making.* Summary report, January 2020. https://www.wwf.org.uk/globalfutures. A collaboration between the World Wildlife Fund and Global Trade Analysis Project and the Natural Capital Project to determine how loss of species and other global change affects the world economy.

Simmons, B. A., et al. 2021. China can help solve the debt and environmental crises. *Science* 371(6528): 468–470. A case for using debt-for-nature swaps for China's Belt and Road Initiative.

Conservation of pristine landscapes and the biodiversity they contain will require the dedication of this and future generations.

An Agenda for the Future

13

As we have seen throughout this book, the causes of the rapid, worldwide decline in biodiversity are no mystery. Ecosystems are destroyed and species are driven to extinction by human resource use, which is propelled by the need of many rural people to simply survive, by the excessive consumption of resources by affluent people and countries, and by the desire to make money (von Weizsäcker and Wijkman 2017). The destruction may be caused by local people in the region, people recently arrived from outside the region, local business interests, large businesses in urban centers, suburban sprawl into rural areas, multinational corporations in other countries, military conflicts, governments, and others. Moreover, people may also be unaware of, or apathetic toward, the impact of human activities on the natural world. Furthermore, although there have never been so many people involved in documenting new species, hundreds of thousands of species are yet to be discovered and may be lost before they are even named (Pimm 2021).

For conservation policies to work, people at all levels of society must believe that the conservation of biodiversity is in their own interest (Manfredo et al. 2016). If conservationists can demonstrate that protecting biodiversity has more value than destroying it, people and their governments may become more willing to preserve biodiversity. This assessment should include not only immediate monetary value but also less tangible aspects, including existence value, option value, and intrinsic value (see Chapter 3). Recent evaluations suggest that reaching international goals for conservation and sustainability is possible but will require major shifts away from "business as usual" practices in production, management, and valuing biodiversity (Tallis et al. 2018).

GLOBAL CHANGE
CONNECTION

13.1 Ongoing Problems and Possible Solutions

LEARNING OBJECTIVES

By the end of this section you should be able to:

13.1.1 Suggest solutions to the major problems facing conservation of biodiversity today.

13.1.2 Identify connections between the conservation of biodiversity and improving the well-being of local people.

There is a consensus among conservation biologists that our efforts to preserve biodiversity face several major problems and that certain changes must be made to policies and practices. We list some of these problems and suggest some solutions next. Note that for the purposes of this text we have simplified the solutions, leaving out many of the intricacies that must be addressed to provide comprehensive, real-world answers.

Problem: Protecting biodiversity is difficult when most of the world's species remain undescribed by scientists and are not known by the general public. Furthermore, most ecosystems lack monitoring to determine how they are changing over time.

Solution: We must train more scientists and enthusiastic amateurs to identify, classify, and monitor species and ecosystems, and we must increase funding in these areas (Pimm 2021). In particular, we need to train more scientists and establish research institutes in developing countries. Citizen scientists can often play an important role in protecting and monitoring biodiversity if they are given some training and guidance by conservation biologists (see Chapter 2). People who are interested in conservation biology should be taught basic skills, such as species identification and environmental monitoring techniques. For example, monitoring efforts by citizen scientists documented the decline of monarch butterflies (*Danaus plexippus*), which led to a petition to list this species as threatened under the Endangered Species Act (Agrawal and Inamine 2018) (**FIGURE 13.1**). Other examples are Project Budburst, which recruits the public to submit phenological data (information on organisms' seasonal cycles) as a way of tracking the impacts of climate change, and iNaturalist, an international program with which amateurs can record occurrences of plants and wildlife (see Figure 2.16). Such citizen science programs are becoming increasingly common and valuable to conservation, especially as we improve our ability to recruit and train citizen scientists and have their data submitted online. Research has shown that training can be very effective in improving the quality of biological data collected by citizen scientists (Gaddis 2018).

GLOBAL CHANGE
CONNECTION

In these types of projects, volunteers not only observe how to conduct science and conservation, but actually perform science and conservation. In this way they can experience a sense of service and ownership of the project and often become advocates. In some cases, this work translates into lasting advances in conservation and can inspire students to pursue careers in the

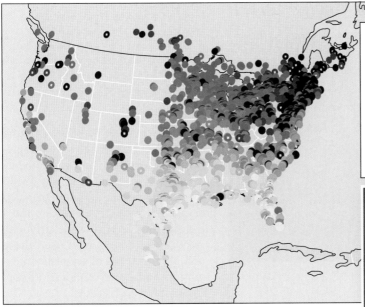

First adult sightings
traveling north
- ○ Before March 1
- ○ March 1–March 14
- ○ March 15–March 28
- ◔ March 29–April 11
- ◑ April 12–April 25
- ◕ April 26–May 9
- ● May 10–May 23
- ● May 24–June 6
- ● June 7–June 20
- ● After June 20

Photo by Ellen Woods

FIGURE 13.1   Each spring, thousands of citizen observers, especially students, contribute their observations to the Journey North website, which tracks the migration of the monarch butterfly, starting from over-wintering sites in central Mexico, moving into the southern United States in March, and arriving in the northern United States and southern Canada in late May and June. There are secondary overwintering sites in California, Florida, and Texas. This monitoring can detect changes in butterfly numbers and migration times. (After Journey North. 2018. https://maps.journeynorth.org/map/?year=2018&map=monarch-adult-first.)

field. Many conservation biologists and ecologists attribute their passion for the field to a transformative outdoor experience such as those provided by these programs.

Educating both the public at large and those whose work directly or indirectly affects endangered species is critically important if we are to stop species extinctions (Struminger et al. 2021). Educational programs can be geared toward specific audiences, such as schoolchildren, senior citizens, or rural people living near national parks (**FIGURE 13.2**). This outreach has often been shown to be effective in increasing positive attitudes toward less-charismatic species and help promote conservation-oriented behaviors (McKinley et al. 2017; Unger and Hickman 2020). Perhaps even more important, conservation practitioners need to be made aware of evidence-based methods for species and resource management, which may require organizations and groups specific to this purpose. These groups, which have been called "evidence bridges" or "boundary organizations," make it easier for scientific information to reach those who can apply it (Kadykalo et al. 2020). For example, in

> Conservation biologists can spread the message of conservation and do better science by educating members of the public and including them in their projects. These citizens then often become advocates for protecting biodiversity.

FIGURE 13.2 Schoolchildren learn how to identify birds and collect data from a guide from Taita Discovery Centre, an outdoor education center in Tsavo, Kenya. Experiences like these connect people to nature and create the next generation of conservation biologists.

the American Southwest, it was found that restoration managers were applying ideas developed and promoted by conservation scientists, largely due to the work of a single nonprofit that produces workshops, conferences, and a website for this purpose (Clark et al. 2019). Practitioners who used these and other sources of scientific information had better restoration outcomes (Sher et al. 2020).

Increasingly, education to improve people's understanding of conservation includes the creative use of digital tools. Information on biodiversity must be made more accessible; this may be accomplished in part through websites dedicated to this purpose, including those with citizen science data, as well as the user-friendly professionally curated data found in digital herbarium databases from around the world, IUCNredlist.org, and plants.usda.gov. The question of how to improve the science, conservation, and education outcomes of citizen science and other educational techniques is an active area of research.

Problem: Many conservation issues are global in scope and involve many countries, making it important but difficult to coordinate conservation actions. Furthermore, international agreements are extremely difficult to enforce, making noncompliance an issue.

GLOBAL CHANGE
CONNECTION

Solution: We must focus on and build upon successful international efforts. Countries are increasingly willing to discuss international conservation issues, as shown by the 2018 conference on climate change (see Chapter 12). Nations are also becoming more willing to sign and implement treaties such as the Convention on Biological Diversity, the United Nations Framework Convention on Climate Change, and the Convention on International Trade in Endangered Species (see Figure 12.7). International conservation efforts are expanding, and further participation in these activities by conservation biologists and the general public should be encouraged. Another positive development is the trend toward establishing transfrontier parks that straddle national borders; these parks benefit wildlife and encourage cooperation between countries (see Figure 12.11).

Citizens and governments of developed countries must also become aware that they bear a direct responsibility for destroying biodiversity through their overconsumption of the world's resources and the specific products they purchase, and that lifestyle changes can benefit biodiversity (Collins et al. 2018). For example, certification programs can help consumers make choices to support sustainable practices, such as for coffee (see Chapter 10), cocoa, wood, and seafood (**FIGURE 13.3A**). In 2018 two of the largest certification programs for cocoa, coffee, and tea—UTZ (a program out of the Netherlands) and the Rainforest Alliance—merged, saving money for the growers, who no longer had to pay for two programs, and simplifying messaging to consumers. These programs typically address not only environmentally sustainable practices but socially responsible ones as well, such as ensuring fair wages and working conditions. Labels on packaging are a marketing tool and help consumers make good choices (**FIGURE 13.3B**).

To encourage the enactment of legislation, conservation professionals must demonstrate how changes in the actions and lifestyles of individuals on the local level, and those of cities and nations on a larger scale, can exert positive influences far beyond the immediate community. Improvements in understanding conservation psychology will help in the design of programs to encourage conservation-minded behaviors.

> Conservation biologists need to support approaches that provide benefits for people and protect biodiversity. One approach is to compensate landowners and local people for the ecosystem services that their land provides.

Problem: Developing countries often want to protect their biodiversity but face great pressure to develop their natural resources to generate needed income.

Solution: Conservation organizations, zoos, aquariums, botanical gardens, natural history museums, governments, and even businesses in developed countries, as well as international organizations such as the United Nations and the World Bank, should continue to provide technical and financial support to developing countries for conservation activities, especially

(A)

(B)

FIGURE 13.3　(A) A farmer in Isla de la Amargura, Caceres, Colombia, harvests fair trade cocoa. (B) Sustainably produced chocolate is available for purchase in an increasing number of stores and includes a variety of certifications, prominently shown as logos on the package. Some brands market themselves specifically as environmentally friendly.

establishing and maintaining protected areas. For example, the Association of Zoos and Aquariums reports contributions of $160 million annually for conservation initiatives (AZA 2021) (see Chapter 8). It is important for these organizations to support the training of conservation biologists in developing countries so local people can become advocates for biodiversity within their own countries (Paknia et al. 2015). Support from developed nations should continue until countries can protect biodiversity with their own resources and personnel. Requiring support from developed nations until then is fair and reasonable since they have the funds to support these parks, drive much of the degradation of biodiversity through consumption, benefit from the exploitation of natural resources in these countries, and make use of biological resources in their agriculture, industry, research programs, zoos, aquariums, botanical gardens, and educational systems. Solutions for economic and social problems in developing countries must be addressed simultaneously, particularly those relating to reducing poverty and ending armed conflicts (see Chapter 12). There is a variety of financial mechanisms to achieve these goals, including direct grants, payments for ecosystem services, debt-for-nature swaps, and trust funds. Individual citizens in developed countries can donate money and participate in organizations and programs that further advance these conservation goals.

Problem: Economic analyses often paint a falsely encouraging picture of development projects that damage the environment. Economic decision-making often fails to include and evaluate ecosystem services and intrinsic values.

Solution: Cost-benefit analyses must evaluate development projects comprehensively by comparing potential project benefits with the full range of costs, including environmental and human costs such as soil erosion, air and water pollution, greenhouse gas production, declines in the availability of natural products and other ecosystem services, and the loss of places for people to live (Alamgir et al. 2017). Local communities and the public at large should be presented with all available information and asked to provide input into the decision process. We must adopt the "polluter pays" principle, in which industries, governments, and individual citizens pay for cleaning up the environmental damage their activities have caused (Zhu and Zhao 2015). And financial subsidies to industries that damage the environment—such as the pesticide, fertilizer, transportation, petrochemical, logging, fishing, and tobacco industries—should end, particularly subsidies to industries that damage human health as well (see Chapter 3). Funds from these perverse subsidies should be redirected to activities that enhance the environment and human well-being, especially to people whose lands provide ecosystem services to the public (see Table 12.2).

Problem: In many areas of the world, disadvantaged people simply trying to survive can be responsible for destroying biodiversity.

Solution: Biodiversity loss from actions of local people can be due to a number of causes, including population growth. This situation can come about in a variety of ways, and the cause should help inform the response. For example, across the world there are Indigenous Peoples who have traditionally minimized their negative impact on the land by being nomadic, but now

GLOBAL CHANGE
CONNECTION

they are more destructive because they have been forced by governments or other factors to settle in one place (Rabinovich et al. 2020) (**FIGURE 13.4**). Programs to address biodiversity loss should thus acknowledge and, when possible, integrate traditions and practices of the people they affect through meaningful input by them (see Chapter 10).

Conservation biologists and conservation organizations increasingly participate in these types of programs for poor rural areas. These agendas often focus on promoting smaller families, establishing reliable food supplies and/or providing training in economically useful skills (Miller et al. 2018). Such efforts should be closely linked to social justice and recognizing basic human rights, especially ownership of the land and the right to a healthy environment (see Chapter 12).

Conservation organizations and businesses should also work together to market green products produced by rural communities, with the profits shared with those communities. Certification programs, such as the ones for coffee and chocolate (see Figure 13.3A), can be a mechanism to ensure that rural communities benefit; for example, products meeting environmental, labor, and developmental standards can be Fair Trade Certified (see Figure 13.3B). More than 1822 organizations in 75 developing countries and representing over 1.8 million farmers and workers have been certified to sell fair trade products, including bananas, tea, cotton, and fresh fruit (Fairtrade International 2021). However, the sources of these products must be carefully investigated to ensure that the practices and goals of the programs are being met. In some cases, investigators have found no meaningful difference

© STR/AFP/Getty Images

FIGURE 13.4 Kazakh nomads herding their livestock across a plain in Altay, in far west China's Xinjiang region. In many regions, nomadic people are being forced to settle in one place, which leads to both poverty and land degradation.

between the practices of the fair trade and conventional sectors. For example, research on the impact of the Forest Stewardship Council (FSC) found no significant impact on deforestation rates in Mexico (Blackman et al. 2018).

At a smaller scale, conservation projects can be created to benefit local economies. For example, the BirdLife International Global Seabird Programme has partnered with the Meme Itumbapo project, a small women's empowerment group in Walvis Bay, Namibia, to produce fishing line that frightens endangered albatross away from baited hooks. This effort has provided income for local people and has been highly effective at reducing the numbers of seabirds killed (RSPB and BirdLife International 2018). In Peru, another women's cooperative begun by conservation biologists helps provide income from an alternative to harvesting endangered frogs (**FIGURE 13.5**). The cooperative's products include hats featuring the endemic frogs, thereby also raising awareness.

Problem: Central governments often make decisions about establishing and managing protected areas with little input from the people and organizations in the affected region. Consequently, local people sometimes feel alienated from conservation projects and do not support them.

Solution: For a conservation project to succeed, the local inhabitants must know they will benefit from it and that their involvement is important (see Chapter 12). To achieve this goal, environmental impact statements, economic forecasts, and other project information should be made publicly available to encourage open discussion at all stages of the project's development and implementation. Local people should be provided with the assistance they need to understand and evaluate the implications of the project. Often they will want to protect biodiversity and associated ecosystem services because they understand that their livelihood, quality of life, and sometimes even their survival depend on protecting the natural environment (Gupta et al. 2015).

Mechanisms should be established to ensure that government agencies, conservation organizations, and local communities and businesses share in access to data and management decisions. In some cases, regional strategies, such as habitat conservation plans or community conservation plans, may have to be developed to reconcile the need for some development (and resulting loss of habitat) with the need to protect species and ecosystems. Conservation biologists working in national parks and other protected areas have a special

Photograph by Richard Reading

FIGURE 13.5 A women's cooperative in Puno, Peru, makes and sells hats featuring the endangered Lake Titicaca frog to sell to tourists. The cooperative was started by conservation biologists with the Denver Zoo seeking to create income streams for local people that did not involve harvesting the animals.

responsibility to engage in meaningful dialogue to explain the purpose and results of their work to nearby communities and school groups and listen to what local people have to say.

Problem: Revenues, business activities, and scientific research associated with national parks and other protected areas may not directly benefit the people living in the surrounding communities.

Solution: People living in local communities often bear the costs but do not receive the benefits of nearby protected areas. Addressing this problem requires developing mechanisms to benefit the local communities. For example, when possible, local people should be trained and employed in protected areas as a way of using local knowledge and providing income (see Figure 9.18). Governments or organizations can also assist local people in developing businesses related to tourism and other park activities. A portion of park revenues should fund local community projects such as schools, clinics, roads, cultural activities, and community businesses—infrastructure that benefits an entire village, town, or region; this establishes a link between conservation programs and the improvement of local lives.

Problem: National parks and conservation areas often have inadequate budgets to pay for conservation activities.

Solution: It is often possible to increase funds for park management by raising rates for admission, lodging, or meals so that they reflect the actual cost of maintaining the area. For example, in the El Nido-Taytay Managed Resource Protected Area, the largest marine protected area in the Philippines, divers and snorkelers are required to pay $4 per person to the park each time they enter the water. Concessions selling goods and services in and around national parks may be required to contribute a percentage of their income to the park's operation.

Problem: Many endangered species and ecosystems occur on private land or on government land managed for timber production, grazing, mining, and other activities. Timber companies that lease forests and ranchers who rent rangeland from the government often damage biodiversity and reduce the productive capacity of the land in the pursuit of short-term profits. Private landowners often regard endangered species on their land as a burden because of restrictions on the use of their habitats.

Solution: More laws should be passed so people are permitted to obtain leases to harvest trees and use rangelands only if they maintain the health of the ecosystem. Governments should also eliminate perverse tax subsidies that encourage exploiting natural resources and should establish payments for land management, especially on private land, that enhance conservation. Additionally, landowners should be educated to protect endangered species and praised publicly for their efforts. It is vital to develop connections among farmers, ranchers, conservation biologists, and hunting and fishing groups because biodiversity, wildlife, and the rural way of life are all intertwined.

We must also seek ways to integrate biodiversity considerations into plans, policies, and practices that depend on, or in some way relate to, the environment, on both the local and the international scales (Redford et al. 2015). To be effective, this requires communication between those who want the ecosystem services and those who are more concerned with the ecosystem itself (QUINTESSENCE Consortium 2016). This approach has been called mainstreaming biodiversity, and it is frequently used in the United Nations and other international treaty bodies. Article 10(a) of the Convention on Biological Diversity (CBD) calls for integrating "as far as possible and as appropriate, the conservation and sustainable use of biological resources into national decision-making." Spurred by such agreements and other, more local, motivators, governments are increasingly incorporating biodiversity into formal laws and standard practices.

Problem: In many countries, governments are inefficient and bound by excessive regulation. Consequently, they are often slow and ineffective at protecting endangered species and ecosystems.

Solution: Local nongovernmental organizations (NGOs) and citizen groups often represent the most effective agents for promoting conservation, especially if they work in concert with government policies, such as economic incentives for conservation (G. Wright et al. 2015). Accordingly, these citizen groups should be encouraged and supported politically, scientifically, and financially. Conservation biologists must educate citizens about local environmental issues and encourage them to act when necessary. Building the capacity of universities, the national media, and NGOs to evaluate, propose, and implement conservation-related policies also effectively encourages national-level action. Individuals, organizations, and businesses should start new foundations to financially support conservation efforts. One of the most important trends in conservation funding and policy is the increased strength of international NGOs, such as the World Wide Fund for Nature (with about 5 million members) and the Royal Society for the Protection of Birds. The number of NGOs has risen dramatically in past decades, and they can substantially influence local conservation programs and environmental policy at the local, national, and international levels.

Problem: Many businesses, banks, and governments remain uninterested in, and unresponsive to, conservation issues.

Solution: Leaders may become more willing to support conservation efforts if they receive additional information about the benefits of more-sustainable practices or perceive strong public support for conservation initiatives (Addison and Bull 2018). In countries with fairly open societies, lobbying and similar efforts may be effective in changing the policies of unresponsive institutions because politicians and other officials generally want to avoid bad publicity. Petitions, rallies, letter-writing campaigns, and economic boycotts all have their place if requests for change are ignored. In many situations, radical groups such as Greenpeace and PETA dominate media attention with dramatic, publicity-grabbing actions, while mainstream conservation organizations follow them to negotiate a compromise. In closed societies, focusing on identifying and educating key leaders is usually a better strategy because of the dangers faced by any opponents of the government. A better understanding of the diverse values that different cultures attribute to biodiversity also can help in promoting sustainable practices.

Conservation biologists often work with environmental activists and nongovernmental organizations to protect biological diversity. This often means finding a balance between doing the best possible science and political advocacy.

13.2 The Role of Conservation Biologists

LEARNING OBJECTIVES

By the end of this section you should be able to:

13.2.1 Evaluate the different roles of conservation biologists.

13.2.2 Identify opportunities to have an impact on conservation in your area.

The problems and solutions outlined here underscore the importance of conservation biologists, who are among the primary participants in conservation efforts. Conservation biology differs from many other scientific disciplines in that it plays an active role in preserving biodiversity in all its forms: species, genetic variability, biological communities, and ecosystem functions. As a result, conservation biologists must balance political advocacy with scientific credibility, maintaining the greatest possible objectivity in their scientific research (Horton et al. 2016). Members of the diverse disciplines that contribute to conservation biology share the common goal of protecting biodiversity in practice, rather than simply investigating and talking about it (Correa Ayram et al. 2016). To accomplish this goal, conservationists must work together with various partners to provide practical solutions to address real-world situations (Owley et al. 2017). Literature reviews suggest that real progress is being made (Godet and Devictor 2018).

Working together is easier said than done, and there is concern among some conservation biologists that disagreements between saving biodiversity for its intrinsic value and emphasizing biodiversity's instrumental value for people (i.e., use values; see Chapter 3) pose a danger for the future of conservation (Piccolo et al. 2018). Reconciling a diversity of viewpoints represents both a challenge and an opportunity for the discipline.

Challenges for conservation biologists

Decisions about park management and species protection increasingly incorporate the ideas and theories of conservation biology (e.g., Natural Resource Stewardship and Science 2015). The need to maintain large parks and protect large populations of endangered species has received widespread attention in both academic and popular literature. The vulnerability of small populations to local extinction, even when they are being carefully protected and managed, and the alarming rates of species extinction and destruction of unique ecosystems worldwide have also been highly publicized. As a result of this publicity, the need to protect biodiversity has entered the political debate and been targeted as a priority for government conservation programs. At the same time, botanical gardens, museums, nature centers, zoos, national parks, and aquariums are reorienting their programs to meet the challenges of protecting biodiversity and embracing the concepts of Earth stewardship (see Chapter 8). The sense of urgency is heightened by the recognition that many endangered species living in cold climates, such as polar bears and penguins, face immediate threats due to a warming climate and the melting of sea ice. What is ultimately required, however, is including the principles of conservation biology into the broader domestic policy arena and economic planning process. An international survey of 758 conservation and environmental leaders working in policy and research found that there was broad agreement about how to incorporate science into policy (Rose et al. 2018). The authors suggested that the main impediment was lack of public support. This highlights that incorporating conservation biology into economic policy and reprioritizing domestic policy goals will require substantial public education and political effort.

Achieving the agenda

Meeting conservation challenges requires conservation biologists to take on several active roles. Conservationists must become more effective *educators*, *leaders*, and *motivators* in the public forum as well as in the classroom (**FIGURE 13.6**). Conservation biologists need to educate as broad a range of people as possible about the problems that stem from a loss of biodiversity, convey a positive message about what needs to be done, and then give examples of some successes (Laurance 2013). The Society for Conservation Biology has made disseminating knowledge the first item in its new code of ethics. Conservation biologists must provide actionable research and leadership that counters the pessimism and passivity so frequently encountered in modern society (Carlson 2013). The International Union for Conservation of Nature demonstrates this approach with their campaign emphasizing "Love. Not Loss," which attempts to capture people's positive feelings toward nature rather than simply convey dry facts about nature's demise (IUCN 2021). Many groups, including anglers, hunters, bird-watchers, hikers, religious groups, and artists, may help conservation efforts once they become aware of the issues or recognize that their self-interest or emotional well-being depends on conservation (Granek et al. 2008). Some evidence suggests that such awareness is increasing worldwide, including in developing countries, such as India (Varma et al. 2015).

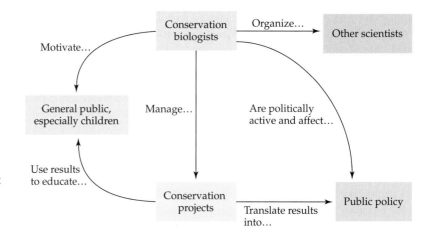

FIGURE 13.6 Conservation biologists must be active in various ways to achieve the goals of conservation biology and to protect biodiversity. Not every conservation biologist can be active in each role, but all the roles are important.

Conservation biologists must reach a wider range of people by speaking to a variety of audiences beyond academia, including at elementary and secondary schools, parks, neighborhood gatherings, and religious gatherings. Conservation biologists must spend more of their time writing articles and editorials for newspapers, magazines, and blogs, as well as speaking on radio, television, and other mass media, in ways the public can understand (Primack and MacKenzie 2019). Social media, such as Instagram, TikTok, YouTube, Facebook, Twitter, and blogs, can be used for effective outreach and communication about conservation issues. Scientists are even communicating with each other this way; at least one study has found that tweeted research papers are more likely to be cited in other manuscripts (Lamb et al. 2018). Data from these approaches can even be used to evaluate how effective scientists are in conveying their message and whether they are reaching their intended audiences (Bombaci et al. 2016) (**FIGURE 13.7**).

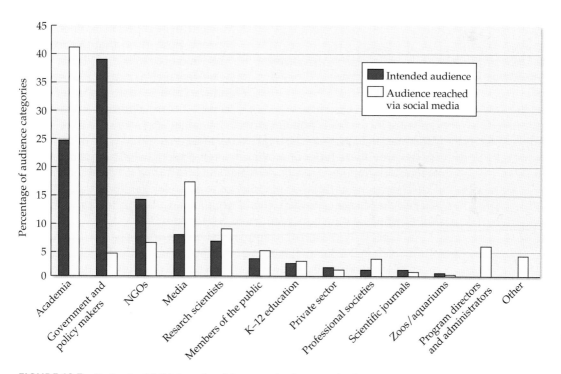

FIGURE 13.7 During the 2013 International Congress for Conservation Biology, efficacy of communication was evaluated by comparing the audience intended by the presenters (as determined via survey) with the audience reached via Twitter. "Audience reached" was determined by counting the senders and recipients of 700 live tweets during the conference and 1711 retweets. The mean number of followers for tweeters and retweeters was 2404, with 40% having over 10,000 followers across 40 countries—thus, considerably more than the 1500 conference attendees. These results suggest that using social media to create media interest may be a good way for scientists to reach the general public, but not necessarily policymakers. (After S. P. Bombaci et al. 2016. *Conserv Biol* 301(1): 216–225. © 2015 Society for Conservation Biology.)

FIGURE 13.8 Oceanographer Sylvia Earle is known throughout the world for her scientific contributions to marine biology as well as her advocacy for conservation. After winning a TED award, she founded a nonprofit to establish new marine protected areas called Mission Blue, about which a documentary was filmed.

Conservation biologists often teach college students and write technical papers addressing conservation issues, but these actions only reach limited audiences: remember that only a few dozen or at most few thousand people read most scientific papers. In contrast, billions of adults watch YouTube videos, listen to podcasts, and read social media daily. Videos such as TED Talks can have strong conservation themes, and many have gone viral via Facebook and Twitter. Dr. Sylvia Earle's talk about marine life conservation is one such example, even leading to a Netflix documentary, *Mission Blue* (**FIGURE 13.8**). Movies can be especially influential for conservation, given that they reach a broad audience (Silk et al. 2018); popular movies such as *Avatar, Finding Dory,* and *The Lorax* have strong conservation themes. Nature conservation documentaries are increasingly popular as well, including *My Octopus Teacher, Racing Extinction,* and *Seaspiracy.* Natural history museums and zoos are often venues for showing these types of movies in large-screen and 3D formats that are especially gripping, a recent example being *Back from the Brink: Saved from Extinction.* Tens of millions of children watch television programs such as *Bill Nye the Science Guy, The Cat in the Hat Knows a Lot About That,* and *Wild Kratts,* which feature both nature and conservation.

An especially powerful example of the role of media was the release of *Blackfish* in 2013, a movie that highlighted the mistreatment of captive orcas (*Orcinus orca*), also known as killer whales, including the questionable methods and practice of capturing them in the wild. The filmmakers worked with scientists and rigorously researched all the information presented. The huge following of the film resulted in legislation proposed in New York and California in 2014 to protect the orcas (A. J. Wright et al. 2015), and Congress unanimously passed an amendment to the US Agriculture Appropriations Act to update regulations concerning captive marine mammals (Charky 2014). Moreover, on March 16, 2016, SeaWorld announced that the orcas then in its care would be the last (CNN 2016). In 2019, Russia announced it would release nearly 100 illegally caught whales, including 10 orcas, after drone footage sparked a criminal investigation (MacFarquhar and Nechepurenko 2019). The whales had been captured in the summer of 2018 and held in pens to sell to Chinese aquariums and marine parks. So long as there is demand, species such as orcas will be at risk, but pressure from conservation biologists and the inspired and informed general public can help mitigate this threat.

The efforts of Merlin Tuttle and Bat Conservation International (BCI) illustrate how public attitudes toward even unpopular species can change. BCI has campaigned throughout the United States and the world to educate people on the importance of bats in ecosystem health, emphasizing their roles as insect eaters, pollinators, and seed dispersers. A valuable part of this effort has involved producing bat photographs and films of exceptional beauty. In Austin, Texas, Tuttle intervened when citizens petitioned the city government to exterminate the hundreds of thousands of Brazilian free-tailed bats (*Tadarida brasiliensis*) that live under a downtown bridge. He and his colleagues convinced people that the bats are both fun to watch and critical in controlling insect populations that feed on important crop plants and bite people. The situation has changed so drastically that the government now protects these bats as a matter of civic pride and practical pest control, and hundreds of citizens and tourists gather every night to watch the swarm of bats emerge from under the bridge on their nightly foraging forays (**FIGURE 13.9**).

Conservation biologists must also become politically active leaders to influence public policy and new laws. Scientists can, and should, participate in politics without sacrificing professional credibility, and they should be willing to acknowledge the role of personal values and funding sources (Garrard et al. 2015). This need for action was evident in the document "World Scientists' Warning to Humanity: A Second Notice," which was signed by 15,364 scientists from 184 countries (Ripple et al. 2017). This call to action has had a broad reach, having been translated into 17 languages and with more than 8000 tweets on Twitter reaching up to 14 million people (Ripple et al. 2018).

© iStock.com/Aneese

FIGURE 13.9 Citizens and tourists gather in the evening to watch Brazilian free-tailed bats emerge from their roosts beneath the Congress Avenue bridge in Austin, Texas.

Involvement in the political process allows conservation biologists to influence the passage of new laws that support the preservation of biodiversity and to argue against legislation that would prove harmful to species or ecosystems. An important first step in this process is joining conservation organizations or mainstream political parties to strengthen them by working within these groups and learning more about the issues. It is important to note that there is also room for people who prefer to work by themselves. Difficulties in getting the US Congress to reauthorize the Endangered Species Act and to ratify the CBD and the United Nations Framework Convention on Climate Change illustrate the need for greater political activism on the part of scientists who understand the implications of failing to take immediate action. Though much of the political process is time-consuming and tedious, it often represents the only way to accomplish major conservation goals, such as acquiring new land for reserves or preventing the overexploitation of old-growth forests. Conservation biologists must master the languages and methods of legal and political processes and form effective alliances with environmental lawyers, citizen groups, and politicians.

An example of the effectiveness of alliances between conservation biologists and other groups is the expansion of the Ikh Nart Nature Reserve in the Dornogobi province of Mongolia. Populations of endangered cinereous vultures (*Aegypius monachus*) (see Chapter 7 opening photo), argali sheep (*Ovis ammon*) (see Figure 7.3), Siberian ibex (*Capra sibirica*), and many other species of animals and plants have been monitored since 2001 by a partnership of conservation biologists at the Denver Zoo, Mongolian National University, Mongolian National Education University, University of Vermont, and Mongolian Academy of Sciences (Ikh Nart Nature Reserve 2021). The scientists communicated the findings of their research to the local government, which responded by doubling the size of the park to about 130,000 ha and restoring some areas that had been badly degraded (Reading 2015). The park has been a source of much scientific information, with nearly 50 peer-reviewed publications in 2020 alone produced by biologists doing research in the park.

A key role for conservation biologists to assume is that of translators, that is, scientists who can take the data and results of conservation science and translate them into legislation, public policy, and management actions (Courchamp et al. 2015). To be effective, conservation biologists must communicate the relevance of their research and demonstrate that their findings are unbiased, while respecting the values and concerns of most or all stakeholders. Conservation scientists have to be aware of the range of issues that may impact their programs and that their programs might affect, and they should be able to speak to general audiences in terms that they can understand. Conservation biologists must take the lead, as their expertise is needed.

Conservation biologists should become motivators, convincing a broad range of people to support conservation efforts and even assisting in formal marketing campaigns to increase awareness (A. J. Wright et al. 2015). Research on attitudes about protected areas found that the extrinsic motivators were less important than the intrinsic ones, suggesting that focusing on the value of nature, social justice, and the well-being of local communities would be an effective strategy (Cetas and Yasué 2017). In general, conservationists must demonstrate to local people that protecting the environment not only saves species and ecosystems, but also improves the long-term health of their families and communities, their own economic well-being, and their quality of life (Morrison 2016). Public discussions, education, and publicity must be a major part of any such program, and they should be delivered in ways that instill trust and a sense of belonging and community. Conservationists should devote careful attention to convincing business leaders and politicians to support conservation efforts. National leaders may be among the most difficult people to convince since they must respond to diverse interests. However, whether their conversion occurs due to reason, sentiment, or professional self-interest, once they join the cause of conservation, these leaders may be in a position to make major contributions.

Finally, and most importantly, conservation biologists must become effective managers and practitioners of conservation projects. They must be willing to walk on the land and go out on the water to find out what is really happening, to get dirty, to knock on doors and talk with local people, and to take risks. Conservation biologists must learn everything they can about the species and ecosystems they are trying to protect, and then make that knowledge available to others in a form that can be readily understood and can affect decision-making.

> The goal of conservation biology is not just to reveal new knowledge, but also to use that knowledge to protect biodiversity. Conservation biologists must learn to demonstrate the practical application of their work.

If conservation biologists are willing to put their ideas into practice and to work with park managers, land use planners, politicians, and local people, progress will follow. Getting the right mix of models, new theories, innovative approaches, and practical examples is necessary for the discipline to succeed. Once this balance is found, conservation biologists, working with energized citizens and government officials, will be in a position to protect the world's biodiversity during this unprecedented era of change.

Summary

- Protecting biodiversity presents us with several challenges. To address these problems, we must change many policies and practices. Changes must occur at the local, national, and international levels and will require action on the part of individuals, conservation organizations, and governments.

- Conservation biologists must demonstrate the practical value of the theories and approaches of their discipline and actively work with all components of society to protect biodiversity and restore the degraded elements of the environment.

- To achieve the long-term goals of conservation biology, practitioners need to become involved in conservation education and the political process.

For Discussion

1. Consider a current conflict over conservation of biodiversity in the news. Does it fit into one or more of the problem-solution pairs provided in this chapter? What are some possible solutions to the conflict, given what you now know?

2. Heather Tallis and her colleagues have outlined "an attainable global vision" (Tallis et al. 2018; see the Suggested Readings) for how to meet international goals for conservation and climate change. Which of their proposals, if any, do you find to be realistic? Why?

3. As a result of studying conservation biology, have you decided to change your lifestyle or your level of political activity? Do you think you can make a difference in the world, and if so, in what way?

Suggested Readings

Blackman, A., L. Goff, and M. R. Planter. 2018. Does eco-certification stem tropical deforestation? Forest Stewardship Council certification in Mexico. *Journal of Environmental Economics and Management* 89: 306–333. It is important to measure the impact of certification programs, especially if they are falling short of expectations.

McKinley, D. C., et al. 2017. Citizen science can improve conservation science, natural resource management, and environmental protection. *Biological Conservation* 208: 15–28. The contribution of citizen scientists can be significant.

Pimm, S. L. 2021. What we need to know to prevent a mass extinction of plant species. *Plants, People, Planet* 3(1): 7–15. A call to action to document the Earth's species before it is too late.

Primack, R. B., A. A. Sher, B. Maas, and V. M. Adams. 2021. Manager characteristics drive conservation success. *Biological Conservation*, 259: 109169. Conservation scientists must consider and measure the human element in conservation and restoration.

Tallis, H. M., et al. 2018. An attainable global vision for conservation and human well-being. *Frontiers in Ecology and the Environment* 16(10): 563–570. Modeling has been used to map out what it would take to meet the demands of a growing human population.

Appendix

Selected Environmental Organizations and Sources of Information

The following are useful printed resources:

Conservation Directory 2017: The Guide to Worldwide Environmental Organizations (2016), published by Carrel Press. This directory lists over 4000 local, national, and international conservation organizations; conservation publications; and more than 18,000 leaders and officials in the field of conservation.

There is No Planet B (2021), by Mike Berners-Lee, published by Cambridge University Press. An overview of how human activities are affecting systems on Earth, using current research and data. It also includes guidance of how we can change our behaviors to improve our future.

Online searches, especially using Google, provide a powerful way to search for information concerning people, organizations, places, and topics. The following are more specialized searchable databases:

Catalogue of Life, 2021 Annual Checklist
www.catalogueoflife.org
The most comprehensive database available of described species on Earth.

Encyclopedia of Life
www.eol.org
Developing resource for species biology.

Global Biodiversity Information Facility
www.gbif.org
Free and open access to biodiversity data.

IUCN Red List
www.iucnredlist.org
Information on all plants, fungi, and animals that have been globally evaluated for conservation status.

USDA Plants List
plants.usda.gov
Data on plants, including distribution, nativity, and wetland status.

Below is a list of some major conservation organizations and resources:

Association of Zoos and Aquariums (AZA)
8403 Colesville Road, Suite 710
Silver Spring, MD 20910 USA
www.aza.org
Preservation and propagation of captive wildlife.

BirdLife International
The David Attenborough Building, 1st floor
Pembroke Street
Cambridge, CB2 3QZ, UK
www.birdlife.org
Determines status, priorities, and conservation plans for birds throughout the world.

Catalogue of Life
www.catalogueoflife.org
An index of the world's known species.

Center for Plant Conservation
15600 San Pasqual Valley Road
Escondido, CA 92027 USA
saveplants.org
Major center for worldwide plant
conservation activities.

CITES Secretariat of Wild Fauna and Flora
International Environment House
11 Chemin des Anémones
CH-1219 Châtelaine, Geneva, Switzerland
www.cites.org
Regulates trade in endangered species.

Conservation International (CI)
2011 Crystal Drive, Suite 600
Arlington, VA 22202 USA
www.conservation.org
Active in international conservation efforts and
developing conservation strategies; home of the
Center for Applied Biodiversity Science.

Convention on Biological Diversity Secretariat
413 Rue Saint-Jacques, Suite 800
Montreal, Quebec, H2Y 1N9, Canada
www.cbd.int
Promotes the goals of the CBD: sustainable
development, biodiversity conservation, and
equitable sharing of genetic resources.

Earthwatch Institute
1380 Soldiers Field Road
Boston, MA 02135 USA
www.earthwatch.org
Clearinghouse for international conservation
projects in which volunteers can work
with scientists.

Employment Opportunities
Various organizations have websites with
environmental and conservation opportunities
and internships throughout the world: www.
webdirectory.com/employment and www.ecojobs.
com, to name a few. A publication of interest is
Careers in the Environment by Mike Fasulo and Paul
Walker, published by McGraw-Hill.

Environmental Defense Fund (EDF)
1875 Connecticut Avenue NW, Suite 600
Washington, DC 20009 USA
www.edf.org
Involved in scientific, legal, and economic issues.

European Center for Nature Conservation (ECNC)
P.O. Box 90154
5000 LG Tilburg, the Netherlands
www.ecnc.org
Provides the scientific expertise that is required for
formulating conservation policy.

Fauna Europeaea
fauna-eu.org
A database on the distribution of 130,000 terrestrial
and freshwater species.

Fauna & Flora International
The David Attenborough Building
Pembroke Street
Cambridge, CB2 3QZ, UK
www.fauna-flora.org
Long-established international conservation body
acting to protect species and ecosystems.

Food and Agriculture Organization of the United Nations (FAO)
Viale delle Terme di Caracalla
00513 Rome, Italy
www.fao.org
A UN agency supporting sustainable agriculture,
rural development, and resource management.

Friends of the Earth
1101 15th Street NW, 11th floor
Washington, DC 20036 USA
www.foe.org
Attention-grabbing organization working to
improve and expand environmental policy.

Global Biodiversity Information Facility
www.gbif.org
An intergovernmental collaboration to facilitate
access to biodiversity data.

Global Environment Facility (GEF) Secretariat
1899 Pennsylvania Avenue NW
Washington, DC 20006 USA
www.thegef.org
Funds international biodiversity and
environmental projects.

Greenpeace International
Ottho Heldringstraat 5
1006 AZ Amsterdam, the Netherlands
www.greenpeace.org/international
Activist organization known for grassroots
efforts and dramatic protests against
environmental damage.

iNaturalist
www.inaturalist.org
An app and database of crowd-sourced species
identifications across the globe.

National Audubon Society

225 Varick Street, 7th floor
New York, NY 10014 USA
www.audubon.org
Involved in wildlife conservation, public education, research, and political lobbying, with emphasis on birds.

National Council for Science and the Environment (NCSE)

740 15th Street NW, Suite 900
Washington, DC 20005 USA
www.ncseonline.org
Works to improve the scientific basis for environmental decision making; their website provides extensive environmental information.

National Wildlife Federation (NWF)

11100 Wildlife Center Drive
Reston, VA 20190 USA
www.nwf.org
Advocates for wildlife conservation. Publishes the *Conservation Directory 2005–2006*, as well as the children's publications *Ranger Rick* and *Your Big Backyard*.

Natural Resources Defense Council (NRDC)

40 West 20th Street, 11th floor
New York, NY 10011 USA
www.nrdc.org
Uses legal and scientific methods to monitor and influence government actions and legislation.

The Nature Conservancy (TNC)

4245 North Fairfax Drive, Suite 100
Arlington, VA 22203 USA
www.nature.org
Emphasizes land preservation.

NatureServe

2511 Richmond Highway, Suite 930
Arlington, VA 22202 USA
www.natureserve.org
Maintains databases of endangered species for North America.

The New York Botanical Garden (NYBG) Institute of Economic Botany (IEB)

International Plant Science Center, the New York Botanical Garden
2900 Southern Boulevard
Bronx, NY 10458 USA
www.nybg.org
Conducts research and conservation programs involving plants that are useful to people.

Ocean Conservancy

1300 19th Street NW, 8th floor
Washington, DC 20036 USA
www.oceanconservancy.org
Focuses on marine wildlife and ocean and coastal habitats.

PREDICTS (Projecting Responses of Ecological Diversity in Changing Terrestrial Systems) database

PREDICTS | Natural History Museum (nhm.ac.uk)
A global database of terrestrial species' responses to human pressures.

Protected Planet

www.protectedplanet.net
Database on protected areas managed by UNEP-WCMC with support from IUCN.

Rainforest Action Network

425 Bush Street, Suite 300
San Francisco, CA 94108 USA
www.ran.org
Works for rainforest conservation and human rights.

Royal Botanic Gardens, Kew

Richmond, Surrey, TW9 3AB, UK
www.kew.org
The famous Kew Gardens are home to a leading botanical research institute and an enormous plant collection.

Sierra Club

2101 Webster Street, Suite 1300
Oakland, CA 94612 USA
www.sierraclub.org
Leading advocate for the preservation of wilderness and open space.

Smithsonian National Zoological Park

3001 Connecticut Avenue NW
Washington, DC 20008 USA
www.nationalzoo.si.edu
The National Zoo and the nearby National Museum of Natural History represent a vast resource of literature, biological materials, and skilled professionals.

Society for Conservation Biology (SCB)

1133 15th Street NW, Suite 300
Washington, DC 20005 USA
www.conbio.org
Leading scientific society for the field. Develops and publicizes new ideas and scientific results through the journal *Conservation Biology* and annual meetings.

Student Conservation Association (SCA)

4245 North Fairfax Drive, Suite 825
Arlington, VA 22203 USA
www.thesca.org
Places volunteers and interns with conservation organizations and public agencies.

United Nations Development Programme (UNDP)
1 United Nations Plaza
New York, NY 10017 USA
www.undp.org
Funds and coordinates international economic
development activities.

United Nations Environment Programme (UNEP)
United Nations Avenue, Gigiri
P.O. Box 30552, 00100
Nairobi, Kenya
www.unep.org
International program of environmental research
and management.

United Nations Environment Programme World Conservation Monitoring Centre (UNEP-WCMC)
219 Huntingdon Road
Cambridge, CB3 0DL, UK
www.unep-wcmc.org
Monitors global wildlife trade, the status of
endangered species, natural resource use, and
protected areas.

United Nations Sustainable Development Goals Tracker
sdg-tracker.org
Presents data on all available indicators from the
global development targets adopted in 2015, to be
achieved by 2030.

United States Fish and Wildlife Service (USFWS)
Department of the Interior
1849 C Street NW
Washington, DC 20240 USA
www.fws.gov
The leading U.S. government agency concerned
with conservation research and management,
with connections to state governments and other
government units, including the National Marine
Fisheries Service, the U.S. Forest Service, and the
Agency for International Development, which is
active in developing nations. The *Conservation
Directory 2005–2006*, mentioned previously, shows
how these units are organized.

Wetlands International
P.O. Box 471
6700 AL Wageningen, the Netherlands
www.wetlands.org
Focus on the conservation and management
of wetlands.

The Wilderness Society
1615 M Street NW
Washington, DC 20036 USA
www.wilderness.org
Devoted to preserving wilderness and wildlife.

Wildlife Conservation Society (WCS)
2300 Southern Boulevard
Bronx, NY 10460 USA
www.wcs.org
Leaders in wildlife conservation and research.

The World Bank
1818 H Street NW
Washington, DC 20433 USA
www.worldbank.org
Multinational bank involved in economic
development; increasingly concerned with
environmental issues.

World Conservation Union (IUCN)
Rue Mauverney 28
1196, Gland, Switzerland
www.iucn.org
Coordinating body for international conservation
efforts. Produces directories of specialists and the
Red List of endangered species.

World Database on Protected Areas (WDPA)
https://www.iucn.org/theme/protected-areas/our-
work/quality-and-effectiveness/world-database-
protected-areas-wdpa
A project of the IUCN; the most comprehensive
global database on protected areas.

World Resources Institute (WRI)
10 G Street NE, Suite 800
Washington, DC 20002 USA
www.wri.org
Produces environmental, conservation, and
development reports.

World Wildlife Fund (WWF)
1250 24th Street NW
Washington, DC 20037 USA
www.wwf.org
Major conservation organization, with
branches throughout the world. Active in national
park management.

The Xerces Society
628 NE Broadway, Suite 200
Portland, OR 97232 USA
www.xerces.org
Focuses on the conservation of insects and
other invertebrates.

Zoological Society of London (ZSL)
Outer Circle, Regent's Park
London, NW1 4RY, UK
www.zsl.org
Center for worldwide activities to preserve nature.

Glossary

Numbers in brackets indicate the chapter[s] in which the term is defined.

A

abiotic Not derived from living things. [2]

acid rain Rainwater that has become acidic due to air pollution. [4]

adaptive management Implementing a management plan and monitoring how well it works, then using the results to adjust the management plan. [9, 11]

Aichi Biodiversity Targets A list of goals to achieve sustainability and the protection of biodiversity developed by the Convention on Biological Diversity (CBD). [12]

Allee effect Inability of a species' social structure to function once a population of that species falls below a certain number or density of individuals. [6]

alleles Different forms of the same gene (e.g., different alleles of the genes for certain blood proteins produce the different blood types found among humans). [2]

alpha diversity The number of different species in a community or specific location; species richness. [2]

amenity value Recreational value of biodiversity, including ecotourism. [3]

anthropogenic Originating from human sources. [5]

arboretum Specialized botanical garden focusing on trees and other woody plants. [8]

artificial incubation Conservation strategy that involves humans taking care of eggs or newborn animals. [8]

artificial insemination Introduction of sperm into a receptive female animal by humans; used to increase the reproductive output particularly of endangered species. [8]

B

background extinction rates The rate of species loss expected to occur in the absence of human impact. [6]

beneficiary value See bequest value. [3]

bequest value The benefit people receive by preserving a resource or species for their children and descendants or future generations and quantified as the amount people are willing to pay for this goal. Also known as beneficiary value. [3]

beta diversity Rate of change of species composition along a gradient or transect. [2]

binomial The unique two-part Latin name taxonomists bestow on a species, such as *Canis lupus* (gray wolf) or *Homo sapiens* (humans). [2]

bioblitz A one-day event in which scientists and citizen scientists perform an intensive biological survey of a designated area in a short time with the goal of documenting all living species in that area. [2]

biocontrol The use of one type of organism, such as an insect, to manage another, undesirable, species, such as an invasive plant. [11]

biocultural restoration Restoring lost ecological knowledge to people to give them an appreciation of the natural world. [11]

biodiversity The complete range of species, biological communities, and their ecosystem interactions and genetic variation within species. Also known as biological diversity. [1]

biodiversity indicators Species or groups of species that provide an estimate of the biodiversity in an area when data on the whole community are unavailable. Also known as surrogate species. [7]

biodiversity offsets See compensatory mitigation. [11]

biological community A group of species that occupies a particular locality. [2]

biological control The use of one living organism to reduce the numbers of another, pest species. See biocontrol. [11]

biological definition of species Among biologists, the most generally used of several definitions of "species." A group of individuals that can potentially breed among themselves in the wild and that do not breed with individuals of other groups. *Compare with* morphological definition of species. [2]

biological diversity See biodiversity. [1]

biomagnification Process whereby toxins become more concentrated in animals at higher levels in the food chain. [4]

biophilia The postulated predisposition in humans to feel an affinity for the diversity of the living world. [1]

biopiracy Collecting and using biological materials for commercial, scientific, or personal use without obtaining the necessary permits. [3]

bioprospecting Collecting biological materials as part of a search for new products. [3]

bioregional management Management system that focuses on a single large ecosystem or a series of linked ecosystems, particularly where they cross political boundaries. [10]

bioremediation The use of an organism to clean up pollutants, such as bacteria that break down the oil in an oil spill or wetland plants that take up agricultural runoff to clean the water. [11]

biosphere reserves Protected areas established as part of a United Nations program to demonstrate the compatibility of biodiversity conservation and sustainable development to benefit local people. [9]

biota A region's plants and animals. [2]

biotic Related to living things. [2]

Bonn Convention Treaty to protect European species, particularly migratory species. Also called the Convention on the Conservation of Migratory Species of Wild Animals. [7]

bushmeat Meat from any wild animal. [3]

C

carnivores An animal species that consumes other animals to survive. Also called a secondary consumer or predator. *Compare with* primary consumers. [2]

carrying capacity The number of individuals or biomass of a species that an ecosystem can support. [2]

census A count of the number of individuals in a population. [7]

co-management Local people working as partners with government agencies and conservation organizations in protected areas. [10]

commodity values *See* direct use values. [3]

common-property resources Natural resources that are not controlled by individuals but collectively owned by society. Also known as open-access resources or common-pool resources. [3]

community-based conservation (CBC) Protection of natural areas and resources that is controlled, owned, and/or managed by the local people; an alternative to government-based conservation. [10]

community conserved areas Protected area managed and sometimes established by local people. [10]

compensatory mitigation When a new site is created or rehabilitated in compensation for a site damaged or destroyed elsewhere. Also known as biodiversity offset. [11]

competition A contest between individuals or groups of animals for resources. Occurs when individuals or a species use a limiting resource in a way that prevents others from using it. [2]

conservation banking A system involving developers paying landowners for the preservation of an endangered species or protected habitat type (or even restoration of a degraded habitat) to compensate for a species or habitat that is destroyed elsewhere. [12]

conservation biology Scientific discipline that draws on diverse fields to carry out research on biodiversity, identify threats to biodiversity, and play an active role in the preservation of biodiversity. [1]

conservation concessions Methods of protecting land whereby a conservation organization pays a government or other landowner to preserve habitat rather than allow an extractive industry to damage the habitat. [12]

conservation corridors Connections between protected areas that allow for dispersal and migration. Also known as habitat corridors or movement corridors. [9]

conservation covenants *See* conservation easements. [12]

conservation development *See* limited development. [12]

conservation easements (CEs) Method of protecting land in which landowners give up the right to develop or build on their property, often in exchange for financial or tax benefit. [10, 12]

conservation genetics The use of genetic information to address issues within conservation biology. [2]

conservation leasing Providing payments to private landowners who actively manage their land for biodiversity protection. [12]

conservation translocations The deliberate placement of organisms to benefit the survival of a population, species, and/or ecosystem. [8]

consumptive use value Value assigned to goods that are collected and consumed locally. [3]

Convention Concerning the Protection of the World Cultural and Natural Heritage *See* World Heritage Convention. [12]

Convention on Biological Diversity (CBD) A treaty that obligates countries to protect the biodiversity within their borders and gives them the right to receive economic benefits from the use of that biodiversity. [12]

Convention on International Trade in Endangered Species (CITES) The international treaty that establishes lists (known as Appendices) of species for which international trade is to be prohibited, regulated, or monitored. [7]

Convention on the Conservation of Migratory Species of Wild Animals (CMS) *See* Bonn Convention. [7]

cost–benefit analysis Comprehensive analysis that compares values gained against the costs of a project or resource use. [3]

cost–effectiveness analysis An alternative to cost–benefit analysis that compares the impact (financial and otherwise) and costs of alternative means of accomplishing an objective, such as the protection of a species. [3]

cross-fostering Conservation strategy in which individuals from a common species raise the offspring of a rare, related species. [8]

cryptic species Two or more species that have similar appearance but that are genetically distinct. [2]

cultural eutrophication Algal blooms and associated impacts caused by excess mineral nutrients released into the water from human activity. [11]

D

debt-for-nature swaps Agreements in which a developing country agrees to fund additional conservation activities in exchange for a conservation organization canceling some of its discounted debt. [12]

decomposers A species that feeds or grows on dead plant and animal material. Also called a detritivore. [2]

deep ecology Philosophy emphasizing biodiversity protection, personal lifestyle changes, and working toward political change. [3]

de-extinction The process of bringing an extinct species back to life in some way using modern genetic technology or through selective breeding. [6]

degazettement Government actions taken to remove the legal status of protected areas. [9]

demographic stochasticity Random variation in birth, death, and reproductive rates in small populations, sometimes causing further decline in population size. Also called demographic variation. [6]

demographic studies Studies in which individuals and populations are monitored over time to determine rates of growth, reproduction, and survival. [7]

demographic variation *See* demographic stochasticity. [6]

described species A species that is officially characterized and identified by science. [2]

detritivores *See* decomposers. [2]

direct use values Value assigned to products, such as timber and animals, that are harvested and directly used by the people who harvest them. Also known as commodity value or private goods. [3]

DNA barcoding A method of species identification using DNA and other genetic technology. [2, 5]

E

Earth Summit An international conference held in 1992 in Rio de Janeiro that resulted in new environmental agreements. Also known as the Rio Summit. [12]

ecocolonialism Practice of governments and conservation organizations disregarding the land rights and traditions of local people in order to establish new conservation areas. [10]

ecological economics Discipline that includes valuations of biodiversity in economic analyses. [3]

ecological footprint The influence a group of people has on both the surrounding environment and locations across the globe as measured by global hectares per person. [4]

ecological resilience A natural ability to recover after disturbance. [11]

ecological restoration Altering a site to reestablish an indigenous ecosystem. [11]

ecological succession A predictable, gradual, and progressive change in species over time. [9]

ecologically extinct A species that has been so reduced in numbers that it no longer has a significant ecological impact on the biological community. *See* functionally extinct. [6]

economic development Economic activity focused on improvements in efficiency and organization but not necessarily on increases in resource consumption. [12]

economic growth Economic activity characterized by increases in the amount of resources used and in the amount of goods and services produced. [12]

economics The study of factors affecting the production, distribution, and consumption of goods and services. [3]

ecosystem A biological community together with its associated physical and chemical environment. [2]

ecosystem diversity The variety of ecosystems present in a place or geographic area. [2]

ecosystem engineers Species that modify the physical structure of an ecosystem. [2]

ecosystem management Large-scale management that often involves multiple stakeholders, the primary goal of which is the preservation of ecosystem components and processes. [1, 10]

ecosystem services Range of benefits provided to people from ecosystems, including flood control, clean water, and reduction of pollution. [3]

ecotourism Tourism, especially in developing countries, focused on viewing unusual and/or especially charismatic biological communities and species that are unique to a country or region. [3]

edge effects Altered environmental and biological conditions at the edges of a fragmented habitat. [4]

effective population size (N_e) The number of breeding individuals in a population. [6]

effect size A statistical term that refers to the magnitude of an impact of interest. [4]

embryo transfer The surgical implantation of embryos into a surrogate mother; used to increase the number of individuals of a rare species, with a common species used as the surrogate mother. [8]

Endangered Species Act (ESA) An important US law passed to protect endangered species and the ecosystems in which they live. [7]

endemic Occurring in a place naturally, without the influence of people (e.g., gray wolves are endemic to Canada). [6]

endemic species Species found in one place and nowhere else (e.g., the many lemur species found only on the island of Madagascar). [2]

endemism The ecological state of being endemic. [7]

environmental DNA (eDNA) DNA shed into the environment by organisms and that is detected with genetic techniques. [2]

environmental economics Discipline that examines the economic impacts of environmental policies and decisions. [3]

environmental ethics Discipline of philosophy that articulates the intrinsic value of the natural world and people's responsibility to protect the environment. [3]

environmental impact assessments Evaluation of a project that considers its possible present and future impacts on the environment. [3]

environmentalism A widespread movement, characterized by political activism, with the goal of protecting the natural environment. [1]

environmental justice Movement that seeks to empower and assist poor and politically weak people in protecting their own environments; their well-being and the protection of biological diversity are enhanced in the process. [3]

environmental stochasticity Random variation in the biological and physical environment. Can increase the risk of extinction in small populations. [6]

eutrophication Process of degradation in aquatic environments caused by nitrogen and phosphorus pollution and characterized by algal blooms and oxygen depletion. [4]

evolution Genetic changes over time in a population that sometimes can result in a new species. [2]

evolutionary definition of species A group of individuals that share unique similarities of their DNA and hence their evolutionary past. [2]

existence value The benefit people receive from knowing that a habitat or species exists and quantified as the amount that people are willing to pay to prevent species from being harmed or going extinct, habitats from being destroyed, and genetic variation from being lost. [3]

ex situ conservation Preservation of species under artificial conditions, such as in zoos, aquariums, and botanical gardens. [8]

extant Presently alive; not extinct. [6]

externalities Hidden costs or benefits that result from an economic activity to individuals or a society not directly involved in that activity. [3]

extinct The condition in which no members of a species are currently living. [6]

extinct in the wild A species no longer found in the wild, but individuals may remain alive in zoos, botanical gardens, or other artificial environments. [6]

extinction cascade A series of linked extinctions whereby the extinction of one species leads to the extinction of one or more other species. [2]

extinction debt When an ecosystem has more species than can be supported by the habitat, following habitat destruction and fragmentation, and some of those species are predicted to go locally extinct. [7]

extinction vortex Tendency of small populations to decline toward extinction. [6]

extirpated Local extinction of a population, even though the species may still exist elsewhere. [6]

extractive reserve Protected area in which sustainable extraction of certain natural products is allowed. [10]

F

fieldwork Observations and data collection in the natural environment. [7]

flagship species A species that captures public attention; aids in conservation efforts, such as establishing a protected area; and may be crucial to ecotourism. [7]

focal species A species that provides a reason for establishing a protected area. [7]

food chains Specific feeding relationships between species at different trophic levels. [2]

food web A network of feeding relationships among species. [2]

founder effect Reduced genetic variability that occurs when a new population is established ("founded") by a small number of individuals. [6]

four Rs Guidelines used by conservation biologists when designing nature reserves: representation, resiliency, redundancy, and reality. [9]

frontier forest Intact blocks of undisturbed forest large enough to support all aspects of biodiversity. [4]

functional diversity The diversity of organisms categorized by their ecological roles or traits rather than their taxonomy. [2]

functionally extinct The state in which a species persists at such reduced numbers that its effects on the other species in its community are negligible. *See* ecologically extinct. [6]

G

gamma diversity The number of species in a large geographic area. [2]

gap analysis Comparing the distribution of endangered species and biological communities with existing and proposed protected areas to determine gaps in protection. [9]

gene flow The movement of genes from one population to another through movement of individuals or gametes. [6]

gene pool The total array of genes and alleles in a population. [2]

genes Units (DNA sequences) on a chromosome that code for specific proteins. Also called loci. [2]

genetically modified organisms (GMOs) Organisms whose genetic code has been altered by scientists using recombinant DNA technology. [5]

genetic diversity The range of genetic variation found within a species. [2]

genetic drift Loss of genetic variation and change in allele frequencies that occur by chance in small populations. [6]

genetic monitoring A method that uses DNA and other genetic techniques to measure presence and abundance species and to detect rates of inbreeding or immigration among populations. [7]

genetic rescue The practice of intentional introduction of genetic variation into the population of a rare species to keep it from extinction. [6]

genetic structure Patterns of genotypes within and among populations. [2]

genome resource banking (GRB) Collecting DNA, eggs, sperm, embryos, and other tissues from species that can be used in breeding programs and scientific research. [8]

genotype Particular combination of alleles that an individual possesses. [2]

geographic information systems (GIS) Computer analyses that integrate and display spatial data; relating in particular to the natural environment, ecosystems, species, protected areas, and human activities. [9]

global change Any alteration of the environment caused by humans that is facilitating changes in ecosystems across the world. [all chapters]

Global Environment Facility (GEF) A large international program involved in funding conservation activities in developing countries. [12]

globalization The increasing interconnectedness of the world's economy. [4]

globally extinct No individuals are presently alive anywhere. [6]

gray literature Written material, especially reports, produced by government agencies and conservation organizations that is not published in scientific journals. [7]

greenhouse effect Warming of the Earth caused by carbon dioxide and other "greenhouse gases" in the atmosphere that allow the sun's radiation to penetrate and warm the Earth but prevent the heat generated by sunlight from re-radiating. Heat is thus trapped near the surface, raising the planet's temperature. [5]

greenhouse gases Gases in the atmosphere, primarily carbon dioxide, that are transparent to sunlight but that trap heat near the Earth's surface. [5]

guild A group of species at the same trophic level that use approximately the same environmental resources. [2]

H

habitat The location or type of environment in which a specific animal or plant species lives. [2]

habitat conservation plans (HCPs) Regional plans that allow development in designated areas while protecting biodiversity in other areas. [7]

habitat corridors *See* conservation corridors. [9]

habitat fragmentation The process whereby a large, continuous area of habitat is both reduced in area and divided into two or more fragments. [4]

habitat islands Intact habitat surrounded by an unprotected matrix of inhospitable terrain. [9]

hard release In the establishment of a new population, when individuals from an outside source are released in a new location without assistance. *Compare with* soft release. [8]

healthy ecosystem Ecosystem in which processes are functioning normally, whether or not there are human influences. [2]

herbivores A species that eats plants or other photosynthetic organisms. Also called a primary consumer. [2]

herbivory Predation on plants. [2]

heterozygous Condition of an individual having two different allele forms of the same gene. [2]

homozygous Condition of an individual having two identical allele forms of the same gene. [2]

hotspots Regions with numerous species, many of which are endemic, that are also under immediate threat from human activity. [7]

hybridize Interbreeding between different species. [2]

hybrids Intermediate offspring resulting from mating between individuals of two different species. [2]

I

in situ conservation Preservation of natural communities and populations of endangered species in the wild. [8]

inbreeding depression Lowered reproduction or production of weak offspring following mating among close relatives or self-fertilization. [6]

indicator species Species used in a conservation plan to identify and often protect a biological community or set of ecosystem processes. [7]

indirect use values Values provided by biodiversity that do not involve harvesting or destroying the resource (such as water quality, soil protection, recreation, and education). Also known as public goods. [3]

insular biogeography A subdiscipline of biogeography devoted to exploring species diversity on island and in isolated natural communities. [6]

integrated conservation development projects (ICDPs) Conservation projects that also provide for the economic needs and welfare of local people. [10]

integrated pest management An approach to controlling undesirable plants or animals that has the goal of minimizing harm to the ecosystem and people, while being cost-effective. [5]

International Union for Conservation of Nature (IUCN) *See* IUCN. [5]

intrinsic value Value of a species and other aspects of biodiversity for their own sake, unrelated to human needs. [3]

introduction program Moving individuals to areas outside their historical range in order to create a new population of an endangered species. [8]

invasive A species that increases in abundance, often because of human changes to the environment, and thereby threatens native species through competition, predation, or by changing ecosystem properties or dynamics. [5]

island biogeography model Formula for the relationship between island size and the number of species living on the island; the model can be used to predict the impact of habitat destruction on species extinctions, viewing remaining habitat as an "island" in the "sea" of a degraded ecosystem. [6]

IUCN International Union for the Conservation of Nature is a major international conservation organization; previously known as The World Conservation Union. [5]

K

keystone resources Any resource in an ecosystem that is crucial to the survival of many species; for example, a watering hole. [2]

keystone species A species that has a disproportionate impact (relative to its numbers or biomass) on the organization of a biological community. Loss of a keystone species may have far-reaching consequences for the community. [2]

L

land ethic Aldo Leopold's philosophy advocating human use of natural resources that is compatible with or even enhances ecosystem health. [1]

land sharing Land use that combines resource use and conservation. [10]

land sparing Land that is protected when other lands are used more intensively. [10]

land trusts Conservation organizations that protect and manage land. [12]

landraces A variety of crop that has unique genetic characteristics; local species that have been adapted by humans over time. [8]

landscape ecology Discipline that investigates patterns of habitat types and their influence on species distribution and ecosystem processes. [9]

legal title The right of ownership of land, recognized by a government and/or judicial system; traditional people often struggle to achieve this recognition. [10]

limited development Compromise involving a landowner, a property developer, and a conservation organization that combines some development with protection of the remaining land. [12]

limiting resource Any requirement for existence whose presence or absence limits a population's size. In the desert, for example, water is a limiting resource. [2]

Living Planet Index A measure of the conservation status of species, based on the IUCN categories. [7]

locally extinct A species that no longer exists in a place where it used to occur, but still exists elsewhere. [6]

loci (singular, locus) *See* genes. [2]

M

management plans A statement of how to protect biodiversity in an area, along with methods for implementation. [9]

market failure Misallocation of resources in which certain individuals or businesses benefit from using a common resource, such as water, the atmosphere, or a forest, but other individuals, businesses, or the society at large bears the cost. [3]

meta-analysis A statistical tool that combines data from multiple sources. [4]

metapopulation Shifting mosaic of populations of the same species linked by some degree of migration; a "population of populations." [7]

microplastics Particles of plastic debris in oceans, the atmosphere, or on land that are less than 5mm long that originate from consumer products and industrial waste. [4]

minimum dynamic area (MDA) Area needed for a population to have a high probability of surviving into the future. [7]

minimum viable population (MVP) Number of individuals necessary to ensure a high probability that a population will survive a certain number of years into the future. [7]

mitigation Process by which a new population or habitat is created to compensate for a habitat damaged or destroyed elsewhere. [8]

morphological definition of species A group of individuals, recognized as a species, that is morphologically, physiologically, or biochemically distinct from other groups. *Compare with* biological definition of species. [2]

morphospecies Individuals that are probably a distinct species based on their appearance but that do not currently have a scientific name. [2]

movement corridors *See* conservation corridors. [9]

multiple-use habitat An area managed to provide a variety of goods and services. [10]

mutations Changes that occur in genes and chromosomes, sometimes resulting in new allele forms and genetic variation. [2]

mutualism When two species benefit each other by their relationship. [2]

N

national environmental fund (NEF) A trust fund or foundation that uses its annual income to support conservation activities. [12]

natural history The ecology and distinctive characteristics of a species. [7]

nonconsumptive use value Value assigned to benefits provided by some aspect of biodiversity that does not involve harvesting or destroying the resource (such as water quality, soil protection, recreation, and education). [3]

nonpoint source pollution Pollution coming from a general area rather than a specific site. [11]

non-use values Values of something that is not presently used; for example, existence value. [3]

normative discipline A discipline that embraces ethical commitment rather than ethical neutrality. [1]

novel ecosystems Ecosystems in which there is a mixture of native and nonnative species coexisting in a community unlike the original or reference site. [11]

O

official development assistance (ODA) International aid by developed countries to developing countries intended to promote economic development. [12]

omnivores Species that eat both plants and animals. [2]

open-access resources Natural resources that are not controlled by individuals but are collectively owned by society. [3]

option value Value of biodiversity in providing possible future benefits for human society (such as new medicines). [3]

P

parasites Organisms that live on or in another organism (host), receiving nutritive benefit while decreasing the fitness of the host, which remains alive. [2]

Paris Accord An agreement made in Paris in 2015 by 195 nations to lower greenhouse gas emissions with the goal of preventing atmospheric temperatures from increasing more than 2°C. [12]

passive restoration Letting an ecosystem recover on its own. [11]

pathogens Disease-causing organisms. [2]

payments for ecosystem services (PES) Direct payments to individual landowners and local communities that protect species or critical ecosystem characteristics. [10, 12]

peer-reviewed The content has been critically evaluated by experts before it is published. [7]

perverse subsidies Government payments or other financial incentives to industries that result in environmentally destructive activities. [3]

phenotype The morphological, physiological, anatomical, and biochemical characteristics of an individual that result from the expression of its genotype in a particular environment. [2]

polymorphic genes Within a population, genes that have more than one form or allele. [2]

population A geographically defined group of individuals of the same species that mate and otherwise interact with one another. *Compare with* metapopulation. [2]

population biology Study of the ecology and genetics of populations, often with a focus on population numbers. [7]

population bottleneck A radical reduction in population size (e.g., following an outbreak of infectious disease), sometimes leading to the loss of genetic variation. [6]

population viability analysis (PVA) Demographic analysis that predicts the probability of a population persisting in an environment for a certain period of time; sometimes linked to various management scenarios. [7]

precautionary principle Principle stating that it may be better to avoid taking a particular action due to the possibility of causing unexpected harm. [3]

predation Act of killing and consuming another organism for food. [2]

predator release hypothesis Hypothesis that attributes the success of invasive species to the absence of specialized natural predators and parasites in their new range. [5]

predators *See* carnivores. [2]

preservationist ethic A belief in the need to preserve wilderness areas for their intrinsic value. [1]

prey An animal that is eaten as food by another species. [2]

primary consumers *See* herbivores. [2]

primary literature Published material that presents the results from original research, which is usually a peer-reviewed article. [7]

primary producers Organisms such as green plants, algae, and seaweeds that obtain their energy directly from the sun via photosynthesis. Also known as autotrophs. [2]

private goods *See* direct use values. [3]

productive use value Value assigned to products that are sold in markets. [3]

protected area A habitat managed primarily or in large part for biodiversity. [9]

public goods Nonconsumptive benefits that belong to society in general, without private ownership. Also known as indirect use values. [3]

R

Ramsar Convention on Wetlands A treaty that promotes the protection of wetlands of international importance. [12]

rapid assessment programs (RAPS) An intensive effort to quickly collect data in order to make decisions. (*See* rapid biodiversity assessments.) [7]

rapid biodiversity assessments Species inventories and vegetation maps made by teams of biologists when urgent decisions must be made on where to establish new protected areas. Also known as *rapid assessment programs* (RAPs). [7]

recombination Mixing of the genes on the two copies of a chromosome that occurs during meiosis (i.e., in the formation of egg and sperm, which contain only one copy of each chromosome). Recombination is an important source of genetic variation. [2]

reconciliation ecology The science of developing urban places in which people and biodiversity can coexist. [10, 11]

recovery criteria Predetermined thresholds (such as numbers of individuals alive in the wild) that signal that an endangered species can be removed from protection under the Endangered Species Act. [7]

Red Data Books Compilations of lists ("Red Lists") of endangered species prepared by the IUCN and other conservation organizations. [7]

Red List criteria Quantitative measures of threats to species based on the probability of extinction. [7]

Red List Index Measure of the conservation status of species based on the IUCN categories. [7]

Red Lists Lists of endangered species prepared by the IUCN. [7]

Reducing Emissions from Deforestation and Forest Degradation (REDD) Program using financial incentives to reduce the emissions of greenhouse gases from deforestation. [12]

reference site Control site that provides goals for restoration in terms of species composition, community structure, and ecosystem processes. [11]

rehabilitation A type of ecological restoration in which a degraded ecosystem is replaced with a different but productive ecosystem type. [11]

reinforcement program Releasing new individuals into an existing population to increase population size and genetic variability. [8]

reintroduction program The release of captive bred or wild-collected individuals at a site within their historical range where the species does not presently occur. [8]

replacement cost approach How much people would have to pay for an equivalent product if what they normally use is unavailable. [3]

representative site Protected area that includes species and ecosystem properties characteristic of a larger area. [7]

resilience The ability of an ecosystem to return to its original state following disturbance. [2]

resistance The ability of an ecosystem to remain in the same state even with ongoing disturbance. [2]

resource conservation ethic Natural resources should be used for the greatest good of the largest number of people for the longest time. [1]

restoration ecology The scientific study of restored populations, communities, and ecosystems. [11]

rewilding Returning species, in particular large mammals to landscape, to approximate their natural condition prior to human impact. [11]

Rio Summit *See* Earth Summit. [12]

S

secondary consumers See carnivores. [2]

secondary invasion When the removal of an invasive species is followed by an invasion by a different species. [11]

secondary literature Published material that reports on the findings of others, such as a literature review or commentary. [7]

seed banks Collections of stored seeds, collected from wild and cultivated plants; used in conservation and agricultural programs. [8]

Shannon diversity index A species diversity index that takes into account the numbers of different species and their relative abundance. [2]

shifting cultivation Farming method in which farmers cut down trees, burn them, plant crops for a few years, and then abandon the site when soil fertility declines. Also called "slash-and-burn" agriculture. [4]

sink populations Populations that receive an influx of new individuals from a source population. [7]

sixth extinction episode The present mass extinction event that is just beginning. [6]

SLOSS debate Controversy concerning the relative advantages of a single large or several small conservation areas. "SLOSS" stands for single *l*arge *o*r *s*everal *s*mall. [9]

soft release In the establishment of a new population, when individuals are given assistance during or after the release to increase the chance of success. *Compare with* hard release. [8]

source populations Established populations from which individuals disperse to new locations. [7]

species–area relationship The number of species found in an area increases with the size of the area (i.e., more species are found on large islands than on small islands). [6]

species diversity The entire range of species found in a particular place. [2]

species richness The number of species found in a community. [2]

stable ecosystems Ecosystems that are able to remain in roughly the same compositional state despite human intervention or stochastic events such as unseasonable weather. [2]

stochasticity Random variation; variation happening by chance. [6]

survey Repeatable sampling method to estimate population size or density, or some other aspect of biodiversity. [7]

sustainable development Economic development that meets present and future human needs without damaging the environment and biodiversity. [1, 12]

symbiotic A mutualistic relationship in which neither of the two species involved can survive without the other. [2]

T

TEK (traditional ecological knowledge) Knowledge of ecosystems acquired by Indigenous People through generations of direct experience with the environment in a particular place. Also known as native science or Indigenous knowledge. [10]

taxonomists Scientists involved in the identification and classification of species. [2]

tertiary consumers The fourth trophic level, in which predators eat other predators. [2]

threatened Species that fall into the endangered or vulnerable to extinction categories in the IUCN system. Under the US Endangered Species Act, refers to species at risk of extinction, but at a lower risk than endangered species. [7]

tragedy of the commons The unregulated use of a public resource that results in its degradation. [3]

transects Lines often designated with measuring tape or permanent markers, along which biological data are collected. [7]

transposable element Segment of DNA capable of changing its location on a chromosome, sometimes resulting in a different trait appearing. [2]

trophic cascade Major changes in vegetation and biodiversity resulting from the loss of a keystone species. [2]

trophic levels Levels of biological communities representing ways in which energy is captured and moved through the ecosystem by the various types of species. *See* primary producers; herbivores; predators; detritivores. [2]

U

umbrella species Protecting an umbrella species results in the protection of other species. [7]

UN Conference on Sustainable Development Held in 2012, this conference linked biodiversity conservation to sustainable development and controlling climate change and emphasized the need for market-based solutions. Also unofficially called the Rio+20. [12]

use values The direct and indirect values provided by some aspect of biodiversity. [3]

urbanization The process of converting land to cities or making an area more urban [4]

W

World Bank International bank established to support economic development in developing countries. [12]

World Heritage Convention (WHC) A treaty that protects cultural and natural areas of international significance. [12]

World Heritage site A cultural or natural area officially recognized as having international significance. [9]

World Summit on Sustainable Development Held in Johannesburg, South Africa, in 2002, this gathering emphasized achieving the social and economic goals of sustainability. [12]

Z

zoning A method of managing protected areas that allows or prohibits certain activities in designated places. [9]

zoonosis A human disease that originated from transmission from another species of animal. [5]

Bibliography

Chapter 1

Beever, E. 2020. Diversity: The roles of optimism in conservation biology. *Conservation Biology* 14: 907–909.

Bevan, E. M., et al. 2019. Comparison of beach temperatures in the nesting range of Kemp's ridley sea turtles in the Gulf of Mexico, Mexico and USA. *Endangered Species Research*, 40: 31–40.

Bevan, E., et al. 2014. In situ nest and hatchling survival at Rancho Nuevo, the primary nesting beach of the Kemp's ridley sea turtle, *Lepidochelys kempii*. *Herpetological Conservation and Biology* 9: 563–577.

Bongaarts, J., B. S. Mensch, and A. K. Blanc. 2017. Trends in the age at reproductive transitions in the developing world: The role of education. *Population studies*, 71(2): 139–154.

Bradshaw, C. J. A., and B. W. Brook. 2014. Human population reduction is not a quick fix for environmental problems. *Proceedings of the National Academy of Sciences USA* 111: 16610–16615.

Bullot, N. J., W. P. Seeley, and S. Davies. 2017. Art and science: A philosophical sketch of their historical complexity and codependence. *Journal of Aesthetics and Art Criticism* 75(4): 453–463.

Burchfield, P. M., C. H. Adams, and J. L. D. Guerrero. 2020. Mexico / United States of America Binational Population Restoration Program U.S. 2020 Report for the Kemp's Ridley Sea Turtle, *Lepidochelys kempii*, on the Coast of Tamaulipas, Mexico. A publication of the Kemp's Ridley Sea Turtle Nest Detection and Enhancement Component of the Sea Turtle Early Restoration Project. Original data courtesy of Mexico's CONANP: Adriana Laura Sarti, Martha Hernandez Lopez, and Hector Hugo Acosta, as well as The Tamaulipas Commission of Parks and Biodiversity's Carlos Alejandro Garza Pena and Abigail Gallegos Barajas.

Burivalova, Z., R. A. Butler, and D. S. Wilcove. 2018. Analyzing Google search data to debunk myths about the public's interest in conservation. *Frontiers in Ecology and the Environment* 16(9): 509–514.

Burlin, N. C. (ed.). 1907. The Indians' Book: An Offering by the American Indians of Indian Lore, Musical and Narrative, to Form a Record of the Songs and Legends of Their Race. Harper and Brothers, New York.

Caillouet, C. W., Jr., et al. 1997. Distinguishing captive-reared from wild Kemp's ridleys. *Marine Turtle Newsletter* 77: 1–6.

Caillouet, C. W., Jr., et al. 2018. Did declining carrying capacity for the Kemp's ridley sea turtle population within the Gulf of Mexico contribute to the nesting setback in 2010–2017? *Chelonian Conservation and Biology* 17: 123–133.

Carson, R. 1962. *Silent Spring*. Houghton Mifflin Company, Boston.

Ceballos, G., et al. 2015. Accelerated modern human-induced species losses: Entering the sixth mass extinction. *Science Advances* 1(5): e1400253.

Cole, T. 1965. Essay on American scenery. *In* J. W. McCoubrey (ed.), *American Art, 1700–1960*, pp. 98–109. Prentice-Hall, Englewood Cliffs, NJ.

Ebbin, S. 2009. Institutional and ethical dimensions of resilience in fishing systems: Perspectives from co-managed fisheries in the Pacific Northwest. *Marine Policy* 33: 264–270.

EIA (US Energy Information Administration). 2018. International energy statistics 2008–2011. Retrieved from www.eia.gov/tools/faqs/faq.cfm?id=85&t=1

Emerson, R. W. 1836. *Nature*. James Monroe and Co., Boston.

Garrard, G. E., et al. 2016. Beyond advocacy: Making space for conservation scientists in public debate. *Conservation Letters* 9(3): 208–212.

Godet, L., and V. Devictor. 2018. What conservation does. *Trends in Ecology & Evolution* 33: 720–730.

Götmark, F., and M. Andersson. 2020. Human fertility in relation to education, economy, religion, contraception, and family planning programs. *BMC Public Health* 20(1): 1–17.

Grove, R. 2002. Climatic fears: Colonialism and the history of environmentalism. *Harvard International Review* 23(4): 50.

Hautier, Y., et al. 2015. Anthropogenic environmental changes affect ecosystem stability via biodiversity. *Science* 348: 336–340.

IUCN (International Union for Conservation of Nature). 2019. The IUCN Red List of Threatened Species. Version 2019-1. Retrieved May 8, 2019 from www.iucnredlist.org

Korngold, J. 2008. God in the Wilderness: Rediscovering the Spirituality of the Great Outdoors with the Adventure Rabbi. Doubleday, New York.

Leopold, A. 1949. *A Sand County Almanac and Sketches Here and There*. Oxford University Press, New York.

Leopold, A. C. 2004. Living with the land ethic. *BioScience* 54: 149–154.

Lindenmayer, D., and M. Hunter. 2010. Some guiding concepts for conservation biology. *Conservation Biology* 24: 1459–1468.

Linnell, J. D., et al. 2015. Framing the relationship between people and nature in the context of European conservation. *Conservation Biology* 29(4): 978–985.

Lutz, W., et al. 2019. Education rather than age structure brings demographic dividend. *Proceedings of the National Academy of Sciences USA* 116(26): 12798–12803.

Martin, T. G., et al. 2017. Timing of protection of critical habitat matters. *Conservation Letters* 10(3): 308–316.

Mathevet, R., F. Bousquet, and C. M. Raymond. 2018. The concept of stewardship in sustainability science and conservation biology. *Biological Conservation* 217: 363–370.

Mazaris, A. D., et al. 2017. Global sea turtle conservation successes. *Science Advances* 3(9): e1600730.

Montes, A. D. N., et al. 2020. Persistent organic pollutants in Kemp's Ridley sea turtle *Lepidochelys kempii* in Playa Rancho Nuevo Sanctuary, Tamaulipas, Mexico. *Science of the Total Environment* 739: 140176.

Muir, J. 1901. *Our National Parks*. Houghton Mifflin, Boston.

Pimm, S. L, et al. 2014. The biodiversity of species and their rates of extinction, distribution, and protection. *Science* 344: 1246752.

Pinchot, G. 1947. *Breaking New Ground*. Harcourt Brace, New York.

Reich, K. J., et al. 2017. δ13C and δ15N in the endangered Kemp's ridley sea turtle *Lepidochelys kempii* after the Deepwater Horizon oil spill. *Endangered Species Research* 33: 281–289.

Roman, J., et al. 2015. Lifting baselines to address the consequences of conservation success. *Trends in Ecology & Evolution* 30(6): 299–302.

SCB (Society for Conservation Biology). 2016. What is SCB? Organizational values. Retrieved from conbio.org/about-scb/who-we-are

Shafer, C. L. 2001. Conservation biology trailblazers: George Wright, Ben Thompson, and Joseph Dixon. *Conservation Biology* 15: 332–344.

Shaver, D. J. 2020. Threats to Kemp's ridley sea turtle (*Lepidochelys kempii* Garman, 1880) nests incubating in situ on the Texas coast. *Herpetology Notes* 13: 907–923.

Singh, S., et al. 2017. Sacred groves: Myths, beliefs, and biodiversity conservation: A case study from Western Himalaya, India. *International Journal of Ecology* 2017: 3828609.

Soulé, M. E. 1985. What is conservation biology? *BioScience* 35: 727–734.

Tilman, D., et al. 2017. Future threats to biodiversity and pathways to their prevention. *Nature* 546(7656): 73.

United Nations. 2019. *World Population Prospects 2019 Highlights*. https://population.un.org/wpp/Publications/Files/WPP2019_Highlights.pdf

Vollset, S. E., et al. 2020. Fertility, mortality, migration, and population scenarios for 195 countries and territories from 2017 to 2100: A forecasting analysis for the Global Burden of Disease Study. *Lancet* 396(10258): 1285–1306.

Wibbels, T., and E. Bevan. 2015. New riddle in the Kemp's ridley saga. *In State of the World's Sea Turtles Report*. Oceanic Society, Washington, DC.

Wibbels, T., and E. Bevan. 2016. A historical perspective of the biology and conservation of the Kemp's ridley sea turtle. [Special issue on the Kemp's ridley sea turtle.] *Gulf of Mexico Science* 33: 129–137.

Wilhere, G. F. 2012. Inadvertent advocacy. *Conservation Biology* 26: 39–46.

Wilson, E. O. 2017. Biophilia and the conservation ethic. *In* D. J. Penn and I. Mysterud (eds.), *Evolutionary Perspectives on Environmental Problems*, pp. 263–272. Routledge, Abingdon, UK.

World Bank. 2014. Energy use (kg of oil equivalent per capita)—United States. Retrieved from https://data.worldbank.org/indicator/EG.USE.PCAP.KG.OE?locations=US&view=chart

WWF (World Wildlife Fund). 2020. *Living Planet Report 2020: Bending the Curve of Biodiversity Loss*. Almond, R. E. A., Grooten, M., and Petersen, T. (eds). WWF, Gland, Switzerland.

Chapter 2

Alva-Basurto, J. C., and J. Arias-Gonzalez. 2014. Modelling the effects of climate change on a Caribbean coral reef food web. *Ecological Modelling* 289: 1–14.

Baker, T. R., et al. 2017. Maximising synergy among tropical plant systematists, ecologists, and evolutionary biologists. *Trends in Ecology & Evolution* 32(4): 258–267.

Barlow, J., et al. 2018. The future of hyperdiverse tropical ecosystems. *Nature* 559(7715): 517.

Beschta, R. L., and W. J. Ripple. 2016. Riparian vegetation recovery in Yellowstone: The first two decades after wolf reintroduction. *Biological Conservation* 198: 93–103.

Bhatta, K. P., J. A. Grytnes, and O. R. Vetaas. 2018. Scale sensitivity of the relationship between alpha and gamma diversity along an alpine elevation gradient in central Nepal. *Journal of Biogeography* 45(4): 804–814.

Bowen, B. W., et al. 2013. The origins of tropical marine biodiversity. *Trends in Ecology & Evolution* 28: 359–366.

Brockman, F. J., and C. J. Murray. 2018. The microbiology of the terrestrial deep subsurface. *In Microbiology of the Terrestrial Deep Subsurface*, pp. 75–102. CRC Press, Boca Raton, FL.

Carbone, C., and J. L. Gittleman. 2002. A common rule for the scaling of carnivore density. *Science* 295: 2273–2276.

Carlson, C. J., et al. 2019. Global estimates of mammalian viral diversity accounting for host sharing. *Nature Ecology & Evolution* 3(7): 1070–1075.

Carstensen, D. W., et al. 2018. Local and regional specialization in plant-pollinator networks. *Oikos* 127(4): 531–537.

Chan, Y. F., et al. 2015. Strains of the morphospecies *Ploeotia costata* (Euglenozoa) isolated from the Western North Pacific (Taiwan) reveal substantial genetic differences. *Journal of Eukaryotic Microbiology* 62: 318–326.

Cilleros, K., et al. 2019. Unlocking biodiversity and conservation studies in high-diversity environments using environmental DNA (eDNA): A test with Guianese freshwater fishes. *Molecular Ecology Resources* 19: 27–46.

Cockle, K. L., and K. Martin. 2015. Temporal dynamics of a commensal network of cavity-nesting vertebrates: Increased diversity during an insect outbreak. *Ecology* 96(4): 1093–1104.

Conservation International and K. J. Caley. 2008. Biological diversity in the Mediterranean Basin. *In* C. J. Cleveland (ed.), *Encyclopedia of Earth.* Environmental Information Coalition, National Council for Science and the Environment, Washington, DC. www.eoearth.org/article/ Biological_diversity_in_the_Mediterranean_Basin

Coogan, S. C., et al. 2018. Functional macronutritional generalism in a large omnivore, the brown bear. *Ecology and Evolution* 8(4): 2365–2376.

Cook, L. M., et al. 2012. Selective bird predation on the peppered moth: The last experiment of Michael Majerus. *Biology Letters* 8(4): 609–612.

Corlett, R. T., and R. B. Primack. 2010. *Tropical Rain Forests: An Ecological and Biogeographical Comparison,* 2nd ed. Blackwell Publishing, Malden, MA.

Daly, N. (National Geographic News). 2021. Nearly 5,000 sea turtles rescued from freezing waters on Texas island. February 19. www.nationalgeographic.com/ animals/article/nearly-5000-sea-turtles-rescued-from-freezing-waters-on-texas-island

Davis, J. (Natural History Museum Science News). 2020. Armoured "slug" among 503 new species described by Museum scientists in 2020. www.nhm.ac.uk/ discover/news/2020/december/503-new-species-described-by-museum-scientists-in-2020.html

Dawud, S. M., et al. 2017. Tree species functional group is a more important driver of soil properties than tree species diversity across major European forest types. *Functional Ecology* 31(5): 1153–1162.

De la Riva, E. G., et al. 2017. The importance of functional diversity in the stability of Mediterranean shrubland communities after the impact of extreme climatic events. *Journal of Plant Ecology* 10(2): 281–293.

Deiner, K., H. Yamanaka, and L. Bernatchez. 2021. The future of biodiversity monitoring and conservation utilizing environmental DNA. *Environmental DNA* 3(1): 3–7.

Delgado-Baquerizo, M., et al. 2018. Ecological drivers of soil microbial diversity and soil biological networks in the Southern Hemisphere. *Ecology* 99(3): 583–596.

Díaz, S., and M. Cabido. 2001. Vive la difference: Plant functional diversity matters to ecosystem processes. *Trends in Ecology & Evolution* 16(11): 646–655.

Domisch, S., et al. 2015. Application of species distribution models in stream ecosystems: The challenges of spatial and temporal scale, environmental predictors and species occurrence data. *Fundamental Applied Limnology* 186(1–2): 45–61.

East, J. L., C. Wilcut, and A. A. Pease. 2017. Aquatic food-web structure along a salinized dryland river. *Freshwater Biology* 62(4): 681–694.

EOL (Encyclopedia of Life). 2021. eol.org.

Fadeeva, N., V. Mordukhovich, and J. Zograf. 2015. New deep-sea large free-living nematodes from marobenthos in the Kuril-Kamchatka trench (the North-West Pacific). *Deep-Sea Research II* 111: 95–103.

Fisher, R., et al. 2015. Species richness on coral reefs and the pursuit of convergent global estimates. *Current Biology* 25: 500–505.

Gagic, V., et al. 2015. Functional identity and diversity of animals predict ecosystem functioning better than species-based indices. *Proceedings of the Royal Society B* 282: 20142620.

Gessner, M. O., et al. 2010. Diversity meets decomposition. *Trends in Ecology & Evolution* 25: 372–380.

Goldberg, C. S., K. M. Strickler, and A. K. Fremier. 2018. Degradation and dispersion limit environmental DNA detection of rare amphibians in wetlands: Increasing efficacy of sampling designs. *Science of the Total Environment* 633: 695–703.

Grassle, J. F. 2001. Marine ecosystems. *In* S. A. Levin (ed.), *Encyclopedia of Biodiversity,* Vol. 4, pp. 13–26. Academic Press, San Diego, CA.

Green, E. J., et al. 2020. Below the canopy: Global trends in forest vertebrate populations and their drivers. *Proceedings of the Royal Society B* 287(1928): 20200533.

Griffin, L. P., et al. 2019. Warming seas increase cold-stunning events for Kemp's ridley sea turtles in the northwest Atlantic. *PLOS One* 14(1): e0211503.

Hale, S. L., and J. L. Koprowski. 2018. Ecosystem-level effects of keystone species reintroduction: A literature review. *Restoration Ecology* 26(3): 439–445.

Healey, A. J., et al. 2018. Cryptic species and parallel genetic structuring in Lethrinid fish: Implications for conservation and management in the southwest Indian Ocean. *Ecology and Evolution* 8(4): 2182–2195.

Henry, A. L., et al. 2021. Invasive tree cover covaries with environmental factors to explain the functional composition of riparian plant communities. *Oecologia,* published online July 31, 2021.

Hollings, T., et al. 2014. Trophic cascades following the disease-induced decline of an apex predator, the Tasmanian devil. *Conservation Biology* 28: 63–75.

Hughes, A. R., et al. 2009. Associations of concern: Declining seagrasses and threatened dependent species. *Frontiers in Ecology and the Environment* 7: 242–246.

Human Microbiome Consortium. 2012. www.human-microbiome.org/

Hunter, M. L., Jr., et al. 2017. Conserving small natural features with large ecological roles: A synthetic overview. *Biological Conservation* 211: 88–95.

Iknayan, K. J., et al. 2014. Detecting diversity: Emerging methods to estimate species diversity. *Trends in Ecology & Evolution* 29(2): 97–106.

IUCN (International Union for Conservation of Nature). 2021. The IUCN Red List index. Retrieved from www.iucnredlist.org/assessment/red-list-index

Jones, C. G., J. H. Lawton, and M. Shachak. 1996. Organisms as ecosystem engineers. *In* F. B. Samson and F. L. Knopf (eds.), *Ecosystem Management*, pp. 130–147. Springer, New York.

Joppa, L. N., D. L. Roberts, and S. L. Pimm. 2011. The population ecology and social behavior of taxonomists. *Trends in Ecology & Evolution* 26: 551–553.

Larsen, B. B., et al. 2017. Inordinate fondness multiplied and redistributed: The number of species on Earth and the new pie of life. *Quarterly Review of Biology* 92(3): 229–265.

Lindenmayer, D. B. 2017. Conserving large old trees as small natural features. *Biological Conservation* 211: 51–59.

Locey, K. J., and Lennon, J. T. 2016. Scaling laws predict global microbial diversity. *Proceedings of the National Academy of Sciences USA* 113(21): 5970–5975.

Louca, S., et al. 2019. A census-based estimate of Earth's bacterial and archaeal diversity. *PLOS Biology* 17(2): e3000106.

Mora, C., et al. 2011. How many species are there on Earth and in the ocean? *PLOS Biology* 9(8): e1001127.

Mori, A. S., K. P. Lertzman, and L. Gustafsson. 2017. Biodiversity and ecosystem services in forest ecosystems: A research agenda for applied forest ecology. *Journal of Applied Ecology* 54(1): 12–27.

Naniwadekar, R., et al. 2015. Reduced hornbill abundance associated with low seed arrival and altered recruitment in a hunted and logged tropical forest. *PLOS One* 10(3): e0120062.

Parks, D. H., et al. 2020. A complete domain-to-species taxonomy for Bacteria and Archaea. *Nature Biotechnology* 38: 1079–1086.

Redford, K. H. 1992. The empty forest. *BioScience* 42(6): 412–422.

Ricketts, T. H., et al. 1999. *Terrestrial Ecoregions of North America: A Conservation Assessment*. Island Press, Washington, DC.

Ripple, W. J., and R. L. Beschta. 2012. Trophic cascades in Yellowstone: The first 15 years after wolf reintroduction. *Biological Conservation* 145: 205–213.

Romero, G. Q., et al. 2015. Ecosystem engineering effects on species diversity across ecosystems: A meta-analysis. *Biological Reviews* 90(3): 877–890.

Rosen, C. E., and N. W. Palm. 2017. Functional classification of the gut microbiota: The key to cracking the microbiota composition code: Functional classifications of the gut microbiota reveal previously hidden contributions of indigenous gut bacteria to human health and disease. *BioEssays* 39(12): 1700032.

Ryan, M. E., et al. 2013. Lethal effects of water quality on threatened California salamanders but not on co-occurring hybrid salamanders. *Conservation Biology* 27: 95–102.

Savage, A. E., et al. 2018. Lost but not forgotten: MHC genotypes predict overwinter survival despite depauperate MHC diversity in a declining frog. *Conservation Genetics* 19(2): 309–322.

Sender, R., S. Fuchs, and R. Milo. 2016. Revised estimates for the number of human and bacteria cells in the body. *PLOS Biology* 14(8): e1002533.

Sher, A. A., and M. Molles. 2021. *Ecology Concepts and Applications*, 9th ed. McGraw Hill Education, New York.

Stork, N. E. 2018. How many species of insects and other terrestrial arthropods are there on Earth? *Annual Review of Entomology* 63: 31–45.

Szczecin'ska, M., et al. 2016. Genetic diversity and population structure of the rare and endangered plant species *Pulsatilla patens* (L.) Mill in east central Europe. *PLOS One* 11(3): e0151730.

Tittensor, D. P., et al. 2010. Global patterns and predictors of marine biodiversity across taxa. *Nature* 466: 1098–1101.

UNEP (United Nations Environment Programme). 2021. Biological diversity in the Mediterranean. Retrieved Sept 18 from www.unep.org/unepmap/resources/factsheets/biological-diversity

Valls, A., M. Coll, and V. Christensen. 2015. Keystone species: Toward an operational concept for marine biodiversity conservation. *Ecological Monographs* 85(1): 29–47.

van't Hof, A. E., et al. 2016. The industrial melanism mutation in British peppered moths is a transposable element. *Nature* 534(7605): 102.

Vidal, M. C., and S. M. Murphy. 2018. Bottom-up vs. top-down effects on terrestrial insect herbivores: A meta-analysis. *Ecology Letters* 21(1): 138–150.

Weiskopf, S. R., et al. 2020. Climate change effects on biodiversity, ecosystems, ecosystem services, and natural resource management in the United States. *Science of the Total Environment* 733: 137782.

Wiens, J. J. 2007. Species delimitations: New approaches for discovering diversity. *Systematic Biology* 56(6): 875–878.

Wimp, G. M., et al. 2011. Do edge responses cascade up or down a multi-trophic food web? *Ecology Letters* 14(9): 863–870.

Xu, W., et al. 2017. Strengthening protected areas for biodiversity and ecosystem services in China. *Proceedings of the National Academy of Sciences USA* 114(7): 1601–1606.

Zellweger, F., et al. 2015. Disentangling the effects of climate, topography, soil and vegetation on stand-scale species richness in temperate forests. *Forest Ecology and Management* 349: 36–44.

Zeng, Y.-X., et al. 2014. Diversity of bacterioplankton in coastal seawaters of Fildes Peninsula, King George Island, Antarctica. *Archives of Microbiology* 196: 137–147.

Zhou, J., et al. 2018. Identification and characterization of microsatellites in *Aconitum reclinatum* (Ranunculaceae), a rare species endemic to North America. *Applications in Plant Sciences* 6(6): e01161.

Chapter 3

ABC News. 2017. Great Barrier Reef '"too big to fail" at $56b, Deloitte Access Economics report says. www.abc.net.au/news/2017-06-26/great-barrier-reef-valued-56b-deloitte/8649936

Akamani, K. 2020. Integrating deep ecology and adaptive governance for sustainable development: Implications for protected areas management. *Sustainability* 12(14): 5757.

Angelsen, A., et al. 2014. Environmental income and rural livelihoods: A global-comparative analysis. *World Development* 64: S12–S28.

Angula, H. N., et al. 2018. Local perceptions of trophy hunting on communal lands in Namibia. *Biological Conservation* 218: 26–31.

Barbier, E. B., et al. 1994. *The Economics of the Tropical Timber Trade*. Earthscan Publications, London.

Bateman, I. J., et al. 2013. Bringing ecosystem services into economic decision-making: Land use in the United Kingdom. *Science* 341: 45–50.

Bearzi, G. 2009. When swordfish conservation biologists eat swordfish. *Conservation Biology* 23: 1–2.

Becker, N., A. Greenfeld, and S. Zemah Shamir. 2018. Cost-benefit analysis of full and partial river restoration: The Kishon River in Israel. *International Journal of Water Resources Development*, https://doi.org/10.1080/07900627.2018.1501349

Bhagwat, S. A., N. Dudley, and S. R. Harrop. 2011. Religious following in biodiversity hotspots: Challenges and opportunities for conservation and development. *Conservation Letters* 4: 234–240.

Bluffstone, R., et al. 2017. Estimated values of carbon sequestration resulting from forest management scenarios. *In* L. Wainger and D. Ervin (eds.), *The Valuation of Ecosystem Services from Farms and Forests: Informing a Systematic Approach to Quantifying Benefits of Conservation Programs*. Report 0114-301. Council on Food, Agricultural and Resource Economics (C-FARE), Washington, DC.

Borsetto, C., and E. M. Wellington. 2017. Bioprospecting soil metagenomes for antibiotics. *In* R. Paterson and N. Lima (eds.), *Bioprospecting*, pp. 113–136. Springer, Cham, Switzerland.

Braunisch, V., et al. 2015. Underpinning the precautionary principle with evidence: A spatial concept for guiding wind power development in endangered species' habitats. *Journal for Nature Conservation* 24: 31–40.

Brienen, R. J., et al. 2015. Long-term decline of the Amazon carbon sink. *Nature* 519(7543): 344.

Buenz, E. J., R. Verpoorte, and B. A. Bauer. 2018. The ethnopharmacologic contribution to bioprospecting natural products. *Annual Review of Pharmacology and Toxicology* 58: 509–530.

Büscher, B., and R. Fletcher. 2017. Destructive creation: Capital accumulation and the structural violence of tourism. *Journal of Sustainable Tourism* 25(5): 651–667.

Chan, H. K., et al. 2015. Improve customs systems to monitor global wildlife trade. *Science* 348(6232): 291–292.

Chaudhary, R., and S. Gautam. 2020. Baseline study for ecotourism development in Nepal: Reflection from Chitwan and Ghorepani. *Journal of Environment Sciences* 6: 98–103.

Chen, W. Y. 2017. Environmental externalities of urban river pollution and restoration: A hedonic analysis in Guangzhou (China). *Landscape and Urban Planning* 157: 170–179.

CNN. 2015. Zimbabwean officials: American man wanted in killing of Cecil the Lion. July 28. www.cnn.com/2015/07/28/africa/zimbabwe-lion-killed/index.html

Corlett, R. T. 2017. Frugivory and seed dispersal by vertebrates in tropical and subtropical Asia: An update. *Global Ecology and Conservation* 11: 1–22.

Costanza, R., et al. 2014. Changes in the global value of ecosystem services. *Global Environmental Change* 26: 152–158.

Crosmary, W. G., S. D. Côté, and H. Fritz. 2015. The assessment of the role of trophy hunting in wildlife conservation. *Animal Conservation* 18(2): 136–137.

Dailymail.com. 2021. Brit who runs a big game trophy hunting in South Africa is offering deals to shoot endangered animals that he insists have become "plentiful" during Covid pandemic. February 14. www.dailymail.co.uk/news/article-9258971/British-trophy-hunter-offering-deals-shoot-plentiful-endangered-animals-South-Africa.html

De Groot, R. S., et al. 2010. Challenges in integrating the concept of ecosystem services and values in landscape planning, management and decision making. *Ecological Complexity* 7(3): 260–272.

Devall, B., and G. Sessions. 1985. *Deep Ecology*. Perigrine Smith, Salt Lake, UT.

Devine, J. A. 2017. Colonizing space and commodifying place: Tourism's violent geographies. *Journal of Sustainable Tourism* 25(5): 634–650.

Di Minin, E., N. Leader-Williams, and C. J. Bradshaw. 2016. Banning trophy hunting will exacerbate biodiversity loss. *Trends in Ecology & Evolution* 31: 99–102.

Donovan, G. H., et al. 2013. The relationship between trees and human health: Evidence from the spread of the emerald ash borer. *American Journal of Preventive Medicine* 44: 139–145.

Ehrlich, P. R., and A. H. Ehrlich. 1982. *Extinction: The Causes and Consequences of the Disappearance of Species*. Gollancz, London.

Ekor, M. 2014. The growing use of herbal medicines: Issues relating to adverse reactions and challenges in monitoring safety. *Frontiers in Pharmacology* 4: 177.

Evaluate Pharma. 2018. World preview 2018, outlook to 2024. Retrieved from http://info.evaluategroup.com/WP2018-EPV.html

Evers, H. G., J. K. Pinnegar, and M. I. Taylor. 2019. Where are they all from? Sources and sustainability in the ornamental freshwater fish trade. *Journal of Fish Biology* 94(6): 909–916.

FAO (Food and Agriculture Organization of the United Nations). 2018a. The State of World Fisheries and Aquaculture 2018: Meeting the Sustainable Development Goals. FAO, Rome.

FAO (Food and Agriculture Organization of the United Nations). 2018b. *Forests and Water Valuation and Payments for Forest Ecosystem Services*. FAO, Rome.

FAO (Food and Agriculture Organization of the United Nations). 2019. Forest product statistics. Retrieved from www.fao.org/forestry/statistics/80938

Fox, M. J. V., and J. D. Erickson. 2020. Design and meaning of the genuine progress indicator: A statistical analysis of the US fifty-state model. *Ecological Economics* 167: 106441.

Gardiner, S., et al. 2010. *Climate Ethics: Essential Readings*. Oxford University Press, New York.

Garnett, K., and D. J. Parsons. 2017. Multi-case review of the application of the precautionary principle in European Union law and case law. *Risk Analysis* 37: 502–516.

Gilstad-Hayden, K., et al. 2015. Research note: Greater tree canopy cover is associated with lower rates of both violent and property crime in New Haven, CT. *Landscape and Urban Planning* 143: 248–253.

Grossi-de-Sá, M. F., et al. 2017. Entomotoxic plant proteins: Potential molecules to develop genetically modified plants resistant to insect-pests. *In* P. Gopalakrishnakone et al. (eds.), *Plant Toxins*, pp. 415–447. Springer, New York

Haines-Young, R., and M. Potschin. 2018. Common International Classification of Ecosystem Services (CICES) V5.1 and Guidance on the Application of the Revised Structure. European Environment Agency. Available from www.cices.eu

Handa, I. T., et al. 2014. Consequences of biodiversity loss for litter decomposition across biomes. *Nature* 509: 218.

Hanley, N., L. P. Dupuy, and E. McLaughlin. 2015. Genuine savings and sustainability. *Journal of Economic Surveys* 29(4): 779–806.

Hardin, G. 1968. The tragedy of the commons. *Science* 162: 1243–1248.

Hautier, Y., et al. 2015. Anthropogenic environmental changes affect ecosystem stability via biodiversity. *Science* 348: 336–340.

Huveneers, C., et al. 2017. The economic value of shark-diving tourism in Australia. *Reviews in Fish Biology and Fisheries* 27(3): 665–680.

Iftekhar, M. S., et al. 2017. How economics can further the success of ecological restoration. *Conservation Biology* 31(2): 261–268.

Ingram, V. J., et al. 2012. Nontimber forest products: Contribution to national economy and strategies for sustainable management. *In* C. de Wasseige, P. de Marcken, and N. Bayolet (eds.), *The Forests of the Congo Basin. State of the Forest 2010*. Office des publications de l'Union Européenne, Luxembourg.

Jiang, Z., and B. Lin. 2014. The perverse fossil fuel subsidies in China: The scale and effects. *Energy* 70: 411–419.

Kilpatrick, A. M., et al. 2017. Conservation of biodiversity as a strategy for improving human health and well-being. *Philosophical Transactions of the Royal Society B* 372(1722): 20160131.

Kremen, C. 2018. The value of pollinator species diversity. *Science* 359(6377): 741–742.

Krisfalusi-Gannon, J., et al. 2018. The role of horseshoe crabs in the biomedical industry and recent trends impacting species sustainability. *Frontiers in Marine Science* 5: 185.

Kubiszewski, I., et al. 2013. Beyond GDP: Measuring and achieving global genuine progress. *Ecological Economics* 93: 57–68.

Lambin, E. F., et al. 2018. The role of supply-chain initiatives in reducing deforestation. *Nature Climate Change* 8: 109–116.

Lawrence, G. B., et al. 2018. Soil base saturation combines with beech bark disease to influence composition and structure of sugar maple–beech forests in an acid rain–impacted region. *Ecosystems* 21(4): 795–810.

Leopold, A. 1949. *A Sand County Almanac and Sketches Here and There*. Oxford University Press, New York.

Leopold, A. 1953. *Round River: From the Journals of Aldo Leopold*. Oxford University Press, New York.

Liu, J., et al. 2018. How does habitat fragmentation affect the biodiversity and ecosystem functioning relationship? *Landscape Ecology* 33: 341–352.

Louv, R. 2005. *Last Child in the Woods: Saving Our Children from Nature-Deficit Disorder*. Algonquin Books, Chapel Hill, NC.

Luck, G. W., et al. 2011. Relations between urban bird and plant communities and human well-being and connection to nature. *Conservation Biology* 25: 816–826.

Luiselli, L., et al. 2020. Bushmeat consumption in large urban centres in West Africa. *Oryx* 54(5): 731–734.

Lyngdoh, S., V. B. Mathur, and B. C. Sinha. 2017. Tigers, tourists and wildlife: Visitor demographics and experience in three Indian tiger reserves. *Biodiversity and Conservation* 26(9): 2187–2204.

MA (Millennium Ecosystem Assessment). 2005. *Ecosystems and Human Well-being.* 4 volumes. Island Press, Washington, DC.

Martin, T. G., et al. 2018. Prioritizing recovery funding to maximize conservation of endangered species. *Conservation Letters* 11(6): e12604.

Martín-López, B., et al. 2014. Trade-offs across value-domains in ecosystem services assessment. *Ecological Indicators* 37: 220–228.

McNeely, J. A., et al. 1990. *Conserving the World's Biological Diversity.* IUCN, World Resources Institute, CI, WWF-US, and the World Bank. Gland, Switzerland, and Washington, DC.

Meadows, D. H., et al. 1972. *The Limits to Growth: A Report for the Club of Rome's Project on the Predicament of Mankind.* Universe Books, New York.

Melin, A. 2016. Biodiversity and Christian ethics: A critical discussion. *In* S. Bergmann, P. M. Scott, M. Jansdotter Samuelsson, and H. Bedford-Strohm (eds.), *Nature, Space and the Sacred,* pp. 95–104. Routledge, Abingdon, UK.

Merckx, T., and H. M. Pereira. 2015. Reshaping agri-environmental subsidies: From marginal farming to large-scale rewilding. *Basic and Applied Ecology* 16: 95–103.

Moat, J., T. W. Gole, and A. P. Davis. 2018. Least Concern to Endangered: Applying climate change projections profoundly influences the extinction risk assessment for wild Arabica coffee. *Global Change Biology* 25(2): 390–403.

Monticini, P. 2010. The ornamental fish trade production and commerce of ornamental fish: Technical-managerial and legislative aspects. *GLOBEFISH Research Programme,* Vol. 102, p. 134. FAO, Rome. Available from https://www.fao.org/in-action/globefish/publications/details-publication/en/c/347680/

Muiños, G., et al. 2015. Frugality and psychological wellbeing: The role of voluntary restriction and the resourceful use of resources. *Psyecology* 6(2): 169–190.

Naess, A. 1989. *Ecology, Community and Lifestyle.* Cambridge University Press, Cambridge, UK.

Naess, A. 2008. *The Ecology of Wisdom: Writings by Arne Naess.* A. Drengson and B. Devall (eds.). Counterpoint, Berkeley, CA.

Narayan, S., et al. 2017. The value of coastal wetlands for flood damage reduction in the northeastern USA. *Scientific Reports* 7(1): 9463.

Nguyen, D. H., S. de Leeuw, and W. E. Dullaert. 2018. Consumer behaviour and order fulfilment in online retailing: A systematic review. *International Journal of Management Reviews* 20: 255–276.

O'Neill, A. 2021. Global gross domestic product (GDP) at current prices from 1985 to 2026. *Statista.* www.statista.com/statistics/268750/global-gross-domestic-product-gdp/

Ostroumov, S. 2017. Aquatic ecosystem service: Improving water quality. Multifunctional role of the biota in water self-purification in marine and freshwater ecosystems. *Caucasus Economic and Social Analysis Journal* 7: 38–41.

Øverland, M., and A. Skrede. 2017. Yeast derived from lignocellulosic biomass as a sustainable feed resource for use in aquaculture. *Journal of the Science of Food and Agriculture* 97(3): 733–742.

Oyebode, O., et al. 2016. Use of traditional medicine in middle-income countries: A WHO-SAGE study. *Health Policy and Planning* 31(8): 984–991.

Paranunzio, R., et al. 2018. New insights in the relation between climate and slope failures at high-elevation sites. *Theoretical and Applied Climatology,* https://doi.org/10.1007/s00704-018-2673-4

Peterson, M. J., et al. 2010. Obscuring ecosystem function with the application of the ecosystem services concept. *Conservation Biology* 24: 113–119.

Phelps, J., and E. L. Webb. 2015. "Invisible" wildlife trades: Southeast Asia's undocumented illegal trade in wild ornamental plants. *Biological Conservation* 186: 296–305.

Potschin-Young, M., et al. 2017. Intermediate ecosystem services: An empty concept? *Ecosystem Services* 27: 124–126.

Powell, K. I., J. M. Chase, and T. M. Knight. 2013. Invasive plants have scale-dependent effects on diversity by altering species-area relationships. *Science* 339: 316–318.

Prescott-Allen, C., and R. Prescott-Allen. 1986. *The First Resource: Wild Species in the North American Economy.* Yale University Press, New Haven, CT.

Reyers, B., et al. 2013. Getting the measure of ecosystem services: A social-ecological approach. *Frontiers in Ecology and the Environment* 11: 268–273.

Robinson, J. G. 2011. Ethical pluralism, pragmatism, and sustainability in conservation practice. *Biological Conservation* 144: 958–965.

Rolston, H., III. 2012. *Environmental Ethics.* Temple University Press, Philadelphia.

Rust, N. A. 2015. Media framing of financial mechanisms for resolving human-predator conflict in Namibia. *Human Dimensions of Wildlife* 20(5): 440–453.

Schebek, L., et al. 2018. Land-use change and CO_2 emissions associated with oil palm expansion in Indonesia by 2020. *In* B. Otjacques, P. Hitzelberger, S. Naumann, and V. Wohlgemuth (eds.), *From Science to Society,* pp. 49–59. Springer, Cham, Switzerland.

Selier, S. A. J., et al. 2014. Sustainability of elephant hunting across international borders in southern Africa: A case study of the greater Mapungubwe Transfrontier Conservation Area. *Journal of Wildlife Management* 78(1): 122–132.

Shutt, K., et al. 2014. Effects of habituation, research and ecotourism on faecal glucocorticoid metabolites in wild western lowland gorillas: Implications for ecotourism. *Biological Conservation* 172: 72–79.

Smith, A. 1909. An inquiry into the nature and causes of the wealth of nations. *In* J. L. Bullock (ed.), *The Harvard Classics*. P. F. Collier and Sons, New York.

Smith, A. C., et al. 2017. How natural capital delivers ecosystem services: A typology derived from a systematic review. *Ecosystem Services* 26: 111–126.

Smith, I. A. 2016. *The Intrinsic Value of Endangered Species*. Routledge, Abingdon, UK.

Sovacool, B. K., J. Kim, and M. Yang. 2021. The hidden costs of energy and mobility: A global meta-analysis and research synthesis of electricity and transport externalities. *Energy Research & Social Science* 72: 101885.

Stenmark, M. 2017. *Environmental Ethics and Policy-Making*. Routledge, Abingdon, UK.

Swanson, F. J., C. Goodrich, and K. D. Moore. 2008. Bridging boundaries: Scientists, creative writers, and the long view of the forest. *Frontiers in Ecology and the Environment* 6: 499–504.

Taylor, P. J., et al. 2018. Economic value of bat predation services: A review and new estimates from macadamia orchards. *Ecosystem Services* 30: 372–381.

TEEB (The Economics of Ecosystems and Biodiversity) and P. Kubar. 2010. *The Economics of Ecosystems and Biodiversity Ecological and Economic Foundations*. Earthscan, London and Washington.

Thompson, R. A., A. J. Lymbery, and S. S. Godfrey. 2018. Parasites at risk: Insights from an endangered marsupial. *Trends in Parasitology* 34: 12–22.

Thoreau, H. D. 1854. *Walden; or, Life in the Woods*. Ticknor and Fields, Boston.

Tietenberg, T. H., and L. Lewis. 2016. *Environmental and Natural Resource Economics*. Routledge, Abingdon, UK.

Tietge, D. J. 2018. Experiencing nature through cable television. *In* K. Rutten, S. Blancke, and R. Soetaert (eds.), *Perspectives on Science and Culture*, pp. 3–18. Purdue University Press, West Lafayette, IN.

Traer, R. 2018. *Doing Environmental Ethics*. Routledge, Abingdon, UK.

USFWS (US Fish and Wildlife Service). 2020a. Reintroduction of a migratory flock of whooping cranes in the eastern United States. Retrieved from www.fws.gov/midwest/whoopingcrane/wcraneqanda.html

USFWS (US Fish and Wildlife Service). 2020b. Sportsmen and women generate nearly $1 billion in conservation funding. Retrieved from www.fws.gov/news/ShowNews.cfm?ref=sportsmen-and-women-generate-nearly-$1-billion-in-conservation-funding-&_ID=36532

Van Ooij, C. 2016. Combining traditional medicine and modern chemistry to fight malaria. *Annals of Translational Medicine* 4(24): 550.

Vanbergen, A. J., and the Insect Pollinator Initiative. 2013. Threats to an ecosystem service: Pressures on pollinators. *Frontiers in Ecology and the Environment* 11: 251–259.

Verified Market Research. 2021. Cyclosporine market size and forecast. Retrieved from www.verifiedmarketresearch.com/product/cyclosporine-market/

Wabnitz, C. C., et al. 2018. Ecotourism, climate change and reef fish consumption in Palau: Benefits, trade-offs and adaptation strategies. *Marine Policy* 88: 323–332.

Wagg, C., et al. 2017. Plant diversity maintains long-term ecosystem productivity under frequent drought by increasing short-term variation. *Ecology* 98(11): 2952–2961.

Wardle, C., et al. 2018. Ecotourism's contributions to conservation: Analysing patterns in published studies. *Journal of Ecotourism*, https://doi.org/10.1080/14724049.2018.1424173

Waswa, F., M. Mcharo, and M. Mworia. 2020. Declining wood fuel and implications for household cooking and diets in tigania Sub-county Kenya. *Scientific African* 8: e00417.

Waterman, C., et al. 2016. Miniaturized cultivation of microbiota for antimalarial drug discovery. *Medicinal Research Reviews* 36(1): 144–168.

WHSRN (Western Hemisphere Shorebird Reserve Network). 2015. *WHSRN: An International Strategy for Saving Shorebirds and Their Habitats*. Manomet Center for Conservation Sciences, Manomet, MA.

Witoszek, N., and M. L. Mueller. 2017. Deep ecology. *Worldviews: Global Religions, Culture, and Ecology* 21(3): 209–217.

Wood, S. A., and M. A. Bradford. 2018. Leveraging a new understanding of how belowground food webs stabilize soil organic matter to promote ecological intensification of agriculture. *In* B. K. Singh (ed.), *Soil Carbon Storage*, pp. 117–136. Academic Press, London.

WTTC (World Travel & Tourism Council). 2019. *Travel & Tourism Economic Impact 2019, World*. WTTC, London.

WWF (World Wildlife Fund). 2018. *Living Planet Report: 2018: Aiming Higher*, M. Grooten and R. E. A. Almond (eds.). WWF, Switzerland.

WWF (World Wildlife Fund). 2020. *Living Planet Report 2020: Bending the Curve of Biodiversity Loss*. R. E. A. Almond, M. Grooten, and T. Petersen (eds.). WWF, Gland, Switzerland.

Yi, B. L. (Public Radio International). 2016. Islamic clerics issue a fatwa against poachers in Indonesia and Malaysia. www.pri.org/stories/2016-01-07/islamic-clerics-issue-fatwa-against-poachers-indonesia-and-malaysia

Chapter 4

Adams, S., and E. K. M. Klobodu. 2017. Urbanization, democracy, bureaucratic quality, and environmental degradation. *Journal of Policy Modeling* 39(6): 1035–1051.

Aguirre-Acosta, N., E. Kowaljow, and R. Aguilar. 2014. Reproductive performance of the invasive tree *Ligustrum lucidum* in a subtropical dry forest: Does habitat fragmentation boost or limit invasion? *Biological Invasions* 16(7): 1397–1410.

Alroy, J. 2017. Effects of habitat disturbance on tropical forest biodiversity. *Proceedings of the National Academy of Sciences USA* 114: 6056–6061.

Alton, L. A., and C. E. Franklin. 2017. Drivers of amphibian declines: Effects of ultraviolet radiation and interactions with other environmental factors. *Climate Change Responses* 4(1): 6.

Alvarado, R., and E. Toledo. 2017. Environmental degradation and economic growth: Evidence for a developing country. *Environment, Development and Sustainability* 19(4): 1205–1218.

Amoroso, R. O., et al. 2018. Bottom trawl fishing footprints on the world's continental shelves. *Proceedings of the National Academy of Sciences USA*, 115(43): E10275–E10282.

Austin, K. G., et al. 2017. Trends in size of tropical deforestation events signal increasing dominance of industrial-scale drivers. *Environmental Research Letters* 12(5): 054009.

Baccini, A., et al. 2017. Tropical forests are a net carbon source based on aboveground measurements of gain and loss. *Science* 358: 230–234.

BBC. 2020. Brazil's Amazon: Deforestation "surges to 12-year high." https://www.bbc.com/news/world-latin-america-55130304

Bebber, D. P., and N. Butt. 2017. Tropical protected areas reduced deforestation carbon emissions by one third from 2000–2012. *Scientific Reports* 7: 14005.

Benjaminsen, T. A., and P. Hiernaux. 2019. From desiccation to global climate change: A history of the desertification narrative in the West African Sahel, 1900–2018. *Global Environment* 12(1): 206–236.

Berners-Lee, M., et al. 2018. Current global food production is sufficient to meet human nutritional needs in 2050 provided there is radical societal adaptation. *Elementa: Science of the Anthropocene*, 6: 52.

Bouwman, H., et al. 2015. First report of the concentrations and implications of DDT residues in chicken eggs from a malaria-controlled area. *Chemosphere* 137: 174–177.

Bradshaw, C. J., and E. Di Minin. 2019. Socio-economic predictors of environmental performance among African nations. *Scientific Reports* 9(1): 1–13

Bryan-Brown, D. N., et al. 2020. Global trends in mangrove forest fragmentation. *Scientific Reports* 10(1): 1–8.

Campbell, M. (Apple Insider). 2018. Apple investing to protect mangrove forest in Colombia to offset carbon emissions. https://appleinsider.com/articles/18/09/13/apple-investing-to-protect-mangrove-forest-in-colombia-to-offset-carbon-emissions

Carbutt, C., W. D. Henwood, and L. A. Gilfedder. 2017. Global plight of native temperate grasslands: Going, going, gone? *Biodiversity and Conservation* 26(12): 2911–2932.

Chávez-Dulanto, P. N., et al. 2020. Increasing the impact of science and technology to provide more people with healthier and safer food. *Food and Energy Security* 10: e259.

Clark, M., and D. Tillman. 2017. Comparative analysis of environmental impacts of agricultural production systems, agricultural input efficiency, and food choice. *Environmental Research Letters* 12(6): 064016.

Comeau, S., et al. 2015. Ocean acidification accelerates dissolution of experimental coral reef communities. *Biogeosciences* 12(2): 365–372.

Çonkar, A. 2020. *Development and Security Challenges in the Sahel Region*. Draft Report. Mediterranean and Middle East Special Group (GSM). NATO Parliamentary Assembly, 2020-06.

Coomes, O. T., and B. C. Miltner. 2017. Indigenous charcoal and biochar production: Potential for soil improvement under shifting cultivation systems. *Land Degradation & Development* 28(3): 811–821.

Cordeiro, N. J., et al. 2015. Forest fragmentation in an African biodiversity hotspot impacts mixed-species bird flocks. *Biological Conservation* 188: 61–71.

Corlett, R. T., and R. B. Primack. 2010. *Tropical Rain Forests: An Ecological and Biogeographical Comparison*, 2nd ed. Blackwell Publishing, Malden, MA.

Crist, E., C. Mora, and R. Engelman. 2017. The interaction of human population, food production, and biodiversity protection. *Science* 356(6335): 260–264.

Curtis, P. G., et al. 2018. Classifying drivers of global forest loss. *Science* 361(6407): 1108–1111.

Daru, B. H., et al. 2013. A global trend towards the loss of evolutionarily unique species in mangrove ecosystems. *PLOS One* 8(6): e66686.

Davidson, N. C. 2014. How much wetland has the world lost? Long-term and recent trends in global wetland area. *Marine and Freshwater Research* 65(10): 934–941.

Díaz-Ortega, G., and E. A. Hernández-Delgado. 2014. Unsustainable land-based source pollution in a climate of change: A roadblock to the conservation and recovery of Elkhorn coral *Acropora palmata* (Lamarck 1816). *Natural Resources* 5(10): 561.

Didham, R. K. 2017. Ecological consequences of habitat fragmentation. *eLS*, https://doi.org/10.1002/9780470015902.a0021904

Dolmen, D., A. G. Finstad, and J. K. Skei. 2018. Amphibian recovery after a decrease in acidic precipitation. *Ambio* 47(3): 355–367.

Donofrio, S., P. Rothrock, and J. Leonard. 2017. Supply change: Tracking corporate commitments to deforestation-free supply chains, 2017. Forest Trends, Washington, DC.

Eagles-Smith, C. A., et al. 2018. Modulators of mercury risk to wildlife and humans in the context of rapid global change. *Ambio* 47(2): 170–197.

Ehrlich, P. R., P. M. Kareiva, and G. C. Daily. 2012. Securing natural capital and expanding equity to rescale civilization. *Nature* 486(7401): 68–73.

Elliott, J. E., and K. H. Elliott. 2013. Tracking marine pollution. *Science* 340: 556–558.

Elrich, P. R., and J. P. Holdren. 1971. Impact of population growth. *Science* 171: 1212–1217.

EPA (US Environmental Protection Agency). 2016. National rivers and streams assessment. Retrieved from www.epa.gov/national-aquatic-resource-surveys/nrsa

Estoque, R. C., et al. 2019. The future of Southeast Asia's forests. *Nature Communications* 10: 1–12.

Fahrig, L. 2017. Ecological responses to habitat fragmentation per se. *Annual Review of Ecology, Evolution, and Systematics* 48: 1–23.

Fahrig, L., et al. 2019. Is habitat fragmentation bad for biodiversity? *Biological Conservation* 230: 179–186.

Fan, J. L., Y. J. Zhang, and B. Wang. 2017. The impact of urbanization on residential energy consumption in China: An aggregated and disaggregated analysis. *Renewable and Sustainable Energy Reviews* 75: 220–233.

FAO (Food and Agriculture Organization of the United Nations). 2011. *The State of the World's Land and Water Resources for Food and Agriculture (SOLAW): Managing Systems at Risk*. FAO , Rome, and Earthscan, London.

Franco, S., V. R. Mandla, and K. R. M. Rao. 2017. Urbanization, energy consumption and emissions in the Indian context: A review. *Renewable and Sustainable Energy Reviews* 71: 898–907.

Geist, H. 2017. *The Causes and Progression of Desertification*. Routledge, Abingdon, UK.

Germanov, E. S., et al. 2018. Microplastics: No small problem for filter-feeding megafauna. *Trends in Ecology & Evolution* 33(4): 227–232.

Global Footprint Network. 2018. National footprint accounts. Retrieved from https://data.footprintnetwork.org

Global Forest Watch. 2020. Dashboard. Retrieved March 5 from www.globalforestwatch.org/dashboards

Guedes-Alonso, R., Z. Sosa-Ferrera, and J. J. Santana-Rodríguez. 2017. Determination of steroid hormones in fish tissues by microwave-assisted extraction coupled to ultra-high performance liquid chromatography tandem mass spectrometry. *Food Chemistry* 237: 1012–1020.

Guo, Q., et al. 2017. Satellite monitoring the spatial-temporal dynamics of desertification in response to climate change and human activities across the Ordos Plateau, China. *Remote Sensing* 9(6): 525.

Haddad, N. M., et al. 2015. Habitat fragmentation and its lasting impact on Earth's ecosystems. *Science Advances* 1(2): e1500052.

Hansen, M. C., et al. 2013. High-resolution global maps of 21st-century forest cover change. *Science* 342(6160): 850–853.

Hansen, M. C., et al. 2020. The fate of tropical forest fragments. *Science Advances* 6(11): eaax8574.

Henderson, K., and M. Loreau. 2019. An ecological theory of changing human population dynamics. *People and Nature* 1(1): 31–43.

Horn, S., et al. 2020. Impact of potential COVID-19 treatment on South African water sources already threatened by pharmaceutical pollution. *Environmental Toxicology and Chemistry* 39: 1305–1306.

Huang, J., et al. 2017. Nitrogen and phosphorus losses and eutrophication potential associated with fertilizer application to cropland in China. *Journal of Cleaner Production* 159: 171–179.

Hughes, T. P., et al. 2017. Coral reefs in the Anthropocene. *Nature* 546(7656): 82.

Hughes, T. P., H. U. I. Huang, and M. A. Young. 2013. The wicked problem of China's disappearing coral reefs. *Conservation Biology* 27(2): 261–269.

Huhta, E., and J. Jokimäki. 2015. Landscape matrix fragmentation effect on virgin forest and managed forest birds: A multi-scale study. *In* J. A. Daniels (ed.), *Advances in Environmental Research*, Vol. 36, pp. 95–111. Nova Science, Hauppauge, New York.

IUCN Freshwater Specialist Group. 2020. Major threats. Retrieved from www.iucnffsg.org/freshwater-fishes/major-threats/

Jakes, A. F., et al. 2018. A fence runs through it: A call for greater attention to the influence of fences on wildlife and ecosystems. *Biological Conservation* 227: 310–318.

Kaplan, S. (Washington Post). 2021. It's wrong to blame overpopulation for "climate change." May 25. www.washingtonpost.com/climate-solutions/2021/05/25/slowing-population-growth-environment/

KC, K. B., et al. 2018. When too much isn't enough: Does current food production meet global nutritional needs?. *PLOS One* 13(10): e0205683.

Khaniabadi, Y. O., et al. 2017. Exposure to PM_{10}, NO_2, and O_3 and impacts on human health. *Environmental Science and Pollution Research* 24(3): 2781–2789.

Kim, K. H., E. Kabir, and S. A. Jahan. 2017. Exposure to pesticides and the associated human health effects. *Science of the Total Environment* 575: 525–535.

Köhler, H. R., and R. Triebskorn. 2013. Wildlife ecotoxicology of pesticides: Can we track effects to the population level and beyond? *Science* 341(6147): 759–765.

Krausmann, F., et al. 2013. Global human appropriation of net primary production doubled in the 20th century. *Proceedings of the National Academy of Sciences USA* 110(25): 10324–10329.

Krauze-Gryz, D., J. Gryz, and J. Goszczyn'ski. 2012. Predation by domestic cats in rural areas of central Poland: An assessment based on two methods. *Journal of Zoology* 288(4): 260–266.

Kroll, G. 2015. An environmental history of roadkill: Road ecology and the making of the permeable highway. *Environmental History* 20(1): 4–28.

Landrigan, P. J., et al. 2018. The Lancet Commission on pollution and health. *Lancet* 391(10119): 462–512.

Lauria, V., et al. 2017. Species distribution models of two critically endangered deep-sea octocorals reveal fishing impacts on vulnerable marine ecosystems in central Mediterranean Sea. *Scientific Reports* 7(1): 8049.

Lavauden, L. 1927. Les forêts du Sahara. *Revue des Eaux et Forêts* 7(65): 329–341.

Loyd, K. A. T., et al. 2013. Quantifying free-roaming domestic cat predation using animal-borne video cameras. *Biological Conservation* 160: 183–189.

Mancini, M. S., et al. 2018. Exploring ecosystem services assessment through Ecological Footprint accounting. *Ecosystem Services* 30: 228–235.

Máñez, K. S., et al. 2014. The Gordian knot of mangrove conservation: Disentangling the role of scale, services and benefits. *Global Environmental Change* 28: 120–128.

Marphatia, A. A., et al. 2020. How much education is needed to delay women's age at marriage and first pregnancy? *Frontiers in Public Health* 7: 396.

Maxwell, S. L., et al. 2016. Biodiversity: The ravages of guns, nets and bulldozers. *Nature* 536(7615): 143–145.

McCarthy, S. 2017. Loss of Great Plains grasslands puts critical ecosystems at risk. www.worldwildlife.org/press-releases/loss-of-great-plains-grasslands-puts-critical-ecosystems-at-risk

McGurn, W. 2018. The population bomb was a dud. *Wall Street Journal* April 3. https://www.wsj.com/articles/the-population-bomb-was-a-dud-1525125341

McKinley, D. C., et al. 2017. Citizen science can improve conservation science, natural resource management, and environmental protection. *Biological Conservation* 208: 15–28.

Ménesguen, A., and G. Lacroix. 2018. Modelling the marine eutrophication: A review. *Science of the Total Environment* 636: 339–354.

Miller-Rushing, A. J., et al. 2019. How does habitat fragmentation affect biodiversity? A controversial question at the core of conservation biology. *Biological Conservation* 232: 271–273.

Mills Busa, J. H. 2013. Deforestation beyond borders: Addressing the disparity between production and consumption of global resources. *Conservation Letters* 6: 192–199.

Mills, E. C., et al. 2018. Forest elephant movement and habitat use in a tropical forest-grassland mosaic in Gabon. *PLOS One* 13(7): e0199387.

Mitsch, W. J., and J. G. Gosselink. 2015. *Wetlands*. John Wiley and Sons, New York.

Morelli, T. L., et al. 2020. The fate of Madagascar's rainforest habitat. *Nature Climate Change* 10: 89–96.

Moreno-Mateos, D., et al. 2012. Structural and functional loss in restored wetland ecosystems. *PLOS Biology* 10(1): 45.

Mukul, S. A., and J. Herbohn. 2016. The impacts of shifting cultivation on secondary forests dynamics in tropics: A synthesis of the key findings and spatiotemporal distribution of research. *Environmental Science and Policy* 55: 167–177.

Nijssen, M. E., M. F. WallisDeVries, and H. Siepel. 2017. Pathways for the effects of increased nitrogen deposition on fauna. *Biological Conservation* 212: 423–431.

Numata, I., et al. 2017. Fire and edge effects in a fragmented tropical forest landscape in the southwestern Amazon. *Forest Ecology and Management* 401: 135–146.

Ostfeld, R. S., et al. 2018. Tick-borne disease risk in a forest food web. *Ecology* 99(7): 1562–1573.

Pandit, R., et al. 2018. Summary for Policymakers of the Assessment Report on Land Degradation and Restoration of the Intergovernmental Science-Policy Platform on Biodiversity and Ecosystem Services. Intergovernmental Platform on Biodiversity and Ecosystem Services, Bonn, Germany.

Peters, R., et al. 2018. Nature divided, scientists united: US-Mexico border wall threatens biodiversity and binational conservation. *BioScience* 68: 740–743.

Pfeifer, M., et al. 2017. Creation of forest edges has a global impact on forest vertebrates. *Nature* 551: 187.

Pikaar, I., et al. 2018. The urgent need to re-engineer nitrogen-efficient food production for the planet. *In* S. Hülsmann and R. Ardakanian (eds.), *Managing Water, Soil and Waste Resources to Achieve Sustainable Development Goals*, pp. 35–69. Springer, Cham, Switzerland.

Ravikumar, A., et al. 2017. Is small-scale agriculture really the main driver of deforestation in the Peruvian Amazon? Moving beyond the prevailing narrative. *Conservation Letters* 10(2): 170–177.

Reynoso, J. A., and G. Williams-Linera. 2017. Herbivory damage on oak seedlings at the edge of cloud forest fragments. *Botanical Sciences* 80: 29–34.

Rockström, J., et al. 2017. Sustainable intensification of agriculture for human prosperity and global sustainability. *Ambio* 46(1): 4–17.

Rossetti, M. R., et al. 2017. Responses of insect herbivores and herbivory to habitat fragmentation: A hierarchical meta-analysis. *Ecology Letters* 20(2): 264–272.

Sayre, N. F., et al. 2013. Earth stewardship of rangelands: Coping with ecological, economic, and political marginality. *Frontiers in Ecology and the Environment* 11: 348–354.

Schaider, L. A., et al. 2014. Pharmaceuticals, perfluorosurfactants, and other organic wastewater compounds in public drinking water wells in a shallow sand and gravel aquifer. *Science of the Total Environment* 468: 384–393.

Scott, S., et al. 2021. Early marriage and early childbearing in South Asia: Trends, inequalities, and drivers from 2005 to 2018. *Annals of the New York Academy of Sciences* 1491(1): 60.

Sharma, S., and S. Chatterjee. 2017. Microplastic pollution, a threat to marine ecosystem and human health: A short review. *Environmental Science and Pollution Research* 24(27): 21530–21547.

Sinha, E., A. M. Michalak, and V. Balaji. 2017. Eutrophication will increase during the 21st century as a result of precipitation changes. *Science* 357(6349): 405–408.

Song, X. P., et al. 2018. Global land change from 1982 to 2016. *Nature* 560(7720): 639–643.

Tatarski, M. (Mongabay). 2021. "Drastic forest development": Vietnam to plant 1 billion trees–but how? https://news.mongabay.com/2021/05/drastic-forest-development-vietnam-to-plant-1-billion-trees-but-how/

Thomas, N., et al. 2017. Distribution and drivers of global mangrove forest change, 1996–2010. *PLOS One* 12(6): e0179302.

Tilman, D., et al. 2017. Future threats to biodiversity and pathways to their prevention. *Nature* 546(7656): 73.

Turcotte, M. M., et al. 2017. The eco-evolutionary impacts of domestication and agricultural practices on wild species. *Philosophical Transactions of the Royal Society B* 372(1712): 20160033.

United Nations Convention to Combat Desertification. 2021. The LDN Target Setting Programme. Retrieved August 10 from www.unccd.int/actions/ldn-target-setting-programme

Valdez, M. (Associated Press). 2018. Washington state builds bridge to keep wildlife off highway. www.apnews.com/36755628e4c44fdabb01cb9599fa5aef

Vieilledent, G., et al. 2018. Combining global tree cover loss data with historical national forest cover maps to look at six decades of deforestation and forest fragmentation in Madagascar. *Biological Conservation* 222: 189–197.

Vivanco, D. F., R. Kemp, and E. van der Voet. 2016. How to deal with the rebound effect? A policy-oriented approach. *Energy Policy* 94: 114–125.

Wackernagel, M., and W. Rees. 1996. *Our Ecological Footprint: Reducing Human Impact on the Earth* (New Catalyst. Bioregional Series). New Society, Gabriola Island, BC.

Wade, L. 2013. Gold's dark side. *Science* 341: 1448–1449.

Walker, R. T., et al. 2019. Avoiding Amazonian catastrophes: Prospects for conservation in the 21st century. *One Earth* 1(2): 202–215.

Weinzettel, J., D. Vačkář, and H. Medková. 2018. Human footprint in biodiversity hotspots. *Frontiers in Ecology and the Environment* 16(8): 447–452.

Weinzettel, J., D. Vačkář, and H. Medková. 2019. Potential net primary production footprint of agriculture: A global trade analysis. *Journal of Industrial Ecology* 23(5):1133–1142.

With, K. A. 2015. How fast do migratory songbirds have to adapt to keep pace with rapidly changing landscapes? *Landscape Ecology* 30(7): 1351–1361.

Wright, C. K., and M. C. Wimberly. 2013. Recent land use change in the Western Corn Belt threatens grasslands and wetlands. *Proceedings of the National Academy of Sciences USA* 110(10): 4134–4139.

WWF (World Wildlife Fund). 2018. *Living Planet Report–2018: Aiming Higher*. M. Grooten and R. E. A. Almond (eds.). WWF, Gland, Switzerland.

WWF (World Wildlife Fund). 2021. Forests the size of France have regrown in the last 20 years, new research shows. May 11. www.worldwildlife.org/press-releases/forests-the-size-of-france-have-regrown-in-the-last-20-years-new-research-shows

Yang, S. L., et al. 2014. Downstream sedimentary and geomorphic impacts of the Three Gorges Dam on the Yangtze River. *Earth-Science Reviews* 138: 469–486.

Zhang, Y., et al. 2020. Atmospheric microplastics: A review on current status and perspectives. *Earth-Science Reviews* 203:103118.

Zhao, Z., et al. 2021. Fire enhances forest degradation within forest edge zones in Africa. *Nature Geoscience* 14(7): 479–483.

Chapter 5

Abatzoglou, J. T., and A. P. Williams. 2016. Impact of anthropogenic climate change on wildfire across western US forests. *Proceedings of the National Academy of Sciences* USA 113: 11770–11775.

Andersson, A. A., et al. 2021. CITES and beyond: Illuminating 20 years of global, legal wildlife trade. *Global Ecology and Conservation* 26: e01455.

Bagla, P. 2010. Hardy cotton-munching pests are latest blow to GM crops. *Science* 327: 1439.

Baiyewu, A. O., et al. 2018. Ethnozoological survey of traditional uses of Temminck's ground pangolin (*Smutsia temminckii*) in South Africa. *Society & Animals* 26(3): 306–325.

Bateman, I. J., et al. 2013. Bringing ecosystem services into economic decision-making: Land use in the United Kingdom. *Science* 341: 45–50.

BBC. January 2021. Japan whale hunting: "By-catch" rule highlighted after minke death. www.bbc.com/news/world-asia-55714815

Beans, C. M., and D. A. Roach. 2015. An invasive plant alters pollinator-mediated phenotypic selection on a native congener. *American Journal of Botany* 102: 50–57.

Bell, R. E., and H. Seroussi. 2020. History, mass loss, structure, and dynamic behavior of the Antarctic Ice Sheet. *Science* 367: 1321-1325.

Benedetti, F., et al. 2018. Investigating uncertainties in zooplankton composition shifts under climate change scenarios in the Mediterranean Sea. *Ecography* 41(2): 345–360.

Burdon, J. J., and J. Zhan. 2020. Climate change and disease in plant communities. *PLOS Biology* 18(11): e3000949.

Burge, C. A., and P. K. Hershberger. 2020. Climate change can drive marine disease. *In* D. C. Behringer, B. R. Silliman, K. D. Lafferty (eds.), *Marine Disease Ecology*. pp. 83–94. Oxford University Press, Oxford Scholarship Online.

Burgess, M. G., S. Polasky, and D. Tilman. 2013. Predicting overfishing and extinction threats in multispecies fisheries. *Proceedings of the National Academy of Sciences USA* 110: 15943–15948.

Bussière, E., L. G. Underhill, and R. Altwegg. 2015. Patterns of bird migration phenology in South Africa suggest Northern Hemisphere climate as the most consistent driver of change. *Global Change Biology* 21(6): 2179–2190.

Byers, J. E. 2021. Marine parasites and disease in the era of global climate change. *Annual Review of Marine Science* 13: 397–420.

Calmy, A., E. Goemaere, and G. Van Cutsem. 2015. HIV and Ebola virus: Two jumped species but not two of a kind. *AIDS* 29(13): 1593–1596.

Campos-Cerqueira, M., et al. 2017. Have bird distributions shifted along an elevational gradient on a tropical mountain? *Ecology and Evolution* 7(23): 9914–9924.

Carey, M. P., et al. 2012. Native invaders: Challenges for science, management, policy, and society. *Frontiers in Ecology and the Environment* 10(7): 373–381.

Castro Galvan, E., M. Maldonado Torres, and M. Pérez Urquiza. 2019. Zea maize reference materials for genetically modified organism detection in Mexico. *Ecology and Evolution*, 9: 12353–12356.

Cazzolla Gatti, R. 2016. Trends in human development and environmental protection. *International Journal of Environmental Studies* 73: 268–276.

CDC (Centers for Disease Control and Prevention). 2021a. Zoonotic diseases. Retrieved from www.cdc.gov/onehealth/basics/zoonotic-diseases.html

CDC (Centers for Disease Control and Prevention). 2021b. Animals and COVID-19; last updated Updated Oct. 5, 2021. Retrieved from www.cdc.gov/coronavirus/2019-ncov/daily-life-coping/animals.html

Chaurasia, A., D. L. Hawksworth, and M. P. de Miranda (eds.). 2020. *GMOs: Implications for Biodiversity Conservation and Ecological Processes*, Vol. 19. Springer Nature, London.

Chen, C., et al. 2020. Global warming and shifts in cropping systems together reduce China's rice production. *Global Food Security* 24: 100359.

Chinn, S., P. S. Hart, and S. Soroka. 2020. Politicization and polarization in climate change news content, 1985–2017. *Science Communication* 42(1): 112–129.

Christian, C., et al. 2013. A review of formal objections to Marine Stewardship Council fisheries certifications. *Biological Conservation* 161: 10–17.

CITES (Convention on International Trade in Endangered Species). 2021. The Appendices. Retrieved September 30, 2021 from https://cites.org/eng/disc/species.php

Climate.gov. 2020. 2020 Arctic sea ice minimum second lowest on record. www.climate.gov/news-features/featured-images/2020-arctic-sea-ice-minimum-second-lowest-record

Coll, M., and E. Wajnberg. 2017. Environmental pest management: A call to shift from a pest-centric to a system-centric approach. *In* M. Coll and E. Wajnberg (eds.), *Environmental Pest Management: Challenges for Agronomists, Ecologists, Economists and Policymakers*, pp. 1–18. Wiley and Sons, Hoboken, NJ.

Cook, J., et al. 2016. Consensus on consensus: A synthesis of consensus estimates on human-caused global warming. *Environmental Research Letters* 11(4): 048002.

Crowley, S. L., S. Hinchliffe, and R. A. McDonald. 2017. Conflict in invasive species management. *Frontiers in Ecology and the Environment* 15(3): 133–141.

Cuthbert, R. N., et al. 2021. Global economic costs of aquatic invasive alien species. *Science of the Total Environment* 775: 145238.

Dangendorf, S., et al. 2019. Persistent acceleration in global sea-level rise since the 1960s. *Nature Climate Change* 9(9): 705–710.

Davenport, C., and E. Lipton (New York Times). 2017. How G.O.P leaders came to view climate change as fake science. June 3. www.nytimes.com/2017/06/03/us/politics/republican-leaders-climate-change.html

Davies, T. E., et al. 2017. Large marine protected areas represent biodiversity now and under climate change. *Scientific Reports* 7(1): 9569.

Davis, M. A. 2009. *Invasion Biology*. Oxford University Press, Oxford, UK.

De la Riva, I., and S. Reichle. 2014. Diversity and conservation of the amphibians of Bolivia. *Herpetological Monographs* 28(1): 46–65.

Dearing, J. A., et al. 2006. Human-environment interactions: Learning from the past. *Regional Environmental Change* 6(1): 1–16.https://doi.org/10.1007/s10113-005-0011-8

Drus, G. M., et al. 2013. The effect of leaf beetle herbivory on the fire behaviour of tamarisk (*Tamarix ramosissima* Lebed.). *International Journal of Wildland Fire* 22(4): 446–458.

Eklund, L., and D. Thompson. 2017. Differences in resource management affects drought vulnerability across the borders between Iraq, Syria, and Turkey. *Ecology and Society* 22(4): 9.

Ellis, E. C., et al. 2010. Anthropogenic transformation of the biomes, 1700 to 2000. *Global Ecology and Biogeography* 19(5): 589–606.

Epanchin-Niell, R. S. 2017. Economics of invasive species policy and management. *Biological Invasions* 19: 3333–3354.

Erisman, B., et al. 2017. Fish spawning aggregations: Where well-placed management actions can yield big benefits for fisheries and conservation. *Fish and Fisheries* 18(1): 128–144.

Esquivel-Muelbert, A., et al. 2019. Compositional response of Amazon forests to climate change. *Global Change Biology* 25(1): 39–56.

Estrada, A., et al. 2017. Impending extinction crisis of the world's primates: Why primates matter. *Science Advances* 3(1): e1600946.

Fairtrade International. 2021. About us. Retrieved September 18 from www.fairtrade.net/about

Fernandes, J. A., et al. 2017. Estimating the ecological, economic and social impacts of ocean acidification and warming on UK fisheries. *Fish and Fisheries* 18(3): 389–411.

Galetti, M., and R. Dirzo. 2013. Ecological and evolutionary consequences of living in a defaunated world. *Biological Conservation* 163: 1–6.

Gallardo, B., et al. 2016. Global ecological impacts of invasive species in aquatic ecosystems. *Global Change Biology* 22(1): 151–163.

Gbashi, S., et al. 2021. Food safety, food security and genetically modified organisms in Africa: A current perspective. *Biotechnology and Genetic Engineering Reviews* 37: 30–63.

Gilmour, J. P., et al. 2013. Recovery of an isolated coral reef system following severe disturbance. *Science* 340: 69–71.

Green, E. J., et al. 2020. Below the canopy: Global trends in forest vertebrate populations and their drivers. *Proceedings of the Royal Society B* 287(1928): 20200533.

Gutiérrez, J. L., and C. Bernstein. 2014. Ecosystem impacts of invasive species. BIOLIEF 2011 2nd World Conference on Biological Invasion and Ecosystem Functioning, Mar del Plata, Argentina, 21–24 November 2011. *Acta Oecologica* 54: 1–138.

Haltinner, K., and D. Sarathchandra. 2018. Climate change skepticism as a psychological coping strategy. *Sociology Compass* 12(6): e12586.

Hannah, L., et al. 2013. Climate change, wine, and conservation. *Proceedings of the National Academy of Sciences USA* 110: 6907–6912.

Heighton, S. P., and P. Gaubert. 2021. A timely systematic review on pangolin research, commercialization, and popularization to identify knowledge gaps and produce conservation guidelines. *Biological Conservation* 256: 109042.

Henson, S. A., et al. 2017. Rapid emergence of climate change in environmental drivers of marine ecosystems. *Nature Communications* 8: 14682.

Hildebrandt, T. B., et al. 2021. Conservation research in times of COVID-19: The rescue of the northern white rhino. *Journal of Applied Animal Ethics Research* 3(1): 16–37. https://doi.org/10.1163/25889567-BJA10009

Hobbs, C. A., et al. 2019. Using DNA barcoding to investigate patterns of species utilisation in UK shark products reveals threatened species on sale. *Scientific Reports* 9(1): 1–10.

Hobbs, R. J., E. S. Higgs, and C. M. Hall. 2013. *Novel Ecosystems: Intervening in the New Ecological World Order.* Wiley-Blackwell, Oxford, UK.

Hönisch, B., et al. 2012. The geological record of ocean acidification. *Science* 335: 1058–1063.

Hulme, P. E., et al. 2018. Integrating invasive species policies across ornamental horticulture supply chains to prevent plant invasions. *Journal of Applied Ecology* 55(1): 92–98.

IPCC (Intergovernmental Panel on Climate Change). 2018. Global Warming of 1.5°C. An IPCC Special Report on the Impacts of Global Warming of 1.5°C above Pre-Industrial Levels and Related Global Greenhouse Gas Emission Pathways, in the Context of Strengthening the Global Response to the Threat of Climate Change, Sustainable Development, and Efforts to Eradicate Poverty. www.ipcc.ch/sr15/

IUCN (International Union for Conservation of Nature). 2016. The IUCN Red List of Threatened Species. Version 2016–1. Retrieved from www.iucnredlist.org

Janssen, J., and S. C. Chng. 2018. Biological parameters used in setting captive-breeding quotas for Indonesia's breeding facilities. *Conservation Biology* 32: 18–25.

JOCI (Joint Ocean Commission Initiative). 2017. Invasive species threaten the Great Lakes economy and ecosystem. https://oceanactionagenda.org/story/invasive-species-threaten-great-lakes-economy-ecosystem/

Johannes, J., et al. 2017. Level of environmental threat posed by horticultural trade in Cactaceae. *Conservation Biology* 31(5): 1066–1075.

Joshua, T. H. (US Department of the Interior). 2020. Interior announces $3.4 million for brown snake control on Guam. www.doi.gov/oia/press/interior-announces-34-million-brown-tree-snake-control-guam

Karalis, D. T., et al. 2020. Genetically modified products, perspectives and challenges. *Cureus* 12(3): e7306.

Keeling, C. D., et al. 1989. A three-dimensional model of atmospheric CO_2 transport based on observed winds: 1. Analysis of observational data. *In* D. H. Peterson (ed.), *Aspects of Climate Variability in the Pacific and the Western Americas*, pp. 165–236. Geophysical Monograph Series, Vol. 55, American Geophysical Union, Washington, DC.

Keledjian, A., et al. (Oceana). 2014. Wasted catch: Unsolved problems in US fisheries. https://oceana.org/reports/wasted-catch-unsolved-problems-us-fisheries

Kerr, J. R., and M. S. Wilson. 2021. Right-wing authoritarianism and social dominance orientation predict rejection of science and scientists. *Group Processes & Intergroup Relations* 24(4): 550–567.

Klein, T., et al. 2019. A nation-wide analysis of tree mortality under climate change: Forest loss and its causes in Israel 1948–2017. *Forest Ecology and Management* 432: 840–849.

Kopf, R. K., al. 2017. Confronting the risks of large-scale invasive species control. *Nature Ecology & Evolution* 1: 0172.

Kuebbing, S. E., L. Souza, and N. J. Sanders. 2014. Effects of co-occurring non-native invasive plant species on old-field succession. *Forest Ecology and Management* 324: 196–204.

Lanzarin, G. A., et al. 2020. Behavioural toxicity of environmental relevant concentrations of a glyphosate commercial formulation-RoundUp® UltraMax-During zebrafish embryogenesis. *Chemosphere* 253: 126636.

LeDee, O. E., et al. 2021. Preparing wildlife for climate change: How far have we come? *Journal of Wildlife Management* 85(1): 7–16.

Lembo, V., et al. 2018. Inter-hemispheric differences in energy budgets and cross-equatorial transport anomalies during the 20th century. *Climate Dynamics*, https://doi.org/10.1007/s00382-018-4572-x

Linde, S. 2020. The politicization of risk: Party cues, polarization, and public perceptions of climate change risk. *Risk Analysis* 40(10): 2002–2018.

Lindsey, R. (Climate.gov). 2021. Earth's hottest month was record hot in 2021. www.climate.gov/news-features/understanding-climate/earths-hottest-month-was-record-hot-2021

Love, S. A., et al. 2018. Does aquatic invasive species removal benefit native fish? The response of gizzard shad (*Dorosoma cepedianum*) to commercial harvest of bighead carp (*Hypophthalmichthys nobilis*) and silver carp (*H. molitrix*). *Hydrobiologia* 817: 403–412.

Magera, A. M., et al. 2013. Recovery trends in marine mammal populations. *PLOS One* 8(10): e77908.

Mainwaring, M. C., et al. 2017. Climate change and nesting behaviour in vertebrates: A review of the ecological threats and potential for adaptive responses. *Biological Reviews* 92(4): 1991–2002.

Marbuah, G., I. M. Gren, and B. McKie. 2014. Economics of harmful invasive species: A review. *Diversity* 6: 500–523.

Matozzo, V., et al. 2018. Ecotoxicological risk assessment for the herbicide glyphosate to non-target aquatic species: A case study with the mussel *Mytilus galloprovincialis*. *Environmental Pollution* 233: 623–632.

Maxwell, S. L., et al. 2016. Biodiversity: The ravages of guns, nets and bulldozers. *Nature News* 536(7615): 143.

Meinhardt, K. A., and C. A. Gehring. 2013. Tamarix and soil ecology. *In* A. Sher and M. Quigley (eds.), *Tamarix: A Case Study of Ecological Change in the American West*. Oxford University Press, New York.

Merow, C., et al. 2017. Climate change both facilitates and inhibits invasive plant ranges in New England. *Proceedings of the National Academy of Sciences USA* 114: E3276–E3284.

Molinos, J. G., et al. 2016. Climate velocity and the future global redistribution of marine biodiversity. *Nature Climate Change* 6(1): 83.

Mollot, G., J. H. Pantel, and T. N. Romanuk. 2017. The effects of invasive species on the decline in species richness: A global meta-analysis. *In* D. A. Bohan and A. J. Dumbrell (eds.), *Advances in Ecological Research*, pp. 61–83. Academic Press, Cambridge, MA.

Morand, S. 2018. Biodiversity and disease transmission. *In* C. J. Hurst (ed.), *The Connections between Ecology and Infectious Disease*, pp. 39–56. Springer, Cham, Switzerland.

Munson, S. M., and A. A. Sher. 2015. Long-term shifts in the phenology of rare and endemic Rocky Mountain plants. *American Journal of Botany* 102(8): 1268–1276.

Murray, D. L., et al. 2017. Continental divide: Climate-mediated fragmentation of species distributions in the boreal forest. *PLOS One* 12(5): e0176706.

Nellemann, C., et al. 2016. The rise of environmental crime: A growing threat to natural resources, peace, development and security. United Nations Environment Programme (UNEP), Gland, Switzerland.

Nelson, E. J., et al. 2013. Climate change's impact on key ecosystem services and the human well-being they support in the US. *Frontiers in Ecology and the Environment* 11: 483–493.

NOAA (National Oceanic and Atmospheric Administration). 2018. Personal communication from the National Marine Fisheries Service, Fisheries Statistics Division.

NOAA (National Oceanic and Atmospheric Administration). 2020. Climate Change: Atmospheric CO_2. www.climate.gov/news-features/understanding-climate/climate-change-atmospheric-carbon-dioxide

NOAA (National Oceanic and Atmospheric Administration). 2021. Climate at a glance: Global time series. Retrieved from www.ncdc.noaa.gov/cag/global/time-series/globe/ocean/ytd/12/1880-2021

Nuñez, T. A., et al. 2013. Connectivity planning to address climate change. *Conservation Biology* 27: 407–416.

Palmer, J. M., et al. 2018. Extreme sensitivity to ultraviolet light in the fungal pathogen causing white-nose syndrome of bats. *Nature Communications* 9(1): 35.

Pardee, G. L., D. W. Inouye, and R. E. Irwin. 2018. Direct and indirect effects of episodic frost on plant growth and reproduction in subalpine wildflowers. *Global Change Biology* 24(2): 848–857.

Pellegrini, A. F., et al. 2018. Fire frequency drives decadal changes in soil carbon and nitrogen and ecosystem productivity. *Nature* 553(7687): 194.

Pérez-Méndez, N., et al. 2020. The economic cost of losing native pollinator species for orchard production. *Journal of Applied Ecology* 57(3): 599–608.

Perry, G., and D. Vice. 2009. Forecasting the risk of brown tree snake dispersal from Guam: A mixed transport-establishment model. *Conservation Biology* 23: 992–1000.

PoachingFacts. 2021. Rhino poaching statistics. Retrieved Sept 30 from www.poachingfacts.com/poaching-statistics/rhino-poaching-statistics

Pyšek, P., et al. 2020. Scientists' warning on invasive alien species. *Biological Reviews* 95(6): 1511–1534.

Quiroga, L. B., et al. 2015. Diet composition of an invasive population of *Lithobates catesbeianus* (American bullfrog) from Argentina. *Journal of Natural History* 49: 1703–1716.

Ramirez, K. S., et al. 2019. Range-expansion effects on the belowground plant microbiome. *Nature Ecology & Evolution* 3(4): 604–611.

Ray, D. K., et al. 2019. Climate change has likely already affected global food production. PLOS One 14(5): e0217148.

Raza, S., et al. 2021. Analysis of the spike proteins suggest pangolin as an intermediate host of COVID-19 (SARS-CoV-2). *International Journal of Agriculture and Biology* 25(3), doi: 10.17957/IJAB/15.1700

Reino, L., et al. 2017. Networks of global bird invasion altered by regional trade ban. *Science Advances* 3(11): e1700783.

Rekker, R. 2021. The nature and origins of political polarization over science. *Public Understanding of Science* 30(4): 352–368.

Rignot, E., et al. 2019. Four decades of Antarctic Ice Sheet mass balance from 1979–2017. *Proceedings of the National Academy of Sciences USA* 116(4): 1095–1103.

Riskas, K. A., M. M. Fuentes, and M. Hamann. 2016. Justifying the need for collaborative management of fisheries bycatch: A lesson from marine turtles in Australia. *Biological Conservation* 196: 40–47.

Ritchie, H., and M. Roser. 2020. CO_2 and greenhouse gas emissions. OurWorldInData.org, https://ourworldindata.org/co2-and-other-greenhouse-gas-emissions

Roberson, L. A., R. A. Watson, and C. J. Klein. 2020. Over 90 endangered fish and invertebrates are caught in industrial fisheries. *Nature Communications* 11(1): 1–8.

Robinson, J. E., and P. Sinovas. 2018. Challenges of analyzing the global trade in CITES-listed wildlife. *Conservation Biology* 32: 1203–1206.

Rose, K. M., D. Brossard, and D. A. Scheufele. 2020. Of society, nature, and health: How perceptions of specific risks and benefits of genetically engineered foods shape public rejection. *Environmental Communication* 14(7): 1017–1031.

Runwall, P. 2021. Why it's so tricky to trace the origin of COVID-19. *National Geographic*. Sept 10. www.nationalgeographic.com/science/article/why-its-so-tricky-to-trace-the-origin-of-covid-19

Russell, J. C., et al. 2015. Predator-free New Zealand: Conservation country. *BioScience* 65(5): 520–525.

Russell, J. C., et al. 2016. Importance of lethal control of invasive predators for island conservation. *Conservation Biology* 30(3): 670–672.

Russell, J. C., et al. 2017. Invasive alien species on islands: Impacts, distribution, interactions and management. *Environmental Conservation* 44: 359–370.

Sanford, E., et al. 2014. Ocean acidification increases the vulnerability of native oysters to predation by invasive snails. *Proceedings of the Royal Society of London B* 281(1778): 20132681.

Santovito, A., M. Audisio, and S. Bonelli. 2020. A micronucleus assay detects genotoxic effects of herbicide exposure in a protected butterfly species. *Ecotoxicology* 29(9): 1390–1398.

Savoca, M. S., et al. 2020. Comprehensive bycatch assessment in US fisheries for prioritizing management. *Nature Sustainability* 3(6): 472–480.

Schmeller, D. S., et al. 2018. People, pollution and pathogens: Global change impacts in mountain freshwater ecosystems. *Science of the Total Environment* 622: 756–763.

Schuur, E. A. G., et al. 2015. Climate change and the permafrost carbon feedback. *Nature* 520(7546): 171–179.

Schuur, T., et al. 2020. Tundra underlain by thawing permafrost persistently emits carbon to the atmosphere over fifteen years of measurements. *Journal of Geophysical Research: Biogeosciences*, e2020JG006044.

Scott, M. (NOAA Climate.gov). 2020. 2020 Arctic sea ice minimum second lowest on record. www.climate.gov/news-features/featured-images/2020-arctic-sea-ice-minimum-second-lowest-record

Seddon, A. W., et al. 2016. Sensitivity of global terrestrial ecosystems to climate variability. *Nature* 531: 229.

Seroussi, H., et al. 2020. ISMIP6 Antarctica: A multi-model ensemble of the Antarctic ice sheet evolution over the 21st century. *The Cryosphere* 14: 3033–3070, https://doi.org/10.5194/tc-14-3033-2020

Shackelford, N., et al. 2013. Finding a middle ground: The native/non-native debate. *Biological Conservation* 158: 55–62.

Siers, S. R., et al. 2018. *Assessment of Brown Tree Snake Activity and Bait Take Following Large-Scale Snake Suppression in Guam*. USDA APHIS Wildlife Services National Wildlife Research Center Rep. QA-2438, Hilo, HI.

Spatz, D. R., et al. 2017. Globally threatened vertebrates on islands with invasive species. *Science Advances* 3: e1603080.

Sunday, J. M., et al. 2017. Ocean acidification can mediate biodiversity shifts by changing biogenic habitat. *Nature Climate Change* 7(1): 81.

Teferra, T. F. 2021. Should we still worry about the safety of GMO foods? Why and why not? A review. *Food Science & Nutrition* 9: 5324–5331.

Terraube, J., and V. Bretagnolle. 2018. Top-down limitation of mesopredators by avian top predators: A call for research on cascading effects at the community and ecosystem scale. *Ibis* 160(3): 693–702.

Tollefson, J. 2018. Global industrial carbon emissions to reach all-time high in 2018. *Nature* 3(369): 370.

Tollefson, J. 2020. How the coronavirus pandemic slashed carbon emissions: In five graphs. *Nature* 582(7811): 158–159.

Treen, K. M. D. I., H. T. Williams, and S. J. O'Neill. 2020. Online misinformation about climate change. *Wiley Interdisciplinary Reviews: Climate Change* 11(5): e665.

Tsatsakis, A. M., et al. 2017. Impact on environment, ecosystem, diversity and health from culturing and using GMOs as feed and food. *Food and Chemical Toxicology* 107: 108–121.

Turcotte, M. M., et al. 2017. The eco-evolutionary impacts of domestication and agricultural practices on wild species. *Philosophical Transactions of the Royal Society B* 372: 20160033.

UNODC (United Nations Office on Drugs and Crime). 2020. United Nations world wildlife crime report 2020. Retrieved from https://www.unodc.org/unodc/en/data-and-analysis/wildlife.html

van Velden, J., K. Wilson, and D. Biggs. 2018. The evidence for the bushmeat crisis in African savannas: A systematic quantitative literature review. *Biological Conservation* 221: 345–356.

Vincent, A. C., et al. 2014. The role of CITES in the conservation of marine fishes subject to international trade. *Fish and Fisheries* 15(4): 563–592.

Vitousek, P. M. 1994. Beyond global warming: Ecology and global change. *Ecology* 75: 1861–1876.

Vitousek, P. M., L. L. Loope, and R. Westbrooks. 2017. Biological invasions as global environmental change. *American Scientist* 84: 218–228.

Waltz, E. 2017. First genetically engineered salmon sold in Canada. *Nature News* 548(7666): 148.

Wildlife Justice Commission. 2019. No rest for pangolins: Trafficking rates spike while crime displaces to new regions. Retrieved from https://wildlifejustice.org/no-rest-for-pangolins-trafficking-rates-spike

WWF (World Wildlife Fund). 2020. *Living Planet Report 2020: Bending the Curve of Biodiversity Loss*. R.E.A. Almond, M. Grooten, and T. Petersen (eds). WWF, Gland, Switzerland.

Xu, H., et al. 2014. Intentionally introduced species: More easily invited than removed. *Biodiversity and Conservation* 23(10): 2637–2643.

Zheng, B., et al. 2020. Satellite-based estimates of decline and rebound in China's CO_2 emissions during COVID-19 pandemic. *Science Advances* 6(49): eabd4998.

Zylberberg, M., et al. 2013. Variation with land use of immune function and prevalence of avian pox in Galapagos finches. *Conservation Biology* 27: 103–112.

Chapter 6

Abascal, F., et al. 2016. Extreme genomic erosion after recurrent demographic bottlenecks in the highly endangered Iberian lynx. *Genome Biology* 17(1): 251.

Allee, W. C. 1931. *Animal Aggregations: A Study in General Sociology*. University of Chicago Press, Chicago, IL.

Allendorf, F. W., and G. Luikart. 2007. *Conservation and the Genetics of Populations*. Blackwell, Oxford, UK.

Arima, E. Y., et al. 2016. Explaining the fragmentation in the Brazilian Amazonian forest. *Journal of Land Use Science* 11(3): 257–277.

Atwood, T. B., et al. 2020. Herbivores at the highest risk of extinction among mammals, birds, and reptiles. *Science Advances* 6(32): eabb8458.

Basset, Y., et al. 2015. The butterflies of Barro Colorado Island, Panama: Local extinction since the 1930s. *PLOS One* 10(8): e0136623.

Beradelli, J., and K. Niemczyk (CBS News). 2021. The Great Dying: Earth's largest-ever mass extinction is a warning for humanity. March 4. www.cbsnews.com/news/great-dying-permian-triassic-extinction-event-warning-humanity/

Bilgmann, K., et al. 2021. Low effective population size in the genetically bottlenecked Australian sea lion is insufficient to maintain genetic variation. *Animal Conservation* 24(5): 847–861.

Bonsall, M. B., et al. 2014. Allee effects and the spatial dynamics of a locally endangered butterfly, the high brown fritillary. *Ecological Applications* 24: 108–120.

Brooks, T. M., et al. 1999. Threat from deforestation to montane and lowland birds and mammals in insular Southeast Asia. *Journal of Animal Ecology* 68(6): 1061–1078.

Broughton, J. M., and E. M. Weitzel. 2018. Population reconstructions for humans and megafauna suggest mixed causes for North American Pleistocene extinctions. *Nature Communications* 9(1): 1–12.

Burkhead, N. M. 2012. Extinction rates in North American freshwater fishes, 1900–2010. *BioScience* 62(9): 798–808.

Butchart, S. H., et al. 2018. Which bird species have gone extinct? A novel quantitative classification approach. *Biological Conservation* 227: 9–18.

Ceballos, G., et al. 2015. Accelerated modern human-induced species losses: Entering the sixth mass extinction. *Science Advances* 1(5): e1400253.

Connor, E. F., and E. D. McCoy. 2001. Species-area relationships. *In* S. A. Levin (ed.), *Encyclopedia of Biodiversity* 5, pp. 397–412. Academic Press, San Diego, CA.

Darwin, C. 1859. The Origin of Species; And, the Descent of Man. Modern Library, New York.

Di Marco, M., et al. 2015. Historical drivers of extinction risk: Using past evidence to direct future monitoring. *Proceedings of the Royal Society B* 282(1813): 20150928.

Dullinger, S., et al. 2013. Europe's other debt crisis caused by the long legacy of future extinctions. *Proceedings of the National Academy of Sciences USA* 110: 7342–7347.

Feng, S., et al. 2019. The genomic footprints of the fall and recovery of the crested ibis. *Current Biology* 29(2): 340–349.

Fisher, D. O., and S. P. Blomberg. 2012. Inferring extinction of mammals from sighting records, threats, and biological traits. *Conservation Biology* 26: 57–67.

Forsman, A. 2014. Effects of genotypic and phenotypic variation on establishment are important for conservation, invasion, and infection biology. *Proceedings of the National Academy of Sciences USA* 111(1): 302–307.

Frankham, R. 2005. Genetics and extinction. *Biological Conservation* 126: 131–140.

Frankham, R. 2015. Genetic rescue of small inbred populations: Meta-analysis reveals large and consistent benefits of gene flow. *Molecular Ecology* 24(11): 2610–2618.

Frankham, R. 2016. Genetic rescue benefits persist to at least the F3 generation, based on a meta-analysis. *Biological Conservation* 195: 33–36.

Frankham, R., C. J. A. Bradshaw, and B. W. Brook. 2014. Genetics in conservation management: Revised recommendations for the 50/500 rules, Red List criteria, and population viability analyses. *Biological Conservation* 170: 56–63.

Frankham, R., J. D. Ballou, and D. A. Briscoe. 2009. *Introduction to Conservation Genetics*, 2nd ed. Cambridge University Press, Cambridge, UK.

Gibson-Reinemer, D. K., K. S. Sheldon, and F. J. Rahel. 2015. Climate change creates rapid species turnover in montane communities. *Ecology and Evolution* 5(12): 2340–2347.

Grayson, D. K. 2016. *Giant Sloths and Sabertooth Cats: Extinct Mammals and the Archaeology of the Ice Age Great Basin*. University of Utah Press, Salt Lake City.

Habel, J. C., and T. Schmitt. 2018. Vanishing of the common species: Empty habitats and the role of genetic diversity. *Biological Conservation* 218: 211–216.

Haddad, N. M., et al. 2015. Habitat fragmentation and its lasting impact on Earth's ecosystems. *Science Advances* 1(2): e1500052.

Hasegawa, K. 2017. Displacement of native white-spotted charr *Salvelinus leucomaenis* by non-native brown trout *Salmo trutta* after resolution of habitat fragmentation by a migration barrier. *Journal of Fish Biology* 90(6): 2475–2479.

Hogg, C. J., et al. 2017. Metapopulation management of an endangered species with limited genetic diversity in the presence of disease: The Tasmanian devil *Sarcophilus harrisii*. *International Zoo Yearbook* 51(1): 137–153.

IPBES (Intergovernmental Science-Policy Platform on Biodiversity and Ecosystem Services). 2019. IPBES global assessment preview. www.ipbes.net/news/ipbes-global-assessment-preview

IUCN (International Union for Conservation of Nature). 2021. The IUCN Red List index. Retrieved from https://www.iucnredlist.org/assessment/red-list-index

IUCN SSC (Species Survival Comission). 2019. *Madagasgar Plant Specialist Group 2019 Report*. www.iucn.org/sites/dev/files/2019_madagascar_plant_sg_report_publication.pdf

Jurikova, H., et al. 2020. Permian–Triassic mass extinction pulses driven by major marine carbon cycle perturbations. *Nature Geoscience* 13(11): 745–750.

Kaiho, K., et al. 2021. Pulsed volcanic combustion events coincident with the end-Permian terrestrial disturbance and the following global crisis. *Geology* 49(3): 289–293.

KopeLion. 2021. Lions in Ngorongoro Crater. Retrieved from https://kopelion.org/lions-ngorongoro-crater

Laikre, L., et al. 2013. Hunting effects on favourable conservation status of highly inbred Swedish wolves. *Conservation Biology* 27: 248–253.

Leidy, R. A., and P. B. Moyle. 2021. Keeping up with the status of freshwater fishes: A California (USA) perspective. *Conservation Science and Practice* 3: e474.

Lucena-Perez, M., et al. 2018. Reproductive biology and genealogy in the endangered Iberian lynx: Implications for conservation. *Mammalian Biology* 89: 7–13.

MacArthur, R. H., and E. O. Wilson. 1967. *The Theory of Island Biogeography*. Princeton University Press, Princeton, NJ.

Mace, G., et al. 2005. Biodiversity. *In* R. Hassan, R. Scholes, and N. Ash (eds.), *Ecosystems and Human Well-Being: Current State and Trends: Findings of the Condition and Trends Working Group*, pp. 77–122. Island Press, Washington, DC.

Maslo, B., et al. 2019. Optimizing conservation benefits for threatened beach fauna following severe natural disturbances. *Science of the Total Environment* 649: 661–671.

McCauley, D. J., et al. 2015. Marine defaunation: Animal loss in the global ocean. *Science* 347(6219): 1255641.

Menges, E. S. 1992. Stochastic modeling of extinction in plant populations. *In* P. L. Fiedler and S. K. Jain (eds.), *Conservation Biology: The Theory and Practice of Nature Conservation, Preservation and Management*, pp. 253–275. Chapman and Hall, New York.

Mikheyev, A. S., et al. 2017. Museum genomics confirms that the Lord Howe Island stick insect survived extinction. *Current Biology* 27(20): 3157–3161.

Miller-Rushing, A. J., and R. B. Primack. 2008. Global warming and flowering times in Thoreau's Concord: A community perspective. *Ecology* 89: 332–341.

Munilla, I., C. Diez, and A. Velando. 2007. Are edge bird populations doomed to extinction? A retrospective analysis of the common guillemot collapse in Iberia. *Biological Conservation* 137: 359–371.

Munson, L., et al. 2008. Climate extremes promote fatal co-infections during canine distemper epidemics in African lions. *PLOS One* 3: e2545.

Nabutanyi, P., and M. J. Wittmann. 2021. Models for eco-evolutionary extinction vortices under balancing selection. *American Naturalist* 197(3): 336–350.

Nagel, R., et al. 2021. Evidence for an Allee effect in a declining fur seal population. *Proceedings of the Royal Society B* 288(1947): 20202882.

Nomoto, H. A., and J. M. Alexander. 2021. Drivers of local extinction risk in alpine plants under warming climate. *Ecology Letters* 24: 1157–1166.

O'Connor, J. E., J. J. Duda, and G. E. Grant. 2015. 1000 dams down and counting. *Science* 348(6234): 496–497.

Palomares, F., et al. 2012. Possible extinction vortex for a population of Iberian lynx on the verge of extirpation. *Conservation Biology* 26: 689–697.

Pimm, S. L., and C. N. Jenkins. 2005. Sustaining the variety of life. *Scientific American* 293(33): 66–73.

Pimm, S. L., and L. N. Joppa. 2015. How many plant species are there, where are they, and at what rate are they going extinct? *Annals of the Missouri Botanical Garden* 100(3): 170–177.

Polishchuk, L. V., et al. 2015. A genetic component of extinction risk in mammals. *Oikos* 124(8): 983–993.

Portman, Z. M., et al. 2018. Local extinction of a rare plant pollinator in southern Utah (USA) associated with invasion by Africanized honey bees. *Biological Invasions* 20(3): 593–606.

Prates, L., and S. I. Perez. 2021. Late Pleistocene South American megafaunal extinctions associated with rise of Fishtail points and human population. *Nature Communications* 12(1): 1–11.

Price, J. P., et al. 2018. Colonization and diversification shape species-area relationships in three

Macaronesian archipelagos. *Journal of Biogeography* 45(9): 2027–2039.

Primack, R. B., A. J. Miller-Rushing, and K. Dharaneeswaran. 2009. Changes in the flora of Thoreau's Concord. *Biological Conservation* 142: 500–508.

Purvis, A., et al. 2000. Predicting extinction risk in declining species. *Proceedings of the Royal Society of London B* 267(1456): 1947–1952.

Quan, N. H., et al. 2018. Conservation of the Mekong Delta wetlands through hydrological management. *Ecological Research* 33(1): 87–103.

Rejmánek, M. 2018. Vascular plant extinctions in California: A critical assessment. *Diversity and Distributions* 24(1): 129–136.

Ribeiro, J., et al. 2016. An integrated trait-based framework to predict extinction risk and guide conservation planning in biodiversity hotspots. *Biological Conservation* 195: 214–223.

Ridding, L. E., et al. 2021. Inconsistent detection of extinction debts using different methods. *Ecography* 44(1): 33–43.

Riverá Ortíz, F. A., et al. 2014. Habitat fragmentation and genetic variability of tetrapod populations. *Animal Conservation* 18(3): 249–258.

Schwitzer, C., et al. 2014. Averting lemur extinctions amid Madagascar's political crisis. *Science* 343: 842–843.

Semper-Pascual, A., et al. 2018. Mapping extinction debt highlights conservation opportunities for birds and mammals in the South American Chaco. *Journal of Applied Ecology* 55(3): 1218–1229.

Sharma, N., M. D. Madhusudan, and A. Sinha. 2014. Local and landscape correlates of primate distribution and persistence in the remnant lowland forests of the Upper Brahmaputra Valley, northeastern India. *Conservation Biology* 28: 95–106.

Štípková, Z., and P. Kindlmann. 2021. Orchid extinction over the last 150 years in the Czech Republic. *Diversity* 13(2): 78.

Swezey, D. S., et al. 2020. Evolved differences in energy metabolism and growth dictate the impacts of ocean acidification on abalone aquaculture. *Proceedings of the National Academy of Sciences USA* 117(42): 26513–26519.

Teixeira, J. C., and C. D. Huber. 2021. The inflated significance of neutral genetic diversity in conservation genetics. *Proceedings of the National Academy of Sciences USA* 118(10): e2015096118.

Tende, T., et al. 2014. Individual identification and genetic variation of lions (*Panthera leo*) from two protected areas in Nigeria. *PLOS One* 9(1): e84288.

Tickner, D., et al. 2020. Bending the curve of global freshwater biodiversity loss: An emergency recovery plan. *BioScience* 70(4): 330–342.

Tilman, D., et al. 1994. Habitat destruction and the extinction debt. *Nature* 371: 65–66.

Todesco, M., et al. 2016. Hybridization and extinction. *Evolutionary Applications* 9(7): 892–908.

Urban, M. C. 2018. Escalator to extinction. *Proceedings of the National Academy of Sciences USA* 115(47): 11871–11873.

Vranckx, G., et al. 2012. Meta-analysis of susceptibility of woody plants to loss of genetic diversity through habitat fragmentation. *Conservation Biology* 26: 228–237.

Wagler, R. 2021. Anthropocene extinction. *Access Science*, https://doi.org/10.1036/1097-8542.039350

Waller, D. M. 2015. Genetic rescue: A safe or risky bet? *Molecular Ecology* 24(11): 2595–2597.

Weiser, E. L., C. E. Grueber, and I. G. Jamieson. 2013. Simulating retention of rare alleles in small populations to assess management options for species with different life histories. *Conservation Biology* 27(2): 335–344.

Willi, Y., et al. 2007. Genetic rescue persists beyond first-generation outbreeding in small populations of a rare plant. *Proceedings of the Royal Society B* 274: 2357–2364.

Wilson, E. O. 1989. Threats to biodiversity. *Scientific American* 261(3): 108–116.

Wood, J. R., et al. 2017. Island extinctions: Processes, patterns, and potential for ecosystem restoration. *Environmental Conservation* 44(4): 348–358.

Woodward, A. 2019. Trump's $5 billion border wall plan could wreak environmental havoc, causing rivers to flood and animals to become "zombie species." *Business Insider*, https://www.businessinsider.in/trumps-5-billion-border-wall-plan-could-wreak-environmental-havoc-causing-rivers-to-flood-and-animals-to-become-zombie-species/articleshow/67446908.cms

Wright, S. 1931. Evolution in Mendelian populations. *Genetics* 16: 97–159.

WWF (World Wildlife Fund) International. 2018. Species directory. https://www.worldwildlife.org/species/directory

Yacine, Y., et al. 2021. Collapse and rescue of evolutionary food webs under global warming. *Journal of Animal Ecology* 90(3): 710–722.

Yoder, A. D., et al. 2018. Neutral theory is the foundation of conservation genetics. *Molecular Biology and Evolution* 35(6): 1322–1326.

Chapter 7

Amel, E., et al. 2017. Beyond the roots of human inaction: Fostering collective effort toward ecosystem conservation. *Science* 356(6335): 275–279.

Arizmendi-Mejía, R., et al. 2015. Combining genetic and demographic data for the conservation of a Mediterranean marine habitat-forming species. *PLOS One* 10(3): e0119585.

Beissinger, S. R. 2015. Endangered species recovery criteria: Reconciling conflicting views. *BioScience* 65(2): 121–122.

BirdLife South Africa. 2017. Bird of the Year 2017: Threats to vultures. Retrieved from www.birdlife.org.za/documents/bird-of-the-year/981-fact-sheet-3-threats-to-vultures-1

Brook, B. W., et al. 2000. Predictive accuracy of population viability analysis in conservation biology. *Nature* 404: 385–387.

Buchalski, M. R., et al. 2015. Genetic population structure of Peninsular bighorn sheep (*Ovis canadensis nelsoni*) indicates substantial gene flow across US-Mexico border. *Biological Conservation* 184: 218–228.

Chaudhary, V., and M. K. Oli. 2020. A critical appraisal of population viability analysis. *Conservation Biology* 34(1): 26–40.

Cindy, B., et al. 2021. Passive eDNA collection enhances aquatic biodiversity analysis. *Communications Biology* 4: 1–12.

Creech, T. G., et al. 2014. Using network theory to prioritize management in a desert bighorn sheep metapopulation. *Landscape Ecology* 29(4): 605–619.

Cruickshank, S. S., et al. 2016. Quantifying population declines based on presence-only records for Red List assessments. *Conservation Biology* 30(5): 1112–1121.

Demery, A. J. C., and M. A. Pipkin. 2021. Safe fieldwork strategies for at-risk individuals, their supervisors and institutions. *Nature Ecology & Evolution* 5(1): 5–9.

Dieter, C. D., and D. J. Schaible. 2014. Distribution and population density of jackrabbits in South Dakota. *Great Plains Research* 24(2): 127–134.

Dubovsky, J. A. 2020. *Status and Harvests of Sandhill Cranes: Mid-Continent, Rocky Mountain, Lower Colorado River Valley and Eastern Populations.* Administrative report. US Fish and Wildlife Service, Lakewood, Colorado.

Elsen, P. R., et al. 2021. Contrasting seasonal patterns of relative temperature and thermal heterogeneity and their influence on breeding and winter bird richness patterns across the conterminous United States. *Ecography* 44(6): 953–965.

Evans, D. M., et al. 2016. Species recovery in the United States: Increasing the effectiveness of the Endangered Species Act. *Issues in Ecology* 20(Winter): 1–28. Available from https://www.fs.usda.gov/treesearch/pubs/50145

Ferreira, C., et al. 2016. The evolution of peer review as a basis for scientific publication: Directional selection towards a robust discipline? *Biological Reviews* 91(3): 597–610.

Flather, C. H., et al. 2011. Minimum viable populations: Is there a "magic number" for conservation practitioners? *Trends in Ecology & Evolution* 26: 307–316.

Forrester, T. D., et al. 2017. Creating advocates for mammal conservation through citizen science. *Biological Conservation* 208: 98–105.

Gray, C. L., et al. 2016. Local biodiversity is higher inside than outside terrestrial protected areas worldwide. *Nature Communications* 7: 12306.

Haase, P., et al. 2018. The next generation of site-based long-term ecological monitoring: Linking essential biodiversity variables and ecosystem integrity. *Science of the Total Environment* 613: 1376–1384.

Hedges, S., et al. 2013. Accuracy, precision, and cost-effectiveness of conventional dung density and fecal DNA based survey methods to estimate Asian elephant (*Elephas maximus*) population size and structure. *Biological Conservation* 159: 101–108.

Henson, P., R. White, and S. P. Thompson. 2018. Improving implementation of the Endangered Species Act: Finding common ground through common sense. *BioScience* 68(11): 861–872.

Himes Boor, G. K. 2014. A framework for developing objective and measurable recovery criteria for threatened and endangered species. *Conservation Biology* 28: 33–43.

Homan, R. N., M. A. Holgerson, and L. M. Biga. 2018. A long-term demographic study of a spotted salamander (*Ambystoma maculatum*) population in central Ohio. *Herpetologica* 74(2): 109–116.

Hunter, R. N., et al. 2018. Surveys of environmental DNA (eDNA): A new approach to estimate occurrence in vulnerable manatee populations. *Endangered Species Research* 35: 101–111.

Ibanez, T., et al. 2017. High endemism and stem density distinguish New Caledonian from other high-diversity rainforests in the Southwest Pacific. *Annals of Botany* 121(1): 25–35.

Jewell, A. 2013. Effect of monitoring technique on quality of conservation science. *Conservation Biology* 27: 501–508.

Jimenez-Lopez, O., et al. 2021. Non-steroidal anti-inflammatory drugs (Nsaids) and their effect on Old World vultures: A scoping review. *Journal of Raptor Research* 55(3): 297–310.

King, T., C. Chamberlan, and A. Courage. 2014. Assessing reintroduction success in long-lived primates through population viability analysis: Western lowland gorillas *Gorilla gorilla gorilla* in Central Africa. *Oryx* 48(02): 294–303.

Kuussaari, M., et al. 2009. Extinction debt: A challenge for biodiversity conservation. *Trends in Ecology & Evolution* 24(10): 564–571.

Ma, B., et al. 2018. Should the endangered status of the giant panda really be reduced? The case of giant panda conservation in Sichuan, China. *Animals* 8(5): 69.

Manlik, O., et al. 2016. The relative importance of reproduction and survival for the conservation of two dolphin populations. *Ecology and Evolution* 6(11): 3496–3512.

Manlik, O., et al. 2018. Demography and genetics suggest reversal of dolphin source-sink dynamics, with implications for conservation. *Marine Mammal Science*, https://doi.org/10.1111/mms.12555

Matthew, E. E., and C. E. Relton. 2021. Training methodology for canine scent detection of a critically endangered lagomorph: A conservation case study. *Journal of Vertebrate Biology* 69(3): 20092.1–14.

McCaffery, R., C. L. Richards-Zawacki, and K. R. Lips. 2015. The demography of *Atelopus* decline: Harlequin frog survival and abundance in central Panama prior to and during a disease outbreak. *Global Ecology and Conservation* 4: 232–242.

Meissen, J. C., S. M. Galatowitsch, and M. W. Cornett. 2017. Assessing long-term risks of prairie seed harvest: What is the role of life-history? *Botany* 95(11): 1081–1092.

Meyer, C., et al. 2016. Range geometry and socio-economics dominate species-level biases in occurrence information. *Global Ecology and Biogeography* 25(10): 1181–1193.

Miller, J. K., et al. 2002. The Endangered Species Act: Dollars and sense? *BioScience* 52: 163–168.

Milligan, B. G., et al. 2018. Disentangling genetic structure for genetic monitoring of complex populations. *Evolutionary Applications* 11(7): 1149–1161.

Nantel, P. L., J. Jones, and C. Drake. 2018. Viability of multiple populations across the range of a species at risk: The case of Pitcher's thistle, *Cirsium pitcheri*, in Canada. *Global Ecology and Conservation* 16: e00445.

Ore, J., et al. 2015. Autonomous aerial water sampling. *In* L. Mejias, P. Corke, and J. Roberts (eds.) *Field and Service Robotics*, pp 137–151. Springer, Switzerland.

Pe'er, G., et al. 2014. Toward better application of minimum area requirements in conservation planning. *Biological Conservation* 170: 92–102.

Phipps, W. L., et al. 2017. Due South: A first assessment of the potential impacts of climate change on Cape vulture occurrence. *Biological Conservation* 210: 16–25.

Pretorius, Y., M. E. Garaï, and L. A. Bates. 2019. The status of African elephant *Loxodonta Africana* populations in South Africa. *Oryx*, 53(4): 757–763.

Reading, R. P., et al. 2019. Home-range size and movement patterns of Hooded Vultures *Necrosyrtes monachus* in southern Africa. *Ostrich* 90(1): 73–77.

Ríos-Saldaña, C. A., M. Delibes-Mateos, and C. C. Ferreira. 2018. Are fieldwork studies being relegated to second place in conservation science? *Global Ecology and Conservation* 14: e00389.

Sæther, B. E., and S. Engen. 2015. The concept of fitness in fluctuating environments. *Trends in Ecology & Evolution* 30(5): 273–281.

Shaffer, M. L. 1981. Minimum population sizes for species conservation. *BioScience* 31: 131–134.

Smith, A. C., and B. P. Edwards. 2021. North American Breeding Bird Survey status and trend estimates to inform a wide range of conservation needs, using a flexible Bayesian hierarchical generalized additive model. *The Condor* 123(1): duaa065.

Stork, N. E., et al. 2017. Consistency of effects of tropical-forest disturbance on species composition and richness relative to use of indicator taxa. *Conservation Biology* 31(4): 924–933.

Swengel, A. B., and S. R. Swengel. 2018. Long-term population monitoring of the Karner blue (*Lepidoptera: Lycaenidae*) in Wisconsin, 1990–2004. *Great Lakes Entomologist* 38(3–4): 11.

Taylor, M. F. J., K. F. Suckling, and J. J. Rachlinski. 2005. The effectiveness of the Endangered Species Act: A quantitative analysis. *BioScience* 55: 360–366.

Terwissen, C., G. Mastromonaco, and D. Murray. 2013. Influence of adrenocorticotrophin hormone challenge and external factors (age, sex, and body region) on hair cortisol concentration in Canada lynx (*Lynx canadensis*). *General and Comparative Endocrinology* 194: 162–167.

Thiollay, J. M. 1989. Area requirements for the conservation of rainforest raptors and game birds in French Guiana. *Conservation Biology* 3: 128–137.

USFWS (US Fish and Wildlife Service). 2013. *Federal and State Endangered and Threatened Species Expenditures Fiscal Year 2013*. www.fws.gov/endangered/esa-library/pdf/2012.EXP.FINAL.pdf

USFWS (US Fish and Wildlife Service). 2020. *Final Report: Bald Eagle Population Size: 2020 Update*. USFWS, Division of Migratory Bird Management, Washington, D.C.

WWF (World Wildlife Fund) Global. 2018. wwf.panda.org.

WWF (World Wildlife Fund). 2020. *Living Planet Report 2020: Bending the Curve of Biodiversity Loss*. R. E. A. Almond, M. Grooten, and T. Petersen (eds.). WWF, Gland, Switzerland

Zepelini, C., et al. 2017. Assessing fishing experts' knowledge to improve conservation strategies for an endangered grouper in the southwestern Atlantic. *Journal of Ethnobiology* 37(3): 478–493.

Chapter 8

Anderson, J. T., and B. H. Song. 2020. Plant adaptation to climate change: Where are we? *Journal of Systematics and Evolution* 58(5): 533–545.

Baker, C. M., M. Bode, and M. A. McCarthy. 2016. Models that predict ecosystem impacts of reintroductions should consider uncertainty and distinguish between direct and indirect effects. *Biological Conservation* 196: 211–212.

Bencatel, J., et al. 2018. Research trends and geographical distribution of mammalian carnivores in Portugal (SW Europe). *PLOS One* 13(11): e0207866.

Bennett, J. R., et al. 2017. Spending limited resources on de-extinction could lead to net biodiversity loss. *Nature Ecology & Evolution* 1(4): 0053.

Berger-Tal, O., et al. 2016. A systematic survey of the integration of behavior into wildlife conservation and management. *Conservation Biology* 30(4): 744–753.

Boast, L. K., et al. 2018. Cheetah translocation and reintroduction programs: Past, present, and future. *In Cheetahs: Biology and Conservation*, pp. 275–289. Academic Press, London.

Bocci, A., et al. 2016. Conservation introduction of the threatened Apennine chamois *Rupicapra pyrenaica ornata*: Post-release dispersal differs between wild-caught and captive founders. *Oryx* 50(1): 128–133.

Canessa, S., et al. 2016. Planning for ex situ conservation in the face of uncertainty. *Conservation Biology* 30(3): 599–609.

Carter, I., J. Foster, and L. Lock. 2017. The role of animal translocations in conserving British wildlife: An overview of recent work and prospects for the future. *EcoHealth* 14: 7–15.

Che-Castaldo, J. P., S. A. Grow, and L. J. Faust. 2018. Evaluating the contribution of North American zoos and aquariums to endangered species recovery. *Scientific Reports* 8(1): 9789.

Che-Castaldo, J., et al. 2019. Patterns in the long-term viability of North American zoo populations. *Zoo Biology* 38(1): 78–94.

Díaz, M., et al. 2018. Independent contributions of threat and popularity to conservation translocations. *Biodiversity and Conservation* 27(6): 1419–1429.

Dolný, A., et al. 2018. How difficult is it to reintroduce a dragonfly? Fifteen years monitoring *Leucorrhinia dubia* at the receiving site. *Biological Conservation* 218: 110–117.

FAO (Food and Agriculture Organization of the United Nations). 2016. *The State of World Fisheries and Aquaculture.* FAO, Rome.

Feldberg, S. 2019. Scientists are getting creative to save this muppet-faced, flightless parrot. *National Geographic*, https://www.nationalgeographic.com/animals/2019/03/endangered-kapako-breeding-technology/

Fisher, M. 2016. Fall, resurrection and uncertainty: An Arabian tale. *Oryx* 50(1): 1–2.

Harding, G., R. A. Griffiths, and L. Pavajeau. 2016. Developments in amphibian captive breeding and reintroduction programs. *Conservation Biology* 30(2): 340–349.

Harrington, L. A., et al. 2013. Conflicting and complementary ethics to animal welfare considerations in reintroductions. *Conservation Biology* 27: 486–500.

He, X., M. L. Johansson, and D. D. Heath. 2016. Role of genomics and transcriptomics in selection of reintroduction source populations. *Conservation Biology* 30(5): 1010–1018.

Hendricks, S. A., et al. 2016. Re-defining historical geographic range in species with sparse records: Implications for the Mexican wolf reintroduction program. *Biological Conservation* 194: 48–57.

Hildebrandt, T. B., et al. 2021. Conservation research in times of COVID-19: The rescue of the northern white rhino. *Journal of Applied Animal Ethics Research* 3(1): 16–37. https://doi.org/10.1163/25889567-BJA10009

Hunter-Ayad, J., et al. 2020. Reintroduction modelling: A guide to choosing and combining models for species reintroductions. *Journal of Applied Ecology* 57(7): 1233–1243.

IUCN/SSC (International Union for Conservation of Nature/Species Survival Commission). 2013. *Guidelines for Reintroductions and Other Conservation Translocations*, Version 1.0. IUCN/SSC, Gland, Switzerland.

Josephson, A. 2018. The economics of zoos. Retrieved from https://smartasset.com/taxes/the-economics-of-zoos

Kolby, J. E., et al. 2015. Rapid response to evaluate the presence of amphibian chytrid fungus (*Batrachochytrium dendrobatidis*) and ranavirus in wild amphibian populations in Madagascar. *PLOS One* 10(6): e0125330.

Liu, H., et al. 2015. Translocation of threatened plants as a conservation measure in China. *Conservation Biology* 29(6): 1537–1551.

Loreno, D. (Fox8). 2021. Scientists announce plans to "de-extinct" the woolly mammoth. September 16. https://fox8.com/news/scientists-announce-plans-to-de-extinct-the-woolly-mammoth/

Lunt, I. D., et al. 2013. Using assisted colonisation to conserve biodiversity and restore ecosystem function under climate change. *Biological Conservation* 157: 172–177.

Malone, E. W., et al. 2018. Which species, how many, and from where: Integrating habitat suitability, population genomics, and abundance estimates into species reintroduction planning. *Global Change Biology* 24(8): 3729–3748.

McCune, L. M. 2018. The protection of Indigenous Peoples' seed rights during ethnobotanical research. *Ethnobiology Letters* 9(1): 67–75.

Minuzzi-Souza, T. T. C., et al. 2016. Vector-borne transmission of *Trypanosoma cruzi* among captive Neotropical primates in a Brazilian zoo. *Parasites and Vectors* 9(1): 1.

Moro, D., et al. 2015. Reintroduction biology of Australian and New Zealand fauna: Progress, emerging themes and future directions. *In* D. P. Armstrong, M. W. Hayward, D. Moro, and P. J. Seddon (eds.), *Advances in Reintroduction Biology of Australian and New Zealand Fauna*, p. 178. CSIRO, Clayton, Australia.

Morris, S. D., et al. 2021. Factors affecting success of conservation translocations of terrestrial vertebrates: A global systematic review. *Global Ecology and Conservation* 28: e01630.

Moss, A., E. Jensen, and M. Gusset. 2017. Probing the link between biodiversity-related knowledge and self-reported pro-conservation behaviour in a global survey of zoo visitors. *Conservation Letters* 10(1): 33–40.

Mounce, R., P. Smith, and S. Brockington. 2017. Ex situ conservation of plant diversity in the world's botanic gardens. *Nature Plants* 3(10): 795.

Odum, R. A. 1995. Zoos: The Ark and a whole lot more. *Zoo Biology* 14(1): 45–47.

Olivieri, I., et al. 2016. Why evolution matters for species conservation: Perspectives from three case studies of plant metapopulations. *Evolutionary Applications* 9(1): 196–211.

Ottewell, K., et al. 2014. Evaluating the success of translocations in maintaining genetic diversity in a threatened mammal. *Biological Conservation* 171: 209–219.

Page, T., and C. Hancock (CNN). 2016. Zebra cousin became extinct 100 years ago. Now, it's back. www.cnn.com/2016/01/25/africa/quagga-project-zebra-conservation-extinct-south-africa/index.html

Parlato, E. H., and D. P. Armstrong. 2013. Predicting post-release establishment using data from multiple reintroductions. *Biological Conservation* 160: 97–104.

Reiter, N., et al. 2016. Orchid re-introductions: An evaluation of success and ecological considerations using key comparative studies from Australia. *Plant Ecology* 217: 81–95.

Resende, P. S., et al. 2020. A global review of animal translocation programs. *Animal Biodiversity and Conservation* 43(2): 221–232.

Rood, S. B., et al. 2015. A twofold strategy for riparian restoration: Combining a functional flow regime and direct seeding to re-establish cottonwoods. *River Research and Applications* 32(5): 836–844.

Royal Botanical Gardens Kew. 2021. Seed collection. Retrieved from https://www.kew.org/science/collections-and-resources/collections/seed-collection

Rueda, C., et al. 2021. Exploratory and territorial behavior in a reintroduced population of Iberian lynx. *Scientific Reports* 11(1): 1–12.

Samojlik, T., et al. 2018. Lessons from Białowieża Forest on the history of protection and the world's first reintroduction of a large carnivore. *Conservation Biology* 32(4): 808–816.

Sampaio, M. B., N. Schiel, and A. da Silva Souto. 2020. From exploitation to conservation: A historical analysis of zoos and their functions in human societies. *Ethnobiology and Conservation*, https://doi.org/10.15451/ec2020-01-9.02-1-32

Sarmento, P., et al. 2019. Spatial organization and social relations in a reintroduced population of endangered Iberian lynx *Lynx pardinus*. *Oryx* 53(2): 344–355.

Schmitz, P., et al. 2015. First steps into the wild: Exploration behavior of European bison after the first reintroduction in western Europe. *PLOS One* 10(11): e0143046.

Seddon, P. J. 2017. The ecology of de-extinction. *Functional Ecology* 31(5): 992–995.

Seddon, P. J., A. Moehrenschlager, and J. Ewen. 2014b. Reintroducing resurrected species: Selecting de-extinction candidates. *Trends in Ecology & Evolution* 29(3): 140–147.

Seddon, P. J., et al. 2014a. Reversing defaunation: Restoring species in a changing world. *Science* 345(6195): 406–412.

Shapiro, B. 2017. Pathways to de-extinction: How close can we get to resurrection of an extinct species? *Functional Ecology* 31(5): 996–1002.

Soorae, P. S. (ed.). 2016. *Global Reintroduction Perspectives: 2016: Case Studies from Around the Globe*. IUCN/SSC Re-introduction Specialist Group & Environment Agency, Abu Dhabi, UAE.

Species360/ZIMS (Zoological Information Management System). 2019. zims.species360.org.

Stringer, A. P., and M. J. Gaywood. 2016. The impacts of beavers *Castor* spp. on biodiversity and the ecological basis for their reintroduction to Scotland, UK. *Mammal Review* 46(4): 270–283.

Traylor-Holzer, K., K. Leus, and K. Bauman. 2018. *Global Integrated Collection Assessment and Planning Workshop for Canids and Hyaenids: Final Report*. IUCN SSC Conservation Planning Specialist Group, Apple Valley, MN.

Traylor-Holzer, K., K. Leus, and K. Bauman. 2019. Integrated Collection Assessment and Planning (ICAP) workshop: Helping zoos move toward the One Plan Approach. *Zoo Biology* 38(1): 95–105.

West, R. S., et al. 2019. Searching for an effective pre-release screening tool for translocations: Can trap temperament predict behaviour and survival in the wild? *Biodiversity and Conservation* 28(1): 229–243.

Wilson, H. B., et al. 2014. Conservation strategies for orangutans: Reintroduction versus habitat preservation and the benefits of sustainably logged forest. *PLOS One* 9(7): e102174.

Yen, S. C., et al. 2015. Residents' attitudes toward reintroduced sika deer in Kenting National Park, Taiwan. *Wildlife Biology* 21(4): 220–226.

Yochim, M. J., and W. R. Lowry. 2016. Creating conditions for policy change in national parks: Contrasting cases in Yellowstone and Yosemite. *Environmental Management* 57(5): 1041–1053.

Zhang, H., et al. 2017. Back into the wild: Apply untapped genetic diversity of wild relatives for crop improvement. *Evolutionary Applications* 10(1): 5–24.

Zhou, Z. K., et al. 2015. Resequencing 302 wild and cultivated accessions identifies genes related to domestication and improvement in soybean. *Nature Biotechnology* 33: 408–414.

Chapter 9

Abesamis, R. A., and G. R. Russ. 2005. Density-dependent spillover from a marine reserve: Long-term evidence. *Ecological Applications* 15: 1798–1812.

Acuña, V., M. Hunter, and A. Ruhí. 2017. Managing temporary streams and rivers as unique rather than second-class ecosystems. *Biological Conservation* 211: 12–19.

Andrello, M., et al. 2015. Extending networks of protected areas to optimize connectivity and population growth rate. *Ecography* 38(3): 273–282.

Baiamonte, G., et al. 2015. Agricultural landscapes and biodiversity conservation: A case study in Sicily (Italy). *Biodiversity and Conservation* 24(13): 3201–3216.

Baker, K. J. M., and T. Warren (Buzzfeed News). 2019. WWF says indigenous people want this park. An internal report says some fear forest ranger "repression." www.buzzfeednews.com/article/katiejmbaker/wwf-eu-messok-dja-fears-repression-ecoguards

Ban, N. C., et al. 2017. Social and ecological effectiveness of large marine protected areas. *Global Environmental Change* 43: 82–91.

Bragina, E. V., et al. 2015. Effectiveness of protected areas in the Western Caucasus before and after the transition to post-socialism. *Biological Conservation* 184: 456–464.

Brockington, D., and D. Wilkie. 2015. Protected areas and poverty. *Philosophical Transactions of the Royal Society B* 370(1681): 20140271.

Burgess, S. C., et al. 2014. Beyond connectivity: How empirical methods can quantify population persistence to improve marine protected-area design. *Ecological Applications* 24: 257–270.

Butchart, S. H. M., et al. 2015. Shortfalls and solutions for meeting national and global conservation targets. *Conservation Letters* 8(5): 329–337.

Carrasco, L., et al. 2021. Global progress in incorporating climate adaptation into land protection for biodiversity since Aichi targets. *Global Change Biology* 27(9): 1788–1801.

Carrizo, S. F., et al. 2017. Critical catchments for freshwater biodiversity conservation in Europe: Identification, prioritisation and gap analysis. *Journal of Applied Ecology* 54(4): 1209–1218.

Corlett, R. T. 2013. Singapore: Half full or half empty? *In* N. S. Sodhi, L. Gibson, and P. H. Raven (eds.), *Conservation Biology: Voices from the Tropics*, pp. 142–147. Wiley-Blackwell, Oxford, UK.

Djossa, C. 2018. These are the world's first national parks. *National Geographic*, www.nationalgeographic.com/travel/national-parks/worlds-first-protected-lands-conservation-yellowstone/

Elliott, T. L., and T. J. Davies. 2019. Phylogenetic attributes, conservation status and geographical origin of species gained and lost over 50 years in a UNESCO Biosphere Reserve. *Biodiversity and Conservation* 28(3): 711–728.

Elsen, P. R., et al. 2020. Keeping pace with climate change in global terrestrial protected areas. *Science Advances*, 6(25), eaay0814.

Fawzi, N. I., V. N. Husna, and J. A. Helms. 2018. Measuring deforestation using remote sensing and its implication for conservation in Gunung Palung National Park, West Kalimantan, Indonesia. *IOP Conference Series: Earth and Environmental Science* 149: 012038.

Fox, H. E., et al. 2012. Reexamining the science of marine protected areas: Linking knowledge to action. *Conservation Letters* 5: 1–10.

Geldmann, J., et al. 2013. Effectiveness of terrestrial protected area in reducing habitat loss and population declines. *Biological Conservation* 161: 230–238.

Geldmann, J., et al. 2018. A global analysis of management capacity and ecological outcomes in terrestrial protected areas. *Conservation Letters* 11(3): e12434.

Gilbert-Norton, L., et al. 2010. A meta-analytic review of corridor effectiveness. *Conservation Biology* 24: 660–668.

Gray, C. L., et al. 2016. Local biodiversity is higher inside than outside terrestrial protected areas worldwide. *Nature Communications* 7: 12306.

Grecchi, R. C., et al. 2017. An integrated remote sensing and GIS approach for monitoring areas affected by selective logging: A case study in northern Mato Grosso, Brazilian Amazon. *International Journal of Applied Earth Observation and Geoinformation* 61: 70–80.

Haddad, N. M., et al. 2015. Habitat fragmentation and its lasting impact on Earth's ecosystems. *Science Advances* 1(2): e1500052.

Hobbs, R. J., et al. 2010. Guiding concepts for park and wilderness stewardship in an era of global environmental change. *Frontiers in Ecology and the Environment* 8: 483–490.

Hunter, M. O., et al. 2015. Structural dynamics of tropical moist forest gaps. *PLOS One* 10(7): e0132144.

Iannella, M., et al. 2018. Coupling GIS spatial analysis and Ensemble Niche Modelling to investigate climate change–related threats to the Sicilian pond turtle *Emys trinacris*, an endangered species from the Mediterranean. *PeerJ* 6: e4969.

IUCN (International Union for Conservation of Nature). 2021. www.iucn.org

IUCN WCPA (International Union for Conservation of Nature, World Commission on Protected Areas). 2021. WCPA Europe. Retrieved from www.iucn.org/commissions/world-commission-protected-areas/regions/wcpa-europe

Jelinski, D. E. 2015. On a landscape ecology of a harlequin environment: The marine landscape. *Landscape Ecology* 30(1): 1–6.

Jiao, Y., et al. 2019. Crises of biodiversity and ecosystem services in Satoyama landscape of Japan: A review on the role of management. *Sustainability* 11(2): 454.

Jones, K. R., et al. 2018. One-third of global protected land is under intense human pressure. *Science* 360(6390): 788–791.

Laguna, E., et al. 2016. Role of micro-reserves in conservation of endemic, rare and endangered plants of the Valencian region (eastern Spain). *Israel Journal of Plant Sciences* 63(4): 320–332.

Langton, M., L. Palmer, and Z. Ma Rhea. 2014. Community-oriented protected areas for indigenous peoples and local communities. *In* S. Stevens (ed.), *Indigenous Peoples, National Parks, and Protected Areas: A New Paradigm Linking Conservation, Culture, and Rights*, pp. 84–107. University of Arizona Press, Tucson, AZ.

Le Roux, D. S., et al. 2016. Enriching small trees with artificial nest boxes cannot mimic the value of large trees for hollow-nesting birds. *Restoration Ecology* 24(2): 252–258.

MacArthur, R. H., and E. O. Wilson. 1967. *The Theory of Island Biogeography*. Princeton University Press, Princeton, NJ.

Magrach, A., A. R. Larrinaga, and L. Santamaría. 2012. Effects of matrix characteristics and interpatch distance on functional connectivity in fragmented temperate rainforests. *Conservation Biology* 26: 238–247.

Mascia, M. B., et al. 2014. Protected area downgrading, downsizing, and degazettement (PADDD) in Africa, Asia, and Latin America and the Caribbean, 1900–2010. *Biological Conservation* 169: 355–361.

Masud, M. M., et al. 2017. Community-based ecotourism management for sustainable development of marine protected areas in Malaysia. *Ocean & Coastal Management* 136: 104–112.

Matthews, T. J., et al. 2016. On the form of species-area relationships in habitat islands and true islands. *Global Ecology and Biogeography* 25(7): 847–858.

McCarthy, M. A., C. J. Thompson, and N. S. G. Williams. 2006. Logic for designing nature reserves for multiple species. *American Naturalist* 167: 717–727.

Middleton, B. A. 2013. Rediscovering traditional vegetation management in preserves: Trading experiences between cultures and continents. *Biological Conservation* 158: 271–279.

Mills, M., et al. 2014. Linking regional planning and local action: Towards using social network analysis in systematic conservation planning. *Biological Conservation* 169: 14–53.

Newmark, W. D. 1995. Extinction of mammal populations in western North American national parks. *Conservation Biology* 9: 512–527.

Nimmo, D. G., et al. 2013. Fire mosaics and reptile conservation in a fire-prone region. *Conservation Biology* 27: 345–353.

Nuñez, T. A., et al. 2013. Connectivity planning to address climate change. *Conservation Biology* 27: 407–416.

Oertli, B. 2018. Freshwater biodiversity conservation: The role of artificial ponds in the 21st century. *Aquatic Conservation: Marine and Freshwater Ecosystems* 28(2): 264–269.

Ogden, L. E. 2015. Do wildlife corridors have a downside? *BioScience* 65(4): 452.

Ojwang, G. O., et al. 2017. *Wildlife Migratory Corridors and Dispersal Areas: Kenya Rangelands and Coastal Terrestrial Ecosystems*. Ministry of Environment and Natural Resources, Nairobi.

Oldekop, J. A., et al. 2015. A global assessment of the social and conservation outcomes of protected areas. *Conservation Biology* 30(1): 133–141.

Pacifici, M., M. Di Marco, and J. E. Watson. 2020. Protected areas are now the last strongholds for many imperiled mammal species. *Conservation Letters* 13(6): e12748.

Packer, C., et al. 2013. Conserving large carnivores: Dollars and fence. *Conservation Letters* 16: 635–641.

Pardini, R., et al. 2005. The role of forest structure, fragment size and corridors in maintaining small mammal abundance and diversity in an Atlantic forest landscape. *Biological Conservation* 12: 253–266.

Pocock, M. J., et al. 2015. Developing and enhancing biodiversity monitoring programmes: A collaborative assessment of priorities. *Journal of Applied Ecology* 52(3): 686–695.

Pristupa, A. O., et al. 2018. Can zoning resolve nature use conflicts? The case of the Numto Nature Park in the Russian Arctic. *Journal of Environmental Planning and Management* 61(10): 1674–1700.

Protected Planet. 2021. www.protectedplanet.net/en/thematic-areas/wdpa?tab=WDPA

Reed, M. G., and P. Abernethy. 2018. Facilitating co-production of transdisciplinary knowledge for sustainability: Working with Canadian Biosphere Reserve Practitioners. *Society & Natural Resources* 31(1): 39–56.

Regos, A., et al. 2016. Predicting the future effectiveness of protected areas for bird conservation in Mediterranean ecosystems under climate change and novel fire regime scenarios. *Diversity and Distributions* 22(1): 83–96.

Richardson, S. J., et al. 2015. Small wetlands are critical for safeguarding rare and threatened plant species. *Applied Vegetation Science* 18(2): 230–241.

Rife, A. N., et al. 2013. When good intentions are not enough: Insights on networks of "paper park" marine protected areas. *Conservation Letters* 6: 200–212.

Rueegger, N. 2017. Artificial tree hollow creation for cavity-using wildlife: Trialling an alternative method to that of nest boxes. *Forest Ecology and Management* 405: 404–412.

Rytwinski, T., et al. 2016. How effective is road mitigation at reducing road-kill? A meta-analysis. *PLOS One* 11: e0166941.

Sayre, R., et al. 2020. An assessment of the representation of ecosystems in global protected areas using new maps of World Climate Regions and World Ecosystems. *Global Ecology and Conservation* 21: e00860.

Scheffer, M., et al. 2015. Creating a safe operating space for iconic ecosystems. *Science* 347(6228): 1317–1319.

Shwartz, A., et al. 2017. Scaling up from protected areas in England: The value of establishing large conservation areas. *Biological Conservation* 212: 279–287.

Simaika, J. P., et al. 2013. Continental-scale conservation prioritization of African dragonflies. *Biological Conservation* 157: 245–254.

Soanes, K., et al. 2013. Movement re-established but not restored: Inferring the effectiveness of road-crossing mitigation for a gliding mammal by monitoring use. *Biological Conservation* 159: 434–441.

Soukhaphon, A., I. G. Baird, and Z. S. Hogan. 2021. The impacts of hydropower dams in the Mekong River Basin: A review. *Water* 13(3): 265.

Soulé, M. E., and D. Simberloff. 1986. What do genetics and ecology tell us about the design of nature reserves? *Biological Conservation* 35: 19–40.

UN Department of Economic and Social Affairs. 2018. 68% of the world population projected to live in urban areas by 2050, says UN. https://www.un.org/development/desa/en/news/population/2018-revision-of-world-urbanization-prospects.html

UNEP-WCMC and IUCN (United Nations Environment Programme World Conservation Monitoring Centre and International Union for Conservation of Nature). 2021. United Nations list of protected areas. Retrieved from www.unep-wcmc.org/resources-and-data/united-nations-list-of-protected-areas

UNEP-WCMC and IUCN (United Nations Environment Programme World Conservation Monitoring Centre and International Union for Conservation of Nature). 2021. *Protected Planet Report 2020*. UNEP-WCMC and IUCN, Cambridge, UK; Gland, Switzerland. https://livereport.protectedplanet.net

UNEP-WCMC, IUCN, and NGS (United Nations Environment Programme World Conservation Monitoring Centre, International Union for Conservation of Nature, and National Geographic Society). 2018. *Protected Planet Report 2018*. Gland, Switzerland, and Washington, DC.

UNESCO (United Nations Educational, Scientific and Cultural Organization). 2021a. Mont Saint Hilaire. Retrieved from www.unesco.org/new/en/natural-sciences/environment/ecological-sciences/biosphere-reserves/europe-north-america/canada/mont-saint-hilaire/

UNESCO (United Nations Educational, Scientific and Cultural Organization). 2021b. Phoenix Islands Protected Area. Retrieved from https://whc.unesco.org/en/list/1325/

Urbanavichus, G. P., and I. N. Urbanavichene. 2017. Contribution to the lichen flora of Erzi Nature Reserve, Republic of Ingushetia, North Caucasus, Russia. *Willdenowia* 47(3): 227–237.

Uwayo, P., et al. 2020. Contribution of Former Poachers for Wildlife Conservation in Rwanda Volcanoes National Park. *Journal of Geoscience and Environment Protection* 8(4): 47–56.

Venter, O., et al. 2018. Bias in protected-area location and its effects on long-term aspirations of biodiversity conventions. *Conservation Biology* 32(1): 127–134.

Watson, J. E., et al. 2014. The performance and potential of protected areas. *Nature* 515: 67–73.

Wikramanayake, E., et al. 2011. A landscape-based conservation strategy to double the wild tiger population. *Conservation Letters* 4: 219–227.

Zomer, R. J., et al. 2015. Projected impact of climate change on the effectiveness of the existing protected area network for biodiversity conservation within Yunnan Province, China. *Biological Conservation* 184: 335–345.

Chapter 10

Altman, I., et al. 2011. A practical approach to implementation of ecosystem-based management: A case study using the Gulf of Maine marine ecosystem. *Frontiers in Ecology and the Environment* 9: 183–189.

Álvarez-Romero, J., et al. 2018. Designing connected marine reserves in the face of global warming. *Global Change Biology* 24: e671–e691.

Baker, J., E. J. Milner-Gulland, and N. Leader-Williams. 2012. Park gazettement and integrated conservation and development as factors in community conflict at Bwindi Impenetrable Forest, Uganda. *Conservation Biology* 26: 160–170.

Ban, N. C., et al. 2013. A social-ecological approach to conservation planning: Embedding social considerations. *Frontiers in Ecology and the Environment* 11: 194–202.

Banerjee, S., et al. 2013. How to sell ecosystem services: A guide for designing new markets. *Frontiers in Ecology and the Environment* 11: 297–304.

Bates, A. E., et al. 2020. COVID-19 pandemic and associated lockdown as a "Global Human Confinement Experiment" to investigate biodiversity conservation. *Biological Conservation* 248: 108665.

Baudron, F., and K. E. Giller. 2014. Agriculture and nature: Trouble and strife? *Biological Conservation* 170: 232–245.

Berkes, F. 1993. Traditional ecological knowledge in perspective. *In* Julian T. Inglis (ed.), *Traditional Ecological Knowledge: Concepts and Cases*, pp. 1–10. International Program on Traditional Ecological Knowledge, Ottawa, Ontario; International Development Research Centre, Ottawa, Canada.

Borrini-Feyerabend, G., et al. 2004. *Sharing Power: Learning by Doing in Co-Management of Natural Resources throughout the World*. IIED and IUCN/CEESP/CMWG, Cenesta, Tehran.

Boydell, S., and H. Holzknecht. 2003. Land—caught in the conflict between custom and commercialism. *Land Use Policy* 20: 203–207.

Braschler, B. 2009. Successfully implementing a citizen-scientist approach to insect monitoring in a resource-poor country. *BioScience* 59: 103–104.

Brooks, J. S. 2017. Design features and project age contribute to joint success in social, ecological, and economic outcomes of community-based conservation projects. *Conservation Letters* 10(1): 23–32.

Carrière, S. M., et al. 2013. Rio +20, biodiversity marginalized. *Conservation* 6: 6–11.

Catano, C. P., et al. 2015. Using scenario planning to evaluate the impacts of climate change on wildlife populations and communities in the Florida Everglades. *Environmental Management* 55(4): 807–823.

Chan, Y. C., et al. 2019. Conserving unprotected important coastal habitats in the Yellow Sea: Shorebird occurrence, distribution and food resources at Lianyungang. *Global Ecology and Conservation* 20: e00724.

Chiu, E. S., et al. 2019. Multiple introductions of domestic cat feline leukemia virus in endangered Florida panthers. *Emerging Infectious Diseases* 25(1): 92.

Corbley, A. (Good News Network). 2020. Five years after turning disused military bases into nature reserves, wolves return and use bases as havens. February 2. www.goodnewsnetwork.org/germany-turns-military-bases-into-nature-reserves/

Criffield, M., et al. 2018. Assessing impacts of intrinsic and extrinsic factors on Florida panther movements. *Journal of Mammalogy* 99(3): 702–712.

CRS (Congressional Research Service). 2020. Federal land ownership: Overview and data. Retrieved from https://crsreports.congress.gov R42346

Cushman, S. A., et al. 2018. Prioritizing core areas, corridors and conflict hotspots for lion conservation in southern Africa. *PLOS One* 13(7): e0196213.

Daru, B. H., and P. C. Le Roux. 2016. Marine protected areas are insufficient to conserve global marine plant diversity. *Global Ecology and Biogeography* 25(3): 324–334.

Davis, A. M., and A. R. Moore. 2016. Conservation potential of artificial water bodies for fish communities on a heavily modified agricultural floodplain. *Aquatic Conservation: Marine and Freshwater Ecosystems* 26(6): 1184–1196.

Di Minin, E., N. Leader-Williams, and C. J. A. Bradshaw. 2016. Banning trophy hunting will exacerbate biodiversity loss. *Trends in Ecology & Evolution* 31(2): 99–102.

Dickman, A. J., E. A. Macdonald, and D. W. Macdonald. 2011. A review of financial instruments to pay for predator conservation and encourage human-carnivore coexistence. *Proceedings of the National Academy of Sciences USA* 108: 13937–13944.

Dinerstein, E., et al. 2013. Enhancing conservation, ecosystem services, and local livelihoods through a wildlife premium mechanism. *Conservation Biology* 27(1): 14–23.

DOD NR Program (US Department of Defense Natural Resources Program). 2016. *Threatened and Endangered Species on DOD Lands: "Enabling the Mission, Defending the Resources."* www.dodworkshops.org/ TE_Species_Fact_Sheet_8–12–2016.pdf

Duchelle, A. E., et al. 2012. Evaluating the opportunities and limitations to multiple use of Brazil nuts and timber in western Amazonia. *Forest Ecology and Management* 268: 39–48.

Earnst, S. L., D. S. Dobkin, and J. A. Ballard. 2012. Changes in avian and plant communities of aspen woodlands over 12 years after livestock removal in the Northwestern Great Basin. *Conservation Biology* 26: 862–871.

Ecosystem Marketplace. 2010. The Katoomba Group. Retrieved from www.ecosystemmarketplace.com

Ellwanger, G., and K. Reiter. 2019. Nature conservation on decommissioned military training areas: German approaches and experiences. *Journal for Nature Conservation* 49: 1–8.

Equator Initiative. 2021. www.equatorinitiative.org

FAO (Food and Agriculture Organization of the United Nations). 2019. FAOSTAT: Coffee. Retrieved from www.fao.org/faostat/en/#search/coffee

Farmer, J. R., et al. 2017. Private landowners, voluntary conservation programs, and implementation of conservation-friendly land management practices. *Conservation Letters* 10(1): 58–66.

FFWCC (Florida Fish and Wildlife Conservation Commission). 2018. *Annual Report on the Research and Management of Florida Panthers: 2017–2018*. Fish and Wildlife Research Institute & Division of Habitat and Species Conservation, Naples, FL.

Fischer, L. K., et al. 2016. Drivers of biodiversity patterns in parks of a growing South American megacity. *Urban Ecosystems* 19(3): 1231–1249.

Florida Natural Areas Inventory. 2019. https://www.fnai. org/conslands/conservation-lands

Galvin, K., T. Beeton, and M. Luizza. 2018. African community-based conservation: A systematic review of social and ecological outcomes. *Ecology and Society* 23(3): 39.

Garnett, S. T., et al. 2018. A spatial overview of the global importance of Indigenous lands for conservation. *Nature Sustainability* 1(7): 369.

Gavin, M. C., et al. 2015. Defining biocultural approaches to conservation. *Trends in Ecology & Evolution* 30(3): 140–145.

Gooden, J., and R. Grenyer. 2019. The psychological appeal of owning private land for conservation. *Conservation Biology* 33(2): 339–350.

Guadilla-Sáez, S., et al. 2019. Biodiversity conservation effectiveness provided by a protection status in temperate forest commons of north Spain. *Forest Ecology and Management* 433: 656–666.

Hanson, T. 2018. Biodiversity conservation and armed conflict: A warfare ecology perspective. *Annals of the New York Academy of Sciences* 1429(1): 50–65.

Hosen, N., H. Nakamura, and Hamzah, A. 2020. Adaptation to climate change: Does traditional ecological knowledge hold the key? *Sustainability* 12(2): 676.

Huffington Post. 2015. Germany to turn 62 military bases into nature sanctuary for birds, beetles and bats. www.huffingtonpost.com/2015/06/19/german-military-bases-nature-reserves_n_7623882.html

Humavindu, M. N., and J. Stage. 2015. Continuous financial support will be needed. *Animal Conservation* 18(1): 18–19.

Hylander, K., et al. 2013. Effects of coffee management on deforestation rates and forest integrity. *Conservation Biology* 27: 1011–1019.

IUCN (International Union for Conservation of Nature). 2016. *Integrating traditional knowledge into species assessments*. Retrieved from www.iucn. org/commissions/commission-environmental-economic-and-social-policy/our-work/specialist-group-sustainable-use-and-livelihoods-suli/events/ iucn-world-conservation-congress-wcc/integrating-traditional

IWGIA (International Work Group for Indigenous Affairs). 2021. www.iwgia.org

Jarvis, D. I., et al. 2016. *Crop Genetic Diversity in the Field and on the Farm: Principles and Applications in Research Practices*. Yale University Press, New Haven, CT.

Kemp, D. R., et al. 2013. Innovative grassland management systems for environmental and livelihood benefits. *Proceedings of the National Academy of Sciences USA* 110: 8369–8374.

King, D. I., et al. 2009. Effects of width, edge and habitat on the abundance and nesting success of scrub-shrub birds in powerline corridors. *Biological Conservation* 142: 2672–2680.

Kinnaird, M. F., and T. G. O'Brien. 2013. Effects of private land use, livestock management, and human tolerance on diversity, distribution, and abundance of large African mammals. *Conservation Biology* 27: 1026–1039.

Koh, L. P., and J. Ghazoul. 2010. Spatially explicit scenario analysis for reconciling agricultural expansion, forest protection, and carbon conservation in Indonesia. *Proceedings of the National Academy of Sciences USA* 107(24): 11140–11144.

Kohler, F., and E. S. Brondizio. 2017. Considering the needs of indigenous and local populations in conservation programs. *Conservation Biology* 31(2): 245–251.

Lapeyre, R. 2015. Wildlife conservation without financial viability? The potential for payments for dispersal areas' services in Namibia. *Animal Conservation* 18(1): 14–15.

Lindsey, P. A., et al. 2013. Benefits of wildlife-based land uses on private lands in Namibia and limitations affecting their development. *Oryx* 47(1): 41–53.

Lindsey, P. A., et al. 2016. Life after Cecil: Channeling global outrage into funding for conservation in Africa. *Conservation Letters* 9(4): 296–301.

Loke, L. H., et al. 2015. Creating complex habitats for restoration and reconciliation. *Ecological Engineering* 77: 307–313.

MA (Millennium Ecosystem Assessment). 2005. *Ecosystems and Human Well-Being.* 4 volumes. Island Press, Washington, DC.

MacKay, A., M. Allard, and M. A. Villard. 2014. Capacity of older plantations to host bird assemblages of naturally regenerated conifer forests: A test at stand and landscape levels. *Biological Conservation* 170: 110–119.

Magle, S. B., et al. 2019. Advancing urban wildlife research through a multi-city collaboration. *Frontiers in Ecology and the Environment* 17(4): 232–239.

Malpai Borderlands Group. 2010. www.malpaiborderlandsgroup.org

Manenti, R., et al. 2020. The good, the bad and the ugly of COVID-19 lockdown effects on wildlife conservation: Insights from the first European locked down country. *Biological Conservation* 249: 108728.

McNeely, J. A. 1989. Protected areas and human ecology: How national parks can contribute to sustaining societies of the twenty-first century. *In* D. Western and M. Pearl (eds.), *Conservation for the Twenty-first Century,* pp. 150–165. Oxford University Press, New York.

Meffert, P. J., and F. Dziock. 2012. What determines occurrence of threatened bird species on urban wastelands? *Biological Conservation* 153: 87–100.

Middleton, B. A. 2013. Rediscovering traditional vegetation management in preserves: Trading experiences between cultures and continents. *Biological Conservation* 158: 271–279.

NACSO (Namibia Association of CBNRM Support Organisations). 2008. *Namibia's Communal Conservancies: A Review of Progress in 2008.* NACSO, Windhoek, Namibia.

NACSO (Namibia Association of CBNRM Support Organisations). 2014. *The State of Community Conservation in Namibia: A Review of Communal Conservancies, Community Forests and Other CBNRM Initiatives (2013 Annual Report).* NACSO, Windhoek, Namibia.

Naidoo, R., et al. 2016. Complementary benefits of tourism and hunting to communal conservancies in Namibia. *Conservation Biology* 30(3): 628–638.

Native Seeds/SEARCH (Southwestern Endangered Aridland Resource Clearing House). 2009. www.nativeseeds.org

Nelson, M. P., et al. 2016. Emotions and the ethics of consequence in conservation decisions: Lessons from Cecil the Lion. *Conservation Letters* 9(4): 302–306.

NPS (US National Park Service). 2009. National Park of American Samoa. Retrieved from www.nps.gov/npsa/index.htm

NPS (US National Park Service). 2021. www.nps.gov

Office of National Marine Sanctuaries. 2021. National Marine Sanctuary of American Samoa. Retrieved from americansamoa.noaa.gov

Phelps, J., et al. 2013. Agricultural intensification escalates future conservation costs. *Proceedings of the National Academy of Sciences USA* 110: 7601–7606.

Philpott, S. M., et al. 2007. Field testing ecological and economic benefits of coffee certification programs. *Conservation Biology* 21: 975–985.

Pikesley, S. K., et al. 2016. Pink sea fans (*Eunicella verrucosa*) as indicators of the spatial efficacy of Marine Protected Areas in southwest UK coastal waters. *Marine Policy* 64: 38–45.

Pitman, R. T., et al. 2017. The conservation costs of game ranching. *Conservation Letters* 10(4): 403–413.

Rai, N. D., and K. S. Bawa. 2013. Inserting politics and history in conservation. *Conservation Biology* 27: 425–428.

Redpath, S. M., et al. 2013. Understanding and managing conservation conflicts. *Trends in Ecology & Evolution* 28: 100–109.

Republic of Namibia MET (Ministry of Environment and Tourism). 2021. Overview of National Parks. Retrieved from www.met.gov.na/national-parks/overview-of-national-parks/292/

Ribeiro, J., et al. 2016. An integrated trait-based framework to predict extinction risk and guide conservation planning in biodiversity hotspots. *Biological Conservation* 195: 214–223.

Riehl, B., H. Zerriffi, and R. Naidoo. 2015. Effects of community-based natural resource management on household welfare in Namibia. *PLOS One* 10(5): e0125531.

Robinson, J. M., et al. 2021. Traditional ecological knowledge in restoration ecology: A call to listen deeply, to engage with, and respect indigenous voices. *Restoration Ecology* 29(4): e13381.

Rodrigues, P., et al. 2018. Coffee management and the conservation of forest bird diversity in southwestern Ethiopia. *Biological Conservation* 217: 131–139.

Roe, D., et al. 2013. Linking biodiversity conservation and poverty reduction: De-polarizing the conservation-poverty debate. *Conservation Letters* 6: 162–171.

Rosenzweig, M. L. 2003. Win-Win Ecology: How the Earth's Species Can Survive in the Midst of Human Enterprise. Oxford University Press, New York.

Sanctuary (The Ministry of Defense Sustainability Magazine). 2018. Environmental project award runner up. *Sanctuary* 47: 6.

Schilthuizen, M. 2019. *Darwin Comes to Town: How the Urban Jungle Drives Evolution*. Picador, New York.

Semeraro, T., et al. 2018. Planning ground based utility scale solar energy as green infrastructure to enhance ecosystem services. *Energy Policy* 117: 218–227.

Stokes, E. J., et al. 2010. Monitoring great ape and elephant abundance at large spatial scales: Measuring effectiveness of a conservation landscape. *PLOS One* 5(4): e10294.

The Guardian. 2020. Race to save 100 whales in Sri Lanka's biggest mass beaching. November 2. www.theguardian.com/environment/2020/nov/02/race-save-pilot-whales-sri-lanka-biggest-mass-beaching

Troupin, D., and Y. Carmel. 2014. Can agro-ecosystems efficiently complement protected area networks? *Biological Conservation* 169: 158–166.

Tuanmu, M. N., et al. 2016. Effects of payments for ecosystem services on wildlife habitat recovery. *Conservation Biology* 30(4): 827–835.

UNEP/MAP (United Nations Environment Programme/Mediterranean Action Plan). 2016. *Countries Agree Ambitious Conservation Measures for Mediterranean*. www.unep.org/Documents.Multilingual/Default.asp?DocumentID=27058&ArticleID=35947&l=en

USFWS (US Fish and Wildlife Service). 2008. *Florida Panther Recovery Plan* (Puma concolor coryi), 3rd revision. www.fws.gov/uploadedFiles/Panther%20Recovery%20Plan.pdf

USFWS (US Fish and Wildlife Service). 2017. Florida panther population estimate updated. https://www.fws.gov/southeast/news/2017/02/florida-panther-population-estimate-updated/

Watson, J. E., et al. 2016. Bolder science needed now for protected areas. *Conservation Biology* 30(2): 243–248.

Weiskopf, S. R., et al. 2019. The conservation value of forest fragments in the increasingly agrarian landscape of Sumatra. *Environmental Conservation* 46: 340–346.

Western, D., R. Groom, and J. Worden. 2009. The impact of subdivision and sedentarization of pastoral lands on wildlife in an African savanna ecosystem. *Biological Conservation* 142: 2538–2546.

World Bank. 2021a. Indigenous Peoples. Retrieved from www.worldbank.org/en/topic/indigenouspeoples

World Bank. 2021b. Namibia. Retrieved from https://data.worldbank.org/country/namibia

Wunder, S. 2013. When payments for environmental services will work for conservation. *Conservation Letters* 6: 230–237.

Zellmer, A. J., et al. 2020. What can we learn from wildlife sightings during the COVID-19 global shutdown? *Ecosphere* 11(8): e03215.

Zhang, L., et al. 2020. Influence of traditional ecological knowledge on conservation of the skywalker hoolock gibbon (*Hoolock tianxing*) outside nature reserves. *Biological Conservation* 241: 108267.

Zielin, S. B., et al. 2016. Ecological investigations to select mitigation options to reduce vehicle-caused mortality of a threatened butterfly. *Journal of Insect Conservation* 20(5): 845–854.

Chapter 11

ACG (Área de Conservación Guanacaste). 2021. www.acguanacaste.ac.cr

Alexander, S., et al. 2016. The relationship between ecological restoration and the ecosystem services concept. *Ecology and Society* 21(1): 34.

Allen, W. H. 1988. Biocultural restoration of a tropical forest: Architects of Costa Rica's emerging Guanacaste National Park plan to make it an integral part of local culture. *BioScience* 38: 156–161.

Alston, J. M., et al. 2019. Reciprocity in restoration ecology: When might large carnivore reintroduction restore ecosystems? *Biological Conservation* 234: 82–89.

AP (Associated Press). 2019. Louisiana hires contractor for wetland restoration project. NewOrleansCityBusiness.com. April 10. https://neworleanscitybusiness.com/blog/2019/04/10/louisiana-hires-contractor-for-wetland-restoration-project/

Archibald, C. L., et al. 2017. Assessing the impact of revegetation and weed control on urban sensitive bird species. *Ecology and Evolution* 7(12): 4200–4208.

Aronson, J. C., et al. 2018. Restoration science does not need redefinition. *Nature Ecology & Evolution* 2(6): 916.

Arthington, A. H., J. M. Bernardo, and M. Ilhéu. 2014. Temporary rivers: Linking ecohydrology, ecological quality and reconciliation ecology. *River Research and Applications* 30(10): 1209–1215.

Baldwin, A. H., R. S. Hammerschlag, and D. R. Cahoon. 2019. Evaluating restored tidal freshwater wetlands. *In* G. M. E. Perillo, E. Wolanski, D. R. Cahoon, and C. S. Hopkinson (eds.), *Coastal Wetlands: An Integrated Ecosystem Approach*, pp. 889–912. Elsevier, Amsterdam, Netherlands.

Barkham, P. 2018. Dutch rewilding experiment sparks backlash as thousands of animals starve. *The Guardian*. April 27. www.theguardian.com/environment/2018/apr/27/dutch-rewilding-experiment-backfires-as-thousands-of-animals-starve

Bateman, H. L., et al. 2014. Indirect effects of biocontrol of an invasive riparian plant (*Tamarix*) alters habitat and reduces herpetofauna abundance. *Biological Invasions* 17: 87–97.

Bayraktarov, E., et al. 2015. The cost and feasibility of marine coastal restoration. *Ecological Applications* 26(4): 1055–1074.

Bean, D., and T. Dudley. 2018. A synoptic review of *Tamarix* biocontrol in North America: Tracking success in the midst of controversy. *BioControl* 63: 361–376.

Bischoff, A., et al. 2018. Hay and seed transfer to re-establish rare grassland species and communities: How important are date and soil preparation? *Biological Conservation* 221: 182–189.

Bloodworth, B. R., et al. 2016. *Tamarisk Beetle (Diorhabda spp.) in the Colorado River Basin: Synthesis of an Expert Panel Forum*. Colorado Mesa University, Grand Junction, CO.

Bradshaw, A. D. 1990. The reclamation of derelict land and the ecology of ecosystems. *In* W. R. Jordan III, M. E. Gilpin, and J. D. Aber (eds.), *Restoration Ecology: A Synthetic Approach to Ecological Research*, pp. 53–74. Cambridge University Press, Cambridge.

Bruce, D. G., et al. 2021. *A Synopsis of Research on the Ecosystem Services Provided by Large-Scale Oyster Restoration in the Chesapeake Bay*. NOAA Technical Memorandum NMFS-OHC-8. https://spo.nmfs.noaa.gov/sites/default/files/TMOHC8.pdf

Bullerjahn, G. S., et al. 2016. Global solutions to regional problems: Collecting global expertise to address the problem of harmful cyanobacterial blooms. A Lake Erie case study. *Harmful Algae* 54: 223–238.

Chazdon, R. L., et al. 2016. Carbon sequestration potential of second-growth forest regeneration in the Latin American tropics. *Science Advances* 2(5): e1501639.

Chen, W. Y. 2017. Environmental externalities of urban river pollution and restoration: A hedonic analysis in Guangzhou (China). *Landscape and Urban Planning* 157: 170–179.

Chimner, R. A., et al. 2017. An overview of peatland restoration in North America: Where are we after 25 years? *Restoration Ecology* 25(2): 283–292.

Clare, S., and I. F. Creed. 2014. Tracking wetland loss to improve evidence-based wetland policy learning and decision making. *Wetlands Ecology and Management* 22: 235–245.

Clark, L. B., et al. 2019. Successful information exchange between restoration science and practice. *Restoration Ecology* 27: 1241–1250.

Cooke, S. J., et al. 2018. Evidence-based restoration in the Anthropocene: From acting with purpose to acting for impact. *Restoration Ecology* 26(2): 201–205.

Cook-Patton, S. C., et al. 2020. Mapping carbon accumulation potential from global natural forest regrowth. *Nature* 585(7826): 545–550.

Cooper, D. J., et al. 2017. Mountain wetland restoration: The role of hydrologic regime and plant

introductions after 15 years in the Colorado Rocky Mountains, USA. *Ecological Engineering* 101: 46–59.

Darrah, A. J., and C. van Riper. 2018. Riparian bird density decline in response to biocontrol of Tamarix from riparian ecosystems along the Dolores River in SW Colorado, USA. *Biological Invasions* 20(3): 709–720.

Davenport, J. 2018. Making the buffalo commons new again: Rangeland restoration and bison reintroduction in the Montana highline. *Great Plains Quarterly* 38(2): 199–225.

DeAngelis, B. M., et al. 2020. Social factors key to landscape-scale coastal restoration: Lessons learned from three US case studies. *Sustainability* 12(3): 869.

Decade on Restoration. 2021. Preventing, halting and reversing the degradation of ecosystems worldwide. Retrieved from www.decadeonrestoration.org

Denning, K. R., and B. L. Foster. 2018. Flower visitor communities are similar on remnant and reconstructed tallgrass prairies despite forb community differences. *Restoration Ecology* 26(4): 751–759.

Derak, M., et al. 2018. A proposed framework for participatory forest restoration in semiarid areas of North Africa. *Restoration Ecology* 26: S18–S25.

Dumroese, R. K., et al. 2015. Considerations for restoring temperate forests of tomorrow: Forest restoration, assisted migration, and bioengineering. *New Forests* 46(5–6): 947–964.

EPA (US Environmental Protection Agency). 2021. Facts and Figures about Materials Waste and Recycling. Retrieved May 26 from www.epa.gov/facts-and-figures-about-materials-waste-and-recycling/national-overview-facts-and-figures-materials

FAO, IUCN CEM, and SER (Food and Agriculture Organization of the United Nations, International Union for Conservation of Nature's Commission on Ecosystem Management, and Society for Ecological Restoration). 2021. *Principles for Ecosystem Restoration to Guide the United Nations Decade 2021–2030*. Rome, FAO.

Foley, M. M., et al. 2017. Dam removal: Listening in. *Water Resources Research* 53(7): 5229–5246.

Glenn, E. P., et al. 2017. Effectiveness of environmental flows for riparian restoration in arid regions: A tale of four rivers. *Ecological Engineering* 106: 695–703.

Goldberg, N., and K. C. Reiss. 2016. Accounting for wetland loss: Wetland mitigation trends in northeast Florida 2006–2013. *Wetlands* 36(2): 373–384.

González, E., A. Masip, and E. Tabacchi. 2016. Poplar plantations along regulated rivers may resemble riparian forests after abandonment: A comparison of passive restoration approaches. *Restoration Ecology* 24(4): 538–547.

González, E., et al. 2015. Restoration of riparian vegetation: A global review of implementation and evaluation approaches in the international, peer-reviewed literature. *Journal of Environmental Management* 158: 85–94.

González, E., et al. 2017. Secondary invasions of noxious weeds associated with control of invasive *Tamarix* are frequent, idiosyncratic and persistent. *Biological Conservation* 213: 106–114.

González, E., et al. 2018. Regeneration of Salicaceae riparian forests in the Northern Hemisphere: A new framework and management tool. *Journal of Environmental Management* 218: 374–387.

Grman, E., T. Bassett, and L. A. Brudvig. 2014. A prairie plant community data set for addressing questions in community assembly and restoration. *Ecology* 95: 2363.

Harris, J. A., et al. 2006. Ecological restoration and global climate change. *Restoration Ecology* 14(2): 170–176.

Hayward, M. W. 2009. Conservation management for the past, present, and future. *Biodiversity and Conservation* 18: 765–775.

Henry, A. L., et al. 2018. Spatial modeling improves understanding patterns of invasive species defoliation by a biocontrol herbivore. *Biological Invasions* 20(12): 3545–3562.

Hertog, I. M., and E. Turnhout. 2018. Ideals and pragmatism in the justification of ecological restoration. *Restoration Ecology* 26: 1221–1229.

Higgs, E. 2017. Novel and designed ecosystems. *Restoration Ecology* 25(1): 8–13.

Hobbs, R. J. 2018. Restoration ecology's silver jubilee: Innovation, debate, and creating a future for restoration ecology. *Restoration Ecology* 26(5): 801–805.

Hobbs, R. J., E. S. Higgs, and C. M. Hall. 2013. *Novel Ecosystems: Intervening in the New Ecological World Order.* Wiley-Blackwell, Oxford, UK.

Holmes, M. A., and G. R. Matlack. 2018. Assembling the forest herb community after abandonment from agriculture: Long-term successional dynamics differ with land-use history. *Journal of Ecology* 106: 2121–2131.

Honold, J., et al. 2016. Restoration in urban spaces: Nature views from home, greenways, and public parks. *Environment and Behavior* 48(6): 796–825.

Hulshof, C. M., and J. S. Powers. 2020. Tropical forest composition and function across space and time: Insights from diverse gradients in Área de Conservación Guanacaste. *Biotropica* 52(6): 1065–1075.

International Joint Commission. 2014. *A Balanced Diet for Lake Erie: Reducing Phosphorus Loadings and Harmful Algal Blooms.* Report of the Lake Erie Ecosystem Priority. International Joint Commission, Washington, Ottawa, and Windsor.

Jakovac, C. C., et al. 2021. The role of land-use history in driving successional pathways and its implications for the restoration of tropical forests. *Biological Reviews*, https://doi.org/10.1111/brv.12694

Janzen, D. H., and W. Hallwachs. 2020. Área de Conservación Guanacaste, northwestern Costa Rica: Converting a tropical national park to conservation via biodevelopment. *Biotropica* 52(6): 1017–1029.

Louisiana Coastal Wetlands Conservation and Restoration Task Force and the Wetlands Conservation and Restoration Authority. 1998. *Coast 2050: Toward a Sustainable Coastal Louisiana.* Louisiana Department of Natural Resources, Baton Rouge, LA.

Martin, D. M. 2017. Ecological restoration should be redefined for the twenty-first century. *Restoration Ecology* 25(5): 668–673.

Mascaro, J., et al. 2012. Novel forests maintain ecosystem processes after the decline of native tree species. *Ecological Monographs* 82(2): 221–228.

McDonald, T., et al. 2016. *International Standards for the Practice of Ecological Restoration–Including Principles and Key Concepts.* Society for Ecological Restoration, Washington, DC.

Meissen, J. C., S. M. Galatowitsch, and M. W. Cornett. 2015. Risks of overharvesting seed from native tallgrass prairies. *Restoration Ecology* 23(6): 882–891.

Miller, J. R., and B. T. Bestelmeyer. 2016. What's wrong with novel ecosystems, really? *Restoration Ecology* 24(5): 577–582.

Moreno-Mateos, D., et al. 2015a. The true loss caused by biodiversity offsets. *Biological Conservation* 192: 552–559.

Moreno-Mateos, D., et al. 2015b. Ecosystem response to interventions: Lessons from restored and created wetland ecosystems. *Journal of Applied Ecology* 52(6): 1528–1537.

Nagler, P. L., et al. 2018. Northern tamarisk beetle (*Diorhabda carinulata*) and tamarisk (*Tamarix* spp.) interactions in the Colorado River basin. *Restoration Ecology* 26(2): 348–359.

Neale, M. W., and E. R. Moffett. 2016. Re-engineering buried urban streams: Daylighting results in rapid changes in stream invertebrate communities. *Ecological Engineering* 87: 175–184.

NYC Parks (New York City Department of Parks & Recreation). 2021. Freshkills Park. Retrieved from www.nycgovparks.org/park-features/freshkills-park

Palmer, M. A., J. B. Zedler, and D. A. Falk. 2016. Ecological theory and restoration ecology. *In* J. B. Zedler, D. A. Falk, and M. A. Palmer (eds.), *Foundations of Restoration Ecology*, pp. 3–26. Island Press, Washington, DC.

Perring, M. P., T. E. Erickson, and P. H. Brancalion. 2018. Rocketing restoration: Enabling the upscaling of ecological restoration in the Anthropocene. *Restoration Ecology* 26(6): 1017–1023.

Primack, R. B., et al. 2021. Manager characteristics drive conservation success. *Biological Conservation* 259: 109169.

Pugh, T. A., et al. 2019. Role of forest regrowth in global carbon sink dynamics. *Proceedings of the National Academy of Sciences USA* 116(10): 4382–4387.

Reiker, J., et al. 2015. Does origin always matter? Evaluating the influence of nonlocal seed provenances for ecological restoration purposes in a widespread and outcrossing plant species. *Ecology and Evolution* 5(23): 5642–5651.

Rohr, J., et al. 2018. The ecology and economics of restoration: When, what, where, and how to restore ecosystems. *Ecology and Society* 23(2): 15.

Saunders, M. I., et al. 2020. Bright spots in coastal marine ecosystem restoration. *Current Biology* 30(24): R1500–R1510.

Scavia, D., et al. 2019. St. Clair–Detroit River system: Phosphorus mass balance and implications for Lake Erie load reduction, monitoring, and climate change. *Journal of Great Lakes Research* 45(1): 40–49.

Seastedt, T. R. 2015. Biological control of invasive plant species: A reassessment for the Anthropocene. *New Phytologist* 205(2): 490–502.

SER (Society of Ecological Resotration). 2021b. A tool for assessing ecosystem recovery: The 5-Star Recovery System in action. www.ser.org/page/SERNews3113

SER (Society of Ecological Restoration). 2021. What is ecological restoration? Retrieved from https://www.ser-rrc.org/what-is-ecological-restoration/

Sher, A. 2013. Introduction to the paradox plant. *In* A. Sher and M. F. Quigley (eds.), *Tamarix: A Case Study of Ecological Change in the American West*, pp. 1–20. Oxford University Press, New York.

Sher, A. A., et al. 2018. Native species recovery after reduction of an invasive tree by biological control with and without active removal. *Ecological Engineering* 111: 167–175.

Sher, A. A., et al. 2020. The human element of restoration success: Manager characteristics affect vegetation recovery following invasive *Tamarix* control. *Wetlands* 40: 1877–1895.

Smith, S., A. A. Sher, and T. A. Grant. 2007. Genetic diversity in restoration materials and the impacts of seed collection in Colorado's restoration plant production industry. *Restoration Ecology* 15: 369–374.

Svenning, J. C., M. Munk, and A. Schweiger. 2018. Trophic rewilding: Ecological restoration of top-down trophic interactions to promote self-regulating biodiverse ecosystems. *Philosophical Transactions of the Royal Society of London B* 373: 20170432.

Thompson, M. S., et al. 2018. Large woody debris "rewilding" rapidly restores biodiversity in riverine food webs. *Journal of Applied Ecology* 55(2): 895–904.

UNEP (United Nations Environment Programme). 2021. Countries commit to restore global land area the size of China. www.unep.org/news-and-stories/story/countries-commit-restore-global-land-area-size-china

Vander Zanden, M. J., et al. 2016. Food web theory and ecological restoration. *In* J. Zedler, D. A. Falk, and M. A. Palmer (eds.), *Foundations of Restoration Ecology*, pp. 301–329. Island Press, Washington, DC.

Wilson, E. O. 1992. *The Diversity of Life*. Harvard University Press, Cambridge.

Chapter 12

Apostolopoulou, E., and W. M. Adams. 2017. Biodiversity offsetting and conservation: Reframing nature to save it. *Oryx* 51(1): 23–31.

Ascensão, F., et al. 2018. Environmental challenges for the Belt and Road Initiative. *Nature Sustainability* 1(5): 206–209.

Barbier, E. B., J. C. Burgess, and T. J. Dean. 2018. How to pay for saving biodiversity. *Science* 360(6388): 486–488.

Barcia, L. 2015. Addis Ababa: Financing the future or financing failure? *Devex*. July 20. https://www.devex.com/news/addis-ababa-financing-the-future-or-financing-failure-86561

Bhattacharya, A., J. Oppenheim, and N. Stern. 2015. *Driving Sustainable Development through Better Infrastructure: Key Elements of a Transformation Program*. Global Economy and Development Working Paper 91. www.brookings.edu/wp-content/uploads/2016/07/07-sustainable-development-infrastructure-v2.pdf

Buchanan, G. M., et al. 2018. The local impacts of World Bank development projects near sites of conservation significance. *Journal of Environment & Development* 27(3): 299–322.

Business Wire. 2015. American burying beetle–Mitigation solutions. USA muddy boggy endangered species bank expansion in Oklahoma. July 30. www.businesswire.com/news/home/20150730006047/en/American-Burying-Beetle-%E2%80%93-Mitigation-Solutions-USA

Carrière, S. M., et al. 2013. Rio +20: Biodiversity marginalized. *Conservation* 6: 6–11.

Cassimon, D., D. Essers, and A. Fauzi. 2014. Indonesia's debt-for-development swaps: Past, present, and future. *Bulletin of Indonesian Economic Studies* 50(1): 75–100.

CBD (Convention on Biological Diversity). 2021. www.cbd.int

Chapman, M., et al. 2020. A payment by any other name: Is Costa Rica's PES a payment for services or a support for stewards? *World Development* 129: 104900.

Clayton, S., and G. Myers. 2015. *Conservation Psychology: Understanding and Promoting Human Care for Nature*. John Wiley & Sons, Hoboken, NJ.

Cohen-Shacham, E., et al. 2015. Using the ecosystem services concept to analyse stakeholder involvement in wetland management. *Wetlands Ecology and Management* 23(2): 241–256.

Costa Rica National Parks. 2021. www.costarica-nationalparks.com

Crees, J. J., et al. 2016. A comparative approach to assess drivers of success in mammalian conservation recovery programs. *Conservation Biology* 30(4): 694–705.

Davenport, C. 2015. Nations approve landmark climate accord in Paris. *New York Times*. December 13. www.nytimes.com/2015/12/13/world/europe/climate-change-accord-paris.html?_r=0

Davenport, C. 2020. What will Trump's most profound legacy be? Possibly climate damage. *New York Times*. www.nytimes.com/2020/11/09/climate/trump-legacy-climate-change.html

Dernbach, J. C., and F. Cheever. 2015. Sustainable development and its discontents. *Transnational Environmental Law* 4(2): 247–287.

Di Minin, E., et al. 2013. Conservation business and conservation planning in a biological diversity hotspot. *Conservation Biology* 27: 808–820.

Dinerstein, E., et al. 2017. An ecoregion-based approach to protecting half the terrestrial realm. *BioScience* 67(6): 534–545.

Dinerstein, E., et al. 2019. A global deal for nature: Guiding principles, milestones, and targets. *Science Advances* 5(4): eaaw2869.

Gallo-Cajiao, E., et al. 2018. Crowdfunding biodiversity conservation. *Conservation Biology* 32(6): 1426–1435.

GEF (Global Environment Facility). 2020. *Annual Performance Report 2020* (Prepared by the Independent Evaluation Office of the GEF). www.thegef.org/sites/default/files/council-meeting-documents/EN_GEF.ME_C58_Inf.01_APR_2020.pdf

GEF (Global Environment Facility). 2020. Investing in our planet. Retrieved from www.thegef.org/gef

Gonçalves, V. K., and M. B. Anselmi. 2018. International civil aviation organization as a climate governance forum: An analysis of the carbon offsetting and reduction. *Revista de Estudos e Pesquisas Avançadas do Terceiro Setor* 1(1): 110–123.

Graves, R. A., et al. 2019. Quantifying the contribution of conservation easements to large-landscape conservation. *Biological Conservation* 232: 83–96.

Hardy, M. J., et al. 2017. Exploring the permanence of conservation covenants. *Conservation Letters* 10(2): 221–230.

Hardy, M. J., et al. 2018. Purchase, protect, resell, repeat: An effective process for conserving biodiversity on private land? *Frontiers in Ecology and the Environment* 16(6): 336–344.

IPBES (Intergovernmental Science-Policy Platform on Biodiversity and Ecosystem Services). 2019. IPES 2019: Global Assessment Report on Biodiversity and Ecosystem Services of the Intergovernmental Science-Policy Platform on Biodiversity and Ecosystem Services. E. S. Brondizio, J. Settele, S. Díaz, and H. T. Ngo (eds.). IPBES Secretariat, Bonn, Germany.

Jackson, B., and J. L. D. Sparks. 2020. Ending slavery by decarbonisation? Exploring the nexus of modern slavery, deforestation, and climate change action via REDD+. *Energy Research & Social Science* 69: 101610.

Kaufmann, E. 2014. Biodiversity conservation aid: Who is getting how much and why? AidData. https://www.aiddata.org/blog/biodiversity-conservation-aid-who-is-getting-how-much-and-why

Knoot, T. G., M. Rickenbach, and K. Silbernagel. 2015. Payments for ecosystem services: Will a new hook net more active family forest owners. *Journal of Forestry* 113(2): 210–218.

Land Trust Alliance. 2021. www.landtrustalliance.org/

Leclère, D., et al. 2020. Bending the curve of terrestrial biodiversity needs an integrated strategy. *Nature* 585(7826): 551–556.

Lin, J., et al. 2014. China's international trade and air pollution in the United States. *Proceedings of the National Academy of Science USA* 111(5): 1736–1741.

Mai, H. J. 2021. U.S. officially rejoins Paris Agreement on climate change. *NPR (National Public Radio)*. https://www.npr.org/2021/02/19/969387323/u-s-officially-rejoins-paris-agreement-on-climate-change

McBride, J. 2018. How does the U.S. spend its foreign aid? Retrieved from Council on Foreign Relations, www.cfr.org/backgrounder/how-does-us-spend-its-foreign-aid

McCarthy, D. P., et al. 2012. Financial costs of meeting global biodiversity conservation targets: Current spending and unmet needs. *Science* 338(6109): 946–949.

Mockrin, M. H., et al. 2017. Balancing housing growth and land conservation: Conservation development preserves private lands near protected areas. *Landscape and Urban Planning* 157: 598–607.

Munawar, S., M. F. Khokhar, and S. Atif. 2015. Reducing emissions from deforestation and forest degradation implementation in northern Pakistan. *International Biodeterioration and Biodegradation* 102: 316–323.

Naeem, S, et al. 2015. Get the science right when paying for nature's services. *Science* 347(6227): 1206–1207.

OECD (Organisation for Economic Co-operation and Development). 2018. *Mainstreaming Biodiversity for Sustainable Development*. OECD, Paris. https://read.oecd-ilibrary.org/environment/mainstreaming-biodiversity-for-sustainable-development_9789264303201-en#page1

Parker, C., et al. 2012. *The Little Biodiversity Finance Book*. Global Canopy Programme, Oxford. http://globalcanopy.org/sites/default/files/documents/resources/LittleBiodiversityFinanceBook_3rd%20edition.pdf

Peace Parks Foundation. 2021. www.peaceparks.org

Pérez-Méndez, N., et al. 2020. The economic cost of losing native pollinator species for orchard production. *Journal of Applied Ecology* 57(3): 599–608.

Phelps, J., et al. 2013. Agricultural intensification escalates future conservation costs. *Proceedings of the National Academy of Sciences USA* 110: 7601–7606.

Ramsar Convention Secretariat. 2021. www.ramsar.org

Rands, M. R., et al. 2010. Biodiversity conservation: Challenges beyond 2010. *Science* 329: 1298–1303.

Reed, S. E., J. A. Hilty, and D. M. Theobald. 2014. Guidelines and incentives for conservation development in local land-use regulations. *Conservation Biology* 28: 258–268.

Ripple, W. J., et al. 2014. Status and ecological effects of the world's largest carnivores. *Science* 343: 1241484.

Robinson, J. G. 2012. Common and conflicting interests in the engagements between conservation organizations and corporations. *Conservation Biology* 26: 967–977.

Roe, D., et al. 2013. Linking biodiversity conservation and poverty reduction: Depolarizing the conservation-poverty debate. *Conservation Letters* 6: 162–171.

Roxburgh, T., et al. 2020. Global Futures: Assessing the Global Economic Impacts of Environmental Change to Support Policy-Making. Summary Report, January 2020. www.wwf.org.uk/globalfutures

Selomane, O., et al. 2015. Towards integrated social-ecological sustainability indicators: Exploring the contribution and gaps in existing global data. *Ecological Economics* 118: 140–146.

Simmons, B. A., et al. 2021. China can help solve the debt and environmental crises. *Science* 371(6528): 468–470.

Sorice, M. G., et al. 2013. Increasing participation in incentive programs for biodiversity conservation. *Ecological Applications* 23: 1146–1155.

State of Nature Partnership. 2019. State of Nature 2019 reports. Retrieved from https://nbn.org.uk/stateofnature2019/reports/

Thondhlana, G., S. Shackleton, and J. Blignaut. 2015. Local institutions, actors, and natural resource governance in Kgalagadi Transfrontier Park and surrounds, South Africa. *Land Use Policy* 47: 121–129.

UN News. 2021. UN launches Decade on Ecosystem Restoration to counter "triple environmental emergency." https://news.un.org/en/story/2021/06/1093362

UNDP (United Nations Development Programme). 2018. *The BIOFIN Workbook 2018: Finance for Nature*. The Biodiversity Finance Initiative, UNDP, New York.

UNESCO (United Nations Educational, Scientific and Cultural Organization). 2021. World Heritage Centre. Retrieved from whc.unesco.org

United Nations. 1993. *Agenda 21: Rio Declaration and Forest Principles*. Post-Rio edition. United Nations Publications, New York.

USFWS (US Fish and Wildlife Service). 2003. *Guidance for the Establishment, Use, and Operation of Conservation Banks*. www.fws.gov/endangered/esa-library/pdf/Conservation_Banking_Guidance.pdf

Waldron, A., et al. 2013. Targeting global conservation funding to limit immediate biodiversity declines. *Proceedings of the National Academy of Sciences USA* 110(29): 12144–12148.

Watanabe, M. E. 2015. The Nagoya Protocol on Access and Benefit Sharing: International treaty poses challenges for biological collections. *BioScience* 65(6): 5430–5450.

World Bank. 2021. www.worldbank.orgWWF (World Wildlife Fund). 2017. WWF recommendations and spatial analysis briefing paper. Retrieved from https://go.nature.com/2v3SwoG

Yue, M. and Nedopil Wang, C. 2021. Debt-for-Nature Swaps: A Triple-Win Solution for Debt Sustainability and Biodiversity Finance in the Belt and Road Initiative (BRI)? https://green-bri.org/wp-content/uploads/2021/01/Yue-2021_Debt-for-nature-swaps-BRI-1.pdf

Chapter 13

Addison, P., and J. W. Bull. 2018. Conservation accord: Corporate incentives. *Science* 360: 1195–1196.

Agrawal, A. A., and H. Inamine. 2018. Mechanisms behind the monarch's decline. *Science* 360: 1294–1296.

Alamgir, M., et al. 2017. Economic, socio-political and environmental risks of road development in the tropics. *Current Biology* 27: R1130–R1140.

AZA (Association of Zoos and Aquariums). 2021. Conservation funding. Retrieved from www.aza.org/conservation-funding

Berrueta, V. M., et al. 2017. Promoting sustainable local development of rural communities and mitigating climate change: The case of Mexico's Patsari improved cookstove project. *Climatic Change* 140: 63–77.

Blackman, A., L. Goff, and M. R. Planter. 2018. Does eco-certification stem tropical deforestation? Forest Stewardship Council certification in Mexico. *Journal of Environmental Economics and Management* 89: 306–333.

Bombaci, S. P., et al. 2016. Using Twitter to communicate conservation science from a professional conference. *Conservation Biology* 30(1): 216–225.

Carlson, T. 2013. The politics of a tree: How a species became national policy. *In* A. Sher and M. F. Quigley (eds.), *Tamarix: A Case Study of Ecological Change in the American West*. Oxford University Press, New York.

Cetas, E. R., and M. Yasué. 2017. A systematic review of motivational values and conservation success in and around protected areas. *Conservation Biology* 31: 203–212.

Charky, N. 2014. Congress attempts to change captivity rules for orcas, marine life. *LA Times*. March 14.

Clark, L., et al. 2019. Information exchange between restoration scientists and managers benefits from a two-way street: A case study in the American Southwest. *Restoration Ecology* 27 (12):41.

CNN. 2016. SeaWorld's orcas will be last generation at parks. March 17. https://www.cnn.com/2016/03/17/us/seaworld-last-generation-of-orcas/index.html

Collins, A., et al. 2018. Learning and teaching sustainability: The contribution of Ecological Footprint calculators. *Journal of Cleaner Production* 174: 1000–1010.

Correa Ayram, C. A., et al. 2016. Habitat connectivity in biodiversity conservation: A review of recent studies and applications. *Progress in Physical Geography* 40: 7–37.

Courchamp, F., et al. 2015. Fundamental ecology is fundamental. *Trends in Ecology & Evolution* 30(1): 9–16.

Gaddis, M. L. 2018. Training citizen scientists for data reliability: A multiple case study to identify themes in current training initiatives. PhD dissertation. University of the Rockies.

Garrard, G. E., et al. 2015. Beyond advocacy: Making space for conservation scientists in public debate. *Conservation Letters*, https://doi.org/10.1111/conl.12193

Godet, L., and V. Devictor. 2018. What conservation does. *Trends in Ecology & Evolution* 33: 720–730.

Granek, E. F., et al. 2008. Engaging recreational fishers in management and conservation: Global case studies. *Conservation Biology* 22: 1125–1134.

Gupta, N., et al. 2015. Assessing recreational fisheries in an emerging economy: Knowledge, perceptions and attitudes of catch-and-release anglers in India. *Fisheries Research* 165: 79–84.

Horton, C., et al. 2016. Credibility and advocacy in conservation science. *Conservation Biology* 30(1): 23–32.

Ikh Nart Nature Reserve. 2021. http://ikhnart.com/

IUCN (International Union for Conservation of Nature). 2021. www.iucn.org

Kadykalo, A. N., et al. 2021. Bridging research and practice in conservation. *Conservation Biology*, https://doi.org/10.1111/cobi.13732

Lamb, C. T., S. L. Gilbert, and A. T. Ford. 2018. Tweet success? Scientific communication correlates with increased citations in *Ecology and Conservation*. *PeerJ* 6: e4564.

Laurance, W. F. 2013. Does research help to safeguard protected areas? *Trends in Ecology & Evolution* 28: 261–266.

MacFarquhar, N., and I. Nechepurenko. 2019. Russia says it will try to free almost 100 whales held in "jail." *New York Times*. April 4. www.nytimes.com/2019/04/04/world/europe/russia-whale-jail.html

Manfredo, M. J., T. L. Teel, and A. M. Dietsch. 2016. Implications of human value shift and persistence for biodiversity conservation. *Conservation Biology* 30(2): 287–296.

McKinley, D. C., et al. 2017. Citizen science can improve conservation science, natural resource management, and environmental protection. *Biological Conservation* 208: 15–28.

Miller, V., et al. 2018. The SOS Pesca project: A multinational and intersectoral collaboration for sustainable fisheries, marine conservation and improved quality of life in coastal communities. *MEDICC Review* 20: 65–70.

Morrison, S. A. 2016. Designing virtuous socio-ecological cycles for biodiversity conservation. *Biological Conservation* 195: 9–16.

Natural Resource Stewardship and Science. 2015. Office of the Associate Director. Strategic framework for National Park Service Research Learning Centers. National Park Service, Natural Resource Stewardship and Science, Washington, DC. https://www.nps.gov/orgs/1778/index.htm

Owley, J., et al. 2017. Climate change challenges for land conservation: Rethinking conservation easements, strategies, and tools. *Denver Law Review* 95: 727.

Paknia, O., H. Rajaei, and A. Koch. 2015. Lack of well-maintained natural history collections and taxonomists in megadiverse developing countries hampers global biodiversity exploration. *Organisms Diversity and Evolution* 15(3): 619–629.

Piccolo, J., et al. 2018. Why conservation scientists should re-embrace their ecocentric roots. *Conservation Biology* 32: 959–961.

Pimm, S. L. 2021. What we need to know to prevent a mass extinction of plant species. *Plants, People, Planet* 3(1):7–15. https://doi.org/10.1002/ppp3.10160

Primack, R. B., et al. 2021. Manager characteristics drive conservation success. *Biological Conservation* 259: 109169.

Primack, R., and C. M. MacKenzie. 2019. Guest post: Strategies for helping your research reach a wider audience. Dynamic Ecology Blog. https://dynamicecology.wordpress.com/2019/04/01/guest-post-strategies-for-helping-your-research-reach-a-wider-audience/

QUINTESSENCE Consortium. 2016. Networking our way to better ecosystem service provision. *Trends in Ecology & Evolution* 31(2): 105–115.

Rabinovich, A., et al. 2020. Protecting the commons: Predictors of willingness to mitigate communal land degradation among Maasai pastoralists. *Journal of Environmental Psychology* 72: 101504.

Reading, R. 2015. *Mongolia Field Program Update*. Association of Zoos and Aquariums (AZA) 2015 mid-year meeting, Columbia, SC. AZA, Silver Spring, MD.

Redford, K. H., et al. 2015. Mainstreaming biodiversity: Conservation for the twenty-first century. *Frontiers in Ecology and Evolution* 3: 137.

Ripple, W. J., et al. 2017. World scientists' warning to humanity: A second notice. *BioScience* 67: 1026–1028.

Ripple, W. J., et al. 2018. The role of scientists' warning in shifting policy from growth to conservation economy. *BioScience* 68(4): 239–240.

Rose, D. C., et al. 2018. The major barriers to evidence-informed conservation policy and possible solutions. *Conservation Letters* 11(5): e12564.

RSPB (Royal Society for the Protection of Birds) and BirdLife International. 2018. *Albatross Task Force Team Highlights 2017–2018*. www.nnf.org.na/images/Progress_Reports/Albatross_Task_Force_Progress_Report_2017-2018-compressed.pdf

Sher, A. A., et al. 2020. The human element of restoration success: Manager characteristics affect vegetation recovery following invasive *Tamarix* control. *Wetlands* 40: 1877–1895.

Silk, M. J., et al. 2018. Considering connections between Hollywood and biodiversity conservation. *Conservation Biology* 32: 597–606.

Struminger, R., et al. 2021. Biological field stations promote science literacy through outreach. *BioScience*, https://doi.org/10.1093/biosci/biab057

Tallis, H. M., et al. 2018. An attainable global vision for conservation and human well-being. *Frontiers in Ecology and the Environment* 16: 563–570.

Unger, S. D., and C.R. Hickman. 2020. A content analysis from 153 years of print and online media shows positive perceptions of the hellbender salamander follow the conservation biology. *Biological Conservation* 246: 108564.

Varma, V., et al. 2015. Perceptions of priority issues in the conservation of biodiversity and ecosystems in India. *Biological Conservation* 187: 201–211.

Von Weizsäcker, E. U., and A. Wijkman. 2017. *Come On! Capitalism, Short-Termism, Population and the Destruction of the Planet*. Springer, New York.

Wright, A. J., et al. 2015. Competitive outreach in the 21st century: Why we need conservation marketing. *Ocean & Coastal Management* 115: 41–48.

Wright, G., et al. 2015. What incentivizes local forest conservation efforts? Evidence from Bolivia. *International Journal of the Commons* 9(1).

Zhu, L. and Y. C. Zhao. 2015. A feasibility assessment of the application of the Polluter-Pays Principle to ship-source pollution in Hong Kong. *Marine Policy* 57: 36–44.

Index

The letter *f* after a page number indicates that the entry is included in a figure; *t* indicates that the entry is included in a table.